# QUANTITATIVE
# APPROXIMATIONS

# QUANTITATIVE APPROXIMATIONS

## George Anastassiou

Department of Mathematical Sciences
The University of Memphis
Memphis, Tennessee

CHAPMAN & HALL/CRC

Boca Raton   London   New York   Washington, D.C.

### Library of Congress Cataloging-in-Publication Data

Anastassiou, George A., 1952-
    Quantitative approximations/George Anastassiou
        p. cm.
    Includes bibliographical references and index.
    ISBN 1-58488-221-2
    1. Approximation theory. 2. Numerical analysis. I. Title.

QA221 .A5365 2000
511'.4--dc21                                                                        00-034615

This book contains information obtained from authentic and highly regarded sources. Reprinted material is quoted with permission, and sources are indicated. A wide variety of references are listed. Reasonable efforts have been made to publish reliable data and information, but the author and the publisher cannot assume responsibility for the validity of all materials or for the consequences of their use.

Apart from any fair dealing for the purpose of research or private study, or criticism or review, as permitted under the UK Copyright Designs and Patents Act, 1988, this publication may not be reproduced, stored or transmitted, in any form or by any means, electronic or mechanical, including photocopying, microfilming, and recording, or by any information storage or retrieval system, without the prior permission in writing of the publishers, or in the case of reprographic reproduction only in accordance with the terms of the licenses issued by the Copyright Licensing Agency in the UK, or in accordance with the terms of the license issued by the appropriate Reproduction Rights Organization outside the UK.

The consent of CRC Press LLC does not extend to copying for general distribution, for promotion, for creating new works, or for resale. Specific permission must be obtained in writing from CRC Press LLC for such copying.

Direct all inquiries to CRC Press LLC, 2000 N.W. Corporate Blvd., Boca Raton, Florida 33431.

**Trademark Notice:** Product or corporate names may be trademarks or registered trademarks, and are used only for identification and explanation, without intent to infringe.

© 2001 by Chapman & Hall/CRC

No claim to original U.S. Government works
International Standard Book Number 1-58488-221-2
Library of Congress Card Number 00-034615
Printed in the United States of America  1  2  3  4  5  6  7  8  9  0
Printed on acid-free paper

*Dedicated to my daughters,
Peggy-Panagiota and Angela*

# Contents

Preface     xiii

1 Introduction     1
    1.1    On Chapter 2: Convergence with Rates of Univariate Neural Network Operators to the Unit Operator . . . . . . . .    1
    1.2    On Chapter 3: Convergence with Rates of Multivariate Neural Network Operators to the Unit Operator . . . . . . . . . . . . . . . . . . . . . . . . . . .    4
    1.3    On Chapter 4: Asymptotic Weak Convergence of Cardaliaguet-Euvrard Neural Network Operators . . . . . . . . . . . . .    7
    1.4    On Chapter 5: Asymptotic Weak Convergence of Squashing Neural Network Operators . . . . . . . . . .    9
    1.5    On Chapter 6: Quantitative Monotone and Probabilistic Wavelet Type Approximation . . . . . . . . .    10
    1.6    On Chapter 7: Quantitative One-Dimensional High Order Wavelet Type Approximation . . . . . . . . .    12
    1.7    On Chapter 8: More on Shape and Probability Preserving One-Dimensional Wavelet Type Operators . . . . . . . . . . . . . . . . . . . . . . . . . . .    14
    1.8    On Chapter 9: Quantitative Multidimensional High Order Wavelet Type Approximation . . . . . . . . .    15
    1.9    On Chapter 10: Rate of Convergence of Probabilistic Discrete Wavelet Approximation . . . . . . .    18
    1.10   On Chapter 11: Asymptotic Nonorthogonal Wavelet Approximation for Deterministic Signals . . . . .    19
    1.11   On Chapter 12: Wavelet Type Differentiated Shift Invariant Integral Operators . . . . . . . . . . . . . .    20
    1.12   On Chapter 13: A Discrete Kac's Formula and Optimal Quantitative Approximation in the Solution of Heat Equation . . . . . . . . . . . . . . . . . .    22

| | | |
|---|---|---|
| 1.13 | On Chapter 14: Quantitative-Asymptotic Expansions for the Probabilistic Representation Formulae for $(C_0)$ $m$-Parameter Operator Semigroups . . . . . . . . . . . . . . . . . . . . | 25 |
| 1.14 | On Chapter 15: Quantitative Probability Limit Theorems over Banach Spaces . . . . . . . . . . . . . . . . . . . . . | 27 |
| 1.15 | On Chapter 16: Quantitative Study of Bias Convergence for Generalized $L$-Statistics . . . . . . . . . . | 30 |
| 1.16 | On Chapter 17: Quantitative Korovkin-Type Results for Vector Valued Functions . . . . . . . . . . . . . . | 33 |
| 1.17 | On Chapter 18: Quantitative $L_p$-Results for Positive Linear Operators . . . . . . . . . . . . . . . . . . | 35 |
| 1.18 | On Chapter 19: On Monotone Approximation Theory . . | 38 |
| 1.19 | On Chapter 20: Comparisons for Local Moduli of Continuity . . . . . . . . . . . . . . . . . . . . . . . . . . . | 41 |
| 1.20 | On Chapter 21: Convergence with Rates of Univariate Singular Integrals to the Unit . . . . . . . . . . | 42 |
| 1.21 | On Chapter 22: About Univariate Ostrowski-Type Inequalities . . . . . . . . . . . . . . . . . . . . . . . | 44 |
| 1.22 | On Chapter 23: About Multidimensional Ostrowski-Type Inequalities . . . . . . . . . . . . . . . . . . | 45 |
| 1.23 | On Chapter 24: General Opial-Type Inequalities for Linear Differential Operators . . . . . . . . . . . . . . . . . . . . . | 47 |
| 1.24 | On Chapter 25: $L_p$-Opial Inequalities Engaging Fractional Derivatives of Functions . . . . . . . . . . . . . . . . . . | 50 |
| 1.25 | On Chapter 26: $L_p$-General Fractional Opial Inequalities . . . . . . . . . . . . . . . . . . . . . . . . . . . | 52 |

# I  On Neural Networks 57

## 2  Convergence with Rates of Univariate Neural Network Operators to the Unit Operator 59
2.1 Convergence with Rates of the Univariate Cardaliaguet–Euvrard Neural Network Operators . . . . . 60
2.2 The Univariate "Squashing Operators" and their Convergence to the Unit with Rates . . . . . . . . . . . . . 76

## 3  Convergence with Rates of Multivariate Neural Network Operators to the Unit Operator 89
3.1 Convergence with Rates of Multivariate Cardaliaguet–Euvrard Neural Network Operators . . . . . 90
3.2 The Multivariate "Squashing Operators" and their Convergence to the Unit with Rates . . . . . . . . . . . . 111

## 4  Asymptotic Weak Convergence of Cardaliaguet–Euvrard Neural Network Operators    121
4.1  Background . . . . . . . . . . . . . . . . . . . . . .  121
4.2  General Result . . . . . . . . . . . . . . . . . . . . .  124
4.3  Supplement . . . . . . . . . . . . . . . . . . . . . .  130

## 5  Asymptotic Weak Convergence of Squashing Neural Network Operators    135
5.1  Background . . . . . . . . . . . . . . . . . . . . . .  135
5.2  General Result . . . . . . . . . . . . . . . . . . . . .  137

# II  On Wavelets    143

## 6  Quantitative Monotone and Probabilistic Wavelet Type Approximation    145
6.1  Approximation by $A_k(f)$ Operators . . . . . . . . . . . . . .  146
6.2  Approximation by $B_k(f)$ Operators . . . . . . . . . . . . . .  156
6.3  More Results on $A_k(f)$ and $B_k(f)$ Operators . . . . . . . .  159

## 7  Quantitative One-Dimensional High Order Wavelet Type Approximation    165
7.1  Estimates and Sharpness . . . . . . . . . . . . . . . . .  165

## 8  More on Shape and Probability Preserving One-Dimensional Wavelet Type Operators    185
8.1  Estimates, Maintenance, and Sharpness . . . . . . . . . . .  186

## 9  Quantitative Multidimensional High Order Wavelet Type Approximation    201
9.1  Estimates and Sharpness . . . . . . . . . . . . . . . . .  201

## 10  Rate of Convergence of Probabilistic Discrete Wavelet Approximation    229
10.1  Approximation by $L_k(f, x)$ Operators . . . . . . . . . . . .  230

## 11  Asymptotic Nonorthogonal Wavelet Approximation for Deterministic Signals    235
11.1  Introduction . . . . . . . . . . . . . . . . . . . . . .  235
11.2  Asymptotic Wavelet Approximation at Resolution $2^{-k}$ . .  237

## 12  Wavelet Type Differentiated Shift-Invariant Integral Operators    259
12.1  General Results . . . . . . . . . . . . . . . . . . . .  259
12.2  Examples . . . . . . . . . . . . . . . . . . . . . . .  269

# III  On Partial Differential Equations 273

## 13  A Discrete Kac's Formula and Optimal Quantitative Approximation in the Solution of Heat Equation 275
13.1 Auxiliary Results . . . . . . . . . . . . . . . . . . . 277
13.2 The Dirichlet Problem: Discrete Case . . . . . . . . . . . 279
13.3 About Approximation on the Grid . . . . . . . . . . . . 293
13.4 On Sharpness of the Error Estimate . . . . . . . . . . . . 299

# IV  On Semigroups 303

## 14  Quantitative Asymptotic Expansions of the Probabilistic Representation Formulae for $(C_0)$ m-Parameter Operator Semigroups 305
14.1 Background . . . . . . . . . . . . . . . . . . . . . 305
14.2 Basic Results . . . . . . . . . . . . . . . . . . . . . 308
14.3 Quantitative Asymptotic Probabilistic Expansions . . . . . 314
14.4 Asymptotic Probabilistic Expansions for
Continuous Type Formulae . . . . . . . . . . . . . . . 323
14.5 Examples . . . . . . . . . . . . . . . . . . . . . . . 327

# V  On Stochastics 333

## 15  Quantitative Probability Limit Theorems over Banach Spaces 335
15.1 Limit Theorems for Martingales in Banach Spaces . . . . . 336
15.1.1 Auxiliary Results . . . . . . . . . . . . . . . 336
15.1.2 Main Theorems . . . . . . . . . . . . . . . . 340
15.2 On Weak Invariance Principle . . . . . . . . . . . . . . 348
15.2.1 Background . . . . . . . . . . . . . . . . . . 348
15.2.2 Central Results . . . . . . . . . . . . . . . . 352

## 16  Quantitative Study of Bias Convergence for Generalized L-Statistics 361
16.1 Background . . . . . . . . . . . . . . . . . . . . . 361
16.2 Quantitative Results for $L$-Statistics with
Standard Weights . . . . . . . . . . . . . . . . . . . 365
16.3 Quantitative Results for $L$-Statistics with
Nonstandard Weights . . . . . . . . . . . . . . . . . 369
16.4 More Conclusions . . . . . . . . . . . . . . . . . . . 379

## VI  On Functional Analysis                                           381

### 17  Quantitative Korovkin-Type Results for Vector Valued Functions    383
  17.1  Background ....................................... 383
  17.2  Auxiliary Results .................................. 385
  17.3  Quantitative General Theorems .................... 396
  17.4  Further Results ................................... 405
  17.5  Applications ...................................... 411

### 18  Quantitative $L_p$-Results for Positive Linear Operators         413
  18.1  Background ....................................... 414
  18.2  Univariate Results ................................ 419
  18.3  Abstract Theory .................................. 433
  18.4  Multidimensional Theory .......................... 438
  18.5  Stochastic Theory ................................ 444

## VII  On Approximation Theory                                          449

### 19  On Monotone Approximation Theory                                 451
  19.1  High Order Quantitative Monotone Approximation with
        Linear Differential Operators ...................... 451
  19.2  Quantitative Monotone Approximation with
        Pseudo-Polynomials ............................... 455
  19.3  Quantitative Bivariate Monotone Approximation ..... 461
  19.4  Quantitative Spline Monotone Approximation
        with Linear Differential Operators ................. 465
        19.4.1  Quantitative Monotone Approximation by
                Polynomial Splines ....................... 465
        19.4.2  Quantitative Monotone Approximation by Periodic
                Polynomial Splines ....................... 469
        19.4.3  Quantitative Monotone Approximation by Discrete
                Polynomial Splines ....................... 471

### 20  Comparisons for Local Moduli of Continuity                       473
  20.1  Univariate Results ................................ 473
  20.2  A Multivariate Result ............................. 479

### 21  Convergence with Rates of Univariate Singular Integrals
    to the Unit                                                         483
  21.1  Background ....................................... 483
  21.2  Quantitative $L^p$-Approximation, $1 \leq p < +\infty$ ....... 485

21.3 Quantitative Uniform Approximation by $Q_{n,\xi}$
Operator . . . . . . . . . . . . . . . . . . . . . . . . . . . . 492

# VIII  On Classical Analysis     495

## 22 About Univariate Ostrowski-Type Inequalities     497
22.1 About Ostrowski's Inequality . . . . . . . . . . . . . . . . 497
22.2 About More General Univariate Ostrowski-Type
Inequalities. . . . . . . . . . . . . . . . . . . . . . . . . . . 499

## 23 About Multidimensional Ostrowski-Type Inequalities     507
23.1 General Results . . . . . . . . . . . . . . . . . . . . . . . . 507

## 24 General Opial-Type Inequalities for Linear Differential Operators     521
24.1 Setting . . . . . . . . . . . . . . . . . . . . . . . . . . . . . 522
24.2 General Results . . . . . . . . . . . . . . . . . . . . . . . . 523
24.3 An Application to Differential Equations . . . . . . . . . . 536

## 25 $L_p$-Opial-Type Inequalities Engaging Fractional Derivatives of Functions     539
25.1 Background . . . . . . . . . . . . . . . . . . . . . . . . . . 539
25.2 General Results . . . . . . . . . . . . . . . . . . . . . . . . 545
25.3 Applications to Differential Equations . . . . . . . . . . . 556

## 26 $L_p$-General Fractional Opial Inequalities     567
26.1 General Results . . . . . . . . . . . . . . . . . . . . . . . . 567

References     585

List of Symbols     599

Index     605

# Preface

This book gives a thorough account of all significant developments in the area of contemporary quantitative mathematics. It shows that quantitative approximation methods penetrate and are applied in so many diverse, important mathematical research areas to the best of the author's knowledge for the first time in a book scheme. These areas include: neural networks, wavelet, partial differential equations, semigroups of operators, probability and statistics, functional analysis, approximation theory, and classical analysis.

The produced/exposed results, inequalities, and asymptotic expansions usually involve some various moduli of smoothness and many times have sharp, that is, optimal results.

Exact, or close to exact, values of errors for all the appearing convergences and differences are found in explicit form all over the book. Also, there are many applications given to the above-listed areas.

The main features of this book are precision and concreteness. That is, all the involved mathematical entities, especially the studied operators, are fully described and specific; in particular, the coefficients appearing there are precisely defined. This monograph contains all of the author's major research results over the past fifteen years on the subjects listed above in the quantitative sense. The exposed results either improve results of others or introduce new trends in research; all, for the first time, in book form.

The proving methods are rather elementary and constructive, that is, making this material accessible to a wide range of graduate students and researchers. This book is intended for use by researchers in the fields of approximation theory, pure mathematics, applied mathematics, differential equations, probability, statistics, numerical analysis, neural networks, wavelets, and engineering. It is also suitable for graduate courses in the above disciplines. Many of the exposed results in a number of chapters come from joint works of the author and the following: X.M. Yu of Southwest Missouri State University; the late S. Cambanis of the University of North Carolina; S. Gal of the University of Oradea, Romania; A. Bendikov of Cornell University; Zhou Mi of Data Stream Co., South Carolina; T. Rychlik of the Polish Academy of Science; and J. Pecaric of the University of Zagreb, Croatia. The author is greatly indebted to these leading researchers in their fields and would like to express his deep gratitude to them. Without these colloborations, this book would not have been as rich and as diverse. Also, the author wishes to thank his typist, Kate MacDougall of Warren, Rhode Island, for doing a great job so punctually.

And last, but not least, the author would like to thank Professor Michiel Hazewinkel of CWI, Amsterdam, the Netherlands, for giving him great encouragement throughout this major project.

George A. Anastassiou

# Chapter 1

## Introduction

To convey some of the essence and format of this monograph to the reader, this chapter briefly presents some of its main results. The proofs of the main results of this monograph are all based on the quantitative approach methods. For the convenience of the reader, the exposed results are numbered as they appear in their respective chapters.

## Part I: On Neural Networks

### 1.1 On Chapter 2: Convergence with Rates of Univariate Neural Network Operators to the Unit Operator

Here we use bell-shaped functions $b$ that are centered and of compact support $[-T, T]$, $T > 0$. Denote $I := \int_{-T}^{T} b(t)dt$, see Definition 2.1.

We study the pointwise convergence with rates over $\mathbf{R}$, to the unit operator, of the Univariate Cardaliaguet–Euvrard neural network operators (see [62]),

$$(F_n(f))(x) := f_n(x) := \sum_{k=-n^2}^{n^2} \frac{f(k/n)}{In^\alpha} \cdot b\left(n^{1-\alpha}\left(x - \frac{k}{n}\right)\right), \qquad (2.1)$$

where $0 < \alpha < 1$, $x \in \mathbf{R}$, and $n \in \mathbf{N}$. Here $f : \mathbf{R} \to \mathbf{R}$ is either continuous and bounded or uniformly continuous.

We present

**THEOREM 2.1.** *Let $x \in \mathbf{R}$, $T > 0$, and $n \in \mathbf{N}$ such that $n \geq \max(T +$*

$|x|$, $T^{-1/\alpha}$). Then it holds

$$|f_n(x) - f(x)| \leq |f(x)| \cdot \left| \sum_{k=\lceil nx-Tn^\alpha \rceil}^{[nx+Tn^\alpha]} \frac{1}{I \cdot n^\alpha} \cdot b\left(n^{1-\alpha} \cdot \left(x - \frac{k}{n}\right)\right) - 1 \right|$$
$$+ \frac{b(0)}{I} \cdot \left(2T + \frac{1}{n^\alpha}\right) \cdot \omega_1\left(f, \frac{T}{n^{1-\alpha}}\right), \qquad (2.5)$$

where $\omega_1$ is the first modulus of continuity of $f$. Inequality (2.5) becomes equality over constant functions.

Here $[\cdot]$, $\lceil \cdot \rceil$ are the integral part and ceiling functions.

Furthermore we get

**THEOREM 2.2.** Let $x \in \mathbf{R}$, $T > 0$, and $n \in \mathbf{N}$ such that $n \geq \max(T + |x|, T^{-1/\alpha})$. Let $f \in C^N(\mathbf{R})$, $N \in \mathbf{N}$, such that $f^{(N)}$ is a uniformly continuous function or $f^{(N)}$ is continuous and bounded. Then it holds

$$|f_n(x) - f(x)| = \left| \sum_{k=-n^2}^{n^2} \frac{f(k/n)}{I \cdot n^\alpha} \cdot b\left(n^{1-\alpha} \cdot \left(x - \frac{k}{n}\right)\right) - f(x) \right|$$

$$\leq |f(x)| \cdot \left| \sum_{k=\lceil nx-Tn^\alpha \rceil}^{[nx+Tn^\alpha]} \frac{1}{I \cdot n^\alpha} \cdot b\left(n^{1-\alpha} \cdot \left(x - \frac{k}{n}\right)\right) - 1 \right|$$

$$+ \frac{b(0)}{I} \cdot \left(2T + \frac{1}{n^\alpha}\right) \cdot \left( \sum_{j=1}^{N} \frac{|f^{(j)}(x)| \cdot T^j}{n^{j \cdot (1-\alpha)} \cdot j!} \right)$$

$$+ \omega_1\left(f^{(N)}, \frac{T}{n^{1-\alpha}}\right) \cdot \frac{T^N}{N! \cdot n^{N \cdot (1-\alpha)}} \cdot \frac{b(0)}{I} \cdot \left(2T + \frac{1}{n^\alpha}\right). \qquad (2.13)$$

Here (2.13) gives us with rates the pointwise convergence of $f_n(x) \to f(x)$, as $n \to +\infty$, $x \in \mathbf{R}$.

We need

**DEFINITION 2.2.** Let the nonnegative function $S : \mathbf{R} \to \mathbf{R}$, $S$ has compact support $[-T, T]$, $T > 0$, and is nondecreasing there and it can be continuous only on either $(-\infty, T]$ or $[-T, T]$. $S$ can have jump discontinuities. We call $S$ the "squashing function" (see also [62]).

# 1. Introduction

Again, $f : \mathbf{R} \to \mathbf{R}$ be either uniformly continuous or continuous and bounded. Assume that

$$I^* := \int_{-T}^{T} S(t)\, dt > 0.$$

For $x \in \mathbf{R}$ we define the "univariate squashing operator"

$$(G_n(f))(x) := \sum_{k=-n^2}^{n^2} \frac{f(k/n)}{I^* \cdot n^\alpha} \cdot S\left(n^{1-\alpha} \cdot \left(x - \frac{k}{n}\right)\right), \qquad (2.26)$$

$0 < \alpha < 1$ and $n \in \mathbf{N}$: $n \geq \max(T + |x|, T^{-1/\alpha})$.

Here we study the pointwise convergence with rates of $(G_n(f))(x) \to f(x)$, as $n \to \infty$, $x \in \mathbf{R}$.

**THEOREM 2.3.** *Under the above terms and assumptions we get*

$$|(G_n(f))(x) - f(x)|$$

$$\leq |f(x)| \cdot \left| \sum_{k=\lceil nx - Tn^\alpha \rceil}^{\lceil nx + Tn^\alpha \rceil} \frac{1}{I^* \cdot n^\alpha} \cdot S\left(n^{1-\alpha} \cdot \left(x - \frac{k}{n}\right)\right) - 1 \right|$$

$$+ \omega_1\left(f, \frac{T}{n^{1-\alpha}}\right) \cdot \frac{S(T)}{I^*} \cdot \left(2T + \frac{1}{n^\alpha}\right). \qquad (2.28)$$

Inequality (2.28) is attained by constant functions.

Moreover we obtain

**THEOREM 2.4.** *Let $x \in \mathbf{R}$, $T > 0$, and $n \in \mathbf{N}$, such that $n \geq \max(T + |x|, T^{-1/\alpha})$. Let $f \in C^N(\mathbf{R})$, $N \in \mathbf{N}$, such that $f^{(N)}$ is a uniformly continuous function or $f^{(N)}$ is continuous and bounded. Then it holds*

$$|(G_n(f))(x) - f(x)|$$

$$= \left| \sum_{k=-n^2}^{n^2} \frac{f(k/n)}{I^* \cdot n^\alpha} \cdot S\left(n^{1-\alpha} \cdot \left(x - \frac{k}{n}\right)\right) - f(x) \right|$$

$$\leq |f(x)| \cdot \left| \sum_{k=\lceil nx-Tn^{\alpha}\rceil}^{[nx+Tn^{\alpha}]} \frac{1}{I^* \cdot n^{\alpha}} \cdot S\left(n^{1-\alpha} \cdot \left(x - \frac{k}{n}\right)\right) - 1 \right|$$

$$+ \frac{S(T)}{I^*} \cdot \left(2T + \frac{1}{n^{\alpha}}\right) \cdot \left(\sum_{j=1}^{N} \frac{|f^{(j)}(x)| \cdot T^j}{j! \cdot n^{j \cdot (1-\alpha)}}\right)$$

$$+ \omega_1\left(f^{(N)}, \frac{T}{n^{1-\alpha}}\right) \cdot \frac{T^N}{N! \cdot n^{N \cdot (1-\alpha)}} \cdot \frac{S(T)}{I^*} \cdot \left(2T + \frac{1}{n^{\alpha}}\right). \quad (2.33)$$

The last is attained by constant functions.

## 1.2 On Chapter 3: Convergence with Rates of Multivariate Neural Network Operators to the Unit Operator

This is a continuation of Chapter 1 to the multidimensional case. Here $b : \mathbf{R}^d \to \mathbf{R}$ ($d \geq 1$) stands for a $d$-dimensional bell-shaped function; see Definition 3.2.

**ASSUMPTION.** Here $b(\vec{x})$ is of compact support $\mathcal{B} := \prod_{i=1}^{d}[-T_i, T_i]$, $T_i > 0$ and it may have jump discontinuities there. Set $I := \int_{\mathcal{B}} b(\vec{x})d\vec{x}$. Notice that $I > 0$. Let $f : \mathbf{R}^d \to \mathbf{R}$ be a continuous and bounded function or a uniformly continuous function.

We study the pointwise convergence with rates over $\mathbf{R}^d$, to the unit operator, of the *multivariate Cardaliaguet-Euvrard neural network operators* (see [62]),

$$(\mathcal{F}_n(f))(\vec{x}) := \sum_{k_1=-n^2}^{n^2} \cdots \sum_{k_d=-n^2}^{n^2} \frac{f(\frac{k_1}{n}, \ldots, \frac{k_d}{n})}{I \cdot n^{\alpha \cdot d}}$$
$$\cdot b\left(n^{1-\alpha} \cdot \left(x_1 - \frac{k_1}{n}\right), \ldots, n^{1-\alpha} \cdot \left(x_d - \frac{k_d}{n}\right)\right). \quad (3.1)$$

where $0 < \alpha < 1$ and $\vec{x} := (x_1, \ldots, x_d) \in \mathbf{R}^d$, $n \in \mathbf{N}$.

We present

## 1. Introduction

**THEOREM 3.1.** *Let $\vec{x} \in \mathbf{R}^d$, then it holds that*

$$|(\mathcal{F}_n(f))(\vec{x}) - f(\vec{x})| \leq |f(\vec{x})| \cdot \left| \sum_{k_1 = \lceil nx_1 - T_1 n^\alpha \rceil}^{[nx_1 + T_1 n^\alpha]} \cdots \sum_{k_d = \lceil nx_d - T_d n^\alpha \rceil}^{[nx_d + T_d n^\alpha]} \frac{1}{I \cdot n^{\alpha \cdot d}} \right.$$

$$\left. \cdot b\left(n^{1-\alpha} \cdot \left(x_1 - \frac{k_1}{n}\right), \ldots, n^{1-\alpha} \cdot \left(x_d - \frac{k_d}{n}\right)\right) - 1 \right|$$

$$+ \frac{b(\vec{0})}{I} \cdot \prod_{i=1}^{d}\left(2T_i + \frac{1}{n^\alpha}\right) \cdot \omega_1\left(f, \frac{T^*}{n^{1-\alpha}}\right), \qquad (3.8)$$

*where $\omega_1$ is the first multidimensional modulus of continuity. Inequality (3.8) is attained by constant functions.*

Furthermore we get

**THEOREM 3.2.** *Let $\vec{x} \in \mathbf{R}^d$, $f \in C^N(\mathbf{R}^d)$, $N \in \mathbf{N}$, such that all of its partial derivatives $f_{\tilde{\alpha}}$ of order $N$, $\tilde{\alpha}: |\tilde{\alpha}| = N$, are uniformly continuous or continuous and bounded. Then it holds*

$$|(\mathcal{F}_n(f))(\vec{x}) - f(\vec{x})| \leq |f(\vec{x})| \cdot \left| \sum_{\vec{k} = \lceil n\vec{x} - \vec{T}n^\alpha \rceil}^{[n\vec{x} + \vec{T}n^\alpha]} \frac{1}{I \cdot n^{\alpha \cdot d}} \right.$$

$$\left. \cdot b\left(n^{1-\alpha} \cdot \left(\vec{x} - \frac{\vec{k}}{n}\right)\right) - 1 \right| + \frac{b(\vec{0})}{I} \cdot \left(\prod_{i=1}^{d}\left(2T_i + \frac{1}{n^\alpha}\right)\right)$$

$$\cdot \left\{ \sum_{j=1}^{N} \frac{(T^*)^j}{j! n^{j(1-\alpha)}} \cdot \left( \left( \sum_{i=1}^{d} \left| \frac{\partial}{\partial x_i} \right| \right)^j f(\vec{x}) \right) \right\}$$

$$+ \frac{(T^*)^N \cdot d^N}{N! n^{N(1-\alpha)}} \cdot \frac{b(\vec{0})}{I} \cdot \left(\prod_{i=1}^{d}\left(2T_i + \frac{1}{n^\alpha}\right)\right)$$

$$\cdot \max_{\tilde{\alpha}: |\tilde{\alpha}| = N} \omega_1\left(f_{\tilde{\alpha}}, \frac{T^*}{n^{1-\alpha}}\right). \qquad (3.27)$$

*Inequality (3.27) is attained by constant functions. Also (3.27) gives us with rates the pointwise convergence of $\mathcal{F}_n(f) \to f$ over $\mathbf{R}^d$, as $n \to +\infty$.*

Additionally we treat the multivariate "squashing operators" and their convergence to the unit with rates.

We need

**DEFINITION 3.4.** *Let the nonnegative function $S : \mathbf{R}^d \to \mathbf{R}$, $d \geq 1$. $S$ has compact support $\mathcal{B} := \prod_{i=1}^{d}[-T_i, T_i]$, $T_i > 0$ and is nondecreasing there for each coordinate. $S$ can be continuous only on either $\prod_{i=1}^{d}(-\infty, T_i]$ or $\mathcal{B}$ and can have jump discontinuities. We call $S$ the multivariate "squashing function" (see also [62]). Suppose that*

$$I^* := \int_{\mathcal{B}} S(\vec{t}) d\vec{t} > 0.$$

Let $f : \mathbf{R}^d \to \mathbf{R}$ be either a uniformly continuous or a continuous and bounded function.

For $\vec{x} \in \mathbf{R}^d$ we define the multivariate "squashing operator"

$$(G_n(f))(\vec{x}) := \sum_{k_1=-n^2}^{n^2} \cdots \sum_{k_d=-n^2}^{n^2} \frac{f\left(\frac{k_1}{n}, \ldots, \frac{k_d}{n}\right)}{I^* \cdot n^{\alpha \cdot d}} \cdot S\left(n^{1-\alpha}\left(x_1 - \frac{k_1}{n}\right), \ldots, n^{1-\alpha}\left(x_d - \frac{k_d}{n}\right)\right), \tag{3.34}$$

where $0 < \alpha < 1$ and $n \in \mathbf{N}$:

$$n \geq \max_{i \in \{1, \ldots, d\}} \{T_i + |x_i|, T_i^{-1/\alpha}\}. \tag{3.5}$$

Here we study the pointwise convergence with rates of $(G_n(f))(\vec{x}) \to f(\vec{x})$, as $n \to +\infty$, $\vec{x} \in \mathbf{R}^d$. This is given by the next result.

**THEOREM 3.3.** *Under the above terms and assumptions we get that*

$$|(G_n(f))(\vec{x}) - f(\vec{x})| \leq |f(\vec{x})| \cdot \left| \sum_{\vec{k} = \lceil n\vec{x} - \vec{T}n^\alpha \rceil}^{[n\vec{x} + \vec{T}n^\alpha]} \frac{1}{I^* \cdot n^{\alpha \cdot d}} \cdot S\left(n^{1-\alpha}\left(\vec{x} - \frac{\vec{k}}{n}\right)\right) - 1 \right| + \frac{S(\vec{T})}{I^*}$$

$$\cdot \prod_{i=1}^{d}\left(2T_i + \frac{1}{n^\alpha}\right) \cdot \omega_1\left(f, \frac{T^*}{n^{1-\alpha}}\right). \tag{3.36}$$

## 1. Introduction

*Inequality (3.36) is attained by constant functions.*

As a related result we give

**THEOREM 3.4.** *Let $\vec{x} \in \mathbf{R}^d$, $f \in C^N(\mathbf{R}^d)$, $N \in \mathbf{N}$, such that all of its partial derivatives $f_{\tilde{\alpha}}$ of order $N$, $\tilde{\alpha}: |\tilde{\alpha}| = N$, are uniformly continuous or continuous and bounded. Then it holds*

$$|(G_n(f))(\vec{x}) - f(\vec{x})| \le |f(\vec{x})| \cdot \left| \sum_{\vec{k}=\lceil n\vec{x}-\vec{T}n^\alpha \rceil}^{\lceil n\vec{x}+\vec{T}n^\alpha \rceil} \frac{1}{I^* \cdot n^{\alpha \cdot d}} \right.$$

$$\left. \cdot S\left(n^{1-\alpha}\left(\vec{x} - \frac{\vec{k}}{n}\right)\right) - 1 \right| + \frac{S(\vec{T})}{I^*} \cdot \left( \prod_{i=1}^d \left(2T_i + \frac{1}{n^\alpha}\right) \right)$$

$$\cdot \left\{ \sum_{j=1}^N \frac{(T^*)^j}{j! n^{j(1-\alpha)}} \cdot \left( \left( \sum_{i=1}^d \left|\frac{\partial}{\partial x_i}\right| \right)^j f(\vec{x}) \right) \right\}$$

$$+ \frac{(T^*)^N \cdot d^N}{N! n^{N(1-\alpha)}} \cdot \frac{S(\vec{T})}{I^*} \cdot \left( \prod_{i=1}^d \left(2T_i + \frac{1}{n^\alpha}\right) \right)$$

$$\cdot \max_{\tilde{\alpha}:|\tilde{\alpha}|=N} \omega_1\left(f_{\tilde{\alpha}}, \frac{T^*}{n^{1-\alpha}}\right). \quad (3.40)$$

*Inequality (3.40) is attained by constant functions. Also (3.40) gives us with rates the pointwise convergence of $G_n(f) \to f$ over $\mathbf{R}^d$, as $n \to +\infty$.*

---

### 1.3 On Chapter 4: Asymptotic Weak Convergence of Cardaliaguet-Euvrard Neural Network Operators

Here $b$ is a bell-shaped function of compact support contained in $[-T, T]$, $T > 0$; or more generally, $b$ can have the property

$$\int_{-\infty}^{\infty} |x|^r b(x)\, dx < \infty \quad \text{for all } r = 0, \ldots, N-1,$$

where $N \in \mathbf{N} : N > \frac{3-2\alpha}{1-\alpha}$, $0 < \alpha < 1$. Set $I = \int_{-\infty}^{\infty} b(x)dx$; see Definition 4.2.

Here we would like to study the class of functions $\Phi_N := \{f \in C^N(\mathbf{R})$, $N \in \mathbf{N}$ such that all $f, f', \ldots, f^{(N)}$ have compact supports contained in $[-T^*, T^*]$, $T^* > 0$, additionally $f^{(N)} \in BV([-T^*, T^*])\}$.

Here we take $N \in \mathbf{N}$ such that

$$N > \frac{3-2\alpha}{1-\alpha}, \quad \text{where } 0 < \alpha < 1.$$

For any $n \in \mathbf{N}$ and $x \in \mathbf{R}$ we consider the Cardaliaguet-Euvrard neural network operators [62]

$$(F_n(f))(x) := \sum_{k=-n^2}^{n^2} \frac{f(\frac{k}{n})}{In^\alpha} b\left(n^{1-\alpha}\left(x - \frac{k}{n}\right)\right). \tag{4.3}$$

In this chapter we study the weak convergence asymptotically of $F_n(f)$ to $f \in \Phi_N$, by taking $n \geq T^*$. More precisely, we present an asymptotic expansion for the error $e_n(f, g)$ of the weak convergence of $F_n(f)$ to $f$, where

$$e_n(f, g) := \int_{-\infty}^{\infty} (F_n(f))(x) g(x) \, dx - \int_{-\infty}^{\infty} f(x) g(x) \, dx, \quad g \in \phi_N. \tag{4.5}$$

We give the following

**THEOREM 4.1.** *For $n \geq T^*$ it holds*

$$e_n(f, g) = \sum_{r=1}^{N-1} \frac{m_r \int_{-T^*}^{T^*} f(t) g^{(r)}(t) \, dt}{Ir! n^{r(1-\alpha)}} + o\left(\frac{1}{n^{(1-\alpha)(N-1)-1}}\right), \tag{4.8}$$

*where*

$$m_r := \int_{-\infty}^{\infty} x^r b(x) \, dx \tag{4.9}$$

*and $N \in \mathbf{N}$ such that*

$$N > \frac{3-2\alpha}{1-\alpha}.$$

## 1.4 On Chapter 5: Asymptotic Weak Convergence of Squashing Neural Network Operators

Here $S$ stands for the "squashing function" of compact support $[-T, T]$, $T > 0$. We assume that

$$J := \int_{-T}^{T} S(x)\,dx > 0. \tag{5.1}$$

Also $\Phi_N$ is the class of functions as in §1.3.

Here we again take $N \in \mathbf{N}$ such that

$$N > \frac{3-2\alpha}{1-\alpha}, \quad \text{where } 0 < \alpha < 1.$$

For any $n \in \mathbf{N}$ and $x \in \mathbf{R}$ we consider the squashing neural network operators

$$(K_n(f))(x) := \sum_{k=-n^2}^{n^2} \frac{f(\frac{k}{n})}{Jn^\alpha} S\left(n^{1-\alpha}\left(x - \frac{k}{n}\right)\right). \tag{5.2}$$

In this chapter we study, asymptotically, the weak convergence of $K_n(f)$ to $f \in \Phi_N$, by taking $n \geq T^*$. More precisely we find an asymptotic expansion for the error $\gamma_n(f, g)$ of the weak convergence of $K_n(f)$ to $f$, where

$$\gamma_n(f, g) := \int_{-\infty}^{\infty} (K_n(f))(x)g(x)\,dx - \int_{-\infty}^{\infty} f(x)g(x)\,dx, \quad g \in \phi_N. \tag{5.4}$$

We present the following

**THEOREM 5.1.** *For* $n \geq T^*$ *it holds that*

$$\gamma_n(f, g) = \sum_{r=1}^{N-1} \frac{m_r \int_{-T^*}^{T^*} f(t)g^{(r)}(t)\,dt}{Jr!\,n^{r(1-\alpha)}} + o\left(\frac{1}{n^{(1-\alpha)(N-1)-1}}\right), \tag{5.7}$$

*where*

$$m_r := \int_{-T}^{T} x^r S(x)\,dx \tag{5.8}$$

and $N \in \mathbf{N}$ such that
$$N > \frac{3-2\alpha}{1-\alpha}.$$

## Part II: On Wavelets

### 1.5 On Chapter 6: Quantitative Monotone and Probabilistic Wavelet Type Approximation

Assume that $\varphi(x)$ is a bounded right-continuous function on $\mathbf{R}$ and has compact support. Define
$$\varphi_{kj}(x) := 2^{\frac{k}{2}}\varphi(2^k x - j) \quad \text{for } k, j \in Z.$$

For $f \in C(\mathbf{R})$, we define
$$A_k(f)(x) := \sum_{j=-\infty}^{\infty} \langle f, \varphi_{kj}\rangle \varphi_{kj}(x) \quad \text{for } k \in Z, \tag{6.1}$$

where
$$\langle f, \varphi_{kj}\rangle := \int_{-\infty}^{\infty} f(t)\varphi_{kj}(t)\, dt.$$

We present

**THEOREM 6.1.** *Assume that $\varphi(x)$ is a bounded function with* $\operatorname{supp} \varphi(x) \subseteq [-a, a]$, $0 < a < +\infty$ *and satisfies the following conditions:*

(i) $\displaystyle\sum_{j=-\infty}^{\infty} \varphi(x-j) \equiv 1$ *on* $\mathbf{R}$;

(ii) *there is a number $b$ such that $\varphi(x)$ is nondecreasing if $x \leq b$ and is nonincreasing if $x \geq b$.*

*Then, for $f \in C(\mathbf{R})$, if $f$ is a nondecreasing function, the linear wavelet type operators $A_k(f)$ defined by (6.1) are also nondecreasing functions on $\mathbf{R}$ and satisfy*
$$|A_k(f)(x) - f(x)| \leq \omega(f, 2^{-k+1}a), \quad x \in \mathbf{R}, \ k \in \mathbf{Z}, \tag{6.6}$$

# 1. Introduction

where $\omega(f,h)$ is the modulus of continuity of $f$. Furthermore, the inequalities (6.6) are sharp.

Next we give

**THEOREM 6.2.** *Assume that $\varphi(x)$ is a bounded right-continuous function with* supp $\varphi(x) \subseteq [-a,a]$, $0 < a < +\infty$, *and satisfies the conditions (i)–(ii) of Theorem 6.1. Let $F(x)$ be a continuous distribution function on $\mathbf{R}$. Then the linear wavelet type operators $A_k(F)$ defined by (6.1) are distribution functions and satisfy*

$$|A_k(F)(x) - F(x)| \leq \omega(F, 2^{-k+1}a), \quad x \in \mathbf{R}, \; k \in \mathbf{Z}.$$

*Furthermore the above inequalities are sharp.*

Now we define another linear wavelet type operator for $f \in C(\mathbf{R})$ and $k \in Z$:

$$B_k(f)(x) = \sum_{j=-\infty}^{\infty} f(2^{-k}j)\varphi(2^k x - j). \tag{6.22}$$

We present

**THEOREM 6.3.** *Assume that $\varphi(x)$ is a bounded function with* supp $\varphi(x) \subseteq [-a,a]$, $0 < a < +\infty$ *and satisfies the conditions (i)–(ii) of Theorem 6.1. Then, for $f \in C(\mathbf{R})$, if $f$ is a nondecreasing function, the linear wavelet type operators $B_k(f)$ defined by (6.22) are also nondecreasing on $\mathbf{R}$ and satisfy*

$$|B_k(f)(x) - f(x)| \leq \omega(f, 2^{-k}a), \quad x \in \mathbf{R}, \; k \in \mathbf{Z}. \tag{6.24}$$

*Furthermore, the inequalities (6.24) are sharp.*

Moreover we have

**THEOREM 6.4.** *Assume that $\varphi(x)$ is a bounded right-continuous function with* supp $\varphi(x) \subseteq [-a,a]$, $0 < a < +\infty$ *and satisfies the conditions (i)–(ii) of Theorem 6.1. Let $F(x)$ be a continuous distribution function on $\mathbf{R}$. Then the linear wavelet type operators $B_k(F)$ defined by (6.22) are distribution functions and satisfy*

$$|B_k(F)(x) - F(x)| \leq \omega(F, 2^{-k}a), \quad x \in \mathbf{R}, \; k \in \mathbf{Z}.$$

*Furthermore the above estimates are sharp.*

## 1.6 On Chapter 7: Quantitative One-Dimensional High Order Wavelet Type Approximation

We obtain

**THEOREM 7.1.** *Let $f \in C^N(\mathbf{R})$, $N \geq 1$, $x \in \mathbf{R}$, and $k \in \mathbf{Z}$. Let $\varphi$ be a bounded function of compact support $\subseteq [-a,a]$, $a > 0$ such that $\sum_{j=-\infty}^{\infty} \varphi(x-j) = 1$ all $x \in \mathbf{R}$. Suppose $\varphi \geq 0$. Put*

$$(B_k(f))(x) := \sum_{j=-\infty}^{\infty} f\left(\frac{j}{2^k}\right) \varphi(2^k x - j).$$

*Then*

$$|(B_k(f))(x) - f(x)| \leq \sum_{i=1}^{N} \frac{|f^{(i)}(x)|}{i!} \frac{a^i}{2^{ki}} + \frac{a^N}{2^{kN} N!} \omega_1\left(f^{(N)}, \frac{a}{2^k}\right), \quad (7.1)$$

*which is attained by a constant function.*

**REMARK 7.1.** i) Given that $f^{(N)}$ is continuous and bounded or uniformly continuous on $\mathbf{R}$, we get that $\omega_1\left(f^{(N)}, \frac{a}{2^k}\right) < \infty$ and $B_k f \to f$, pointwise over $\mathbf{R}$, as $k \to \infty$.

Also it holds

**THEOREM 7.2.** *Same assumptions as in Theorem 7.1. Additionally suppose that $\varphi$ is Lebesgue measurable (then $\int_{-\infty}^{+\infty} \varphi(x) dx = 1$). Define*

$$\varphi_{kj}(x) := 2^{k/2} \varphi(2^k x - j) \quad \text{all } k, j \in \mathbf{Z},$$

$$\langle f, \varphi_{kj} \rangle := \int_{-\infty}^{\infty} f(t) \varphi_{kj}(t) dt,$$

*and*

$$(A_k(f))(x) := \sum_{j=-\infty}^{\infty} \langle f, \varphi_{kj} \rangle \varphi_{kj}(x)$$

$$= \sum_{j=-\infty}^{\infty} \left( \int_{-\infty}^{\infty} f\left(\frac{u}{2^k}\right) \varphi(u-j) du \right) \varphi(2^k x - j).$$

1. Introduction

*Then it holds*

$$|(A_k(f))(x) - f(x)| \le \sum_{i=1}^{N} \frac{|f^{(i)}(x)|}{i!} \frac{a^i}{2^{i(k-1)}} + \frac{a^N}{N! 2^{N(k-1)}} \omega_1\left(f^{(N)}, \frac{a}{2^{k-1}}\right), \quad (7.5)$$

*which is attained by constants.*

As above $A_k \to I$, $k \to \infty$, pointwise.
Furthermore we get

**THEOREM 7.3.** *Same assumptions as in Theorem 7.1. Define*

$$C_k(f)(x) := \sum_{j=-\infty}^{\infty} \gamma_{kj}(f) \varphi(2^k x - j)$$

$$:= \sum_{j=-\infty}^{\infty} \left(2^k \int_{2^{-k}j}^{2^{-k}(j+1)} f(t) dt\right) \varphi(2^k x - j).$$

*That is,*

$$\gamma_{kj}(f) := 2^k \int_{2^{-k}j}^{2^{-k}(j+1)} f(t) dt = 2^k \int_{0}^{2^{-k}} f\left(t + \frac{j}{2^k}\right) dt.$$

*Then* ($x \in \mathbf{R}$, $k \in \mathbf{Z}$)

$$|(C_k(f))(x) - f(x)| \le \sum_{i=1}^{N} \frac{|f^{(i)}(x)|(a+1)^i}{i! \, 2^{ki}} + \frac{(a+1)^N}{2^{kN} N!} \omega_1\left(f^{(N)}, \frac{a+1}{2^k}\right), \quad (7.19)$$

*which is attained by constants.*

Again, as above, $C_k \to I$, $k \to \infty$, pointwise.
One obtains that

$$|(C_k(f))(x) - f(x)| \le \omega_1\left(f, \frac{\alpha+1}{2^k}\right), \quad f \in C(\mathbf{R}), x \in \mathbf{R}, k \in \mathbf{Z}. \quad (7.37)$$

We prove that (7.37) is sharp over convex $f \in C(\mathbf{R})$.

## 1.7 On Chapter 8: More on Shape and Probability Preserving One-Dimensional Wavelet Type Operators

Let $f \in C^N(\mathbf{R})$, $N \geq 0$, $x \in \mathbf{R}$, $k, j \in \mathbf{Z}$. Let $\varphi$ be a bounded function of compact support $\subseteq [-a, a]$, $a > 0$ such that $\sum_{j=-\infty}^{\infty} \varphi(x-j) = 1$ all $x \in \mathbf{R}$. Also assume there exists $b \in \mathbf{R}$ such that $\varphi$ is nondecreasing for $x \leq b$ and $\varphi$ is nonincreasing for $x \geq b$ (that is, $\varphi \geq 0$).

Define

$$C_k(f)(x) := \sum_{j=-\infty}^{\infty} \gamma_{kj}(f)\varphi(2^k x - j)$$
$$:= \sum_{j=-\infty}^{\infty} \left( 2^k \int_{2^{-k}j}^{2^{-k}(j+1)} f(t)dt \right) \varphi(2^k x - j), \quad (8.1)$$

i.e.,

$$\gamma_{kj}(f) := 2^k \int_{2^{-k}j}^{2^{-k}(j+1)} f(t)dt = 2^k \int_0^{2^{-k}} f\left(t + \frac{j}{2^k}\right) dt. \quad (8.2)$$

Also define

$$D_k(f)(x) := \sum_{j=-\infty}^{\infty} \delta_{kj}(f)\varphi(2^k x - j), \quad (8.3)$$

where

$$\delta_{kj}(f) := \sum_{r=0}^{n} w_r f\left(\frac{j}{2^k} + \frac{r}{2^k n}\right), \quad n \in \mathbf{N}, w_r \geq 0, \sum_{r=0}^{n} w_r = 1. \quad (8.4)$$

We give

**THEOREM 8.1.** *Let $\varphi$ as in Lemma 8.3. Additionally suppose that $\varphi$ is right continuous. Let $f \in C(\mathbf{R})$ be a probability distribution function. Then, for any $k \in \mathbf{Z}$, we have that $C_k(f)$, $D_k(f)$ are probability distribution functions.*

Also we mention

**PROPOSITION 7.1.** *Let $f \in C(\mathbf{R})$, $x \in \mathbf{R}$, $k \in \mathbf{Z}$. Let $\varphi$ be a bounded function of compact support $\subseteq [-a, a]$, $a > 0$ such that $\sum_{j=-\infty}^{\infty} \varphi(x-j) = 1$*

## 1. Introduction

on **R**. Also there exists $b \in \mathbf{R}$ such that $\varphi$ is nondecreasing for $x \leq b$ and $\varphi$ is nonincreasing for $x \geq b$. Then it holds

$$|C_k(f)(x) - f(x)| \leq \omega_1\left(f, \frac{a+1}{2^k}\right), \tag{8.9}$$

and

$$|D_k(f)(x) - f(x)| \leq \omega_1\left(f, \frac{a+1}{2^k}\right). \tag{8.10}$$

Optimality is presented in

**THEOREM 8.2.** *Inequalities (8.9) and (8.10) are sharp (in fact attained).*

As a related result we present

**PROPOSITION 8.2.** *Let $f \in C(\mathbf{R})$ which is convex. Let $\varphi$ as in Lemma 8.4. Then $C_k(f)(x), D_k(f)(x)$ are convex, any $k \in \mathbf{Z}$.*

Also it holds

**PROPOSITION 8.3.** *Let $\varphi$ as in Lemma 8.5. Let $f \in C(\mathbf{R})$ which is 3th-convex. Then $C_k(f)(x)$ and $D_k(f)(x)$ are 3th-convex, any $k \in \mathbf{Z}$.*

### 1.8 On Chapter 9: Quantitative Multidimensional High Order Wavelet Type Approximation

We establish

**THEOREM 9.1.** *Let $f \in C^N(\mathbf{R}^r)$, $N$ and $r \geq 1$; $\vec{x} \in \mathbf{R}^r$ and $k \in \mathbf{Z}$. Let $\varphi \geq 0$ be a bounded function on $\mathbf{R}^r$ of compact support $\subseteq \prod_{i=1}^r [-a_i, a_i]$, $0 < a_i < +\infty$, $a := \max(a_1, \ldots, a_r)$. Suppose that*

$$\sum_{j_1=-\infty}^{\infty} \cdots \sum_{j_r=-\infty}^{\infty} \varphi(x_1 - j_1, \ldots, x_r - j_r) = 1, \quad \text{all } \vec{x} := (x_1, \ldots, x_r) \in \mathbf{R}^r, \tag{9.1}$$

*in short*

$$\sum_{\vec{j}=-\infty}^{\infty} \varphi(\vec{x} - \vec{j}) = 1, \quad \text{all } \vec{x} \in \mathbf{R}^r, \tag{9.1'}$$

where $\vec{j} := (j_1, \ldots, j_r)$. Put

$$B_k(f)(x_1, \ldots, x_r) := \sum_{j_1=-\infty}^{\infty} \cdots \sum_{j_r=-\infty}^{\infty} f\left(\frac{j_1}{2^k}, \ldots, \frac{j_r}{2^k}\right) \cdot \varphi(2^k x_1 - j_1, \ldots, 2^k x_r - j_r), \quad \text{any } k \in \mathbf{Z}, \tag{9.2}$$

all $(x_1, \ldots, x_r) \in \mathbf{R}^r$; in short

$$B_k(f)(\vec{x}) = \sum_{\vec{j}=-\infty}^{\infty} f\left(\frac{\vec{j}}{2^k}\right) \varphi(2^k \vec{x} - \vec{j}), \quad \text{any } k \in \mathbf{Z}, \text{ all } \vec{x} \in \mathbf{R}^r. \tag{9.2'}$$

Here we further suppose, that all of the partial derivatives of $f$ of order $N$, denoted by

$$f_{\tilde{\alpha}} := \frac{\partial^{\tilde{\alpha}} f}{\partial x^{\tilde{\alpha}}} \left(\tilde{\alpha} := (\alpha_1, \ldots, \alpha_r), \alpha_i \in \mathbf{Z}^+, i = 1, \ldots, r : |\tilde{\alpha}| = \sum_{i=1}^{r} \alpha_i = N\right),$$

are continuous and bounded or uniformly continuous on $\mathbf{R}^r$.
Then it holds

$$|(B_k(f))(\vec{x}) - f(\vec{x})| \leq \sum_{j=1}^{N} \frac{a^j}{j! 2^{kj}} \left(\left(\sum_{i=1}^{r} \left|\frac{\partial}{\partial x_i}\right|\right)^j f(\vec{x})\right) + \frac{a^N r^N}{N! 2^{kN}} \max_{\tilde{\alpha}: |\tilde{\alpha}|=N} \omega_1\left(f_{\tilde{\alpha}}, \frac{a}{2^k}\right), \quad \text{any } k \in \mathbf{Z}, \tag{9.3}$$

which is attained by constant functions, where $\omega_1$ is appropriate multidimensional first modulus of continuity.

**REMARK 9.1.** i) Clearly here $B_k f \to f$ pointwise over $\mathbf{R}^r$, as $k \to \infty$.

Also it holds

**THEOREM 9.2.** *Here we consider $f$ and $\varphi$ as in Theorem 9.1 and use the same notations. Additionally suppose that $\varphi$ is Lebesque measurable. Define*

$$A_k(f)(\vec{x}) := \sum_{\vec{j}=-\infty}^{\infty} \alpha_{k\vec{j}}(f) \varphi(2^k \vec{x} - \vec{j}), \tag{9.9}$$

where

$$\alpha_{k\vec{j}}(f) := \int_{\mathbf{R}^r} f\left(\frac{\vec{u}}{2^k}\right) \varphi(\vec{u} - \vec{j}) d\vec{u}, \quad k \in \mathbf{Z}. \tag{9.10}$$

*1. Introduction*                                                                                                                    17

*Then for any $k \in \mathbf{Z}$ we get*

$$|A_k(f)(\vec{x}) - f(\vec{x})| \leq \sum_{j=1}^{N} \frac{a^j}{j! 2^{(k-1)j}} \left( \left( \sum_{i=1}^{r} \left| \frac{\partial}{\partial x_i} \right| \right)^j f(\vec{x}) \right) \\
+ \frac{a^N r^N}{N! 2^{(k-1)\mathbf{N}}} \max_{\tilde{\alpha}: |\tilde{\alpha}|=N} \omega_1 \left( f_{\tilde{\alpha}}, \frac{a}{2^{k-1}} \right), \quad (9.11)$$

*which is attained by constant functions.*

**REMARK 9.2.** i) *Clearly here $A_k f \to f$ pointwise on $\mathbf{R}^r$, as $k \to \infty$.*

Furthermore we get

**THEOREM 9.3.** *Here are same assumptions and notations as in Theorem 9.1. Put*

$$\gamma_{k\vec{j}}(f) := 2^{rk} \int_{2^{-k}\vec{j}}^{2^{-k}(\vec{j}+\vec{1})} f(\vec{t}) d\vec{t} = 2^{rk} \int_{\vec{0}}^{2^{-k}} f\left( \vec{t} + \frac{\vec{j}}{2^k} \right) d\vec{t}, \quad (9.19)$$

*and*

$$C_k(f)(\vec{x}) := \sum_{\vec{j}=-\infty}^{\infty} \gamma_{k\vec{j}}(f) \varphi(2^k \vec{x} - \vec{j}), \quad (9.20)$$

*all $\vec{x} \subset \mathbf{R}^r$ and $k \subset \mathbf{Z}$. Then it holds*

$$|C_k(f)(\vec{x}) - f(\vec{x})| \leq \sum_{j=1}^{N} \frac{(a+1)^j}{j! 2^{kj}} \left( \left( \sum_{i=1}^{r} \left| \frac{\partial}{\partial x_i} \right| \right)^j f(\vec{x}) \right) \\
+ \frac{(a+1)^N r^N}{N! 2^{kN}} \max_{\tilde{\alpha}: |\tilde{\alpha}|=N} \omega_1 \left( f_{\tilde{\alpha}}, \frac{a+1}{2^k} \right), \quad (9.21)$$

*which is attained by constant functions.*

**REMARK 9.3.** i) *Clearly here $C_k f \to f$ pointwise on $\mathbf{R}^r$, as $k \to \infty$.*

Specializing we find

**PROPOSITION 9.1.** *Let $f \in C(\mathbf{R}^r)$, $r \geq 1$, $a := \max(a_1, \ldots, a_r)$, $k \in \mathbf{Z}$, $\vec{x} \in \mathbf{R}^r$.*

*i) Under the notations and assumptions of Theorem 9.2 we get*

$$|A_k(f)(\vec{x}) - f(\vec{x})| \leq \omega_1 \left( f, \frac{a}{2^{k-1}} \right). \quad (9.36)$$

ii) Under the notations and assumptions of Theorems 9.1 and 9.3 we have
$$|C_k(f)(\vec{x}) - f(\vec{x})| \leq \omega_1\left(f, \frac{a+1}{2^k}\right). \tag{9.37}$$

Inequalities (9.36), (9.37) are proved to be sharp!

## 1.9 On Chapter 10: Rate of Convergence of Probabilistic Discrete Wavelet Approximation

Let $\Phi(x) := \begin{cases} 1, & x \geq 0 \\ 0, & x < 0 \end{cases}$ and consider

$$\mathcal{F} := \left\{ F \text{ distribution functions on } \mathbf{R} : \\ \int_{-\infty}^{+\infty} |F(x) - \Phi(x)|^p \, dx < +\infty, \quad 1 \leq p < +\infty \right\} \tag{10.1}$$

For any distribution function $F \in \mathcal{F}$, we define

$$L_0(F, x) := \sum_{j=-\infty}^{\infty} (C_j - C_{j-1})\Phi(x - j) \tag{10.3}$$

where

$$C_j := \int_j^{j+1} F(t)\, dt, \quad j \in Z \tag{10.4}$$

and

$$L_k(F, x) := L_0(F(2^{-k}\cdot), 2^k x), \quad k \in Z. \tag{10.5}$$

We present

**THEOREM 10.1.** *Let $F(x)$ be a distribution function in $\mathcal{F}$ and $1 \leq p < +\infty$. Then $L_k(F, x)$ defined by (10.3) and (10.5) are distribution functions such that*
$$\|L_k(F, x) - F(x)\|_p \leq 2\omega(F, 2^{-k})_p, \tag{10.11}$$
*where $\omega(\cdot, \cdot)_p$ is the $L_p$-modulus of continuity.*

## 1.10 On Chapter 11: Asymptotic Nonorthogonal Wavelet Approximation for Deterministic Signals

We present the following result.

**THEOREM 11.1.** *Let $f$ be such that $f, f', f'', \ldots, f^{(n)}$ are functions in $L_1(\mathbf{R})$ and $f^{(n)} \in BV(\mathbf{R}) \cap C(\mathbf{R})$, $n \geq 2$. Let $\varphi$ be a Lebesgue measurable bounded function with support $\varphi \subseteq [-a, a]$, $0 < a < +\infty$, such that*

$$\sum_{j=-\infty}^{+\infty} \varphi(x-j) = 1, \quad \forall x \in \mathbf{R}. \tag{11.1}$$

Put

$$(B_k f)(x) := \sum_{j=-\infty}^{+\infty} f\left(\frac{j}{2^k}\right) \cdot \varphi(2^k x - j) \tag{11.2}$$

and

$$e_k^2(f) := \int_{-\infty}^{+\infty} (f(x) - (B_k f)(x))^2 \cdot dx, \quad k \in \mathbf{Z}. \tag{11.3}$$

Set

$$\beta := \begin{cases} [2a], & \text{if } 2a \neq \text{integer}, \\ 2a - 1, & \text{if } 2a \text{ is an integer}, \end{cases} \tag{11.4}$$

where $[\,]$ is the integral part function. Put

$$\ell := \begin{cases} \frac{n-1}{2}, & n \text{ is odd}, \\ \frac{n-2}{2}, & n \text{ is even} \end{cases} = \lceil \frac{n}{2} \rceil - 1, \tag{11.5}$$

where $\lceil \cdot \rceil$ is the ceiling of the number. Call

$$\rho_\gamma(f) := \frac{(-1)^\gamma}{(2\gamma)!} \cdot \int_{-\infty}^{+\infty} (f^{(\gamma)}(t))^2 dt, \quad \gamma = 1, \ldots, \ell. \tag{11.6}$$

Set ($\gamma = 1, \ldots, \ell$)

$$\lambda_\gamma(\varphi) := \int_{-\infty}^{+\infty} \varphi(u) \cdot \left\{ -2 \cdot u^{2\cdot\gamma} + \sum_{q=1}^{\beta} q^{2\cdot\gamma} \cdot (\varphi(u+q) + \varphi(u-q)) \right\} \cdot du. \tag{11.7}$$

Then, for $k \in \mathbf{N}$, we have that

$$e_k^2(f) = \sum_{\gamma=1}^{\ell^*} \frac{\rho_\gamma(f) \cdot \lambda_\gamma(\varphi)}{4^{k \cdot \gamma}} + o\left(\frac{1}{2^{k \cdot (n-2)}}\right), \qquad (11.8)$$

where

$$\ell^* := \begin{cases} \ell, & \text{if } n \text{ is even,} \\ \ell - 1, & \text{if } n \text{ is odd.} \end{cases}$$

We see that $e_k(f) \to 0$ as $k \to +\infty$, i.e., for $n \geq 2$ we get that $B_k(f) \to f$ as $k \to +\infty$ in $\mathcal{L}_2$-norm.

## 1.11 On Chapter 12: Wavelet Type Differentiated Shift Invariant Integral Operators

Let $X := C_U(\mathbf{R})$ be the space of uniformly continuous real-valued functions on $\mathbf{R}$ and $C(\mathbf{R})$ the space of continuous functions from $\mathbf{R}$ into itself. We use the $p$th modulus of smoothness

$$\omega_p(f; \delta) = \sup_{0 < h \leq \delta} \|\Delta_h^p f\|_\infty,$$

where

$$\Delta_h^p f(t) := \sum_{i=0}^{p} (-1)^{p-i} \binom{p}{i} f(t + ih)$$

is the forward finite difference. Let $\{l_k\}_{k \in \mathbf{Z}}$ be a sequence of positive linear operators from $C(\mathbf{R})$ into itself such that

$$l_k(f)(x) = l_0(f(2^{-k} \cdot))(x), \quad x \in \mathbf{R}, \quad f \in C(\mathbf{R}). \qquad (12.1)$$

Suppose that

$$l_0 : C^{(i)}(\mathbf{R}) \to C^{(i)}(\mathbf{R}), \quad i \in \mathbf{N} \text{ fixed}, \qquad (12.2)$$

where $C^{(i)}(\mathbf{R})$ is the space of $i$-times continuously differentiable functions.
In this chapter we consider only $f \in C^{(i)}(\mathbf{R})$ such that $f^{(i)} \in C_U(\mathbf{R})$. Let $\varphi$ be a continuous real-valued function of compact support $\subseteq [-a, a]$.

# 1. Introduction

$\varphi \geq 0$ and
$$\int_{-\infty}^{+\infty} \varphi(x-u)du = 1, \quad x \in \mathbf{R}. \tag{12.3}$$

Let $\{L_k\}_{k \in \mathbf{Z}}$ be the sequence of linear operators acting on $C^{(i)}(\mathbf{R})$ and defined as follows

$$L_0(f)(x) := \int_{-\infty}^{+\infty} l_0(f)(u)\varphi(x-u)du, \tag{12.5}$$

and
$$L_k(f)(x) := L_0(f(2^{-k}\cdot))(2^k x), \quad x \in \mathbf{R}, \ k \in \mathbf{Z}, \tag{12.6}$$

**ASSUMPTION.** For fixed $a > 0$, $f \in C^{(i)}(\mathbf{R})$, ($i \in \mathbf{N}$ fixed) such that $f^{(i)} \in X$, let us suppose that

$$\sup_{\substack{u,y \in \mathbf{R} \\ |u-y| \leq \alpha}} |l_0^{(i)}(f)(u) - f^{(i)}(y)| \leq \omega_1\left(f^{(i)}; \frac{ma+n}{2^r}\right), \tag{12.10}$$

where $m \in \mathbf{N}$, $n \in \mathbf{Z}_+$, $r \in \mathbf{Z}$.

We present

**THEOREM 12.1.** *For any $f \in C^{(i)}(\mathbf{R})$ such that $f^{(i)} \in X$, $i \in \mathbf{N}$ fixed, suppose that*

$$|\Delta_h^p l_0^{(i)}(f)(x-u)| \leq \omega_p(f^{(i)}; h), \quad \forall\, x, u \in \mathbf{R},\ h > 0, \tag{12.11}$$

*where $p \in \mathbf{N}$ is fixed and $\Delta_h^p$ is the pth forward finite difference operator. Then*
$$\omega_p(L_k^{(i)}(f); \delta) \leq \omega_p(f^{(i)}; \delta), \quad \forall\, \delta > 0,\ k \in \mathbf{Z}. \tag{12.12}$$

Next we define the operators

$$L_{k,j}(f)(x) = \int_{-\infty}^{+\infty} \ell_k(f)(2^k x - ju)\varphi(u)du, \quad k \in \mathbf{Z},\ j \in \mathbf{N}. \tag{12.13}$$

We prove that $L_{k,j}^{(i)}$ are shift invariant operators under a mild assumption: see Theorem 12.3.

We give the global smoothness preservation property of $L_{k,j}^{(i)}$.

**THEOREM 12.4.** *Assume that (12.11) holds, for $f \in C^{(i)}(\mathbf{R})$, $f^{(i)} \in X$, $i \in \mathbf{N}$ fixed. Then it holds*

$$\omega_p(L_{k,j}^{(i)}(f); \delta) \le \omega_p(f^{(i)}; \delta), \quad k \in \mathbf{Z}, \ j \in \mathbf{N}, \ \delta > 0. \tag{12.16}$$

Inequality (12.16) is sharp when $p = 1$, under a mild assumption on $\ell_0$, see Theorem 12.10.

Convergence with rates of $L_{k,j}^{(i)}$ operators is established:

**THEOREM 12.5.** *Let $f \in C^{(i)}(\mathbf{R})$, $f^{(i)} \in X$, $i \in \mathbf{N}$ fixed. Suppose that (12.10) holds. Then*

$$|L_{k,j}^{(i)}(f)(x) - f^{(i)}(x)| \le \omega_1\left(f^{(i)}; \frac{mja+n}{2^{k+r}}\right), \tag{12.17}$$

*where $m, j \in \mathbf{N}$, $n \in \mathbf{Z}_+$, $k, r \in \mathbf{Z}$.*

Several examples follow.

# Part III: On Partial Differential Equations

## 1.12 On Chapter 13: A Discrete Kac's Formula and Optimal Quantitative Approximation in the Solution of Heat Equation

Let $\Omega$ be the open unit cube in $\mathbf{R}^l$, $l \ge 1$ and $\dot{\Omega} := \Omega \times I$ be an "interval" in space-time $\dot{\mathbf{R}}^l = \mathbf{R}^l \times \mathbf{R}$, where $I := (0, T)$, $T > 0$. Denote by $\dot{\Delta} = \frac{1}{2}\Delta - \partial_t$ the heat operator, where $\Delta$ stands for the Laplacian operator in $\mathbf{R}^l$. Given a continuous function $\lambda \ge 0$ on $\dot{\Omega}$, consider the Dirichlet problem in $\dot{\Omega}$:

$$\dot{\Delta} u(\dot{x}) - \lambda(\dot{x})u(\dot{x}) = -f(\dot{x}), \quad \dot{x} \in \dot{\Omega},$$

$$\lim_{\dot{x} \to \dot{y}} u(\dot{x}) = \varphi(\dot{y}), \quad \forall \dot{y} \in \partial\dot{\Omega} - \{\dot{x} = (x, t) : t = T\}.$$

It is a known fact [103] that the solution $u$ of this equation is also ex-

## 1. Introduction

pressed by

$$u(\dot{x}) := E_{\dot{x}}\left\{\int_0^{\tau_{\dot{\Omega}}} M_s \cdot f(\dot{x}(s))\,ds + M_{\tau_{\dot{\Omega}}} \cdot \varphi(\dot{x}(\tau_{\dot{\Omega}}))\right\}, \qquad (13.10)$$

where $E$ stands for the expectation, and $\tau_{\dot{\Omega}}$ is the first exit time of the space-time Brownian motion $\dot{x}(\cdot,\omega)$, $\dot{x}(0,\omega) = \dot{x}$, of an open regular set $\dot{\Omega} \subset \dot{R}^l$. Then, according to [103], the function $u(\dot{x})$, $\dot{x} \in \dot{\Omega}$, gives a solution of the following Dirichlet problem

$$\left.\begin{aligned}\dot{\Delta} u(\dot{x}) - \lambda(\dot{x})u(\dot{x}) &= -f(\dot{x}), & \dot{x} &\in \dot{\Omega}, \\ \lim_{\dot{x}\to\dot{y}} u(\dot{x}) &= \varphi(\dot{y}), & \dot{y} &\in \partial\dot{\Omega}.\end{aligned}\right\} \qquad (13.11)$$

Here we consider a discrete analog of the Kac's formula and then its application to approximation on a uniform grid.

In particular we consider the following Dirichlet problem on the interval $\dot{\Omega} = \{\dot{x} \in \dot{R}^l : 0 < x_i < 1,\ 0 < t < \infty,\ i = 1,\ldots,l\}$:

$$\left.\begin{aligned}\dot{\Delta} u(\dot{x}) - \lambda(\dot{x})u(\dot{x}) &= -f(\dot{x}), & \dot{x} &\in \dot{\Omega}, \\ u(\dot{x}) &= \varphi(\dot{x}), & \dot{x} &\in \partial\dot{\Omega},\end{aligned}\right\} \qquad (13.22)$$

with $\varphi$ and $\lambda \geq 0$ continuous functions and $f$ a locally bounded Hölder function. Here we restrict our treatment to problem (13.22) for which $u \in C^{(2)}(\overline{\dot{\Omega}})$.

Let $h := 1/n$ with $n \in \mathbf{N}$ and $k := h^2/l$. An approximate solution $u_h$, defined on the grid $\overline{\dot{\Omega}}_h = \dot{\Omega}_h \cup \partial\dot{\Omega}_h$, where

$$\overline{\dot{\Omega}}_h := \{\dot{x} = (x_1,\ldots,x_l,t) \in \dot{R}^l : x_i = k_i h,\ t = jk,$$

$$k_i = 0,\ldots,n,\ i = 1,\ldots,l,\ j = 0,1,\ldots\},$$

and

$$\dot{\Omega}_h := \overline{\dot{\Omega}}_h \cap \dot{\Omega},\quad \partial\dot{\Omega}_h := \overline{\dot{\Omega}}_h \setminus \dot{\Omega}_h,$$

is gotten as the solution of the discrete counterpart to (13.22)

$$\left.\begin{aligned}\dot{\Delta}_h u_h(\dot{x}) - \lambda(\dot{x})u_h(\dot{x}) &= -f(\dot{x}), & \dot{x} &\in \dot{\Omega}_h, \\ u_h(\dot{x}) &= \varphi(\dot{x}), & \dot{x} &\in \partial\dot{\Omega}_h.\end{aligned}\right\} \qquad (13.23)$$

Here the "discrete" parabolic Laplacian $\dot{\Delta}_h$ is given by

$$\dot{\Delta}_h u(\dot{x}) := \frac{1}{2} h^{-2} \left( \sum_{k=1}^{l} u(\dot{x} \pm he_k - k \cdot e) - 2lu(\dot{x}) \right),$$

where $k = k(h) := h^2/l$.

Next the rate of convergence of $u_h \to u$ as $h \downarrow 0$ can be measured by the partial moduli of continuity of $f : \omega_1$ and $\omega_{2,i}$, $i = 1, \ldots, l$, where

$$\omega_1(\delta, f; \overline{\dot{\Omega}}) := \sup\{|f(\dot{x} + \theta e) - f(\dot{x})| : \dot{x}, \dot{x} + \theta e \in \overline{\dot{\Omega}}, |\theta| < \delta\},$$

$$\omega_{2,i}(\delta, f; \overline{\dot{\Omega}}) := \sup\{|f(\dot{x} + \theta e_i) - 2f(\dot{x}) + f(\dot{x} - \theta e_i)| :$$

$$\dot{x}, \dot{x} \pm \theta e_i \in \overline{\dot{\Omega}}, |\theta| < \delta\}.$$

Namely we have

**THEOREM 13.10.** *Assume that the solution $u$ of the problem (13.22) is in $C^{(2)}(\overline{\dot{\Omega}})$, then for the solution $u_h$ of the problem (13.23) the following inequality holds valid*

$$\|u_h - u\|_{\overline{\dot{\Omega}}_{h,T}} \leq \min\left\{T, \frac{1}{4}\right\} \left[ \frac{1}{4} \sum_{i=1}^{l} \omega_{2,i}(h, \partial^2_{x_i} u; \overline{\dot{\Omega}}_{h,T}) \right.$$

$$\left. + \frac{1}{2} \sum_{i=1}^{l} \omega_1(k, \partial^2_{x_i} u; \overline{\dot{\Omega}}_{h,T}) + \omega_1(k, \partial_t u; \overline{\dot{\Omega}}_{h,T}) \right].$$
(13.27)

Here $\dot{\Omega}_{h,T} := \{\dot{x} \in \dot{\Omega}_h : t \leq T\}$. Inequality (13.27) is sharp.

# Part IV: On Semigroups

## 1.13 On Chapter 14: Quantitative-Asymptotic Expansions for the Probabilistic Representation Formulae for $(C_0)$ $m$-Parameter Operator Semigroups

For terms, definitions, and notations please see §14.1. In this chapter $T$ stands for the semigroup, $E$ for the expectation, and $\omega$ is a special modulus of continuity.

We prove

**THEOREM 14.1.** *Let $X = (X_{01}, \ldots, X_{0m})$ be an $\mathbf{R}_+^m$-valued random vector with $E(X) = x = (x_1, \ldots, x_m)$ and there exists a $\delta > 0$ such that $\Psi_{\tilde{X}}^*(\delta) < \infty$. Let further that $X_i \overset{i.i.d.}{\sim} X$, $i = 1, 2, \ldots$, and denote $S_n = \sum_{i=1}^n X_i$. Assume $g \in D^{2k}$ for some fixed $k \in \mathbf{N}$. Then for sufficiently large $n$ we obtain*

$$\left\| \{E[T(X/n)]\}^n g - T(x)g - \sum_{j=2}^{2k} \frac{1}{j!} E\left[ \left( \left( \frac{1}{n} S_n - x \right) \bullet A \right)^j \right] T(x)g \right\|$$

$$\leq \frac{K}{n^k} \omega\left(T, 2k, g; \frac{1}{\sqrt{n}}\right), \tag{14.18}$$

*where $K$ is a constant independent of $n$.*

Also we obtain

**THEOREM 14.2.** *Let $N$ be a $\mathbf{Z}_+$-valued random variable with $E(N) = \eta$, $\eta > 0$, and let $Y_l \overset{i.i.d.}{\sim} Y$, $l = 1, 2, \ldots$, where $Y = (Y_{01}, \ldots, Y_{0m})$ is an $\mathbf{R}_+^m$-valued random vector independent of $N$ with $E(Y) = \gamma = (\gamma_1, \ldots, \gamma_m)$. Suppose that there exists a $\delta > 0$ such that*

$$\Psi_N(\Psi_{\tilde{Y}}^*(\delta)) < \infty.$$

*Let further, $X_i \overset{i.i.d.}{\sim} X$, $i = 1, 2, \ldots$, where $X = \sum_{j=1}^N Y_j$, and assume*

$g \in D^{2k}$ for some fixed $k \in \mathbf{Z}_+$. Then for sufficiently large $n$ we obtain

$$\left\| \{\Psi_N(E[T(Y/n)])\}^n g - T(\eta\gamma)g \right.$$
$$\left. - \sum_{j=2}^{2k} \frac{1}{j!} E\left[ \left( \left( \frac{1}{n} \sum_{i=1}^{n} X_i - \eta\gamma \right) \bullet A \right)^j \right] T(\eta\gamma)g \right\| \quad (14.25)$$
$$\leq \frac{K}{n^k} \omega\left( T, 2k, g; \frac{1}{\sqrt{n}} \right),$$

where $K$ is a constant independent of $n$.

And furthermore we get

**THEOREM 14.3.** *Let $N = (N_1, \ldots, N_m)$ be a $\mathbf{Z}_+^m$-valued random vector with $E(N) = \eta = (\eta_1, \ldots, \eta_m)$. For each $r \in \{1, \ldots, m\}$, let $\{Y_{lr}\}_{l=1}^{\infty}$ be a sequence of i.i.d. $\mathbf{R}_+$-valued random variables distributed as $Y$, which is a fixed $\mathbf{R}_+$-valued random variable with $E(Y) = \gamma$. $N$ and $Y_{lr}$'s ($l = 1, 2, \ldots, r = 1, \ldots, m$) are assumed to be altogether independent. Also suppose that there exists a $\delta > 0$ such that*

$$\Psi_{\tilde{N}}(\Psi_Y^*(\delta)) < \infty.$$

*Let further $X_i \overset{i.i.d.}{\sim} X$, $i = 1, 2, \ldots$, where*

$$X := \left( \sum_{l_1}^{N_1} Y_{l_1 1}, \ldots, \sum_{l_m}^{N_m} Y_{l_m m} \right).$$

*Assume $g \in D^{2k}$ for some fixed $k \in \mathbf{Z}_+$. Then for sufficiently large $n$ we find*

$$\left\| \left\{ E\left[ T\left( \sum_{l_1=1}^{N_1} \frac{1}{n} Y_{l_1 1}, \ldots, \sum_{l_m=1}^{N_m} \frac{1}{n} Y_{l_m m} \right) \right] \right\}^n g - T(\gamma\eta)g \right.$$
$$\left. - \sum_{j=2}^{2k} \frac{1}{j!} E\left[ \left( \left( \frac{1}{n} \sum_{i=1}^{n} X_i - \gamma\eta \right) \bullet A \right)^j \right] T(\gamma\eta)g \right\| \quad (14.26)$$
$$\leq \frac{K}{n^k} \omega\left( T, 2k, g; \frac{1}{\sqrt{n}} \right),$$

*where $K$ is a constant independent of $n$.*

Several examples follow.

# Part V: On Stochastics

## 1.14  On Chapter 15: Quantitative Probability Limit Theorems over Banach Spaces

Here we give limit theorems for martingales in Banach spaces and we discuss the weak invariance principle. For terms, definitions, and notations, please see §§15.1.1 and 15.2.1.

We present

**THEOREM 15.1.** Let $(X_i, \mathcal{F}_i)_{i \in \mathbf{Z}_+}$ be a Martingale Difference Sequence (MDS), $Z$ be a $\varphi$-decomposable r.v. with $E(Z) = 0$, and $E(\|Z\|_B^r) < \infty$, $r \in \mathbf{N} - \{1\}$. Assume that $E(\|X_i\|_B^r) < \infty$ $\forall i \in \mathbf{N}$; also suppose that

$$E(X_i^v \mid \mathcal{F}_{i-1}) = E(\mathbf{Z}_i^v) \quad a.s. \quad (1 \le |v| \le r - 1, i \in \mathbf{N}). \tag{15.21}$$

Let $f \in C_B^{r-1}$. Then it holds

$$|E(f(\varphi(n) \cdot S_n)) - E(f(Z))| \le \omega_1(f^{(r-1)}, h) \cdot h^{r-1} \cdot \left\{ \frac{\varphi^{r-1}(n)}{r!} \right.$$

$$\left. + \frac{(\varphi(n))^{(r-1)^2/r}}{2 \cdot (r-1)!} + \frac{(\varphi(n))^{((r^2 - 3r + 2)/r)}}{8 \cdot (r-2)!} \right\}, \tag{15.22}$$

- where

$$h := \left( \varphi(n) \cdot E \left( \left( \sum_{i=1}^{n} (\|X_i\|_B + \|Z_i\|_B) \right)^r \right) \right)^{1/r}. \tag{15.23}$$

We also give the weak law of large numbers for MDS on $B$-spaces with rates when $\|(\|x\|_B)'\|_{\mathcal{L}_1} \le A$, $x \in B - \{0\}$.

**THEOREM 15.2.** Let $(X_i, \mathcal{F}_i)_{i \in \mathbf{Z}_+}$ be an MDS, and $E(\|X_i\|_B^2) < \infty$, $i \in \mathbf{N}$. Let $f \in C_B^1$. Then it holds

$$|E(f(\varphi(n) \cdot S_n)) - f(0)| \le \omega_1(f', h) \cdot \frac{h}{2} \cdot \left\{ \varphi(n) + \sqrt{\varphi(n)} + \frac{1}{4} \right\}. \tag{15.40}$$

where

$$h = \left( \varphi(n) \cdot E\left( \left( \sum_{i=1}^{n} \|X_i\|_B \right)^2 \right) \right)^{1/2}. \tag{15.41}$$

Furthermore we give the central limit theorem for martingales in Banach spaces with a countable basis.

**THEOREM 15.3.** *Let $(X_i, \mathcal{F}_i)_{i \in \mathbf{Z}_+}$ be an MDS, $R$ be the covariance functional of any mean zero Gaussian r.v., and $r \in \mathbf{N} - \{1\}$. Assume $E(\|X_i\|_B^r) < \infty$, $i \in \mathbf{N}$, and suppose that there exists a sequence of $a_i > 0$ such that*

$$E(X_i^v \mid \mathcal{F}_{i-1}) = a_i^{|v|} \cdot E(X_R^v) \tag{15.42}$$

*a.s.* $(1 \leq |v| \leq r-1, i \in \mathbf{N})$. *Let $f \in C_B^{r-1}$. Put*

$$A_n := \left( \sum_{i=1}^{n} a_i^2 \right)^{1/2}. \tag{15.43}$$

*Then it holds*

$$|E(f(A_n^{-1} \cdot S_n)) - E(f(X_r))| \leq \omega_1(f^{(r-1)}, h) \cdot h^{r-1}$$
$$\cdot \left\{ \frac{A_n^{1-r}}{r!} + \frac{A_n^{-(r-1)^2/r}}{2 \cdot (r-1)!} + \frac{A_n^{-((r^2-3r+2)/r)}}{8 \cdot (r-2)!} \right\}. \tag{15.44}$$

Also we study the *weak invariance principle*, that is, the convergence of the distributions of the random polygonal lines

$$S_n(\omega, t) := n^{-1/2} \cdot \left( \sum_{i=0}^{[nt]} X_i(\omega) + (nt - [nt]) \cdot X_{[nt]+1}(\omega) \right) \tag{15.46}$$

to the Wiener-measure $W_R$ on $C = C([0, 1])$, the space of all real-valued, continuous functions on $[0, 1]$ with the supremum-norm.

We give

**THEOREM 15.4.** *Let $(X_i)_{i \in \mathbf{Z}_+}$ be a sequence of identically distributed real r.v.s with mean 0 and variance 1, and let $(a_{n_i})_{1 \leq i \leq n} > 0$, $n \in \mathbf{N}$ and $A_n := (\sum_{i=1}^{n} a_{ni}^2)^{1/2}$. Suppose that*

$$\zeta_r := E(|X_1|^r) < \infty \tag{15.64}$$

1. Introduction

for $r \in \mathbf{N} - \{1\}$, and for the r.f.s $(X_{ni})_{1 \le i \le n}$, $n \in \mathbf{N}$ defined in (15.47) we assume

$$E(X_{ni}^{\nu} \mid \mathcal{A}_{ni}) = a_{ni}^{j} \cdot A_n^{-j} \cdot E(W^{\nu}) \quad a.s. \tag{15.65}$$

for all $|\nu| = j$, $1 \le j \le r-1$, $1 \le i \le n$, and $n \in \mathbf{N}$. Here

$$\mathcal{A}_{ni} := \mathcal{A}(X_1, \ldots, X_{i-1}, W_{n,i+1}, \ldots, W_{nn}) \tag{15.66}$$

(the sub-$\sigma$-algebra generated by $\{X_1, \ldots, X_{i-1}, W_{n,i+1}, \ldots, W_{nn}\}$). Let $f \in C_C^{r-1}$. Then $P_{S_n}$ converges weakly to the Wiener-measure $W_R$, with rates given from

$$|E(f(S_n(t))) - E(f(W(t)))| \le \omega_1(f^{(r-1)}, h)$$
$$\cdot \left\{ \frac{1}{r!} + \left( \frac{n^{((3-r)/2)} \cdot \zeta_r^{((r-1)/r)} + A_n^{1-r} \cdot (\sum_{i=1}^n a_{ni}^{r-1}) \cdot l_r^{((r-1)/r)}}{2 \cdot (r-1)!} \right) \right.$$
$$\left. + \frac{h}{8(r-2)!} \cdot \left\{ n^{((4-r)/2)} \cdot \zeta_r^{((r-2)/r)} + A_n^{2-r} \left( \sum_{i=1}^n a_{ni}^{r-2} \right) \cdot l_r^{((r-2)/r)} \right\} \right\}, \tag{15.67}$$

where

$$h := h_n := n^{(2-r)/2} \cdot \zeta_r + A_n^{-r} \cdot \left( \sum_{i=1}^n a_{ni}^r \right) \cdot l_r < \infty. \tag{15.68}$$

Let $X^*$ be a standard normally distributed r.v., then we prove

**THEOREM 15.5.** *Let $(X_i)_{i \in \mathbf{N}}$ be a sequence of identically distributed, possibly dependent, real r.v.s for which $E(X_1) = 0$, $\mathrm{Var}(X_1) = 1$, and $\zeta_r := E(|X_1|^r) < \infty$, for $r \in \mathbf{N} - \{1\}$.*

*(i) Suppose that $(X_i)_{i \in \mathbf{N}}$ is dependent from below and that*

$$E(X_i^j \mid \mathcal{E}_{i-1}) = E(X^{*j}) \quad a.s. \ (1 \le j \le r-1, i \in \mathbf{N}), \tag{15.85}$$

where

$$\mathcal{E}_{i-1} := \mathcal{A}(X_1, \ldots, X_{i-1}).$$

*Let $f : \mathbf{R} \to \mathbf{R}$ for which $f^{(j)}$ is bounded and continuous on $\mathbf{R}$ for all*

$0 \leq j \leq r-1$. Then it holds

$$\left| E\left( f\left( \frac{\sum_{i=1}^{n} X_i}{\sqrt{n}} \right) \right) - E(f(X^*)) \right|$$
$$\leq \omega_1(f^{(r-1)}, h) \cdot \left\{ \frac{1}{r!} + \frac{n^{((3-r)/2)} \cdot (\zeta_r^{((r-1)/r)} + l_r^{((r-1)/r)})}{2 \cdot (r-1)!} \right.$$
$$\left. + \frac{h}{8 \cdot (r-2)!} \cdot n^{((4-r)/2)} \cdot (\zeta_r^{((r-2)/r)} + l_r^{((r-2)/r)}) \right\}. \tag{15.86}$$

Here
$$l_r = E(|X^*|^r) < \infty$$

and
$$h := h_n = n^{((2-r)/2)} \cdot (\zeta_r + l_r).$$

(ii) Assume that the sequence $(X_i)_{i \in \mathbf{N}}$ is dependent from above and that

$$E(X_i^j \mid F_{i+1}^*) = E(X^{*j}) \quad a.s. \quad (1 \leq j \leq r-1, i \in \mathbf{N}), \tag{15.87}$$

where
$$F_{i+1}^* := \mathcal{A}(X_{i+1}, \ldots, X_n)$$

and $f$ as in part (i). Then (15.86) is true again. The RHS (15.86) converges to zero as $n \to \infty$, when $r \geq 5$.

## 1.15 On Chapter 16: Quantitative Study of Bias Convergence for Generalized $L$-Statistics

For a detailed explanation of terms, definitions, and notations, please see Chapter 16 itself.

Here let $X_n$, $n \in \mathbf{N}$, be a sequence of independent identically distributed (i.i.d.) random variables with a common distibution function $F$. For a sample of size $n : X_1, \ldots, X_n$; the sequence of order statistics is denoted by $X_{1:n}, \ldots, X_{n:n}$. A sequence $L_n$, $n \in \mathbf{N}$, of (generalized) $L$-statistics is defined by

$$L_n = \sum_{i=1}^{n} c_{i,n} g(X_{i:n}), \quad n \in \mathbf{N}, \tag{16.1}$$

## 1. Introduction

where $g$ is a score function. For a weight function $f : [0,1] \to \mathbf{R}$ we define the $L$-estimate sequence

$$L_n(f,g) = \sum_{i=1}^{n} l_{i,n} g(X_{i:n}) = n^{-1} \sum_{i=1}^{n} f\left(\frac{i-1}{n-1}\right) g(X_{i:n}), \quad n \in \mathbf{N} \setminus \{1\}. \tag{16.4}$$

We also need to mention the limit functional

$$\mu(f,g,F) = \mathrm{E}_F[g(X_1) f(F(X_1))] = \int_0^1 g(F^{-1}(y)) f(y)\, dy. \tag{16.6}$$

Let

$$(B_n f)(x) = \sum_{k=0}^{n} \binom{n}{k} x^k (1-x)^{n-k} f\left(\frac{k}{n}\right) \tag{16.11}$$

the $n$th Bernstein operator for $f \in C[0,1]$.

We give

**THEOREM 16.1.** *For an i.i.d. sequence $X_i$, $i \in \mathbf{N}$, with a common distribution function $F$ and $\mathrm{E}|g(X_1)| < \infty$, and $L$-statistics defined by (16.4), we get*

$$\mathrm{E}_F L_n(f,g) = \int_0^1 g(F^{-1}(y))(B_{n-1} f)(y)\, dy, \tag{16.12}$$

*If $f \in C^1[0,1]$, then for $n \in \mathbf{N} \setminus \{1\}$, it holds*

$$|\mathrm{E}_F L_n(f,g) - \mu(f,g,F)| \le \frac{25}{32}(n-1)^{-1/2} \omega_1\left(f', \frac{1}{4}(n-1)^{-1/2}\right) \mathrm{E}_F |g(X_1)|.$$

Also we present

**THEOREM 16.2.** *Under the assumptions and notations of Theorem 16.1, with $f \in C[0,1]$, we have for all $n \in \mathbf{N} \setminus \{1\}$*

$$|\mathrm{E}_F L_n(f,g) - \mu(f,g,F)| \le \|B_{n-1} f - f\|_\infty \mathrm{E}_F |g(X_1)|$$
$$\le \begin{cases} \frac{4306 + 837\sqrt{6}}{5832} \omega_1\left(f, \frac{1}{\sqrt{n-1}}\right) \\ \frac{35}{32} \omega_2\left(f, \frac{1}{\sqrt{n-1}}\right) \end{cases} \mathrm{E}_F |g(X_1)|. \tag{16.13}$$

*Furthermore, the first estimate in (16.13) is optimal.*

Above $\omega_2$ is the second modulus of smoothness.

Furthermore we have

**THEOREM 16.3.** *All notations and assumptions here are as in Theorem 16.2. For all $n \in \mathbf{N} \setminus \{1\}$, we obtain*

$$|E_F L_n(f,g) - \mu(f,g,F)| \leq \begin{Bmatrix} 2K(f,(n-1)^{-1}), \\ C\omega_2^\varphi(f,(n-1)^{-1/2}) \end{Bmatrix} E_F|g(X_1)|,$$

where $K, \omega_2^\varphi$ are the Ditzian-Totik K-functional and second modulus of smoothness, respectively.

Next we would like to mention the Kantorovich operators

$$(K_n f)(x) = (n+1) \sum_{k=0}^{n} \left( \int_{k/(n+1)}^{(k+1)/(n+1)} f(t)\,dt \right) p_{n,k}(x), \quad n \in \mathbf{N} \cup \{0\},$$
(16.14)

for either $f \in L_p[0,1]$ or $f \in C[0,1]$ in cases $1 \leq p < \infty$ and $p = \infty$, respectively. Here

$$p_{n,k}(x) := \binom{n}{k} x^k (1-x)^{n-k}.$$

We give

**THEOREM 16.7.** *Let $X_i$, $i \in \mathbf{N}$, be an i.i.d. sequence of random variables. We follow the assumptions and notation of Remark 16.2 and suppose that $\mathrm{E}|g(X_1)|^q < \infty$ for $q = p/(p-1)$ and $p > 1$. Then it holds*

$$E_F K_n(f,g) = \int_0^1 g(F^{-1}(y))(K_{n-1}f)(y)\,dy$$

*and*

$$|E_F K_n(f,g) - \mu(f,g,F)| \leq \|K_{n-1}f - f\|_p (E_F|g(X_1)|^q)^{1/q}, \quad n \in \mathbf{N} \setminus \{1\}.$$

Here the $L$-estimate sequence is

$$K_n(f,g) = \sum_{i=1}^{n} k_{i,n} g(X_{i:n}) = \sum_{i=1}^{n} \left[ \int_{(i-1)/n}^{i/n} f(t)\,dt \right] g(X_{i:n}), \quad n \in \mathbf{N}.$$
(16.5)

Consequently we get

1. Introduction

**THEOREM 16.8.** *Under the assumptions and notations of Theorem 16.7, we obtain*

$$|E_F K_n(f,g) - \mu(f,g,F)|$$

$$\leq \begin{cases} CK^*(f,(n-1)^{-1/2})_p \\ C^*[\omega_2^\varphi(f,(n-1)^{-1/2})_p + \omega_1(f,(n-1)^{-1})_p] \end{cases} (E_F|g(X_1)|^q)^{1/q},$$

*for $1 < p < \infty$ and universal constants $C, C^* > 0$. Furthermore, if $p = \infty$, then*

$$|E_F K_n(f,g) - \mu(f,g,F)| \leq \|K_{n-1}f - f\|_\infty E_F|g(X_1)|$$

$$\leq \begin{cases} CK^*(f,(n-1)^{-1/2})_\infty \\ C^*[\omega_2^\varphi(f,(n-1)^{-1/2})_\infty + \omega_1(f,(n-1)^{-1})_\infty] \\ \frac{M}{n-1}\left[\int_{(n-1)^{-1/2}}^{1/2} \omega_2^\varphi(f,t)_\infty t^{-3}\,dt + E_0(f)_\infty\right] \end{cases} E_F|g(X_1)|,$$

*where $M > 0$ and $E_0(f)_\infty := \inf_{c \in \mathbf{R}} \|f - c\|_\infty$.*

Here $K^*$ is the Gonska-Zhou $K$-functional and $\omega_2^\varphi(\cdot,\cdot)_p$ the second modulus of smoothness with weight function $\varphi(x) = \sqrt{x(1-x)}$, with respect to the $L_p$ norm. And $\omega_1(\cdot,\cdot)_p$ is the first $L_p$-modulus of continuity.

We also present quantitative results for $L$-statistics with nonstandard weights where we involve the Bernstein-Durrmeyer operators, etc.

# Part VI: On Functional Analysis

## 1.16 On Chapter 17: Quantitative Korovkin-Type Results for Vector Valued Functions

Let $X$ be a normed vector space, $Y$ be a Banach lattice, $M \subset X$ is a compact and convex subset. Consider the space of continuous functions from $M$ into $Y$, denoted by $C(M,Y)$, and the space of bounded functions $B(M,Y)$. Here we study the rate of the uniform convergence of lattice homomorphisms $T : C(M,Y) \to C(M,Y)$ or $T : C(M,Y) \to B(M,Y)$ to the unit operator $I$.

In our main results we assume that $X$ is a Banach space.

Our presented inequalities are of Shisha-Mond type, i.e., of Korovikin type. More precisely, we find upper bounds to $|||Tf - f|||_\infty$, $f \in C(M, Y)$, and $|||TP - P|||_\infty$, $P \in C^n(M, Y)$, $n \in \mathbf{N}$, (space of $n$-times continuously Fréchet differentiable functions), where $||| \cdot |||_\infty$ is the supremum norm in $C(M, Y)$ or $B(M, Y)$. These inequalities involve the modulus of continuity $\omega_1$ of $f$ or $P^{(n)}$.

We present

**THEOREM 17.1.** *Let $M$ be a compact and convex subset of $(X, \|\cdot\|)$ and $(Y, \|\cdot\|, <)$ is a Banach lattice. Let $T$ be a lattice homomorphism from $C(M, Y)$ into itself, and $f \in C(M, Y)$. Then it holds*

$$|||Tf - f|||_\infty \leq |||((T(f(x_0))) - f)(x_0)|||_{\infty, x_0} \\ + \omega_1(f, |||(T(\|x - x_0\| \cdot i))(x_0)|||_{\infty, x_0}) \cdot (1 + |||T(i)|||_\infty), \quad (17.19)$$

*where $||| \cdot |||_\infty$, $||| \cdot |||_{\infty, x_0}$ are the supremum norms taken over $M$ and over all $x_0 \in M$, respectively, and $i \in Y^+$ is such that $\|i\| = 1$.*

Furthermore we obtain

**THEOREM 17.2.** *Let $X$ be a Banach space, $Y$ be a Banach lattice, $M$ be a compact and convex subset of $X$, $h > 0$, and $T$ is a lattice homomorphism from $C(M, Y)$ into itself. Consider a function $P$ from $X$ into $Y$ such that $P\big|_M \in C^n(M, Y)$, $n \in \mathbf{N}$. Then it holds*

$$|||TP - P|||_\infty$$

$$\leq |||(T(P(x_0)))(x_0) - P(x_0)|||_{\infty, x_0}$$

$$+ \sum_{k=1}^{n} \frac{1}{k!} \cdot |||(T(P^{(k)}(x_0)(x - x_0)^k))(x_0)|||_{\infty, x_0}$$

$$+ \omega_1(P^{(n)}, h) \cdot \left[ \frac{1}{(n+1)! \cdot h} \cdot |||(T(\|x - x_0\|^{n+1} i))(x_0)|||_{\infty, x_0} \right.$$

$$+ \frac{1}{2 \cdot n!} \cdot |||(T(\|x - x_0\|^n \cdot i))(x_0)|||_{\infty, x_0}$$

$$\left. + \frac{h}{8 \cdot (n-1)!} \cdot |||(T(\|x - x_0\|^{n-1}, i))(x_0)|||_{\infty, x_0} \right]. \quad (17.23)$$

# 1. Introduction

Here $i \in Y^+$ is such that $\|i\| = 1$, and $\||\cdot|\|_\infty$, $\||\cdot|\|_{\infty, x_0}$ are the supremum norms taken over $M$ and over all $x_0 \in M$, respectively.

Other applications follow.

## 1.17 On Chapter 18: Quantitative $L_p$-Results for Positive Linear Operators

A sample of these results follows:

**THEOREM 18.1.** *Consider the positive linear operator*

$$L : C([a,b]) \longrightarrow C([a,b]).$$

*Define* $(n \in \mathbf{N})$

$$D_n := \|(L(|t - \cdot|^n))(\cdot)\|_\infty^{1/n}, \tag{18.8}$$

*where* $\|\cdot\|_\infty$ *is the supremum norm. Let* $f \in C^n([a,b])$. *Then we get that*

$$\|Lf - f\|_\infty \le \|f\|_\infty \cdot \|L1 - 1\|_\infty + \sum_{k=1}^n \frac{\|f^{(k)}\|_\infty}{k!} \cdot \|(L(t-\cdot)^k)(\cdot)\|_\infty$$

$$+ \omega_1(f^{(n)}, D_n) \cdot D_n^{n-1} \left( \frac{(b-a)}{(n+1)!} + \frac{D_n}{2 \cdot n!} + \frac{D_n^2}{8 \cdot (b-a) \cdot (n-1)!} \right). \tag{18.9}$$

An improved result for $f \in C^1([a,b])$ with respect to $\|\cdot\|_\infty$ comes next.

**THEOREM 18.2.** *Let* $L \not\equiv 0$ *be a positive linear operator from* $C([a,b])$ *into itself. Set*

$$\rho := \|(L(t-x_0)^2)(x_0)\|_\infty^{1/2}, \tag{18.16}$$

*and consider* $r > 0$. *Let* $f \in C^1([a,b])$. *Then it holds*

$$\|Lf - f\|_\infty - \|f\|_\infty \cdot \|L1 - 1\|_\infty - \|f'\|_\infty \cdot \|(L((t-x_0))(x_0)\|_\infty$$

$$\le \begin{cases} \frac{1}{8 \cdot r} \cdot (2 + \sqrt{\|L(1)\|_\infty} \cdot r)^2 \cdot \omega_1(f', r \cdot \rho) \cdot \rho, & \text{if } r \le \frac{2}{\sqrt{\|L(1)\|_\infty}}; \\ \sqrt{\|L(1)\|_\infty} \cdot \omega_1(f', r \cdot \rho) \cdot \rho, & \text{if } r > \frac{2}{\sqrt{\|L(1)\|_\infty}}. \end{cases}$$
$$\tag{18.17}$$

Furthermore we get

**THEOREM 18.4.** *Let $([a,b], \mathcal{B}, \mu)$, $a \neq b$, be a measure space, where $\mathcal{B}$ is the Borel $\sigma$-algebra on $[a,b]$ and $\mu$ is a positive finite measure on $[a,b]$. (Note that $C([a,b]) \subset L_p([a,b], \mathcal{B}, \mu)$, any $p > 1$.) Here $\|\cdot\|_p$ stands for the related $L_p$ norm with respect to $\mu$. Let $p, q > 1$ such that $\frac{1}{p} + \frac{1}{q} = 1$. Consider the positive linear operator*

$$L : C([a,b]) \longrightarrow C([a,b]).$$

Set ($n \in \mathbf{N}$)
$$h_n := \|(L(|t - \cdot|^n))(\cdot)\|_p^{1/n}. \tag{18.29}$$

Let $f \in C^n([a,b])$. *Then it holds*

$$\|Lf - f\|_p \leq \|f\|_{p \cdot q} \cdot \|L1 - 1\|_{p^2} + \sum_{k=1}^{n} \frac{1}{k!} \cdot \|f^{(k)}\|_{p \cdot q} \cdot \|L(t - \cdot)^k(\cdot)\|_{p^2} + \omega_1(f^{(n)}, h_n)$$

$$\cdot h_n^{n-1} \cdot \left[ \frac{(b-a)}{(n+1)!} + \frac{h_n}{2 \cdot n!} + \frac{h_n^2}{4 \cdot (n-1)! \cdot (b-a)} \right]. \tag{18.30}$$

An abstract result follows.

**THEOREM 18.7.** *Let $M$ be a nonempty compact convex subset of the real normed vector space $(V, \|\cdot\|)$. Let $(M, \mathcal{B}, \mu)$ be a measure space, where $\mathcal{B}$ is the Borel $\sigma$-algebra on $M$ and $\mu$ is a positive finite measure on $M$. (Note that $C(M) \subset L_p(M, \mathcal{B}, \mu)$, any $1 \leq p < +\infty$.) Consider the positive linear operator*

$$L : C(M) \longrightarrow C(M).$$

*Assume*
$$\begin{aligned} (L(1))(x_0) &> 0, & \forall x_0 \in M, \\ (L(\|x - x_0\|^r))(x_0) &> 0, & \forall x_0 \in M, \ r > 1. \end{aligned} \tag{18.61}$$

*Call*
$$F(x_0) := \left( \frac{(L(\|x - x_0\|^r))(x_0)}{(L(1))(x_0)} \right)^{1/r}, \quad \forall x_0 \in M. \tag{18.62}$$

Let $f \in C(M)$. *Then it holds*

$$\|Lf - f\|_p \leq \|f\|_\infty \cdot \|L1 - 1\|_p + \omega_1(f; r \cdot \|F\|_p) \cdot \|L(1)\|_\infty \cdot \left[ (\mu(M))^{1/p} + \frac{1}{r} \right]. \tag{18.63}$$

## 1. Introduction

A multidimensional result comes next.

**THEOREM 18.8.** *Take $Q \subset \mathbf{R}^k$ of the form $Q := \{\mathbf{x} \in \mathbf{R}^k : \|\mathbf{x}\| \leq 1\}$, where $\|\cdot\|$ is the $\ell_1$-norm in $\mathbf{R}^k$, $k \geq 1$. Let $(Q, \mathcal{B}, \mu)$ be a measure space, where $\mathcal{B}$ is the Borel $\sigma$-algebra on $Q$ and $\mu$ is a positive finite measure on $Q$. (Note that $C(Q) \subset L_p(Q, \mathcal{B}, \mu)$, $1 \leq p < +\infty$). Consider the positive linear operator*

$$L : C(Q) \longrightarrow C(Q),$$

*such that*

$$(L(1))(\mathbf{x}_0) = 1, \quad \forall \mathbf{x}_0 \in Q. \tag{18.69}$$

*Suppose that*

$$h_1 := \|(L(\|\mathbf{x} - \mathbf{x}_0\|))(\mathbf{x}_0)\|_\infty > 0, \tag{18.70}$$

$$h_2 := \|(L(\|\mathbf{x} - \mathbf{x}_0\|))(\mathbf{x}_0)\|_p > 0, \quad 1 \leq p < +\infty. \tag{18.71}$$

*(Notice that $h_1, h_2 < +\infty$.)*

*Let $f \in C^n(Q)$, $n \geq 2$, and assume that each of its nth partial derivatives $f_\alpha$ has a modulus of continuity $\omega_1(f_\alpha, h_i) \leq w_i$, where $w_i$, $i = 1, 2$ are given positive numbers.*

*Set*

$$g_{(\mathbf{x}, \mathbf{x}_0)}(t) := f(\mathbf{x}_0 + t \cdot (\mathbf{x} - \mathbf{x}_0)), \quad t \geq 0, \ \mathbf{x}_0, \mathbf{x} \in Q. \tag{18.72}$$

*Then we have the following estimates*

*(i)*

$$\|Lf - f\|_\infty \leq \sum_{j=1}^n \frac{1}{j!} \cdot \|(L(g^{(j)}_{(\mathbf{x}, \mathbf{x}_0)}(0)))(\mathbf{x}_0)\|_\infty$$
$$+ \omega_1 \cdot \left( \frac{2^n}{(n+1)!} + \frac{h_1 \cdot 2^{n-2}}{n!} + \frac{h_1^2 \cdot 2^{n-5}}{(n-1)!} \right), \tag{18.73}$$

*(ii)*

$$\|Lf - f\|_p \leq \sum_{j=1}^n \frac{1}{j!} \cdot \|(L(g^{(j)}_{(\mathbf{x}, \mathbf{x}_0)}(0)))(\mathbf{x}_0)\|_p$$
$$+ \omega_2 \cdot \left( \frac{2^n}{(n+1)!} + \frac{h_2 \cdot 2^{n-2}}{n!} + \frac{h_2^2 \cdot 2^{n-5}}{(n-1)!} \right). \tag{18.74}$$

# Part VII: On Approximation Theory

## 1.18 On Chapter 19: On Monotone Approximation Theory

We give (here $I = [-1,1]$)

**THEOREM 19.2.** *Let $h$, $\nu$, $r$ be integers, $0 \leq h \leq \nu \leq r$ and let $f \in C^r(I)$, with $f^{(r)}$ having modulus of smoothness $\omega_s(f^{(r)}, \delta)$ there, $s \geq 1$. Let $\alpha_j(x)$, $j = h, h+1, \ldots, \nu$ be real functions, defined and bounded on $I$ and suppose $\alpha_k(x)$ is either $\geq \alpha > 0$ or $\leq \beta < 0$ throughout $I$.*

*Take the operator*

$$L = \sum_{j=h}^{\nu} \alpha_j(x) \cdot \left[\frac{d^j}{dx^j}\right]$$

*and assume throughout $I$,*

$$L(f) \geq 0. \tag{19.3}$$

*Then, for every integer $n \geq \max(4(r+1), r+s)$, there exists a real polynomial $Q_n(x)$ of degree $\leq n$ such that*

$$L(Q_n) \geq 0 \text{ throughout } I,$$

*and*

$$\|f^{(k)} - Q_n^{(k)}\|_\infty \leq \frac{C}{n^{r-\nu}} \omega_s\left(f^{(r)}, \frac{1}{n}\right), \quad 0 \leq k \leq h. \tag{19.4}$$

*Moreover, we get*

$$\|f^{(k)} - Q_n^{(k)}\|_\infty \leq \frac{C}{n^{r-k}} \omega_s\left(f^{(r)}, \frac{1}{n}\right), \quad h+1 \leq k \leq r. \tag{19.5}$$

*where $C$ is a constant independent of $f$ and $n$.*

Next let again $I = [-1,1]$, $n, m \in \mathbf{Z}_+$, $P_\theta$ denote the space of algebraic polynomials of degree $\leq \theta$.

Consider the tensor product spaces $P_n \otimes C(I)$, $C(I) \otimes P_m$ and their sum $P_n \otimes C(I) + C(I) \otimes P_m$, that is

$$P_n \otimes C(I) + C(I) \otimes P_m$$

## 1. Introduction

$$= \left\{ \sum_{i=0}^{n} x^i A_i(y) + \sum_{j=0}^{m} B_j(x) y^j; A_i, B_j \in C(I), x, y \in I \right\},$$

the space of pseudopolynomials.

We prove (here $f^{(k,\ell)}$ denotes a mixed partial derivative)

**THEOREM 19.4.** *Let $h_1, h_2, \nu_1, \nu_2, r, p$ be integers, $0 \leq h_1 \leq \nu_1 \leq r$, $0 \leq h_2 \leq \nu_2 \leq p$ and let $f \in C^{r,p}(I^2)$, with $f^{(r,p)}$ having a mixed modulus of smoothness $\omega_{s,q}(f^{(r,p)}; x, y)$ there, $s, q \in \mathbf{N}$. Let $\alpha_{ij}(x,y)$, $i = h_1, h_1+1, \ldots \nu_1$; $j = h_2, h_2+1, \ldots, \nu_2$ be real-valued functions, defined and bounded in $I^2$ and suppose $\alpha_{h_1 h_2}$ is either $\geq \alpha > 0$ or $\leq \beta < 0$ throughout $I^2$. Take the operator*

$$L = \sum_{i=h_1}^{\nu_1} \sum_{j=h_2}^{\nu_2} \alpha_{ij}(x,y) \frac{\partial^{i+j}}{\partial x^i \partial y^j}$$

*and assume, throughout $I^2$, that*

$$L(f) \geq 0. \qquad (19.11)$$

*Then, for any integers $n, m$ with $n \geq \max\{4(r+1), r+s\}$, $m \geq \max\{4(p+1), p+q\}$, there exists a pseudopolynomial*

$$Q_{n,m} \in (P_n \otimes C(I) + C(I) \otimes P_m)$$

*such that $L(Q_{n,m}) \geq 0$ throughout $I^2$ and*

$$\|f^{(k,l)} - Q_{n,m}^{(k,l)}\|_\infty \leq \frac{C}{n^{r-\nu_1} m^{p-\nu_2}} \cdot \omega_{s,q}\left(f^{(r,p)}; \frac{1}{n}, \frac{1}{m}\right) \qquad (19.12)$$

*for all $(0,0) \leq (k,l) \leq (h_1, h_2)$. Moreover we get*

$$\|f^{(k,l)} - Q_{n,m}^{(k,l)}\|_\infty \leq \frac{C}{n^{r-k} m^{p-l}} \cdot \omega_{s,q}\left(f^{(r,p)}; \frac{1}{n}, \frac{1}{m}\right) \qquad (19.13)$$

*for all $(h_1+1, h_2+1) \leq (k,l) \leq (r,p)$. Also (19.13) is valid whenever $0 \leq k \leq h_1$, $h_2+1 \leq l \leq p$, or $h_1+1 \leq k \leq r$, $0 \leq l \leq h_2$. Here $C$ is a constant independent of $f$ and $n, m$. It depends only on $r, p, s, q, L$.*

Next we need

**DEFINITION 19.1.** (D.D. Stancu [164]). Let $f \in C([0,1]^2)$, $[0,1]^2 = [0,1] \times [0,1]$, where $(x_1, y_1)$, $(x_2, y_2) \in [0,1]^2$ and $\delta_1, \delta_2 \geq 0$. The first modulus of continuity of $f$ is defined as follows:

$$\omega(f, \delta_1, \delta_2) = \sup_{\substack{|x_1-x_2|\leq \delta_1 \\ |y_1-y_2|\leq \delta_2}} |f(x_1, y_1) - f(x_2, y_2)|.$$

We establish

**THEOREM 19.6.** Let $h_1, h_2, \nu_1, \nu_2, r, p$ be integers, $0 \leq h_1 \leq \nu_1 \leq r$, $0 \leq h_2 \leq \nu_2 \leq p$ and let $f \in C^{r,p}([0,1]^2)$. Let $\alpha_{ij}(x,y)$, $i = h_1, h_1+1, \ldots, \nu_1$; $j = h_2, h_2+1, \ldots, \nu_2$ be real-valued functions, defined and bounded in $[0,1]^2$ and assume $\alpha_{h_1 h_2}$ is either $\geq \alpha > 0$ or $\leq \beta < 0$ throughout $[0,1]^2$. Take the operator

$$L = \sum_{i=h_1}^{\nu_1} \sum_{j=h_2}^{\nu_2} \alpha_{ij}(x,y) \partial^{i+j}/\partial x^i \partial y^j$$

and assume that throughout $[0,1]^2$,

$$L(f) \geq 0. \tag{19.18}$$

Then for integers $m, n$ with $m > r$, $n > p$, there exists a polynomial $Q_{m,n}(x,y)$ of degree $(m,n)$ such that $L(Q_{m,n}(x,y)) \geq 0$ throughout $[0,1]^2$ and

$$\|f^{(k,l)} - Q_{m,n}^{(k,l)}\|_\infty \leq \frac{P_{m,n}(L,f)}{(h_1-k)!(h_2-l)!} + M_{m,n}^{k,l}(f), \tag{19.19}$$

all $(0,0) \leq (k,l) \leq (h_1, h_2)$. Furthermore we obtain

$$\|f^{(k,l)} - Q_{m,n}^{(k,l)}\|_\infty \leq M_{m,n}^{k,l}(f), \tag{19.20}$$

for all $(h_1+1, h_2+1) \leq (k,l) \leq (r,p)$. Also (19.20) is valid whenever $0 \leq k \leq h_1$, $h_2+1 \leq l \leq p$ or $h_1+1 \leq k \leq r$, $0 \leq l \leq h_2$. Here

$$M_{m,n}^{k,l} \equiv M_{m,n}^{k,l}(f) \equiv t(k,l) \cdot \omega\left(f^{(k,l)}; \frac{1}{\sqrt{m-k}}, \frac{1}{\sqrt{n-l}}\right)$$

$$+ \max\left\{\frac{k(k-1)}{m}, \frac{l(l-1)}{n}\right\} \cdot \|f^{(k,l)}\|_\infty$$

1. Introduction

and
$$P_{m,n} \equiv P_{m,n}(L,f) \equiv \sum_{i=h_1}^{\nu_1} \sum_{j=h_2}^{\nu_2} l_{ij} \cdot M_{m,n}^{i,j},$$

where $t$ is a positive real-valued function on $\mathbf{N}^2$ and

$$l_{ij} \equiv \sup_{(x,y)\in[0,1]^2} |\alpha_{h_1 h_2}^{-1}(x,y) \cdot \alpha_{ij}(x,y)| < \infty.$$

## 1.19 On Chapter 20: Comparisons for Local Moduli of Continuity

Here we use the following notion

**DEFINITION 20.1.** Let $f \in C([a,b])$ and $x_0 \in [a,b]$. Then the first local modulus of continuity of $f$ at $x_0$ is defined as follows:

$$\omega_1(f, x_0, h) := \sup_{\substack{x\in[a,b] \\ |x-x_0|\leq h}} |f(x) - f(x_0)|, \qquad (20.1)$$

where $0 \leq h \leq b - a$.

It follows the main result.

**THEOREM 20.1.** Let $f \in C^N([a,b])$, $1 \leq n \leq N$, $x_0 \in (a,b)$ fixed, and $0 \leq h \leq b-a$. Then it holds

$$\omega_1(f, x_0, h) \leq \sum_{k=1}^n \frac{|f^{(k)}(x_0)|}{k!} \cdot h^k + \frac{h^n}{n!} \cdot \omega_1(f^{(n)}, x_0, h). \qquad (20.4)$$

Also we need

**DEFINITION 20.2.** Let $f \in C([a,b])$ and $x_0 \in [a,b]$. Then the second local modulus of continuity of $f$ at $x_0$ is given by

$$\omega_2(f, x_0, h) := \sup_{\substack{\text{all } y: a\leq x_0\pm y\leq b \\ \text{with } |y|\leq h}} |f(x_0+y) + f(x_0-y) - 2f(x_0)|, \qquad (20.9)$$

where
$$0 \leq h \leq \frac{b-a}{2}.$$

Another main result follows.

**THEOREM 20.2.** *Let $f \in C^N([a,b])$, $N \in \mathbb{N}$ and $n$ be even such that $2 \leq n \leq N$, $x_0 \in (a,b)$ fixed. Then we get*

$$\omega_2(f, x_0, h) \leq 2 \cdot \sum_{\rho=1}^{n/2} \frac{|f^{(2\rho)}(x_0)|}{(2\rho)!} \cdot h^{2\rho} + \frac{h^n}{n!} \cdot \omega_2(f^{(n)}, x_0, h), \qquad (20.10)$$

*where*

$$0 \leq h \leq (b-a)/2.$$

## 1.20 On Chapter 21: Convergence with Rates of Univariate Singular Integrals to the Unit

Here for $\xi > 0$ we study the Jackson-type generalizations of Picard, Poisson-Cauchy, and Gauss-Weierstrass singular integrals, respectively,

$$P_{n,\xi}(f;x) = -\frac{1}{2\xi} \sum_{k=1}^{n+1} (-1)^k \binom{n+1}{k} \int_{-\infty}^{+\infty} f(x+kt) e^{-|t|/\xi} dt,$$

$$Q_{n,\xi}(f;x) = \frac{1}{-\left(\frac{2}{\xi}\right) \tan^{-1}\left(\frac{\pi}{\xi}\right)} \sum_{k=1}^{n+1} (-1)^k$$

$$\cdot \binom{n+1}{k} \int_{-\pi}^{\pi} \frac{f(x+kt)}{t^2 + \xi^2} dt,$$

and

$$W_{n,\xi}(f;x) = -\frac{1}{2C(\xi)} \sum_{k=1}^{n+1} (-1)^k \binom{n+1}{k} \int_{-\pi}^{\pi} f(x+kt) e^{-t^2/\xi^2} dt,$$

$$C(\xi) = \int_0^\pi e^{-t^2/\xi^2} dt.$$

## 1. Introduction

We prove

**THEOREM 21.1.** *Here consider* $X = L^1(\mathbf{R})$ *(for* $P_{n,\xi}$*),* $X = L^1_{2\pi}(\mathbf{R})$ *(for* $W_{n,\xi}, Q_{n,\xi}$*),* $\xi > 0$, $n \in \mathbf{N}$, $f \in X$. *Then it holds*

$$\|f - P_{n,\xi}\|_X \leq \left[\sum_{k=0}^{n+1} \binom{n+1}{k} k!\right] \omega_{n+1}(f;\xi)_X, \quad \xi > 0; \quad (21.1)$$

$$\|f - W_{n,\xi}(f)\|_X \leq \left[1/\int_0^\pi e^{-u^2} du\right] \left[\int_0^{+\infty} (u+1)^{n+1} e^{-u^2} du\right]$$
$$\omega_{n+1}(f;\xi)_{L^1_{2\pi}(\mathbf{R})}, \quad 0 < \xi \leq 1; \quad (21.2)$$

$$\|f - Q_{n,\xi}(f)\|_X \leq K(n,\xi)\omega_{n+1}(f;\xi)_{L^1_{2\pi}(\mathbf{R})}, \quad \xi > 0, \quad (21.3)$$

*where*

$$K(n,\xi) = \left[1/\tan^{-1}\frac{\pi}{\xi}\right] \int_0^{\pi/\xi} \frac{(u+1)^{n+1}}{u^2+1} du.$$

Also we establish

**THEOREM 21.2.** *Let us take* $X = L^p(\mathbf{R})$ *(for* $P_{n,\xi}$*),* $X = L^p_{2\pi}(\mathbf{R})$ *(for* $W_{n,p}, Q_{n,\xi}$*),* $0 < \xi \leq 1$, $n \in \mathbf{N}$, $1 < p < +\infty$, $\frac{1}{p} + \frac{1}{q} = 1$, $f \in X$. *Then it holds*

$$\|f - P_{n,\xi}(f)\|_X \leq (2/q)^{1/q} \|g\|_{L^p(\mathbf{R}_+)} \omega_{n+1}(f;\xi)_X,$$

*where* $g(u) = (u+1)^{n+1} e^{-u/2}$;

$$\|f - W_{n,\xi}(f)\|_X \leq \left(\sqrt{\frac{\pi}{2q}}\right)^{1/q} \frac{1}{\int_0^\pi e^{-u^2} du} \|h\|_{L^p(\mathbf{R}_+)} \omega_{n+1}(f;\xi)_X,$$

*where* $h(u) = (u+1)^{n+1} e^{-u^2/2}$;

$$\|f - Q_{n,\xi}(f)\|_X \leq K_p(n,\xi)\omega_{n+1}(f;\xi)_{L^p_{2\pi}(\mathbf{R})},$$

*where*

$$K_p(n,\xi) = \left[\frac{1}{\tan^{-1}\frac{\pi}{\xi}} \int_0^{\pi/\xi} (u+1)^{(n+1)p} \frac{1}{u^2+1} du\right]^{1/p}.$$

And furthermore

**THEOREM 21.3.** *For $0 < \xi \leq 1$, $n \in \mathbf{N}$, $f \in X = C_{2\pi}(\mathbf{R})$, it holds*

$$\|f - Q_{n,\xi}(f)\|_X \leq K(n,\xi)\omega_{n+1}(f;\xi)_X,$$

*where $K(n,\xi)$ is given by Theorem 21.1.*

## Part VIII: On Classical Analysis

### 1.21 On Chapter 22: About Univariate Ostrowski-Type Inequalities

Ostrowski's inequality (see Ostrowski [136]) has as follows:

$$\left| \frac{1}{b-a} \cdot \int_a^b f(y)dy - f(x) \right| \leq \left( \frac{1}{4} + \frac{(x - \frac{a+b}{2})^2}{(b-a)^2} \right) \cdot (b-a) \cdot \|f'\|_\infty, \quad (22.1)$$

where $f \in C^1([a,b])$, $x \in [a,b]$. Inequality (22.1) is sharp because the function in ( ) cannot be replaced by a smaller one. One can easily see that

$$\left( \frac{1}{4} + \frac{(x - \frac{a+b}{2})^2}{(b-a)^2} \right) \cdot (b-a) = \frac{(x-a)^2 + (b-x)^2}{2 \cdot (b-a)}. \quad (22.2)$$

We give a different proof for (22.1) in

**THEOREM 22.1.** *Let $f \in C^1([a,b])$, $x \in [a,b]$. Then it holds*

$$\left| \frac{1}{b-a} \cdot \int_a^b f(y)dy - f(x) \right| \leq \left( \frac{(x-a)^2 + (b-x)^2}{2 \cdot (b-a)} \right) \cdot \|f'\|_\infty. \quad (22.3)$$

*Inequality (22.3) is sharp. In particular the optimal function is*

$$f^*(y) := |y - x|^\alpha \cdot (b-a), \quad \alpha > 1. \quad (22.4)$$

Generalizing we get

**THEOREM 22.2.** Let $f \in C^{n+1}([a,b])$, $n \in \mathbf{N}$ and $x \in [a,b]$ be fixed, such that $f^{(k)}(x) = 0$, $k = 1, \ldots, n$. Then it holds

$$\left| \frac{1}{b-a} \cdot \int_a^b f(y)dy - f(x) \right| \leq \frac{\|f^{(n+1)}\|_\infty}{(n+2)!} \cdot \left( \frac{(x-a)^{n+2} + (b-x)^{n+2}}{b-a} \right). \tag{22.13}$$

Inequality (22.13) is sharp. In particular, when $n$ is odd it is attained by $f^*(y) := (y-x)^{n+1} \cdot (b-a)$, while when $n$ is even the optimal function is

$$\tilde{f}(y) := |y-x|^{n+\alpha} \cdot (b-a), \quad \alpha > 1.$$

Furthermore it holds

**THEOREM 22.3.** Let $f \in C^{n+1}([a,b])$, $n \in \mathbf{N}$ such that $f^{(k)}((a+b)/2) = 0$, all $k$ even $\in \{1, \ldots, n\}$. Then it holds

$$\left| \frac{1}{b-a} \cdot \int_a^b f(y)dy - f\left(\frac{a+b}{2}\right) \right| \leq \frac{\|f^{(n+1)}\|_\infty}{(n+2)!} \cdot \frac{(b-a)^{n+1}}{2^{n+1}}. \tag{22.18}$$

Inequality (22.18) is sharp. More precisely, when $n$ is odd it is attained by $f^*(y) := (y - \frac{a+b}{2})^{n+1} \cdot (b-a)$, while when $n$ is even the optimal function is

$$\tilde{f}(y) := \left| y - \frac{a+b}{2} \right|^{n+\alpha} \cdot (b-a), \quad \alpha > 1.$$

## 1.22 On Chapter 23: About Multidimensional Ostrowski-Type Inequalities

We prove the basic result

**THEOREM 23.1.** Let $f \in C^1(\prod_{i=1}^k [a_i, b_i])$, where $a_i < b_i$; $a_i, b_i \in \mathbf{R}$, $i = 1, \ldots, k$, and let $\vec{x}_0 := (x_{01}, \ldots, x_{0k}) \in \prod_{i=1}^k [a_i, b_i]$ be fixed. Then it holds

$$\left| \frac{1}{\prod_{i=1}^k (b_i - a_i)} \int_{a_1}^{b_1} \cdots \int_{a_i}^{b_i} \cdots \int_{a_k}^{b_k} f(z_1, \ldots, z_k) dz_1 \cdots dz_k - f(\vec{x}_0) \right|$$

$$\leq \sum_{i=1}^{k} \left( \frac{(x_{0i} - a_i)^2 + (b_i - x_{0i})^2}{2(b_i - a_i)} \right) \left\| \frac{\partial f}{\partial z_i} \right\|_{\infty}. \tag{23.1}$$

Inequality (23.1) is sharp, here the optimal function is

$$f^*(z_1, \ldots, z_k) := \sum_{i=1}^{k} |z_i - x_{0i}|^{\alpha_i}, \quad \alpha_i > 1. \tag{23.2}$$

Generalizing we obtain

**THEOREM 23.2.** *Let $Q$ be a compact and convex subset of $\mathbf{R}^k$, $k \geq 1$. Let $f \in C^{n+1}(Q)$, $n \in \mathbf{N}$ and $\vec{x}_0 \in Q$ be fixed such that all partial derivatives $f_\alpha := \frac{\partial^\alpha f}{\partial z^\alpha}$, where $\alpha = (\alpha_1, \ldots, \alpha_k)$, $\alpha_i \in \mathbf{Z}^+$, $i = 1, \ldots, k$, $|\alpha| = \sum_{i=1}^{k} \alpha_i = j$, $j = 1, \ldots, n$ fulfill $f_\alpha(\vec{x}_0) = 0$. Then it holds*

$$\left| \frac{1}{\mathrm{Vol}(Q)} \int_Q f(\vec{z}) \, d\vec{z} - f(\vec{x}_0) \right| \\ \leq \frac{D_{n+1}(f)}{(n+1)! \mathrm{Vol}(Q)} \int_Q (\|\vec{z} - \vec{x}_0\|_{\ell_1})^{n+1} \, d\vec{z}, \tag{23.7}$$

where

$$D_{n+1}(f) := \max_{\alpha : |\alpha| = n+1} \|f_\alpha\|_{\infty} \tag{23.8}$$

and

$$\|\vec{z} - \vec{x}_0\|_{\ell_1} := \sum_{i=1}^{k} |z_i - x_{0i}|. \tag{23.9}$$

Other interesting related results are in

**PROPOSITION 23.1.** *Let $f \in C^{n+1}([a_1, b_1] \times [a_2, b_2])$, $n \in \mathbf{N}$, where $a_1 < b_1$, $a_2 < b_2$; $a_1, a_2, b_1, b_2 \in \mathbf{R}$ and let $\vec{x}_0 = (x_{01}, x_{02}) \in [a_1, b_1] \times [a_2, b_2]$ be fixed. We suppose here that all partial derivatives $f_\alpha := \frac{\partial^\alpha f}{\partial z^\alpha}$, where $\alpha = (\alpha_1, \alpha_2)$, $\alpha_1, \alpha_2 \in \mathbf{Z}^+$, $|\alpha| = \alpha_1 + \alpha_2 = j$, $j = 1, \ldots, n$ fulfill $f_\alpha(\vec{x}_0) = 0$. Then it holds*

$$\left| \frac{1}{(b_1 - a_1)(b_2 - a_2)} \int_{a_1}^{b_1} \int_{a_2}^{b_2} f(z_1, z_2) \, dz_1 dz_2 - f(\vec{x}_0) \right|$$

$$\leq \sum_{\ell=0}^{n+1} \left\{ \frac{[(x_{01} - a_1)^{n+2-\ell} + (b_1 - x_{01})^{n+2-\ell}][(x_{02} - a_2)^{\ell+1} + (b_2 - x_{02})^{\ell+1}]}{(n+2-\ell)!(\ell+1)!(b_1 - a_1)(b_2 - a_2)} \right\}$$

## 1. Introduction

$$\cdot \left\| \frac{\partial^{n+1} f}{\partial z_1^{n+1-\ell} \partial z_2^{\ell}} \right\|_{\infty}. \tag{23.22}$$

*Inequality (23.22) is sharp.*

And

**PROPOSITION 23.2.** *Let $f \in C^2(\prod_{i=1}^{3}[a_i, b_i])$, where $a_i < b_i$, $i = 1, 2, 3$; $a_i, b_i \in \mathbf{R}$ and let $\vec{x}_0 = (x_{01}, x_{02}, x_{03}) \in \prod_{i=1}^{3}[a_i, b_i]$ is fixed. We suppose here that $\frac{\partial f}{\partial z_i}(\vec{x}_0) = 0$; $i = 1, 2, 3$. Then it holds*

$$\left| \frac{1}{\prod_{i=1}^{3}(b_i - a_i)} \int_{a_1}^{b_1} \int_{a_2}^{b_2} \int_{a_3}^{b_3} f(z_1, z_2, z_3) \, dz_1 dz_2 dz_3 - f(\vec{x}_0) \right| \tag{23.27}$$

$$\leq \left\{ \sum_{i=1}^{3} \left( \frac{(x_{0i} - a_i)^3 + (b_i - x_{0i})^3}{6(b_i - a_i)} \right) \left\| \frac{\partial^2 f}{\partial z_i^2} \right\|_{\infty} \right.$$

$$+ \sum_{i=1}^{2} \frac{((x_{0i} - a_i)^2 + (b_i - x_{0i})^2)((x_{0,i+1} - a_{i+1})^2 + (b_{i+1} - x_{0,i+1})^2)}{4(b_i - a_i)(b_{i+1} - a_{i+1})}$$

$$\cdot \left\| \frac{\partial^2 f}{\partial z_i \partial z_{i+1}} \right\|_{\infty}$$

$$+ \frac{((x_{03} - a_3)^2 + (b_3 - x_{03})^2)((x_{01} - a_1)^2 + (b_1 - x_{01})^2)}{4(b_3 - a_3)(b_1 - a_1)} \left\| \frac{\partial^2 f}{\partial z_3 \partial z_1} \right\|_{\infty} \right\}.$$

*Inequality (23.27) is sharp and is attained by*

$$f^*(z_1, z_2, z_3) = \sum_{i=1}^{3} (z_i - x_{0i})^2. \tag{23.28}$$

---

## 1.23 On Chapter 24: General Opial-Type Inequalities for Linear Differential Operators

Here we use notions from [111], pp. 145–154. Let $I$ be a closed interval of $\mathbf{R}$. Let $\alpha_i(x)$, $i = 0, 1, \ldots, n-1$ ($n \in \mathbf{N}$), $h(x)$ be continuous functions on $I$ and let $L = D^n + \alpha_{n-1}(x)D^{n-1} + \cdots + \alpha_0(x)$ be a fixed linear differential operator on $C^n(I)$. Let $y_1(x), \ldots, y_n(x)$ be a set of linear independent solutions to $Ly = 0$. Here the associated Green's function for $L$ is given by

$$H(x,t) := \begin{vmatrix} y_1(t) \cdots y_n(t) \\ y_1'(t) \cdots y_n'(t) \\ \vdots \\ y_1^{(n-2)}(t) \cdots y_n^{(n-2)}(t) \\ y_1(x) \cdots y_n(x) \end{vmatrix} \bigg/ \begin{vmatrix} y_1(t) \cdots y_n(t) \\ y_1'(t) \cdots y_n'(t) \\ \vdots \\ y_1^{(n-2)}(t) \cdots y_n^{(n-2)}(t) \\ y_1^{(n-1)}(t) \cdots y_n^{(n-1)}(t) \end{vmatrix}$$

which is a continuous function on $I^2$.

Take fixed $x_0 \in I$, then

$$y(x) = \int_{x_0}^{x} H(x,t) h(t)\, dt, \quad \text{all } x \in I$$

is the unique solution of the initial value problem

$$Ly = h; \quad y^{(i)}(x_0) = 0, \quad i = 0, 1, \ldots, n-1.$$

We give

**THEOREM 24.1.** *Let $x \geq x_0$; $x_0, x \in I$ and $r > 1$, $\alpha, \beta > 0$, $r > \alpha$ and let continuous functions $p(w) > 0$ and $q(w) \geq 0$ on $I$. Then it holds*

$$\int_{x_0}^{x} q(w)|y(w)|^{\beta}|(Ly)(w)|^{\alpha}\, dw \leq K \left[\int_{x_0}^{x} p(w)|(Ly)(w)|^r\, dw\right]^{(\frac{\alpha+\beta}{r})}. \quad (24.1)$$

*Here*

$$K := \left(\frac{\alpha}{\alpha+\beta}\right)^{\alpha/r} \cdot \left[\int_{x_0}^{x} (q^r(w) p^{-\alpha}(w))^{\frac{1}{(r-\alpha)}} (P_1(w))^{\frac{\beta(r-1)}{r-\alpha}}\, dw\right]^{\frac{(r-\alpha)}{r}}, \quad (24.2)$$

*and*

$$P_1(w) := \int_{x_0}^{w} (p(t))^{-\frac{1}{(r-1)}} |H(w,t)|^{\frac{r}{r-1}}\, dt.$$

The counterpart of the previous result comes next.

**THEOREM 24.2.** *Let $x \leq x_0$; $x, x_0 \in I$ and $r > 1$, $\alpha, \beta > 0$, $r > \alpha$. And let continuous functions $p(w) > 0$ and $q(w) \geq 0$ on $I$. Then it holds*

$$\int_{x}^{x_0} q(w)|y(w)|^{\beta}|(Ly)(w)|^{\alpha}\, dw \leq K \left[\int_{x}^{x_0} p(w)|(Ly)(w)|^r\, dw\right]^{(\frac{\alpha+\beta}{r})}. \quad (24.11)$$

# 1. Introduction

*Here*

$$K := \left(\frac{\alpha}{\alpha+\beta}\right)^{\alpha/r} \left[\int_x^{x_0} (q^r(w)p^{-\alpha}(w))^{\frac{1}{(r-\alpha)}} (P_1(w))^{\frac{\beta(r-1)}{r-\alpha}} dw\right]^{\frac{(r-\alpha)}{r}} \quad (24.12)$$

*and*

$$P_1(w) := \int_w^{x_0} (p(t))^{-\frac{1}{(r-1)}} |H(w,t)|^{\frac{r}{r-1}} dt.$$

Furthermore in the next we consider functions $p(w)$, $q(w)$ that are nonnegative and Lebesgue measurable over $I$, points $x_0, x \in I$ and real numbers $\alpha, \beta, r \neq 0$. Suppose that $\left(\frac{\alpha+\beta}{\alpha}\right) > 0$. Here either $x_0 \leq w \leq x$ or $x \leq w \leq x_0$.

We set

$$P_1(w) := \left|\int_{x_0}^w (p(t))^{-\frac{1}{r-1}} |H(w,t)|^{\frac{r}{r-1}} dt\right|, \quad (24.28)$$

$$z(w) := \left|\int_{x_0}^w p(t)|h(t)|^r dt\right|, \quad (24.29)$$

$$QY := \left|\int_{x_0}^x q(w)|y(w)|^\beta |h(w)|^\alpha dw\right|, \quad (24.30)$$

$$PQ := \left|\int_{x_0}^x (q^r(w)p^{-\alpha}(w))^{\frac{1}{r-\alpha}} (P_1(w))^{\frac{\beta(r-1)}{(r-\alpha)}} dw\right|, \quad (24.31)$$

*and*

$$P_1 := P_1(x), \quad z := z(x). \quad (24.32)$$

**THEOREM 24.3.** Let $x > x_0$. Suppose that $H(w,t) \geq 0$ for $x_0 \leq t \leq w$, $w \in I$. Also suppose that: either $r > 1$, $\beta > 0$, $0 < \alpha < r$, or $r < \alpha < 0$, $\beta < 0$, or $-\alpha < \beta < 0$, $0 < \alpha < r < 1$, and $P_1(w)$ exists for all $w \in [x_0, x]$, $PQ < \infty$, $z < \infty$. Then it holds

$$\int_{x_0}^x q(w)|y(w)|^\beta |(Ly)(w)|^\alpha dw \leq K \left[\int_{x_0}^x p(w)|(Ly)(w)|^r dw\right]^{\frac{(\alpha+\beta)}{r}}, \quad (24.33)$$

*where*

$$K := \left(\frac{\alpha}{\alpha+\beta}\right)^{\alpha/r} (PQ)^{\frac{(r-\alpha)}{r}}. \quad (24.34)$$

The counterpart of the previous result follows.

**THEOREM 24.4.** Let $x > x_0$. Suppose again that $H(w,t) \geq 0$ for $x_0 \leq t \leq w$, $w \in I$. Also suppose that: either

$$\left\{\begin{array}{l} \beta > 0,\ 0 < r < \min(\alpha, 1),\ or\ \alpha < 0 < r < 1, \\ 0 < \beta < -\alpha\ and\ P_1(w)\ exists\ for\ all \\ w \in [x_0, x],\ PQ < \infty,\ QY < \infty; \end{array}\right\} \quad (24.43)$$

or

$$\left\{\begin{array}{l} \beta < 0,\ \alpha < 0,\ r > 1,\ or\ 1 < r < \alpha,\ -\alpha < \beta < 0, \\ and\ P_1(w),\ z(w)\ exists\ for\ all\ w \in [x_0, x], \\ PQ < \infty,\ QY < \infty; \end{array}\right\} \quad (24.44)$$

or

$$\left\{\begin{array}{l} \beta > 0,\ r < 0 < \alpha,\ or\ \alpha < r < 0,\ 0 < \beta < -\alpha, \\ and\ z(w)\ exists\ for\ all\ w \in [x_0, x], \\ PQ < \infty,\ QY < \infty. \end{array}\right\} \quad (24.45)$$

Then it holds

$$\int_{x_0}^{x} q(w)|y(w)|^{\beta} |(Ly)(w)|^{\alpha}\, dw \geq K \left[\int_{x_0}^{x} p(w)|(Ly)(w)|^{r}\, dw\right]^{(\frac{\alpha+\beta}{r})}, \quad (24.46)$$

where

$$K := \left(\frac{\alpha}{\alpha+\beta}\right)^{\alpha/r} (PQ)^{(\frac{r-\alpha}{r})}. \quad (24.47)$$

Finally, an application for proving uniqueness of solution of a general initial value problem ends the chapter.

---

## 1.24 On Chapter 25: $L_p$-Opial Inequalities Engaging Fractional Derivatives of Functions

Here we follow all terms and notations as in Chapter 25. A sample of results follows:

**THEOREM 25.2.** Let $\nu, \gamma \geq 1$ such that $\nu - \gamma \geq 1$ and $f \in C_{x_0}^{\nu}([a,b])$ with $f^{(i)}(x_0) = 0$, $i = 0, 1, \ldots, n-1$, $n := [\nu]$. Here $x, x_0 \in [a,b] : x \geq x_0$. Let $p, q > 1$ such that $\frac{1}{p} + \frac{1}{q} = 1$. Then it holds

$$\int_{x_0}^{x} |(D_{x_0}^{\gamma} f)(w)|\, |(D_{x_0}^{\nu} f)(w)|\, dw$$

## 1. Introduction

$$\le \frac{(x-x_0)^{\frac{p\nu-p\gamma-p+2}{p}}}{(\sqrt[q]{2})\Gamma(\nu-\gamma)((p\nu-p\gamma-p+1)(p\nu-p\gamma-p+2))^{1/p}}$$

$$\cdot \left(\int_{x_0}^x |(D_{x_0}^\nu f)(w)|^q dw\right)^{2/q}. \qquad (25.17)$$

Here $D_{x_0}^\gamma f$, $D_{x_0}^\nu f$ are fractional derivatives of $f$ of orders $\gamma, \nu$, respectively.

Also it holds

**THEOREM 25.3.** *Let $\nu, \gamma \ge 1$ such that $\nu - \gamma \ge 1$ and $f \in C_{x_0}^\nu([a,b])$ with $f^{(i)}(x_0) = 0$, $i = 0, 1, \ldots, n-1$, $n := \lceil \nu \rceil$. Here $x, x_0 \in [a,b] : x > x_0$. Suppose that $(D_{x_0}^\nu f)(t) \ne 0$ and of fixed sign over $[x_0, b]$. Let $p, q$ such that $0 < p < 1$ and $\frac{1}{p} + \frac{1}{q} = 1$. Then it holds*

$$\int_{x_0}^x |D_{x_0}^\gamma f|(w) \cdot |D_{x_0}^\nu f|(w) dw$$

$$\ge \frac{(x-x_0)^{\frac{p\nu-p\gamma-p+2}{p}}}{\sqrt[q]{2}\,\Gamma(\nu-\gamma)((p\nu-p\gamma-p+1)(p\nu-p\gamma-p+2))^{1/p}}$$

$$\cdot \left(\int_{x_0}^x (|D_{x_0}^\nu f|(w))^q dw\right)^{2/q}. \qquad (25.20)$$

The following inequality involves fractional derivatives of three different orders.

**THEOREM 25.4.** *Let $\gamma, k, \nu \ge 1$ such that $\nu - k > 1$, $\nu - \gamma \ge 1$ and $\gamma = k+1$, also $f \in C_{x_0}^\nu([a,b])$ with $f^{(i)}(x_0) = 0$, $i = 0, 1, \ldots, n-1$, $n := \lceil \nu \rceil$. Here $x, x_0 \in [a, b] : x \ge x_0$. Let $p, q > 1$ such that $\frac{1}{p} + \frac{1}{q} = 1$. Then it holds*

$$\int_{x_0}^x |(D_{x_0}^k f)(w)| \, |(D_{x_0}^\gamma f)(w)| dw \le \frac{(x-x_0)^{\frac{2(p\nu-pk-p+1)}{p}}}{2(\Gamma(\nu-k))^2(p\nu-pk-p+1)^{2/p}}$$

$$\cdot \left(\int_{x_0}^x |(D_{x_0}^\nu f)(w)|^q dw\right)^{2/q}. \qquad (25.22)$$

The counterpart of Theorem 25.4 follows.

**THEOREM 25.5.** *Let $\gamma, k, \nu \ge 1$ such that $\nu - k > 1$, $\nu - \gamma \ge 1$ and $\gamma = k+1$, also $f \in C_{x_0}^\nu([a,b])$ with $f^{(i)}(x_0) = 0$, $i = 0, 1, \ldots, n-1$, $n := \lceil \nu \rceil$.*

Here $x, x_0 \in [a,b] : x > x_0$. Suppose that $(D_{x_0}^{\nu} f)(t) \not\equiv 0$ and of fixed sign over $[x_0, b]$. Let $p, q$ such that $0 < p < 1$ and $\frac{1}{p} + \frac{1}{q} = 1$. Then it holds

$$\int_{x_0}^{x} |D_{x_0}^{k} f|(w) |D_{x_0}^{\gamma} f|(w) dw \geq \frac{(x-x_0)^{\frac{2(p\nu - pk - p + 1)}{p}}}{2(\Gamma(\nu-k))^2 (p\nu - pk - p + 1)^{2/p}}$$

$$\cdot \left( \int_{x_0}^{x} |D_{x_0}^{\nu} f|^q(w) \, dw \right)^{2/q}. \tag{25.30}$$

Applications to fractional differential equations are also presented.

⋮

## 1.25 On Chapter 26: $L_p$-General Fractional Opial Inequalities

This continues the material of Chapter 25.
We present

**THEOREM 26.1.** Let $\gamma_i \geq 1$, $\nu \geq 2$ such that $\nu - \gamma_i \geq 1$; $i = 1, \ldots, \ell$ and $f \in C_{x_0}^{\nu}([a,b])$ with $f^{(j)}(x_0) = 0$, $j = 0, 1, \ldots, n-1$, $n := \lceil \nu \rceil$. Here $x, x_0 \in [a,b] : x \geq x_0$. Let $q_1, q_2 > 0$ continuous functions on $[a,b]$ and $r_i > 0 : \sum_{i=1}^{\ell} r_i = r$. Let $s_1, s_1' > 1: \frac{1}{s_1} + \frac{1}{s_1'} = 1$ and $s_2, s_2' > 1: \frac{1}{s_2} + \frac{1}{s_2'} = 1$, and $p > s_2$. Furthermore suppose that

$$Q_1 := \left( \int_{x_0}^{x} (q_1(w))^{s_1} dw \right)^{1/s_1'} < +\infty \tag{26.1}$$

and

$$Q_2 := \left( \int_{x_0}^{x} (q_2(w))^{-s_2'/p} dw \right)^{r/s_2'} < +\infty. \tag{26.2}$$

Call $\sigma := \frac{p - s_2}{p s_2}$. Then it holds

$$\int_{x_0}^{x} q_1(w) \prod_{i=1}^{\ell} |(D_{x_0}^{\gamma_i}(f))(w)|^{r_i} dw \leq Q_1 Q_2$$

$$\cdot \prod_{i=1}^{\ell} \left\{ \frac{\sigma^{r_i \sigma}}{(\Gamma(\nu - \gamma_i))^{r_i} (\nu - \gamma_i - 1 + \sigma)^{r_i \sigma}} \right\}$$

## 1. Introduction

$$\frac{(x-x_0)^{(\sum_{i=1}^{\ell}(\nu-\gamma_i-1)r_i+\sigma r+\frac{1}{s_1})}}{((\sum_{i=1}^{\ell}(\nu-\gamma_i-1)r_is_1)+rs_1\sigma+1)^{1/s_1}}$$

$$\cdot \left(\int_{x_0}^{x} q_2(w)|(D_{x_0}^\nu f)(w)|^p dw\right)^{r/p}. \tag{26.3}$$

Here $D_{x_0}^{\gamma_i} f$, $D_{x_0}^\nu f$ are fractional derivatives of $f$ of orders $\gamma_i$, $\nu$, respectively.

The counterpart of the last theorem follows

**THEOREM 26.2.** Let $\gamma_i \geq 1$, $\nu \geq 2$ such that $\nu - \gamma_i \geq 1$; $i = 1, \ldots, \ell$ and $f \in C_{x_0}^\nu([a,b])$ with $f^{(j)}(x_0) = 0$, $j = 0, 1, \ldots, n-1$, $n := [\nu]$. Here $x, x_0 \in [a,b] : x > x_0$. Let $q_1, q_2 > 0$ continuous functions on $[a,b]$ and $r_i > 0$: $\sum_{i=1}^{\ell} r_i = r$. Let $0 < s_1, s_2 < 1$ and $s_1', s_2' < 0$ such that $\frac{1}{s_1} + \frac{1}{s_1'} = 1$, $\frac{1}{s_2} + \frac{1}{s_2'} = 1$. Assume that $(D_{x_0}^\nu f)(t)$ is never zero and of fixed sign over $[x_0, b]$. Furthermore suppose that

$$Q_1 := \left(\int_{x_0}^{x} (q_1(w))^{s_1'} dw\right)^{1/s_1'} < +\infty \tag{26.5}$$

and

$$Q_2 := \left(\int_{x_0}^{x} (q_2(w))^{-s_2'} dw\right)^{r/s_2'} < +\infty. \tag{26.6}$$

Set $\lambda := \frac{s_1 s_2}{s_1 s_2 - 1}$. Then it holds

$$\int_{x_0}^{x} q_1(w) \left(\prod_{i=1}^{\ell} (|D_{x_0}^{\gamma_i}(f)|(w))^{r_i}\right) dw$$

$$\geq \frac{Q_1 Q_2}{\prod_{i=1}^{\ell}\{(\Gamma(\nu-\gamma_i))^{r_i} \cdot ((\nu-\gamma_i-1)s_2^2 s_1 + 1)^{(\frac{r_i}{s_2^2 s_1})}\}}$$

$$\cdot \frac{(x-x_0)^{\{(\sum_{i=1}^{\ell} r_i((\nu-\gamma_i-1)s_1+s_2^{-2}))+1\}/s_1}}{\{(\sum_{i=1}^{\ell} r_i((\nu-\gamma_i-1)s_1+s_2^{-2}))+1\}^{1/s_1}}$$

$$\cdot \left(\int_{x_0}^{x} q_2^{\lambda s_2}(w)|(D_{x_0}^\nu f)(w)|^{\lambda s_2} dw\right)^{r/\lambda s_2}. \tag{26.7}$$

A quite sharp fractional Opial type inequality follows.

**THEOREM 26.3.** Let $k \geq 1$, $\gamma \geq 2$, $\nu \geq 3$ such that $\nu - \gamma \geq 1$, $\nu - k \geq 1$, $\gamma - k \geq 1$ and $f \in C_{x_0}^{\nu}([a,b])$ with $f^{(i)}(x_0) = 0$, $i = 1, \ldots, n-1$, $n := [\nu]$. Here $x, x_0 \in [a,b] : x \geq x_0$. Let $p, q > 1$: $\frac{1}{p} + \frac{1}{q} = 1$. Then it holds

$$\int_{x_0}^{x} |D_{x_0}^{\gamma} f|(w) \cdot |D_{x_0}^{k} f|(w) dw \tag{26.14}$$

$$\leq \frac{2^{-1/p}(x-x_0)^{(2\nu - k - \gamma - 1 + \frac{2}{q})} \cdot \left(\int_{x_0}^{x} |D_{x_0}^{\nu} f(w)|^p dw\right)^{2/p}}{\Gamma(\nu-k)\Gamma(\nu-\gamma+1)((\nu-\gamma)q+1)^{1/q}(2\nu q - kq - \gamma q - q + 2)^{1/q}}.$$

The following is a fractional integral Opial type result involving derivatives of generalized Riemann-Liouville integral. Let $x_0 \in [a,b] \subset \mathbf{R}$, $f \in C([a,b])$ and the generalized Riemann-Liouville integral

$$(J_{\nu}^{x_0} f)(w) := \frac{1}{\Gamma(\nu)} \int_{x_0}^{w} (w-t)^{\nu-1} f(t) dt, \tag{26.19}$$

all $x_0 \leq w \leq b$, and $\nu > 1$. Let $k \in \mathbf{N}$ then

$$(J_{\nu}^{x_0} f)^{(k)}(w) = (J_{\nu-k}^{x_0} f)(w). \tag{26.20}$$

**THEOREM 26.5.** Let $f \in C([a,b])$ such that $|f|$ is decreasing. Let $\nu \geq 2$, $n := [\nu]$ and $\ell_i \in \mathbf{N}$ such that $\nu - \ell_i \geq 1$, $i = 1, \ldots, k$, $\ell_i \in \{1, \ldots, n-1\}$, $1 \leq k \leq n-1$, $\ell := \sum_{i=1}^{k} \ell_i$. Here $p, q > 1$: $\frac{1}{p} + \frac{1}{q} = 1$ and $x, x_0 \in [a,b]$ such that $x \geq x_0$. Then it holds

$$\int_{x_0}^{x} \prod_{i=1}^{k} |(J_{\nu-\ell_i}^{x_0} f)(w)| dw \leq \frac{\left(\int_{x_0}^{x} |f(t)|^{qk} dt\right)^{1/q}}{\prod_{i=1}^{k} \Gamma(\nu-\ell_i)}$$
$$\cdot \frac{p(x-x_0)^{\left(\frac{1+pk\nu - p\ell}{p}\right)}}{(pk\nu - p\ell - pk + 1)^{1/p}(pk\nu - p\ell - pk + p + 1)}. \tag{26.21}$$

Finally, the counterpart of Theorem 26.5 for fractional derivatives comes next.

**THEOREM 26.6.** Let $[x_0, x] \subset [a,b]$, $f \in C_{x_0}^{\nu}([a,b])$, $\nu \geq 2$, $n := [\nu]$, $\gamma_i \geq 1$ such that $\nu - \gamma_i \geq 1$, all $i = 1, \ldots, k$, $k \in \mathbf{N} - \{1\}$. Here suppose that $f^{(j)}(x_0) = 0$, $j = 0, 1, \ldots, n-1$. Set $\gamma := \sum_{i=1}^{k} \gamma_i$. Let $p, q > 1$ such that

## 1. Introduction

$\frac{1}{p} + \frac{1}{q} = 1$. Furthermore suppose that $|(D_{x_0}^{\nu}f)(t)|$ is decreasing on $[x_0, x]$. Then it holds

$$\int_{x_0}^{x} \prod_{i=1}^{k} |(D_{x_0}^{\gamma_i}f)(w)| dw \qquad (26.28)$$

$$\leq \frac{p(x-x_0)^{\frac{(1+k\nu p - \gamma p)}{p}} \left(\int_{x_0}^{x} |D_{x_0}^{\nu}f(t)|^{kq} dt\right)^{1/q}}{\prod_{i=1}^{k} \Gamma(\nu - \gamma_i) \cdot (k\nu p - \gamma p - kp + 1)^{1/p} \cdot (k\nu p - \gamma p - kp + p + 1)}.$$

# Part I
# On Neural Networks

# Chapter 2

## Convergence with Rates of Univariate Neural Network Operators to the Unit Operator

In this chapter we present a complete study of determination of the rate of convergence to the unit operator of specific well-known *univariate neural network* operators, namely, of Cardaliaguet–Euvrard and "Squashing" univariate operators. This is given by *Jackson type inequalities* that involve the *modulus of continuity* of the engaged function or its derivative. Here we follow the basic study done by the author [17].

The *univariate Cardaliaguet–Euvrard* (2.1) *operators* were first introduced and studied extensively in [62], where the authors, among many other interesting things, proved that these operators converge uniformly on compacta, to the unit over continuous and bounded functions. The *univariate squashing operator* (2.26) was motivated and inspired by the *"squashing function"* and generated Theorem 6 of [62]. The work in [62] is qualitative where the used bell-shaped function is general. However, the work presented here, though greatly motivated by [62], is quantitative and the used bell-shaped and "squashing" functions are of compact support. We give a series of inequalities giving close upper bounds to the errors in approximating the unit operator by the above univariate neural network induced operators. All involved constants are well determined. These are mainly *pointwise estimates* involving the first modulus of continuity of the engaged continuous function or its derivatives. We also present some $L_p$-related estimates.

## 2.1 Convergence with Rates of the Univariate Cardaliaguet–Euvrard Neural Network Operators

We use the following (see [62]).

**DEFINITION 2.1** *A function* $b : \mathbf{R} \to \mathbf{R}$ *is said to be* bell-shaped *if* $b$ *belongs to* $L^1$ *and its integral is nonzero, if it is nondecreasing on* $(-\infty, a)$ *and nonincreasing on* $[a, +\infty)$, *where* $a$ *belongs to* $\mathbf{R}$. *In particular* $b(x)$ *is a nonnegative number and at* $a$ $b$ *takes a global maximum; it is the* center *of the bell-shaped function. A bell-shaped function is said to be* centered *if its center is zero. The function* $b(x)$ *may have jump discontinuities. In this chapter we consider only centered bell-shaped functions of compact support* $[-T, T]$, $T > 0$. *Denote* $I := \int_{-T}^{T} b(t)\, dt$. *Notice that* $I > 0$.

**EXAMPLES 2.1**
(1) $b(x)$ can be the *characteristic* function over $[-1, 1]$.
(2) $b(x)$ can be the *hat* function over $[-1, 1]$, i.e.,

$$b(x) = \begin{cases} 1 + x, & -1 \leq x \leq 0, \\ 1 - x, & 0 < x \leq 1, \\ 0, & \text{elsewhere.} \end{cases}$$

Here we consider functions $f : \mathbf{R} \to \mathbf{R}$ that are either continuous and bounded, or uniformly continuous.

In this section we study the pointwise convergence with rates over the real line, to the unit operator, of the *univariate Cardaliaguet–Euvrard neural network operators* (see [62]),

$$(F_n(f))(x) := f_n(x) := \sum_{k=-n^2}^{n^2} \frac{f(k/n)}{I n^\alpha} \cdot b\left(n^{1-\alpha} \cdot \left(x - \frac{k}{n}\right)\right), \quad (2.1)$$

where $0 < \alpha < 1$ and $x \in \mathbf{R}$, $n \in \mathbf{N}$. The terms in the sum (2.1) can be nonzero iff

$$\left| n^{1-\alpha} \left( x - \frac{k}{n} \right) \right| \leq T, \text{ i.e., } \left| x - \frac{k}{n} \right| \leq \frac{T}{n^{1-\alpha}}$$

iff

$$nx - T n^\alpha \leq k \leq nx + T n^\alpha. \quad (2.2)$$

## 2. Rates of Univariate Neural Network Operators

To have the desired order

$$-n^2 \leq nx - Tn^\alpha \leq nx + Tn^\alpha \leq n^2, \tag{2.3}$$

it is sufficient enough to suppose that

$$n \geq T + |x|. \tag{2.4}$$

When $x \in [-T, T]$ it is enough to suppose $n \geq 2T$ which implies (2.3).

### PROPOSITION 2.1
Let $a \leq b$, $a, b \in \mathbf{R}$. Let $\mathrm{card}(k)$ $(\geq 0)$ be the maximum number of integers contained in $[a, b]$. Then

$$\max(0, (b-a)-1) \leq \mathrm{card}(k) \leq (b-a)+1.$$

**PROOF**  Let the consecutive $k_1, \ldots, k_l \in \mathbf{Z}$ be in $[a, b]$, $l = \mathrm{card}(k) \geq 0$, such that $k_1 < k_2 < \cdots < k_l$. Here $k_{i+1} - k_i = 1$, $i = 1, \ldots, l-1$; $0 \leq k_1 - a \leq 1$, $0 \leq b - k_l \leq 1$. Hence

$$l - 1 \leq b - a = (k_1 - a) + (b - k_l) + \sum_{i=1}^{l-1}(k_{i+1} - k_i) \leq l + 1.$$

Thus

$$(b-a) - 1 \leq l.$$

And

$$l \leq (b-a) + 1.$$

∎

*Note.* We would like to establish a lower bound on $\mathrm{card}(k)$ over the interval $[nx - Tn^\alpha, nx + Tn^\alpha]$. From Proposition 2.1 we find that

$$\mathrm{card}(k) \geq \max(2Tn^\alpha - 1, 0).$$

We obtain $\mathrm{card}(k) \geq 1$, if

$$2Tn^\alpha - 1 \geq 1 \quad \text{iff} \quad n \geq T^{-1/\alpha}.$$

So, to have the desired order (2.3) and card$(k) \geq 1$ over $[nx - Tn^\alpha, nx + Tn^\alpha]$, we need to take

$$n \geq \max(T + |x|, T^{-1/\alpha}). \qquad (2.4)^*$$

Also notice that card$(k) \to +\infty$, as $n \to +\infty$. We set $b^* := b(0)$ the maximum of $b(x)$.

Denote by $[\cdot]$ the integral part of a number and by $\lceil \cdot \rceil$ its ceiling. Next we present the first main result.

**THEOREM 2.1**
 *Let $x \in \mathbf{R}$, $T > 0$, and $n \in \mathbf{N}$ such that $n \geq \max(T + |x|, T^{-1/\alpha})$. Then*

$$|f_n(x) - f(x)| \leq |f(x)| \cdot \left| \sum_{k=\lceil nx-Tn^\alpha \rceil}^{[nx+Tn^\alpha]} \frac{1}{I \cdot n^\alpha} \cdot b\left(n^{1-\alpha} \cdot \left(x - \frac{k}{n}\right)\right) - 1 \right|$$

$$+ \frac{b^*}{I} \cdot \left(2T + \frac{1}{n^\alpha}\right) \cdot \omega_1\left(f, \frac{T}{n^{1-\alpha}}\right), \qquad (2.5)$$

*where $\omega_1$ is the first modulus of continuity of $f$. Inequality (2.5) becomes equality for constant functions.*

**PROOF** Notice that

$$\sum_{k=\lceil nx-Tn^\alpha \rceil}^{[nx+Tn^\alpha]} \frac{1}{I \cdot n^\alpha} \cdot b\left(n^{1-\alpha} \cdot \left(x - \frac{k}{n}\right)\right) \leq \frac{b^*}{I \cdot n^\alpha} \cdot \sum_{k=\lceil nx-Tn^\alpha \rceil}^{[nx+Tn^\alpha]} 1$$

$$\leq \frac{b^*}{I \cdot n^\alpha} \cdot (2Tn^\alpha + 1) = \frac{b^*}{I} \cdot \left(2T + \frac{1}{n^\alpha}\right). \qquad (2.6)$$

Next we estimate

$$\Delta := |f_n(x) - f(x)| = \left| \sum_{k=-n^2}^{n^2} \frac{f(k/n)}{I \cdot n^\alpha} \cdot b\left(n^{1-\alpha} \cdot \left(x - \frac{k}{n}\right)\right) - f(x) \right|$$

$$= \left| \sum_{k=-n^2}^{\lceil nx-Tn^\alpha \rceil - 1} \frac{f(k/n)}{I \cdot n^\alpha} \cdot b\left(n^{1-\alpha} \cdot \left(x - \frac{k}{n}\right)\right) \right.$$

2. Rates of Univariate Neural Network Operators 63

$$+ \sum_{k=[nx+Tn^\alpha]+1}^{n^2} \frac{f(k/n)}{I \cdot n^\alpha} \cdot b\left(n^{1-\alpha} \cdot \left(x - \frac{k}{n}\right)\right)$$

$$+ \sum_{k=\lceil nx-Tn^\alpha\rceil}^{[nx+Tn^\alpha]} \frac{f(k/n)}{I \cdot n^\alpha} \cdot b\left(n^{1-\alpha} \cdot \left(x - \frac{k}{n}\right)\right) - f(x)\Bigg|$$

$$= \left| \sum_{k=\lceil nx-Tn^\alpha\rceil}^{[nx+Tn^\alpha]} \frac{f(k/n)}{I \cdot n^\alpha} \cdot b\left(n^{1-\alpha} \cdot \left(x - \frac{k}{n}\right)\right) - f(x) \right|.$$

The last comes by the compact support $[-T, T]$ of $b$ and (2.2); i.e., we have that

$$\Delta = \left| \sum_{k=\lceil nx-Tn^\alpha\rceil}^{[nx+Tn^\alpha]} \frac{f(k/n)}{I \cdot n^\alpha} \cdot b\left(n^{1-\alpha} \cdot \left(x - \frac{k}{n}\right)\right) - f(x) \right|$$

$$= \Bigg| \sum_{k=\lceil nx-Tn^\alpha\rceil}^{[nx+Tn^\alpha]} \frac{f(k/n)}{I \cdot n^\alpha} \cdot b\left(n^{1-\alpha} \cdot \left(x - \frac{k}{n}\right)\right)$$

$$- f(x) \cdot \sum_{k=\lceil nx-Tn^\alpha\rceil}^{[nx+Tn^\alpha]} \frac{1}{I \cdot n^\alpha} \cdot b\left(n^{1-\alpha} \cdot \left(x - \frac{k}{n}\right)\right)$$

$$+ f(x) \cdot \sum_{k=\lceil nx-Tn^\alpha\rceil}^{[nx+Tn^\alpha]} \frac{1}{I \cdot n^\alpha} \cdot b\left(n^{1-\alpha} \cdot \left(x - \frac{k}{n}\right)\right) - f(x) \Bigg|$$

$$\leq |f(x)| \cdot \left| \sum_{k=\lceil nx-Tn^\alpha\rceil}^{[nx+Tn^\alpha]} \frac{1}{I \cdot n^\alpha} \cdot b\left(n^{1-\alpha} \cdot \left(x - \frac{k}{n}\right)\right) - 1 \right|$$

$$+ \left| \sum_{k=\lceil nx-Tn^\alpha\rceil}^{[nx+Tn^\alpha]} \frac{(f(k/n) - f(x))}{I \cdot n^\alpha} \cdot b\left(n^{1-\alpha} \cdot \left(x - \frac{k}{n}\right)\right) \right|$$

$$\leq |f(x)| \cdot \left| \sum_{k=\lceil nx-Tn^\alpha \rceil}^{[nx+Tn^\alpha]} \frac{1}{I \cdot n^\alpha} \cdot b\left(n^{1-\alpha} \cdot \left(x - \frac{k}{n}\right)\right) - 1 \right|$$

$$+ \sum_{k=\lceil nx-Tn^\alpha \rceil}^{[nx+Tn^\alpha]} \frac{\omega_1(f, |k/n - x|)}{I \cdot n^\alpha} \cdot b\left(n^{1-\alpha} \cdot \left(x - \frac{k}{n}\right)\right).$$

Therefore

$$|f_n(x) - f(x)| \leq |f(x)| \cdot \left| \sum_{k=\lceil nx-Tn^\alpha \rceil}^{[nx+Tn^\alpha]} \frac{1}{I \cdot n^\alpha} \cdot b\left(n^{1-\alpha} \cdot \left(x - \frac{k}{n}\right)\right) - 1 \right|$$

$$+ \omega_1\left(f, \frac{T}{n^{1-\alpha}}\right) \cdot \left( \sum_{k=\lceil nx-Tn^\alpha \rceil}^{[nx+Tn^\alpha]} \frac{1}{I \cdot n^\alpha} \cdot b\left(n^{1-\alpha} \cdot \left(x - \frac{k}{n}\right)\right) \right).$$

Using now the upper bound (2.6) we get (2.5). ∎

We will use

**LEMMA 2.1**
*We have that*

$$S_n(x) := \sum_{k=\lceil nx-Tn^\alpha \rceil}^{[nx+Tn^\alpha]} \frac{1}{I \cdot n^\alpha} \cdot b\left(n^{1-\alpha} \cdot \left(x - \frac{k}{n}\right)\right) \to 1,$$

*pointwise, as $n \to +\infty$, where $x \in \mathbf{R}$.*

**REMARK 2.1** Using Lemma 2.1 now, from inequality (2.5) we obtain that $f_n(x) \to f(x)$, pointwise with rates, as $n \to +\infty$, where $x \in \mathbf{R}$. ∎

**PROOF** (of Lemma 2.1). Here $b(x)$ is nondecreasing over $[-T, 0]$ and nonincreasing over $[0, T]$.
(i) Case of $\lceil nx \rceil + 1 \leq k \leq [nx + Tn^\alpha]$; i.e., $nx < nx + 1 \leq \lceil nx \rceil + 1 \leq k \leq [nx + Tn^\alpha]$. Let

$$nx \leq k - 1 \leq t \leq k \Rightarrow x - \frac{k}{n} \leq x - \frac{t}{n} \leq x - \frac{(k-1)}{n} < 0.$$

## 2. Rates of Univariate Neural Network Operators

Hence

$$b\left(n^{1-\alpha}\left(x-\frac{k}{n}\right)\right) \leq b\left(n^{1-\alpha}\left(x-\frac{t}{n}\right)\right)$$

$$\leq b\left(n^{1-\alpha}\left(x-\frac{(k-1)}{n}\right)\right).$$

Thus

$$b\left(n^{1-\alpha}\left(x-\frac{k}{n}\right)\right) \leq \int_{k-1}^{k} b\left(n^{1-\alpha}\left(x-\frac{t}{n}\right)\right) dt.$$

Now let $nx < k \leq t \leq k+1$. Therefore

$$x - \frac{(k+1)}{n} \leq x - \frac{t}{n} \leq x - \frac{k}{n} < 0.$$

So that

$$b\left(n^{1-\alpha}\left(x-\frac{(k+1)}{n}\right)\right) \leq b\left(n^{1-\alpha}\left(x-\frac{t}{n}\right)\right)$$

$$\leq b\left(n^{1-\alpha}\left(x-\frac{k}{n}\right)\right).$$

Then

$$\int_{k}^{k+1} b\left(n^{1-\alpha}\left(x-\frac{t}{n}\right)\right) \cdot dt \leq b\left(n^{1-\alpha}\left(x-\frac{k}{n}\right)\right).$$

We have established that

$$\int_{k}^{k+1} b\left(n^{1-\alpha}\left(x-\frac{t}{n}\right)\right) \cdot dt \leq b\left(n^{1-\alpha}\left(x-\frac{k}{n}\right)\right)$$

$$\leq \int_{k-1}^{k} b\left(n^{1-\alpha}\left(x-\frac{t}{n}\right)\right) \cdot dt, \quad (2.7)$$

for $\lceil nx \rceil + 1 \leq k \leq \lceil nx + Tn^{\alpha} \rceil$, where $k$ is an integer.

(ii) Case of $\lceil nx - Tn^{\alpha} \rceil \leq k \leq \lceil nx \rceil - 1$; i.e., $k < k+1 \leq \lceil nx \rceil < nx$. From $k - 1 \leq t \leq k < nx$ we have that

$$x - \frac{(k-1)}{n} \geq x - \frac{t}{n} \geq x - \frac{k}{n} > 0.$$

Hence

$$0 < b\left(n^{1-\alpha}\left(x - \frac{(k-1)}{n}\right)\right) \le b\left(n^{1-\alpha}\left(x - \frac{t}{n}\right)\right) \le b\left(n^{1-\alpha}\left(x - \frac{k}{n}\right)\right).$$

So that

$$\int_{k-1}^{k} b\left(n^{1-\alpha}\left(x - \frac{t}{n}\right)\right) \cdot dt \le b\left(n^{1-\alpha}\left(x - \frac{k}{n}\right)\right).$$

Let $k \le t \le k+1 < nx$. Therefore

$$x - \frac{k}{n} \ge x - \frac{t}{n} \ge x - \frac{(k+1)}{n} > 0.$$

Thus

$$b\left(n^{1-\alpha}\left(x - \frac{k}{n}\right)\right) \le b\left(n^{1-\alpha}\left(x - \frac{t}{n}\right)\right) \le b\left(n^{1-\alpha}\left(x - \frac{(k+1)}{n}\right)\right).$$

Then

$$b\left(n^{1-\alpha}\left(x - \frac{k}{n}\right)\right) \le \int_{k}^{k+1} b\left(n^{1-\alpha}\left(x - \frac{t}{n}\right)\right) \cdot dt.$$

We have found that

$$\int_{k-1}^{k} b\left(n^{1-\alpha}\left(x - \frac{t}{n}\right)\right) \cdot dt \le b\left(n^{1-\alpha}\left(x - \frac{k}{n}\right)\right)$$

$$\le \int_{k}^{k+1} b\left(n^{1-\alpha}\left(x - \frac{t}{n}\right)\right) \cdot dt, \quad (2.8)$$

for $\lceil nx - Tn^{\alpha} \rceil \le k \le \lceil nx \rceil - 1$, where $k$ is an integer.

It is clear that

$$S_n^3(x) := \frac{1}{I \cdot n^{\alpha}} \cdot b\left(n^{1-\alpha}\left(x - \frac{\lceil nx \rceil}{n}\right)\right) \to 0,$$

$$S_n^4(x) := \frac{1}{I \cdot n^{\alpha}} \cdot b\left(n^{1-\alpha}\left(x - \frac{\lceil nx \rceil}{n}\right)\right) \to 0, \quad (2.9)$$

as $n \to +\infty$. Set

$$S_n^1(x) := \sum_{k=\lceil nx \rceil+1}^{[nx+Tn^\alpha]} \frac{1}{I \cdot n^\alpha} \cdot b\left(n^{1-\alpha} \cdot \left(x - \frac{k}{n}\right)\right),$$

$$S_n^2(x) := \sum_{k=\lceil nx-Tn^\alpha \rceil}^{[nx]-1} \frac{1}{I \cdot n^\alpha} \cdot b\left(n^{1-\alpha} \cdot \left(x - \frac{k}{n}\right)\right). \tag{2.10}$$

So that
$$S_n(x) := S_n^1(x) + S_n^2(x) + S_n^3(x) + S_n^4(x). \tag{2.11}$$

From (2.7) we get that

$$\frac{1}{I \cdot n^\alpha} \cdot \int_{\lceil nx \rceil+1}^{[nx+Tn^\alpha]+1} b\left(n^{1-\alpha}\left(x - \frac{t}{n}\right)\right) \cdot dt$$

$$\leq S_n^1(x) \leq \frac{1}{I \cdot n^\alpha} \cdot \int_{\lceil nx \rceil}^{[nx+Tn^\alpha]} b\left(n^{1-\alpha}\left(x - \frac{t}{n}\right)\right) dt.$$

That is

$$\frac{1}{I} \cdot \int_{n^{1-\alpha}(x-([nx+Tn^\alpha]+1)/n)}^{n^{1-\alpha}(x-(\lceil nx \rceil+1)/n)} b(t)\, dt$$

$$\leq S_n^1(x) \leq \frac{1}{I} \cdot \int_{n^{1-\alpha}(x-[nx+Tn^\alpha]/n)}^{n^{1-\alpha}(x-\lceil nx \rceil/n)} b(t)\, dt. \tag{2.12}$$

But we see that

$$\lim_{n \to +\infty}\left(\frac{nx - \lceil nx \rceil - 1}{n^\alpha}\right) = \lim_{n \to +\infty}\left(\frac{nx - \lceil nx \rceil}{n^\alpha}\right) = 0,$$

also that

$$\lim_{n \to +\infty}\left(\frac{nx - [nx+Tn^\alpha] - 1}{n^\alpha}\right) = \lim_{n \to +\infty}\left(\frac{nx - [nx+Tn^\alpha]}{n^\alpha}\right) = -T.$$

Thus we obtain
$$S_n^1(x) \to \frac{1}{I} \cdot \int_{-T}^{0} b(t)\, dt, \quad \text{as } n \to +\infty.$$

Likewise from (2.8) we find that

$$\frac{1}{I \cdot n^\alpha} \cdot \int_{\lceil nx - Tn^\alpha \rceil - 1}^{[nx]-1} b\left(n^{1-\alpha}\left(x - \frac{t}{n}\right)\right) \cdot dt$$

$$\leq S_n^2(x) \leq \frac{1}{I \cdot n^\alpha} \cdot \int_{\lceil nx - Tn^\alpha \rceil}^{[nx]} b\left(n^{1-\alpha}\left(x - \frac{t}{n}\right)\right) \cdot dt.$$

That is

$$\frac{1}{I} \cdot \int_{n^{1-\alpha} \cdot (x - ([nx]-1)/n)}^{n^{1-\alpha} \cdot (x - (\lceil nx - Tn^\alpha \rceil - 1)/n)} b(t)\, dt \leq S_n^2(x)$$

$$\leq \frac{1}{I} \cdot \int_{n^{1-\alpha} \cdot (x - [nx]/n)}^{n^{1-\alpha} \cdot (x - \lceil nx - Tn^\alpha \rceil /n)} b(t)\, dt.$$

Because

$$\lim_{n \to +\infty} \left(\frac{nx - \lceil nx - Tn^\alpha \rceil + 1}{n^\alpha}\right) = \lim_{n \to +\infty} \left(\frac{nx - \lceil nx - Tn^\alpha \rceil}{n^\alpha}\right) = T,$$

and

$$\lim_{n \to +\infty} \left(\frac{nx - [nx] + 1}{n^\alpha}\right) = \lim_{n \to +\infty} \left(\frac{nx - [nx]}{n^\alpha}\right) = 0,$$

we get that
$$S_n^2(x) \to \frac{1}{I} \cdot \int_0^T b(t)\, dt, \quad \text{as } n \to +\infty.$$

We have proved now that
$$\lim_{n \to +\infty} S_n(x) = 1, \quad x \in \mathbf{R}.$$

∎

Now we present the second main result.

## 2. Rates of Univariate Neural Network Operators

**THEOREM 2.2**
Let $x \in \mathbf{R}$, $T > 0$, and $n \in \mathbf{N}$ such that $n \geq \max(T + |x|, T^{-1/\alpha})$. Let $f \in C^N(\mathbf{R})$, $N \in \mathbf{N}$, such that $f^{(N)}$ is a uniformly continuous function or $f^{(N)}$ is continuous and bounded. Then

$$|f_n(x) - f(x)| = \left| \sum_{k=-n^2}^{n^2} \frac{f(k/n)}{I \cdot n^\alpha} \cdot b\left(n^{1-\alpha} \cdot \left(x - \frac{k}{n}\right)\right) - f(x) \right|$$

$$\leq |f(x)| \cdot \left| \sum_{k=\lceil nx-Tn^\alpha \rceil}^{\lfloor nx+Tn^\alpha \rfloor} \frac{1}{I \cdot n^\alpha} \cdot b\left(n^{1-\alpha} \cdot \left(x - \frac{k}{n}\right)\right) - 1 \right|$$

$$+ \frac{b^*}{I} \cdot \left(2T + \frac{1}{n^\alpha}\right) \cdot \left( \sum_{j=1}^{N} \frac{|f^{(j)}(x)| \cdot T^j}{n^{j \cdot (1-\alpha)} \cdot j!} \right)$$

$$+ \omega_1\left(f^{(N)}, \frac{T}{n^{1-\alpha}}\right) \cdot \frac{T^N}{N! \cdot n^{N \cdot (1-\alpha)}} \cdot \frac{b^*}{I} \cdot \left(2T + \frac{1}{n^\alpha}\right). \quad (2.13)$$

**REMARK 2.2** Inequality (2.13) is attained by constant functions. Also notice that as $n \to +\infty$, we have the R.H.S. (2.13) $\to 0$ (by Lemma 2.1, etc.), therefore L.H.S. (2.13) $\to 0$. That is, (2.13) gives us with rates the pointwise convergence of $f_n(x) \to f(x)$, as $n \to +\infty$, $x \in \mathbf{R}$. ∎

**PROOF** (of Theorem 2.2). Notice that $b$ here is of compact support $[-T, T]$ and all assumptions are as stated earlier. From Taylor's formula we see that

$$f\left(\frac{k}{n}\right) = \sum_{j=0}^{N} \frac{f^{(j)}(x)}{j!} \cdot \left(\frac{k}{n} - x\right)^j$$

$$+ \int_x^{k/n} (f^{(N)}(t) - f^{(N)}(x)) \cdot \frac{((k/n) - t)^{N-1}}{(N-1)!} \cdot dt.$$

So that

$$\frac{f(k/n) \cdot b(n^{1-\alpha}(x - k/n))}{I \cdot n^\alpha} = \sum_{j=0}^{N} \frac{f^{(j)}(x)}{j!} \cdot \left(\frac{k}{n} - x\right)^j$$

$$\cdot \frac{b(n^{1-\alpha} \cdot (x - k/n))}{I \cdot n^\alpha} + \frac{b(n^{1-\alpha} \cdot (x - k/n))}{I \cdot n^\alpha}$$

$$\cdot \int_x^{k/n} (f^{(N)}(t) - f^{(N)}(x)) \cdot \frac{(k/n - t)^{N-1}}{(N-1)!} \cdot dt.$$

Therefore

$$(F_n(f))(x) - f(x) = \sum_{k=\lceil nx - Tn^\alpha \rceil}^{\lceil nx + Tn^\alpha \rceil} \frac{f(k/n) \cdot b(n^{1-\alpha} \cdot (x - k/n))}{I \cdot n^\alpha} - f(x)$$

$$= -f(x) + \sum_{j=0}^{N} \frac{f^{(j)}(x)}{j!}$$

$$\cdot \left( \sum_{k=\lceil nx - Tn^\alpha \rceil}^{\lceil nx + Tn^\alpha \rceil} \frac{(k/n - x)^j}{I \cdot n^\alpha} \cdot b(n^{1-\alpha} \cdot (x - k/n)) \right) + \mathcal{R} =: \otimes,$$

where

$$\mathcal{R} := \sum_{k=\lceil nx - Tn^\alpha \rceil}^{\lceil nx + Tn^\alpha \rceil} \frac{b(n^{1-\alpha} \cdot (x - k/n))}{I \cdot n^\alpha}$$

$$\cdot \int_x^{k/n} (f^{(N)}(t) - f^{(N)}(x)) \cdot \frac{(k/n - t)^{N-1}}{(N-1)!} \cdot dt, \quad (2.14)$$

$$\otimes = -f(x) + f(x) \cdot \left( \sum_{k=\lceil nx - Tn^\alpha \rceil}^{\lceil nx + Tn^\alpha \rceil} \frac{1}{I \cdot n^\alpha} \cdot b \left( n^{1-\alpha} \cdot \left( x - \frac{k}{n} \right) \right) \right)$$

$$+ \sum_{j=1}^{N} \frac{f^{(j)}(x)}{j!} \cdot \left( \sum_{k=\lceil nx - Tn^\alpha \rceil}^{\lceil nx + Tn^\alpha \rceil} \frac{(k/n - x)^j}{I \cdot n^\alpha} \cdot b \left( n^{1-\alpha} \cdot \left( x - \frac{k}{n} \right) \right) \right) + \mathcal{R}.$$

Hence

$$(F_n(f))(x) - f(x) = f(x) \cdot \left( \sum_{k=\lceil nx - Tn^\alpha \rceil}^{\lceil nx + Tn^\alpha \rceil} \frac{1}{I \cdot n^\alpha} \cdot b \left( n^{1-\alpha} \cdot \left( x - \frac{k}{n} \right) \right) - 1 \right)$$

$$+ \sum_{j=1}^{N} \frac{f^{(j)}(x)}{j!} \cdot \left( \sum_{k=\lceil nx-Tn^{\alpha} \rceil}^{\lceil nx+Tn^{\alpha} \rceil} \frac{(k/n-x)^j}{I \cdot n^{\alpha}} \cdot b\left(n^{1-\alpha} \cdot \left(x - \frac{k}{n}\right)\right) \right) + \mathcal{R}.$$

So that

$$|(F_n(f))(x) - f(x)|$$

$$\leq |f(x)| \cdot \left| \sum_{k=\lceil nx-Tn^{\alpha} \rceil}^{\lceil nx+Tn^{\alpha} \rceil} \frac{1}{I \cdot n^{\alpha}} \cdot b\left(n^{1-\alpha} \cdot \left(x - \frac{k}{n}\right)\right) - 1 \right|$$

$$+ \sum_{j=1}^{N} \frac{|f^{(j)}(x)|}{j!}$$

$$\cdot \left( \sum_{k=\lceil nx-Tn^{\alpha} \rceil}^{\lceil nx+Tn^{\alpha} \rceil} \frac{T^j}{n^{j \cdot (1-\alpha)} \cdot I \cdot n^{\alpha}} \cdot b\left(n^{1-\alpha} \cdot \left(x - \frac{k}{n}\right)\right) \right) + |\mathcal{R}|.$$

Furthermore (by (2.6))

$$|(F_n(f))(x) - f(x)|$$

$$\leq |f(x)| \cdot \left| \sum_{k=\lceil nx-Tn^{\alpha} \rceil}^{\lceil nx+Tn^{\alpha} \rceil} \frac{1}{I \cdot n^{\alpha}} \cdot b\left(n^{1-\alpha} \cdot \left(x - \frac{k}{n}\right)\right) - 1 \right|$$

$$+ \sum_{j=1}^{N} \frac{|f^{(j)}(x)|}{j!} \cdot \frac{T^j}{n^{j \cdot (1-\alpha)}} \cdot \frac{b^*}{I} \cdot \left(2T + \frac{1}{n^{\alpha}}\right) + |\mathcal{R}|$$

$$= |f(x)| \cdot \left| \sum_{k=\lceil nx-Tn^{\alpha} \rceil}^{\lceil nx+Tn^{\alpha} \rceil} \frac{1}{I \cdot n^{\alpha}} \cdot b\left(n^{1-\alpha} \cdot \left(x - \frac{k}{n}\right)\right) - 1 \right|$$

$$+ \frac{b^*}{I} \cdot \left(2T + \frac{1}{n^{\alpha}}\right) \cdot \left( \sum_{j=1}^{N} \frac{T^j \cdot |f^{(j)}(x)|}{n^{j \cdot (1-\alpha)} \cdot j!} \right) + |\mathcal{R}|. \qquad (2.15)$$

In the following we estimate

$$|\mathcal{R}| = \left| \sum_{k=\lceil nx-Tn^\alpha \rceil}^{\lceil nx+Tn^\alpha \rceil} \frac{b(n^{1-\alpha} \cdot (x - k/n))}{I \cdot n^\alpha} \right.$$

$$\left. \cdot \int_x^{k/n} (f^{(N)}(t) - f^{(N)}(x)) \cdot \frac{(k/n - t)^{N-1}}{(N-1)!} \cdot dt \right|$$

$$\leq \sum_{k=\lceil nx-Tn^\alpha \rceil}^{\lceil nx+Tn^\alpha \rceil} \frac{b(n^{1-\alpha} \cdot (x - k/n))}{I \cdot n^\alpha}$$

$$\cdot \left| \int_x^{k/n} (f^{(N)}(t) - f^{(N)}(x)) \cdot \frac{(k/n - t)^{N-1}}{(N-1)!} \cdot dt \right|$$

$$\leq \sum_{k=\lceil nx-Tn^\alpha \rceil}^{\lceil nx+Tn^\alpha \rceil} \frac{b(n^{1-\alpha} \cdot (x - k/n))}{I \cdot n^\alpha} \cdot \gamma \leq (*),$$

where

$$\gamma := \left| \int_x^{k/n} |f^{(N)}(t) - f^{(N)}(x)| \cdot \frac{|k/n - t|^{N-1}}{(N-1)!} \cdot dt \right| \quad (2.16)$$

$$(*) \leq \sum_{k=\lceil nx-Tn^\alpha \rceil}^{\lceil nx+Tn^\alpha \rceil} \frac{b(n^{1-\alpha} \cdot (x - k/n))}{I \cdot n^\alpha} \cdot \varphi, \quad (2.17)$$

with

$$\varphi := \omega_1\left(f^{(N)}, \frac{T}{n^{1-\alpha}}\right) \cdot \frac{T^N}{N! \cdot n^{N \cdot (1-\alpha)}}. \quad (2.18)$$

This last part of the inequality (2.17) comes from the next:
(i) Let $x \leq k/n$, then

$$\gamma = \int_x^{k/n} |f^{(N)}(t) - f^{(N)}(x)| \cdot \frac{|k/n - t|^{N-1}}{(N-1)!} \cdot dt$$

$$\leq \int_x^{k/n} \omega_1(f^{(N)}, |t - x|) \cdot \frac{|k/n - t|^{N-1}}{(N-1)!} \cdot dt$$

## 2. Rates of Univariate Neural Network Operators

$$\leq \omega_1\left(f^{(N)}, \left|x - \frac{k}{n}\right|\right) \cdot \int_x^{k/n} \frac{(k/n - t)^{N-1}}{(N-1)!} \cdot dt$$

$$\leq \omega_1\left(f^{(N)}, \frac{T}{n^{1-\alpha}}\right) \cdot \frac{(k/n - x)^N}{N!} \leq \omega_1\left(f^{(N)}, \frac{T}{n^{1-\alpha}}\right) \cdot \frac{T^N}{N! \cdot n^{N \cdot (1-\alpha)}};$$

i.e., when $x \leq k/n$ we find

$$\gamma \leq \omega_1\left(f^{(N)}, \frac{T}{n^{1-\alpha}}\right) \cdot \frac{T^N}{N! \cdot n^{N \cdot (1-\alpha)}}. \tag{2.19}$$

(ii) Let $x \geq k/n$, then

$$\gamma = \left|\int_{k/n}^x |f^{(N)}(t) - f^{(N)}(x)| \cdot \frac{|t - k/n|^{N-1}}{(N-1)!} \cdot dt\right|$$

$$= \int_{k/n}^x |f^{(N)}(t) - f^{(N)}(x)| \cdot \frac{(t - k/n)^{N-1}}{(N-1)!} \cdot dt$$

$$\leq \int_{k/n}^x \omega_1(f^{(N)}, |t - x|) \cdot \frac{(t - k/n)^{N-1}}{(N-1)!} \cdot dt$$

$$\leq \omega_1\left(f^{(N)}, \left|x - \frac{k}{n}\right|\right) \cdot \int_{k/n}^x \frac{(t - k/n)^{N-1}}{(N-1)!} \cdot dt$$

$$= \omega_1\left(f^{(N)}, \left|x - \frac{k}{n}\right|\right) \cdot \frac{(x - k/n)^N}{N!}$$

$$\leq \omega_1\left(f^{(N)}, \frac{T}{n^{1-\alpha}}\right) \cdot \frac{T^N}{N! \cdot n^{N \cdot (1-\alpha)}}.$$

Hence, in both cases we have found that

$$\gamma \leq \omega_1\left(f^{(N)}, \frac{T}{n^{1-\alpha}}\right) \cdot \frac{T^N}{N! \cdot n^{N \cdot (1-\alpha)}}. \tag{2.20}$$

Consequently, from (2.6), (2.17), and (2.20) we get

$$|\mathcal{R}| \leq \omega_1\left(f^{(N)}, \frac{T}{n^{1-\alpha}}\right) \cdot \frac{T^N}{N! \cdot n^{N \cdot (1-\alpha)}} \cdot \frac{b^*}{I} \cdot \left(2T + \frac{1}{n^\alpha}\right). \quad (2.21)$$

Finally, from (2.15) and (2.21) we conclude inequality (2.13). ∎

**COROLLARY 2.1**
Let $b(x)$ be a centered bell-shaped continuous function on **R** of compact support $[-T, T]$. Let $x \in [-T^*, T^*]$, $T^* > 0$, and $n \in \mathbf{N}$ be such that $n \geq \max(T + T^*, T^{-1/\alpha})$, $0 < \alpha < 1$. Consider $p \geq 1$. Then it holds

$$\|f_n - f\|_{p, [-T^*, T^*]} \leq \|f\|_{\infty, [-T^*, T^*]}$$

$$\cdot \left\| \sum_{k=\lceil nx - Tn^\alpha \rceil}^{\lceil nx + Tn^\alpha \rceil} \frac{1}{I \cdot n^\alpha} \cdot b\left(n^{1-\alpha} \cdot \left(x - \frac{k}{n}\right)\right) - 1 \right\|_{p, [-T^*, T^*]}$$

$$+ \omega_1\left(f, \frac{T}{n^{1-\alpha}}\right) \cdot \frac{b^*}{I} \cdot \left(2T + \frac{1}{n^\alpha}\right) \cdot 2^{1/p} \cdot T^{*1/p}, \quad (2.22)$$

where $I := \int_{-T}^{T} b(t)\, dt$.

Inequality (2.22) is attained by constant functions. Also from (2.22) we get the $L_p$ convergence of $f_n$ to $f$ with rates.

**PROOF** From Theorem 2.1 (2.5) we have that

$$|f_n(x) - f(x)|$$

$$\leq \|f\|_{\infty, [-T^*, T^*]} \cdot \left| \sum_{k=\lceil nx - Tn^\alpha \rceil}^{\lceil nx + Tn^\alpha \rceil} \frac{1}{I \cdot n^\alpha} \cdot b\left(n^{1-\alpha} \cdot \left(x - \frac{k}{n}\right)\right) - 1 \right|$$

$$+ \frac{b^*}{I} \cdot \left(2T + \frac{1}{n^\alpha}\right) \cdot \omega_1\left(f, \frac{T}{n^{1-\alpha}}\right). \quad (2.23)$$

Inequality (2.22) now comes by integration of (2.23) and the properties of

## 2. Rates of Univariate Neural Network Operators

the $L_p$-norm. From (2.6) we get that

$$\sum_{k=\lceil nx-Tn^\alpha \rceil}^{\lfloor nx+Tn^\alpha \rfloor} \frac{1}{I \cdot n^\alpha} \cdot b\left(n^{1-\alpha} \cdot \left(x - \frac{k}{n}\right)\right) \le \frac{b^*}{I} \cdot (2T+1).$$

Hence

$$\left| \sum_{k=\lceil nx-Tn^\alpha \rceil}^{\lfloor nx+Tn^\alpha \rfloor} \frac{1}{I \cdot n^\alpha} \cdot b\left(n^{1-\alpha} \cdot \left(x - \frac{k}{n}\right)\right) - 1 \right|^p$$

$$\le \left(\frac{b^*}{I} \cdot (2T+1) + 1\right)^p =: M > 0,$$

for all $n \in \mathbf{N}$, and all $x \in [-T^*, T^*]$. Also, from Lemma 2.1, we obtain that

$$\lim_{n \to +\infty} \left| \sum_{k=\lceil nx-Tn^\alpha \rceil}^{\lfloor nx+Tn^\alpha \rfloor} \frac{1}{I \cdot n^\alpha} \cdot b\left(n^{1-\alpha} \cdot \left(x - \frac{k}{n}\right)\right) - 1 \right|^p = 0,$$

all $x \in [-T^*, T^*]$. Now it is clear from the bounded convergence theorem that

$$\lim_{n \to \infty} \left\| \sum_{k=\lceil nx-Tn^\alpha \rceil}^{\lfloor nx+Tn^\alpha \rfloor} \frac{1}{I \cdot n^\alpha} \cdot b\left(n^{1-\alpha} \cdot \left(x - \frac{k}{n}\right)\right) - 1 \right\|_{p, [-T^*, T^*]} = 0.$$

∎

**COROLLARY 2.2**

*Let $b(x)$ be a centered bell-shaped continuous function on $\mathbf{R}$ of compact support $[-T, T]$. Let $x \in [-T^*, T^*]$, $T^* > 0$, and $n \in \mathbf{N}$ be such that $n \ge \max(T + T^*, T^{-1/\alpha})$, $0 < \alpha < 1$. Consider $p \ge 1$. Then we get*

$$\|f_n - f\|_{p, [-T^*, T^*]} \le \|f\|_{\infty, [-T^*, T^*]}$$

$$\cdot \left\| \sum_{k=\lceil nx-Tn^\alpha \rceil}^{\lfloor nx+Tn^\alpha \rfloor} \frac{1}{I \cdot n^\alpha} \cdot b\left(n^{1-\alpha} \cdot \left(x - \frac{k}{n}\right)\right) - 1 \right\|_{p, [-T^*, T^*]}$$

$$+ \frac{b*}{I} \cdot \left(2T + \frac{1}{n^\alpha}\right) \cdot \left(\sum_{j=1}^{N} \frac{T^j \cdot \|f^{(j)}\|_{p,[-T^*,T^*]}}{n^{j\cdot(1-\alpha)} \cdot j!}\right)$$

$$+ \omega_1\left(f^{(N)}, \frac{T}{n^{1-\alpha}}\right) \cdot \frac{2^{1/p} \cdot T^N \cdot T^{*1/p}}{N! \cdot n^{N\cdot(1-\alpha)}} \cdot \frac{b*}{I} \cdot \left(2T + \frac{1}{n^\alpha}\right), \quad (2.24)$$

where $N \geq 1$.

The last is attained by constant functions. Here from (2.24) we again obtain the $L_p$ convergence of $f_n$ to $f$ with rates.

**PROOF** From Theorem 2.2 (2.13) we see that

$$|f_n(x) - f(x)|$$

$$\leq \|f\|_{\infty,[-T^*,T^*]} \cdot \left|\sum_{k=\lceil nx-Tn^\alpha \rceil}^{\lceil nx+Tn^\alpha \rceil} \frac{1}{I \cdot n^\alpha} \cdot b\left(n^{1-\alpha} \cdot \left(x - \frac{k}{n}\right)\right) - 1\right|$$

$$+ \frac{b*}{I} \cdot \left(2T + \frac{1}{n^\alpha}\right) \cdot \left(\sum_{j=1}^{n} \frac{|f^{(j)}(x)| \cdot T^j}{n^{j\cdot(1-\alpha)} j!}\right)$$

$$+ \omega_1\left(f^{(N)}, \frac{T}{n^{1-\alpha}}\right) \cdot \frac{T^N}{N! \cdot n^{N\cdot(1-\alpha)}} \cdot \frac{b*}{I} \cdot \left(2T + \frac{1}{n^\alpha}\right). \quad (2.25)$$

Inequality (2.24) now comes by integration of (2.25) and the properties of the $L_p$-norm. ∎

## 2.2 The Univariate "Squashing Operators" and their Convergence to the Unit with Rates

We use

**DEFINITION 2.2** *Let the nonnegative function $S : \mathbf{R} \to \mathbf{R}$, $S$ has compact support $[-T,T]$, $T > 0$, and is nondecreasing there and it can be*

## 2. Rates of Univariate Neural Network Operators

continuous only on either $(-\infty, T]$ or $[-T, T]$. $S$ can have jump discontinuities. We call $S$ the "squashing function" (see also [62]).

Let $f : \mathbf{R} \to \mathbf{R}$ be either uniformly continuous or continuous and bounded. Assume that

$$I^* := \int_{-T}^{T} S(t)\,dt > 0.$$

Clearly
$$\max_{x \in [-T,T]} S(x) = S(T).$$

For $x \in \mathbf{R}$ we define the "univariate squashing operator"

$$(G_n(f))(x) := \sum_{k=-n^2}^{n^2} \frac{f(k/n)}{I^* \cdot n^\alpha} \cdot S\left(n^{1-\alpha} \cdot \left(x - \frac{k}{n}\right)\right), \qquad (2.26)$$

$0 < \alpha < 1$ and $n \in \mathbf{N}$: $n \geq \max(T + |x|, T^{-1/\alpha})$. It is easy to see that

$$(G_n(f))(x) = \sum_{k=\lceil nx - Tn^\alpha \rceil}^{[nx+Tn^\alpha]} \frac{f(k/n)}{I^* \cdot n^\alpha} \cdot S\left(n^{1-\alpha} \cdot \left(x - \frac{k}{n}\right)\right). \qquad (2.27)$$

Here we study the pointwise convergence with rates of $(G_n(f))(x) \to f(x)$, as $n \to +\infty$, $x \in \mathbf{R}$.

**THEOREM 2.3**
Under the above terms and assumptions we get

$$|(G_n(f))(x) - f(x)|$$

$$\leq |f(x)| \cdot \left| \sum_{k=\lceil nx-Tn^\alpha \rceil}^{[nx+Tn^\alpha]} \frac{1}{I^* \cdot n^\alpha} \cdot S\left(n^{1-\alpha} \cdot \left(x - \frac{k}{n}\right)\right) - 1 \right|$$

$$+ \omega_1\left(f, \frac{T}{n^{1-\alpha}}\right) \cdot \frac{S(T)}{I^*} \cdot \left(2T + \frac{1}{n^\alpha}\right). \qquad (2.28)$$

Inequality (2.28) is attained by constant functions.

**PROOF** We observe that

$$|(G_n f)(x) - f(x)|$$

$$= \left| \sum_{k=\lceil nx-Tn^\alpha \rceil}^{[nx+Tn^\alpha]} \frac{f(k/n)}{I^* \cdot n^\alpha} \cdot S\left(n^{1-\alpha} \cdot \left(x - \frac{k}{n}\right)\right) - f(x) \right|$$

$$= \left| \sum_{k=\lceil nx-Tn^\alpha \rceil}^{[nx+Tn^\alpha]} \frac{(f(k/n) - f(x))}{I^* \cdot n^\alpha} \cdot S\left(n^{1-\alpha} \cdot \left(x - \frac{k}{n}\right)\right) \right.$$

$$\left. + f(x) \cdot \sum_{k=\lceil nx-Tn^\alpha \rceil}^{[nx+Tn^\alpha]} \frac{1}{I^* \cdot n^\alpha} \cdot S\left(n^{1-\alpha} \cdot \left(x - \frac{k}{n}\right)\right) - f(x) \right|$$

$$\leq \sum_{k=\lceil nx-Tn^\alpha \rceil}^{[nx+Tn^\alpha]} \frac{|f(k/n) - f(x)|}{I^* \cdot n^\alpha} \cdot S\left(n^{1-\alpha} \cdot \left(x - \frac{k}{n}\right)\right)$$

$$+ |f(x)| \cdot \left| \sum_{k=\lceil nx-Tn^\alpha \rceil}^{[nx+Tn^\alpha]} \frac{1}{I^* \cdot n^\alpha} \cdot S\left(n^{1-\alpha} \cdot \left(x - \frac{k}{n}\right)\right) - 1 \right|$$

$$\leq \omega_1\left(f, \frac{T}{n^{1-\alpha}}\right) \cdot \sum_{k=\lceil nx-Tn^\alpha \rceil}^{[nx+Tn^\alpha]} \frac{1}{I^* \cdot n^\alpha} \cdot S\left(n^{1-\alpha} \cdot \left(x - \frac{k}{n}\right)\right)$$

$$+ |f(x)| \cdot \left| \sum_{k=\lceil nx-Tn^\alpha \rceil}^{[nx+Tn^\alpha]} \frac{1}{I^* \cdot n^\alpha} \cdot S\left(n^{1-\alpha} \cdot \left(x - \frac{k}{n}\right)\right) - 1 \right|.$$

Hence

$$|(G_n f)(x) - f(x)|$$

$$\leq |f(x)| \cdot \left| \sum_{k=\lceil nx-Tn^\alpha \rceil}^{[nx+Tn^\alpha]} \frac{1}{I^* \cdot n^\alpha} \cdot S\left(n^{1-\alpha} \cdot \left(x - \frac{k}{n}\right)\right) - 1 \right|$$

## 2. Rates of Univariate Neural Network Operators

$$+ \omega_1\left(f, \frac{T}{n^{1-\alpha}}\right) \cdot \frac{S(T)}{I^* \cdot n^\alpha} \cdot \sum_{k=\lceil nx-Tn^\alpha \rceil}^{\lceil nx+Tn^\alpha \rceil} 1.$$

Using in the last that

$$\sum_{k=\lceil nx-Tn^\alpha \rceil}^{\lceil nx+Tn^\alpha \rceil} 1 \leq (2Tn^\alpha + 1)$$

we have proved (2.28). ∎

**LEMMA 2.2**
We have that

$$D_n(x) := \sum_{k=\lceil nx-Tn^\alpha \rceil}^{\lceil nx+Tn^\alpha \rceil} \frac{1}{I^* \cdot n^\alpha} \cdot S\left(n^{1-\alpha} \cdot \left(x - \frac{k}{n}\right)\right) \to 1,$$

pointwise, as $n \to +\infty$, where $x \in \mathbf{R}$.

**REMARK 2.3**   Using Lemma 2.2 now, from inequality (2.28) we get $(G_n(f))(x) \to f(x)$, pointwise with rates, as $n \to +\infty$, where $x \in \mathbf{R}$. ∎

**PROOF**   (of Lemma 2.2).
(i) Case of $\lceil nx \rceil + 1 \leq k \leq \lceil nx + Tn^\alpha \rceil$, i.e.,

$$nx < nx + 1 \leq \lceil nx \rceil + 1 \leq k \leq \lceil nx + Tn^\alpha \rceil.$$

Consider $t$,
$$nx \leq k - 1 \leq t \leq k,$$
that is,
$$x - \frac{k}{n} \leq x - \frac{t}{n} \leq x - \frac{(k-1)}{n} < 0.$$

Because $S$ is nondecreasing we observe that

$$0 \leq S\left(n^{1-\alpha} \cdot \left(x - \frac{k}{n}\right)\right) \leq S\left(n^{1-\alpha} \cdot \left(x - \frac{t}{n}\right)\right)$$

$$\leq S\left(n^{1-\alpha} \cdot \left(x - \frac{(k-1)}{n}\right)\right).$$

Thus
$$S\left(n^{1-\alpha}\cdot\left(x-\frac{k}{n}\right)\right) \leq \int_{k-1}^{k} S\left(n^{1-\alpha}\cdot\left(x-\frac{t}{n}\right)\right)\cdot dt.$$

Now, let $nx < k \leq t \leq k+1$, then
$$x - \frac{(k+1)}{n} \leq x - \frac{t}{n} \leq x - \frac{k}{n} < 0.$$

Hence
$$0 \leq S\left(n^{1-\alpha}\cdot\left(x-\frac{(k+1)}{n}\right)\right) \leq S\left(n^{1-\alpha}\cdot\left(x-\frac{t}{n}\right)\right)$$
$$\leq S\left(n^{1-\alpha}\cdot\left(x-\frac{k}{n}\right)\right).$$

Consequently we get that
$$\int_{k}^{k+1} S\left(n^{1-\alpha}\cdot\left(x-\frac{t}{n}\right)\right) dt \leq S\left(n^{1-\alpha}\cdot\left(x-\frac{k}{n}\right)\right);$$

i.e., we have found that
$$\int_{k}^{k+1} S\left(n^{1-\alpha}\cdot\left(x-\frac{t}{n}\right)\right) dt \leq S\left(n^{1-\alpha}\cdot\left(x-\frac{k}{n}\right)\right)$$
$$\leq \int_{k-1}^{k} S\left(n^{1-\alpha}\cdot\left(x-\frac{t}{n}\right)\right)\cdot dt, \qquad (2.29)$$

for any integer $k$, $\lceil nx \rceil + 1 \leq k \leq \lfloor nx + Tn^\alpha \rfloor$.

(ii) Case of $\lceil nx - Tn^\alpha \rceil \leq k \leq \lceil nx \rceil - 1$. Here
$$k < k+1 \leq \lceil nx \rceil < nx.$$

Consider $t$,
$$k-1 \leq t \leq k < nx,$$

then
$$x - \frac{(k-1)}{n} \geq x - \frac{t}{n} \geq x - \frac{k}{n} > 0.$$

## 2. Rates of Univariate Neural Network Operators

Since $S$ is nondecreasing, we obtain

$$S\left(n^{1-\alpha} \cdot \left(x - \frac{(k-1)}{n}\right)\right) \geq S\left(n^{1-\alpha} \cdot \left(x - \frac{t}{n}\right)\right)$$

$$\geq S\left(n^{1-\alpha} \cdot \left(x - \frac{k}{n}\right)\right) \geq 0.$$

Therefore

$$\int_{k-1}^{k} S\left(n^{1-\alpha} \cdot \left(x - \frac{t}{n}\right)\right) \cdot dt \geq S\left(n^{1-\alpha} \cdot \left(x - \frac{k}{n}\right)\right).$$

Let $k \leq t \leq k+1 < nx$, then

$$x - \frac{k}{n} \geq x - \frac{t}{n} \geq x - \frac{(k+1)}{n} > 0.$$

Consequently

$$S\left(n^{1-\alpha} \cdot \left(x - \frac{k}{n}\right)\right) \geq S\left(n^{1-\alpha} \cdot \left(x - \frac{t}{n}\right)\right)$$

$$\geq S\left(n^{1-\alpha} \cdot \left(x - \frac{(k+1)}{n}\right)\right) \geq 0.$$

That is,

$$S\left(n^{1-\alpha} \cdot \left(x - \frac{k}{n}\right)\right) \geq \int_{k}^{k+1} S\left(n^{1-\alpha} \cdot \left(x - \frac{t}{n}\right)\right) \cdot dt.$$

We have established that

$$\int_{k}^{k+1} S\left(n^{1-\alpha} \cdot \left(x - \frac{t}{n}\right)\right) \cdot dt \leq S\left(n^{1-\alpha} \cdot \left(x - \frac{k}{n}\right)\right)$$

$$\leq \int_{k-1}^{k} S\left(n^{1-\alpha} \cdot \left(x - \frac{t}{n}\right)\right) \cdot dt, \quad (2.30)$$

for any integer $k$,

$$\lceil nx - Tn^{\alpha} \rceil \leq k \leq \lceil nx \rceil - 1.$$

Notice that for specific $k^*$ we obtain

$$0 \le \frac{1}{I^* \cdot n^\alpha} \cdot S\left(n^{1-\alpha} \cdot \left(x - \frac{k^*}{n}\right)\right) \le \frac{S(T)}{I^* \cdot n^\alpha} \to 0,$$

as $n \to +\infty$.

Thus

$$D_n^3(x) := \frac{1}{I^* \cdot n^\alpha} \cdot S\left(n^{1-\alpha} \cdot \left(x - \frac{\lceil nx \rceil}{n}\right)\right) \to 0$$

and

$$D_n^4(x) := \frac{1}{I^* \cdot n^\alpha} \cdot S\left(n^{1-\alpha} \cdot \left(x - \frac{\lceil nx \rceil}{n}\right)\right) \to 0,$$

as $n \to +\infty$. Denote by

$$D_n^1(x) := \sum_{k=\lceil nx \rceil+1}^{\lceil nx+Tn^\alpha \rceil} \frac{1}{I^* \cdot n^\alpha} \cdot S\left(n^{1-\alpha} \cdot \left(x - \frac{k}{n}\right)\right)$$

and

$$D_n^2(x) := \sum_{k=\lceil nx-Tn^\alpha \rceil}^{\lceil nx \rceil-1} \frac{1}{I^* \cdot n^\alpha} \cdot S\left(n^{1-\alpha} \cdot \left(x - \frac{k}{n}\right)\right).$$

So we can represent

$$D_n(x) = D_n^1(x) + D_n^2(x) + D_n^3(x) + D_n^4(x).$$

From (2.29) we get

$$\frac{1}{I^* \cdot n^\alpha} \cdot \int_{\lceil nx \rceil+1}^{\lceil nx+Tn^\alpha \rceil+1} S\left(n^{1-\alpha} \cdot \left(x - \frac{t}{n}\right)\right) \cdot dt$$

$$\le D_n^1(x) \le \frac{1}{I^* \cdot n^\alpha} \cdot \int_{\lceil nx \rceil}^{\lceil nx+Tn^\alpha \rceil} S\left(n^{1-\alpha} \cdot \left(x - \frac{t}{n}\right)\right) \cdot dt.$$

So that

$$\frac{1}{I^*} \cdot \int_{n^{1-\alpha} \cdot (x - (\lceil nx+Tn^\alpha \rceil+1)/n)}^{n^{1-\alpha} \cdot (x - (\lceil nx \rceil+1)/n)} S(t) \cdot dt$$

$$\le D_n^1(x) \le \frac{1}{I^*} \cdot \int_{n^{1-\alpha} \cdot (x - \lceil nx+Tn^\alpha \rceil/n)}^{n^{1-\alpha} \cdot (x - \lceil nx \rceil/n)} S(t) \, dt.$$

Because
$$\lim_{n\to+\infty}\left(\frac{nx-\lceil nx\rceil-1}{n^\alpha}\right)=\lim_{n\to+\infty}\left(\frac{nx-\lceil nx\rceil}{n^\alpha}\right)=0$$

and
$$\lim_{n\to+\infty}\left(\frac{nx-\lceil nx+Tn^\alpha\rceil-1}{n^\alpha}\right)=\lim_{n\to+\infty}\left(\frac{nx-\lceil nx+Tn^\alpha\rceil}{n^\alpha}\right)=-T,$$

we see that
$$\lim_{n\to+\infty}D_n^1(x)=\frac{1}{I^*}\cdot\int_{-T}^{0}S(t)\cdot dt. \qquad (2.31)$$

Likewise we get from (2.30) that
$$\frac{1}{I^*\cdot n^\alpha}\cdot\int_{\lceil nx-Tn^\alpha\rceil}^{\lceil nx\rceil}S\left(n^{1-\alpha}\cdot\left(x-\frac{t}{n}\right)\right)\cdot dt$$
$$\le D_n^2(x)\le\frac{1}{I^*\cdot n^\alpha}\cdot\int_{\lceil nx-Tn^\alpha\rceil-1}^{\lceil nx\rceil-1}S\left(n^{1-\alpha}\cdot\left(x-\frac{t}{n}\right)\right)\cdot dt.$$

In other words,
$$\frac{1}{I^*}\cdot\int_{n^{1-\alpha}\cdot(x-\lceil nx\rceil/n)}^{n^{1-\alpha}\cdot(x-\lceil nx-Tn^\alpha\rceil/n)}S(t)\cdot dt\le D_n^2(x)$$
$$\le\frac{1}{I^*}\cdot\int_{n^{1-\alpha}\cdot(x-(\lceil nx\rceil-1)/n)}^{n^{1-\alpha}\cdot(x-(\lceil nx-Tn^\alpha\rceil-1)/n)}S(t)\cdot dt.$$

Because
$$\lim_{n\to+\infty}\left(\frac{nx-\lceil nx\rceil}{n^\alpha}\right)=\lim_{n\to+\infty}\left(\frac{nx-\lceil nx\rceil+1}{n^\alpha}\right)=0,$$

and
$$\lim_{n\to+\infty}\left(\frac{nx-\lceil nx-Tn^\alpha\rceil}{n^\alpha}\right)=\lim_{n\to+\infty}\left(\frac{nx-\lceil nx-Tn^\alpha\rceil+1}{n^\alpha}\right)=T,$$

we obtain that
$$\lim_{n\to+\infty} D_n^2(x) = \frac{1}{I^*} \cdot \int_0^T S(t) \cdot dt. \qquad (2.32)$$

Finally, from all the above, the representation of $D_n(x)$, and (2.31), (2.32), we conclude that $\lim_{n\to+\infty} D_n(x) = 1$, pointwise, $x \in \mathbf{R}$. ∎

As a related result we present

**THEOREM 2.4**
Let $x \in \mathbf{R}$, $T > 0$, and $n \in \mathbf{N}$, such that $n \geq \max(T + |x|, T^{-1/\alpha})$. Let $f \in C^N(\mathbf{R})$, $N \in \mathbf{N}$, such that $f^{(N)}$ is a uniformly continuous function or $f^{(N)}$ is continuous and bounded. Then

$$|(G_n(f))(x) - f(x)|$$

$$= \left| \sum_{k=-n^2}^{n^2} \frac{f(k/n)}{I^* \cdot n^\alpha} \cdot S\left(n^{1-\alpha} \cdot \left(x - \frac{k}{n}\right)\right) - f(x) \right|$$

$$\leq |f(x)| \cdot \left| \sum_{k=\lceil nx-Tn^\alpha \rceil}^{\lfloor nx+Tn^\alpha \rfloor} \frac{1}{I^* \cdot n^\alpha} \cdot S\left(n^{1-\alpha} \cdot \left(x - \frac{k}{n}\right)\right) - 1 \right|$$

$$+ \frac{S(T)}{I^*} \cdot \left(2T + \frac{1}{n^\alpha}\right) \cdot \left( \sum_{j=1}^N \frac{|f^{(j)}(x)| \cdot T^j}{j! \cdot n^{j \cdot (1-\alpha)}} \right)$$

$$+ \omega_1\left(f^{(N)}, \frac{T}{n^{1-\alpha}}\right) \cdot \frac{T^N}{N! \cdot n^{N \cdot (1-\alpha)}} \cdot \frac{S(T)}{I^*} \cdot \left(2T + \frac{1}{n^\alpha}\right). \quad (2.33)$$

Inequality (2.33) is attained by constant functions. Furthermore (2.33) gives us with rates (say, Lemma 2.2), the pointwise convergence of $G_n(f)(x) \to f(x)$, as $n \to +\infty$, $x \in \mathbf{R}$.

**PROOF**   As in the proof of Theorem 2.2, we observe that

$$\frac{f(k/n)}{I^* \cdot n^\alpha} \cdot S\left(n^{1-\alpha} \cdot \left(x - \frac{k}{n}\right)\right) = \sum_{j=0}^N \frac{f^{(j)}(x)}{j!} \cdot \left(\frac{k}{n} - x\right)^j$$

## 2. Rates of Univariate Neural Network Operators

$$\cdot \frac{S(n^{1-\alpha} \cdot (x - k/n))}{I^* \cdot n^\alpha} + \frac{S(n^{1-\alpha} \cdot (x - k/n))}{I^* \cdot n^\alpha}$$

$$\cdot \int_x^{k/n} (f^{(N)}(t) - f^{(N)}(x)) \cdot \frac{(k/n - t)^{N-1}}{(N-1)!} \cdot dt.$$

Hence

$$(G_n(f))(x) - f(x)$$

$$= \sum_{k=\lceil nx - Tn^\alpha \rceil}^{\lfloor nx + Tn^\alpha \rfloor} f\left(\frac{k}{n}\right) \cdot \frac{S(n^{1-\alpha} \cdot (x - k/n))}{I^* \cdot n^\alpha} - f(x)$$

$$= -f(x) + \sum_{j=0}^{N} \frac{f^{(j)}(x)}{j!}$$

$$\cdot \left( \sum_{k=\lceil nx - Tn^\alpha \rceil}^{\lfloor nx + Tn^\alpha \rfloor} \frac{(k/n - x)^j}{I^* \cdot n^\alpha} \cdot S\left(n^{1-\alpha} \cdot \left(x - \frac{k}{n}\right)\right) \right) + \mathcal{R}^* =: \otimes.$$

Here

$$\mathcal{R}^* := \sum_{k=\lceil nx - Tn^\alpha \rceil}^{\lfloor nx + Tn^\alpha \rfloor} \frac{S(n^{1-\alpha} \cdot (x - k/n))}{I^* \cdot n^\alpha}$$

$$\cdot \int_x^{k/n} (f^{(N)}(t) - f^{(N)}(x)) \cdot \frac{(k/n - t)^{Nn-1}}{(N-1)!} \cdot dt.$$

That is

$$\otimes = f(x) \cdot \left( \sum_{k=\lceil nx - Tn^\alpha \rceil}^{\lfloor nx + Tn^\alpha \rfloor} \frac{1}{I^* \cdot n^\alpha} \cdot S\left(n^{1-\alpha} \cdot \left(x - \frac{k}{n}\right)\right) - 1 \right) + \sum_{j=1}^{N} \frac{f^{(j)}(x)}{j!}$$

$$\cdot \left( \sum_{k=\lceil nx - Tn^\alpha \rceil}^{\lfloor nx + Tn^\alpha \rfloor} \frac{(k/n - x)^j}{I^* \cdot n^\alpha} \cdot S\left(n^{1-\alpha} \cdot \left(x - \frac{k}{n}\right)\right) \right) + \mathcal{R}^*.$$

Consequently

$$|(G_n(f))(x) - f(x)|$$

$$\leq |f(x)| \cdot \left| \sum_{k=\lceil nx-Tn^\alpha \rceil}^{[nx+Tn^\alpha]} \frac{1}{I^* \cdot n^\alpha} \cdot S\left(n^{1-\alpha} \cdot \left(x - \frac{k}{n}\right)\right) - 1 \right|$$

$$+ \sum_{j=1}^{N} \frac{|f^{(j)}(x)|}{j!}$$

$$\cdot \left( \sum_{k=\lceil nx-Tn^\alpha \rceil}^{[nx+Tn^\alpha]} \frac{(T^j/n^{j \cdot (1-\alpha)})}{I^* \cdot n^\alpha} \cdot S\left(n^{1-\alpha} \cdot \left(x - \frac{k}{n}\right)\right) \right) + |\mathcal{R}^*|$$

$$\leq \text{ (by (2.6) for } S\text{)}$$

$$|f(x)| \cdot \left| \sum_{k=\lceil nx-Tn^\alpha \rceil}^{[nx+Tn^\alpha]} \frac{1}{I^* \cdot n^\alpha} \cdot S\left(n^{1-\alpha} \cdot \left(x - \frac{k}{n}\right)\right) - 1 \right|$$

$$+ \sum_{j=1}^{N} \frac{|f^{(j)}(x)|}{j!} \cdot \frac{T^j}{n^{j \cdot (1-\alpha)}} \cdot \frac{S(T)}{I^*} \cdot \left(2T + \frac{1}{n^\alpha}\right) + |\mathcal{R}^*|.$$

Therefore

$$|(G_n(f))(x) - f(x)|$$

$$\leq |f(x)| \cdot \left| \sum_{k=\lceil nx-Tn^\alpha \rceil}^{[nx+Tn^\alpha]} \frac{1}{I^* \cdot n^\alpha} \cdot S\left(n^{1-\alpha} \cdot \left(x - \frac{k}{n}\right)\right) - 1 \right|$$

$$+ \frac{S(T)}{I^*} \cdot \left(2T + \frac{1}{n^\alpha}\right) \cdot \left( \sum_{j=1}^{N} \frac{|f^{(j)}(x)| \cdot T^j}{j! \cdot n^{j \cdot (1-\alpha)}} \right) + |\mathcal{R}^*|. \quad (2.34)$$

## 2. Rates of Univariate Neural Network Operators

Next we estimate

$$|\mathcal{R}^*| \le \sum_{k=\lceil nx - Tn^\alpha \rceil}^{\lceil nx + Tn^\alpha \rceil} \frac{S(n^{1-\alpha} \cdot (x - k/n))}{I^* \cdot n^\alpha}$$

$$\cdot \left| \int_x^{k/n} (f^{(N)}(t) - f^{(N)}(x)) \cdot \frac{(k/n - t)^{N-1}}{(N-1)!} \cdot dt \right|$$

$$\le \sum_{k=\lceil nx - Tn^\alpha \rceil}^{\lceil nx + Tn^\alpha \rceil} \frac{S(n^{1-\alpha} \cdot (x - k/n))}{I^* \cdot n^\alpha} \cdot \gamma =: \otimes.$$

Here

$$\gamma := \left| \int_x^{k/n} |f^{(N)}(t) - f^{(N)}(x)| \cdot \frac{|k/n - t|^{N-1}}{(N-1)!} \cdot dt \right|$$

$$\le \omega_1 \left( f^{(N)}, \frac{T}{n^{1-\alpha}} \right) \cdot \frac{T^N}{N! \cdot n^{N \cdot (1-\alpha)}}, \quad \text{(by (2.20))}.$$

That is

$$\otimes \le \sum_{k=\lceil nx - Tn^\alpha \rceil}^{\lceil nx + Tn^\alpha \rceil} \frac{S(n^{1-\alpha} \cdot (x - k/n))}{I^* \cdot n^\alpha} \cdot \omega_1 \left( f^{(N)}, \frac{T}{n^{1-\alpha}} \right) \cdot \frac{T^N}{N! n^{N \cdot (1-\alpha)}}.$$

Finally, using (2.6) with the function $S$ instead of $b$ we get

$$|\mathcal{R}^*| \le \omega_1 \left( f^{(N)}, \frac{T}{n^{1-\alpha}} \right) \cdot \frac{T^N}{N! \cdot n^{N \cdot (1-\alpha)}} \cdot \frac{S(T)}{I^*} \cdot \left( 2T + \frac{1}{n^\alpha} \right). \quad (2.35)$$

It is now clear that (2.34) and (2.35) imply (2.33). ∎

**REMARK 2.4** (i) The maps $F_n$, $G_n$, $n \in \mathbf{N}$, are positive linear operators.

(ii) Let $x \in [-T, T]$, $S \in C^r([-T, T])$, $r \in \mathbf{N}$, such that $S^{(r)}(x) \ge 0$ over

$[-T, T]$. Let $f \geq 0$ and $n \in \mathbf{N}$, $n \geq \max(2T, T^{-1/\alpha})$, $0 < \alpha < 1$. Then

$$(G_n(f))^{(r)}(x) = \sum_{k=\lceil nx - Tn^\alpha \rceil}^{[nx+Tn^\alpha]} \frac{f(k/n)}{I^*} \cdot n^{r-(r+1)\alpha} \cdot S^{(r)}\left(n^{1-\alpha} \cdot \left(x - \frac{k}{n}\right)\right)$$

$$\geq 0,$$

over $[-T, T]$. ∎

# Chapter 3

## Convergence with Rates of Multivariate Neural Network Operators to the Unit Operator

In this chapter we give a complete study of determination of the rate of convergence to the unit operator of specific well-known *multivariate neural network operators*, namely of Cardaliaguet–Euvrard and "Squashing" multivariate operators. This is given by *multidimensional Jackson type inequalities* that involve the *multidimensional modulus of continuity* of the engaged *multivariate function* or its partial derivative of specific order. Here we follow the basic study done by the author [30].

The multivariate *Cardaliaguet–Euvrard* (3.1) *operators* were first introduced and studied thoroughly in [62], where the authors, among many other interesting things, proved that these multivariate operators converge uniformly on compacta to the unit—over continuous and bounded multivariate functions. Our *multivariate "squashing operator"* (3.34) was motivated and inspired by the "*squashing function*" and the related Theorem 6 of [62].

The work in [62] is qualitative where the used multivariate bell-shaped function is general. However, the work presented here, though greatly motivated by [62], is quantitative and the used multivariate bell-shaped and "squashing" functions are of compact support.

This chapter is the continuation Chapter 2 in the multidimensional case. We produce a set of multivariate inequalities giving close upper bounds to the errors in approximating the unit operator by the above multidimensional neural network induced operators. All constants appearing there are well determined. These are mainly pointwise estimates involving the first multivariate modulus of continuity of the engaged multivariate continuous function or its partial derivatives of certain fixed order. We also present some $L_p$-related estimates.

## 3.1 Convergence with Rates of Multivariate Cardaliaguet–Euvrard Neural Network Operators

We use the following (see [62]):

**DEFINITION 3.1** *A function* $b: \mathbf{R} \to \mathbf{R}$ *is said to be bell-shaped if* $b$ *belongs to* $L^1$ *and its integral is nonzero, if it is nondecreasing on* $(-\infty, a)$ *and nonincreasing on* $[a, +\infty)$, *where* $a$ *belongs to* $\mathbf{R}$. *In particular* $b(x)$ *is a nonnegative number and at* $a$, $b$ *takes a global maximum; it is the* center *of the bell-shaped function. A bell-shaped function is said to be* centered *if its center is zero.*

**DEFINITION 3.2** *(See [62].) A function* $b: \mathbf{R}^d \to \mathbf{R}$ $(d \geq 1)$ *is said to be a* $d$-*dimensional bell-shaped function if it is integrable and its integral is not zero, and if for all* $i = 1, \ldots, d$,

$$t \to b(x_1, \ldots, t, \ldots, x_d)$$

*is a centered bell-shaped function, where* $\vec{x} := (x_1, \ldots, x_d) \in \mathbf{R}^d$ *arbitrary.*

**EXAMPLE 3.1**

(From [62].) Let $b$ be a centered bell-shaped function over $\mathbf{R}$, then $(x_1, \ldots, x_d) \to b(x_1) \ldots b(x_d)$ is a $d$-dimensional bell-shaped function.

**ASSUMPTION** Here $b(\vec{x})$ is of compact support $\mathcal{B} := \prod_{i=1}^{d}[-T_i, T_i]$, $T_i > 0$ and it may have jump discontinuities there. Set $I := \int_{\mathcal{B}} b(\vec{x}) d\vec{x}$. Note that $I > 0$. Let $f: \mathbf{R}^d \to \mathbf{R}$ be a continuous and bounded function or a uniformly continuous function.

In this section we study the pointwise convergence with rates over $\mathbf{R}^d$, to the unit operator, of the *multivariate Cardaliaguet–Euvrard neural network operators* (see [62]),

$$(\mathcal{F}_n(f))(\vec{x}) := \sum_{k_1=-n^2}^{n^2} \cdots \sum_{k_d=-n^2}^{n^2} \frac{f(\frac{k_1}{n}, \ldots, \frac{k_d}{n})}{I \cdot n^{\alpha \cdot d}}$$

$$\cdot b\left(n^{1-\alpha}\left(x_1 - \frac{k_1}{n}\right), \ldots, n^{1-\alpha} \cdot \left(x_d - \frac{k_d}{n}\right)\right), \quad (3.1)$$

## 3. Rates of Multivariate Neural Network Operators

where $0 < \alpha < 1$ and $\vec{x} := (x_1, \ldots, x_d) \in \mathbf{R}^d$, $n \in \mathbf{N}$. Clearly $\mathcal{F}_n$ is a positive linear operator.

The terms in the multiple sum (3.1) can be nonzero iff simultaneously

$$\left| n^{1-\alpha} \cdot \left( x_i - \frac{k_i}{n} \right) \right| \le T_i, \quad \text{all } i = 1, \ldots, d$$

i.e., $\left| x_i - \frac{k_i}{n} \right| \le \frac{T_i}{n^{1-\alpha}}$, all $i = 1, \ldots, d$, iff

$$nx_i - T_i \cdot n^\alpha \le k_i \le nx_i + T_i n^\alpha, \quad \text{all } i = 1, \ldots, d. \tag{3.2}$$

To have the order

$$-n^2 \le nx_i - T_i n^\alpha \le k_i \le nx_i + T_i \cdot n^\alpha \le n^2, \tag{3.3}$$

we need $n \ge T_i + |x_i|$, all $i = 1, \ldots, d$. So (3.3) is true when we consider

$$n \ge \max_{i \in \{1, \ldots, d\}} (T_i + |x_i|). \tag{3.4}$$

When $\vec{x} \in B$ in order to have (3.3) it is enough to suppose that $n \ge 2T^*$, where $T^* := \max\{T_1, \ldots, T_d\} > 0$. Take

$$\tilde{I}_i := [nx_i - T_i n^\alpha, nx_i + T_i n^\alpha], \quad i = 1, \ldots, d, \ n \in \mathbf{N}.$$

The length of $\tilde{I}_i$ is $2T_i n^\alpha$. By Proposition 2.1 we obtain that the cardinality of $k_i \in \mathbf{Z}$ that belong to $\tilde{I}_i := \text{card}(k_i) \ge \max(2T_i n^\alpha - 1, 0)$, any $i \in \{1, \ldots, d\}$. In order to have $\text{card}(k_i) \ge 1$ we need $2T_i n^\alpha - 1 \ge 1$ iff $n \ge T_i^{-1/\alpha}$, any $i \in \{1, \ldots, d\}$.

Therefore, a sufficient condition for causing the order (3.3) along with the interval $\tilde{I}_i$ to contain at least one integer for all $i = 1, \ldots, d$ is that

$$n \ge \max_{i \in \{1, \ldots, d\}} (T_i + |x_i|, T_i^{-1/\alpha}). \tag{3.5}$$

Clearly as $n \to +\infty$ we get that $\text{card}(k_i) \to +\infty$, all $i = 1, \ldots, d$. Also notice that $\text{card}(k_i)$ equals to the cardinality of integers in $[\lceil nx_i - T_i n^\alpha \rceil, \lceil nx_i + T_i n^\alpha \rceil]$ for all $i = 1, \ldots, d$.

We set $b^* := b(\vec{o})$ the maximum of $b(\vec{x})$. From now on in this chapter we will assume (3.5). Therefore

$$(\mathcal{F}_n(f))(\vec{x}) = \sum_{k_1=\lceil nx_1-T_1 n^\alpha\rceil}^{[nx_1+T_1 n^\alpha]} \cdots \sum_{k_d=\lceil nx_d-T_d n^\alpha\rceil}^{[nx_d+T_d n^\alpha]} \frac{f\left(\frac{k_1}{n},\ldots,\frac{k_d}{n}\right)}{I \cdot n^{\alpha \cdot d}}$$

$$\cdot b\left(n^{1-\alpha}\cdot\left(x_1-\frac{k_1}{n}\right),\ldots,n^{1-\alpha}\cdot\left(x_d-\frac{k_d}{n}\right)\right), \quad (3.6)$$

all $\vec{x} := (x_1,\ldots,x_d) \in \mathbf{R}^d$, where

$$I = \int_{-T_1}^{T_1} \cdots \int_{-T_d}^{T_d} b(x_1,\ldots,x_d) dx_1 \cdots dx_d.$$

Denote by $\|\cdot\|_\infty$ the maximum norm on $\mathbf{R}^d$, $d \geq 1$. So if $\left|n^{1-\alpha}\cdot\left(x_i - \frac{k_i}{n}\right)\right| \leq T_i$, all $i = 1,\ldots,d$, we find that

$$\left\|\left(\vec{x}-\frac{\vec{k}}{n}\right)\right\|_\infty \leq \frac{T^*}{n^{1-\alpha}}$$

where $\vec{k} := (k_1,\ldots,k_d)$.

**DEFINITION 3.3** Let $f : \mathbf{R}^d \to \mathbf{R}$. We call

$$\omega_1(f,h) := \sup_{\substack{\text{all } \vec{x},\vec{y}: \\ \|\vec{x}-\vec{y}\|_\infty \leq h}} |f(\vec{x})-f(\vec{y})|, \quad (3.7)$$

where $h > 0$, the first multidimensional modulus of continuity of $f$.

Next we present the first main result.

**THEOREM 3.1**
Let $\vec{x} \in \mathbf{R}^d$, then it holds that

$$|(\mathcal{F}_n(f))(\vec{x}) - f(\vec{x})| \leq |f(\vec{x})| \cdot \left| \sum_{k_1=\lceil nx_1-T_1 n^\alpha\rceil}^{[nx_1+T_1 n^\alpha]} \cdots \sum_{k_d=\lceil nx_d-T_d n^\alpha\rceil}^{[nx_d+T_d n^\alpha]} \frac{1}{I \cdot n^{\alpha \cdot d}}\right.$$

## 3. Rates of Multivariate Neural Network Operators

$$\cdot b\left(n^{1-\alpha}\cdot\left(x_1-\frac{k_1}{n}\right),\ldots,n^{1-\alpha}\cdot\left(x_d-\frac{k_d}{n}\right)\right)-1\bigg|$$

$$+\frac{b^*}{I}\cdot\prod_{i=1}^{d}\left(2T_i+\frac{1}{n^\alpha}\right)\cdot\omega_1\left(f,\frac{T^*}{n^{1-\alpha}}\right). \tag{3.8}$$

*Inequality (3.8) is attained by constant functions.*

**PROOF**    Notice that

$$\xi_1:=\frac{1}{I\cdot n^{\alpha\cdot d}}\cdot\sum_{k_1=\lceil nx_1-T_1n^\alpha\rceil}^{\lfloor nx_1+T_1n^\alpha\rfloor}\cdots\sum_{k_d=\lceil nx_d-T_dn^\alpha\rceil}^{\lfloor nx_d+T_dn^\alpha\rfloor}1$$

$$\leq\frac{1}{I\cdot n^{\alpha\cdot d}}\cdot\sum_{k_1=\lceil nx_1-T_1n^\alpha\rceil}^{\lfloor nx_1+T_1n^\alpha\rfloor}\cdots\sum_{k_{d-1}=\lceil nx_{d-1}-T_{d-1}n^\alpha\rceil}^{\lfloor nx_{d-1}+T_{d-1}n^\alpha\rfloor}(2T_dn^\alpha+1)$$

$$\leq\frac{1}{I\cdot n^{\alpha\cdot d}}\cdot\prod_{i=1}^{d}(2T_in^\alpha+1).$$

That is

$$\xi_1\leq\frac{1}{I}\cdot\prod_{i=1}^{d}\left(2T_i+\frac{1}{n^\alpha}\right). \tag{3.9}$$

In the following we estimate ($\vec{T}:=(T_1,\ldots,T_d)$)

$$|(\mathcal{F}_n(f))(\vec{x})-f(\vec{x})|\stackrel{(3.6)}{=}\Bigg|\sum_{k_1=\lceil nx_1-T_1n^\alpha\rceil}^{\lfloor nx_1+T_1n^\alpha\rfloor}\cdots\sum_{k_d=\lceil nx_d-T_dn^\alpha\rceil}^{\lfloor nx_d+T_dn^\alpha\rfloor}\frac{f\left(\frac{k_1}{n},\ldots,\frac{k_d}{n}\right)}{I\cdot n^{\alpha\cdot d}}$$

$$\cdot b\left(n^{1-\alpha}\cdot\left(x_1-\frac{k_1}{n}\right),\ldots,n^{1-\alpha}\cdot\left(x_d-\frac{k_d}{n}\right)\right)-f(\vec{x})\Bigg|$$

$$=\Bigg|\sum_{\vec{k}=\lceil n\vec{x}-\vec{T}n^\alpha\rceil}^{\lfloor n\vec{x}+\vec{T}n^\alpha\rfloor}\frac{f\left(\frac{\vec{k}}{n}\right)}{I\cdot n^{\alpha\cdot d}}\cdot b\left(n^{1-\alpha}\cdot\left(\vec{x}-\frac{\vec{k}}{n}\right)\right)-f(\vec{x})\Bigg|$$

$$= \left| \sum_{\vec{k}=\lceil n\vec{x}-\vec{T}n^\alpha \rceil}^{\lfloor n\vec{x}+\vec{T}n^\alpha \rfloor} \frac{\left(f\left(\frac{\vec{k}}{n}\right) - f(\vec{x})\right)}{I \cdot n^{\alpha \cdot d}} \cdot b\left(n^{1-\alpha} \cdot \left(\vec{x} - \frac{\vec{k}}{n}\right)\right) \right.$$

$$\left. + f(\vec{x}) \cdot \sum_{\vec{k}=\lceil n\vec{x}-\vec{T}n^\alpha \rceil}^{\lfloor n\vec{x}+\vec{T}n^\alpha \rfloor} \frac{1}{I \cdot n^{\alpha \cdot d}} \cdot b\left(n^{1-\alpha} \cdot \left(\vec{x} - \frac{\vec{k}}{n}\right)\right) - f(\vec{x}) \right|$$

$$\leq \sum_{\vec{k}=\lceil n\vec{x}-\vec{T}n^\alpha \rceil}^{\lfloor n\vec{x}+\vec{T}n^\alpha \rfloor} \frac{\left|f\left(\frac{\vec{k}}{n}\right) - f(\vec{x})\right|}{I \cdot n^{\alpha \cdot d}} \cdot b\left(n^{1-\alpha} \cdot \left(\vec{x} - \frac{\vec{k}}{n}\right)\right) + |f(\vec{x})|$$

$$\cdot \left| \sum_{\vec{k}=\lceil n\vec{x}-\vec{T}n^\alpha \rceil}^{\lfloor n\vec{x}+\vec{T}n^\alpha \rfloor} \frac{1}{I \cdot n^{\alpha \cdot d}} \cdot b\left(n^{1-\alpha} \cdot \left(\vec{x} - \frac{\vec{k}}{n}\right)\right) - 1 \right|$$

$$\leq \sum_{\vec{k}=\lceil n\vec{x}-\vec{T}n^\alpha \rceil}^{\lfloor n\vec{x}+\vec{T}n^\alpha \rfloor} \frac{\omega_1\left(f, \left\|\frac{\vec{k}}{n} - \vec{x}\right\|_\infty\right)}{I \cdot n^{\alpha \cdot d}} \cdot b\left(n^{1-\alpha} \cdot \left(\vec{x} - \frac{\vec{k}}{n}\right)\right) + |f(\vec{x})|$$

$$\cdot \left| \sum_{\vec{k}=\lceil n\vec{x}-\vec{T}n^\alpha \rceil}^{\lfloor n\vec{x}+\vec{T}n^\alpha \rfloor} \frac{1}{I \cdot n^{\alpha \cdot d}} \cdot b\left(n^{1-\alpha} \cdot \left(\vec{x} - \frac{\vec{k}}{n}\right)\right) - 1 \right|.$$

Thus,

$$|(\mathcal{F}_n(f))(\vec{x}) - f(\vec{x})| \leq \omega_1\left(f, \frac{T^*}{n^{1-\alpha}}\right) \cdot b^* \cdot \xi_1 + |f(\vec{x})|$$

$$\cdot \left| \sum_{k_1=\lceil nx_1-T_1 n^\alpha \rceil}^{\lfloor nx_1+T_1 n^\alpha \rfloor} \cdots \sum_{k_d=\lceil nx_d-T_d \cdot n^\alpha \rceil}^{\lfloor nx_d+T_d \cdot n^\alpha \rfloor} \frac{1}{I \cdot n^{\alpha \cdot d}} \right.$$

$$\left. \cdot b\left(n^{1-\alpha} \cdot \left(x_1 - \frac{k_1}{n}\right), \ldots, n^{1-\alpha} \cdot \left(x_d - \frac{k_d}{n}\right)\right) - 1 \right|. \quad (3.10)$$

Inserting (3.9) into (3.10) we obtain (3.8). ∎

We use

**LEMMA 3.1**
It holds true that ($\vec{x} \in \mathbf{R}^d$)

$$S_n(\vec{x}) := \sum_{k_1 = \lceil nx_1 - T_1 n^\alpha \rceil}^{\lceil nx_1 + T_1 n^\alpha \rceil} \cdots \sum_{k_d = \lceil nx_d - T_d n^\alpha \rceil}^{\lceil nx_d + T_d n^\alpha \rceil} \frac{1}{I \cdot n^{\alpha \cdot d}}$$

$$\cdot b\left(n^{1-\alpha} \cdot \left(x_1 - \frac{k_1}{n}\right), \ldots, n^{1-\alpha} \cdot \left(x_d - \frac{k_d}{n}\right)\right) \to 1,$$

pointwise, as $n \to +\infty$.

**REMARK 3.1** By Lemma 3.1 and inequality (3.8) we get that $(\mathcal{F}_n(f))(\vec{x}) \to f(\vec{x})$, pointwise with rates, as $n \to +\infty$, where $\vec{x} \in \mathbf{R}^d$, $d \geq 1$. ∎

To prove Lemma 3.1 we will need

**LEMMA 3.2**
(i) Let $t_i \leq t'_i \leq 0$, $i = 1, 2, \ldots, d$, then

$$b(t_1, t_2, \ldots, t_d) \leq b(t'_1, t'_2, \ldots, t'_d).$$

(ii) Let $0 \leq t_i \leq t'_i$, $i = 1, 2, \ldots, d$, then

$$b(t_1, t_2, \ldots, t_d) \geq b(t'_1, t'_2, \ldots, t'_d).$$

**PROOF** (i) Notice that

$$b(t_1, t_2, \ldots, t_d) \leq b(t'_1, t_2, \ldots, t_d) \leq b(t'_1, t'_2, t_3, \ldots, t_d)$$

$$\leq b(t'_1, t'_2, t'_3, \ldots, t_d) \leq b(t'_1, t'_2, t'_3, \ldots, t'_d).$$

(ii) Similar proof. ∎

In general we observe

**LEMMA 3.3**
(i) Let $t_i \leq t'_i \leq 0$, where $i \in \{1, 2, \ldots, d\}$, $1 \leq |i| \leq d$, then

$$b(t_1, \ldots, t_i, \ldots, t_d) \leq b(t_1, \ldots, t'_i, \ldots, t_d).$$

(ii) Let $0 \leq t_i \leq t'_i$, where $i \in \{1, 2, \ldots, d\}$; $1 \leq |i| \leq d$, then

$$b(t_1, \ldots, t_i, \ldots, t_d) \geq b(t_1, \ldots, t'_i, \ldots, t_d).$$

**PROOF**   Similar to Lemma 3.2.   ∎

That is, monotonicity of $b$ is true for some or all coordinates.

**PROOF**   (of Lemma 3.1).
(i) Case of $k_i \in \mathbf{Z}$: $\lceil nx_i \rceil + 1 \leq k_i \leq \lceil nx_i + T_i n^\alpha \rceil$, all $i = 1, 2, \ldots, d$. Here

$$nx_i < nx_i + 1 \leq \lceil nx_i \rceil + 1 \leq k_i \leq \lceil nx_i + T_i n^\alpha \rceil.$$

Let $t_i$ such that $nx_i \leq k_1 - 1 \leq t_i \leq k_i$. Thus

$$x_i - \frac{k_i}{n} \leq x_i - \frac{t_i}{n} \leq x_i - \frac{(k_i - 1)}{n} < 0, \quad \text{all } i = 1, \ldots, d.$$

By Lemma 3.2 we get

$$b\left(n^{1-\alpha} \cdot \left(\vec{x} - \frac{\vec{k}}{n}\right)\right) \leq b\left(n^{1-\alpha} \cdot \left(\vec{x} - \frac{\vec{t}}{n}\right)\right)$$

$$\leq b\left(n^{1-\alpha} \cdot \left(\vec{x} - \frac{(\vec{k} - 1)}{n}\right)\right),$$

where $\vec{k} := (k_1, k_2, \ldots, k_d)$. So that

$$b\left(n^{1-\alpha} \cdot \left(\vec{x} - \frac{\vec{k}}{n}\right)\right) \leq \int_{\vec{k}-1}^{\vec{k}} b\left(n^{1-\alpha} \cdot \left(\vec{x} - \frac{\vec{t}}{n}\right)\right) \cdot d\vec{t}.$$

Next let $t_i$ such that $nx_i < k_i \leq t_i \leq k_i + 1$. Hence

$$x_i - \frac{(k_i + 1)}{n} \leq x_i - \frac{t_i}{n} \leq x_i - \frac{k_i}{n} < 0, \quad \text{all } i = 1, \ldots, d.$$

Again, by Lemma 3.2 we obtain

$$b\left(n^{1-\alpha} \cdot \left(\vec{x} - \frac{(\vec{k}+1)}{n}\right)\right) \leq b\left(n^{1-\alpha} \cdot \left(\vec{x} - \frac{\vec{t}}{n}\right)\right)$$

## 3. Rates of Multivariate Neural Network Operators

$$\leq b\left(n^{1-\alpha} \cdot \left(\vec{x} - \frac{\vec{k}}{n}\right)\right).$$

Consequently

$$\int_{\vec{k}}^{\overrightarrow{k+1}} b\left(n^{1-\alpha} \cdot \left(\vec{x} - \frac{\vec{t}}{n}\right)\right) d\vec{t} \leq b\left(n^{1-\alpha} \cdot \left(\vec{x} - \frac{\vec{k}}{n}\right)\right).$$

We have established that

$$\int_{\vec{k}}^{\overrightarrow{k+1}} b\left(n^{1-\alpha} \cdot \left(\vec{x} - \frac{\vec{t}}{n}\right)\right) d\vec{t} \leq b\left(n^{1-\alpha} \cdot \left(\vec{x} - \frac{\vec{k}}{n}\right)\right)$$

$$\leq \int_{\overrightarrow{k-1}}^{\vec{k}} b\left(n^{1-\alpha} \cdot \left(\vec{x} - \frac{\vec{t}}{n}\right)\right) d\vec{t} \quad (3.11)$$

for any $\vec{k} := (k_1, \ldots, k_d) \in \mathbf{Z}^d$ such that

$$\lceil nx_i \rceil + 1 \leq k_i \leq \lceil nx_i + T_i n^\alpha \rceil,$$

for all $i = 1, 2, \ldots, d$.

(ii) Case of $k_i \in \mathbf{Z}$: $\lceil nx_i - T_i n^\alpha \rceil \leq k_i \leq \lceil nx_i \rceil - 1$, all $i = 1, 2, \ldots, d$.
Here we have

$$k_i \leq k_i + 1 \leq \lceil nx_i \rceil < nx_i.$$

Let $t_i: k_i - 1 \leq t_i \leq k_i \leq nx_i$, then

$$x_i - \frac{(k_i - 1)}{n} \geq x_i - \frac{t_i}{n} \geq x_i - \frac{k_i}{n} > 0.$$

Hence (Lemma 3.2) we obtain that

$$b\left(n^{1-\alpha} \cdot \left(\vec{x} - \frac{(\overrightarrow{k-1})}{n}\right)\right) \leq b\left(n^{1-\alpha} \cdot \left(\vec{x} - \frac{\vec{t}}{n}\right)\right)$$

$$\leq b\left(n^{1-\alpha} \cdot \left(\vec{x} - \frac{\vec{k}}{n}\right)\right).$$

Consequently

$$\int_{\vec{k}-1}^{\vec{k}} b\left(n^{1-\alpha}\cdot\left(\vec{x}-\frac{\vec{t}}{n}\right)\right) d\vec{t} \leq b\left(n^{1-\alpha}\cdot\left(\vec{x}-\frac{\vec{k}}{n}\right)\right).$$

Now let $t_i$ such that
$$k_i \leq t_i \leq k_i + 1 < nx_i.$$

Then
$$x_i - \frac{k_i}{n} \geq x_i - \frac{t_i}{n} \geq x_i - \frac{(k_i+1)}{n} > 0.$$

Hence (Lemma 3.2)

$$b\left(n^{1-\alpha}\cdot\left(\vec{x}-\frac{\vec{k}}{n}\right)\right) \leq b\left(n^{1-\alpha}\cdot\left(\vec{x}-\frac{\vec{t}}{n}\right)\right)$$

$$\leq b\left(n^{1-\alpha}\cdot\left(\vec{x}-\frac{\overrightarrow{(k+1)}}{n}\right)\right),$$

and

$$b\left(n^{1-\alpha}\cdot\left(\vec{x}-\frac{\vec{k}}{n}\right)\right) \leq \int_{\vec{k}}^{\overrightarrow{k+1}} b\left(n^{1-\alpha}\cdot\left(\vec{x}-\frac{\vec{t}}{n}\right)\right) d\vec{t}.$$

We have established that

$$\int_{\vec{k}-1}^{\vec{k}} b\left(n^{1-\alpha}\cdot\left(\vec{x}-\frac{\vec{t}}{n}\right)\right) d\vec{t} \leq b\left(n^{1-\alpha}\cdot\left(\vec{x}-\frac{\vec{k}}{n}\right)\right)$$

$$\leq \int_{\vec{k}}^{\overrightarrow{k+1}} b\left(n^{1-\alpha}\cdot\left(\vec{x}-\frac{\vec{t}}{n}\right)\right) d\vec{t} \quad (3.12)$$

for any $\vec{k} := (k_1, \ldots, k_d) \in \mathbf{Z}^d$ such that

$$\lceil nx_i - T_i n^\alpha \rceil \leq k_i \leq [nx_i] - 1$$

for all $i = 1, 2, \ldots, d$. Notice also that, as $n \to +\infty$, we obtain

$$S_n^3(\vec{x}) := \frac{1}{I \cdot n^{\alpha \cdot d}} \cdot b\left(n^{1-\alpha}\cdot\left(\vec{x}-\frac{[n\vec{x}]}{n}\right)\right) \to 0$$

## 3. Rates of Multivariate Neural Network Operators

and
$$S_n^4(\vec{x}) := \frac{1}{I \cdot n^{\alpha \cdot d}} \cdot b\left(n^{1-\alpha} \cdot \left(\vec{x} - \frac{\lceil n\vec{x} \rceil}{n}\right)\right) \to 0. \tag{3.13}$$

Set
$$S_n^1(\vec{x}) := \sum_{\vec{k} = \lceil \vec{nx} \rceil + \vec{1}}^{\lceil nx + \vec{T} n^\alpha \rceil} \frac{1}{I \cdot n^{\alpha \cdot d}} \cdot b\left(n^{1-\alpha} \cdot \left(\vec{x} - \frac{\vec{k}}{n}\right)\right),$$

$$S_n^2(\vec{x}) := \sum_{\vec{k} = \lceil nx - \vec{T} n^\alpha \rceil}^{\lceil \vec{nx} \rceil - \vec{1}} \frac{1}{I \cdot n^{\alpha \cdot d}} \cdot b\left(n^{1-\alpha} \cdot \left(\vec{x} - \frac{\vec{k}}{n}\right)\right), \tag{3.14}$$

and
$$S_n^5(\vec{x}) := \text{the sum of all other possible mixed subsums}$$
$$\tilde{S}_n^j(\vec{x}), \quad j = 1, \ldots, 4^d.$$

Then
$$S_n(\vec{x}) = S_n^1(\vec{x}) + S_n^2(\vec{x}) + S_n^3(\vec{x}) + S_n^4(\vec{x}) + S_n^5(\vec{x}). \tag{3.15}$$

(iii) Case of *only* some $i \in \{1, \ldots, d\}$ such that $\lceil nx_i \rceil + 1 \le k_i \le \lceil nx_i + T_i n^\alpha \rceil$, here $1 \le |i| < d$. Call $\vec{k}^* := (k_i)$, $\vec{t}^* := (t_i)$. Similarly to (3.11) (use of Lemma 3.3) we find that

$$\int_{\vec{k}^*}^{\overrightarrow{\vec{k}^*+1}} b\left(n^{1-\alpha} \cdot \left(x_{d-i} - \frac{k_{d-i}}{n}, x_i - \frac{t_i}{n}\right)\right) d\vec{t}^* \le b\left(n^{1-\alpha} \cdot \left(\vec{x} - \frac{\vec{k}}{n}\right)\right)$$

$$\le \int_{\overrightarrow{\vec{k}^*-1}}^{\vec{k}^*} b\left(n^{1-\alpha} \cdot \left(x_{d-i} - \frac{k_{d-i}}{n}, x_i - \frac{t_i}{n}\right)\right) d\vec{t}^*. \tag{3.16}$$

(iv) Case of *only* some $i \in \{1, \ldots, d\}$ such that $\lceil nx_i - T_i n^\alpha \rceil \le k_i \le \lceil nx_i \rceil - 1$, here $1 \le |i| < d$. Call $\vec{k}^* := (k_i)$, $\vec{t}^* := (t_i)$. Similar to (3.12) (use of Lemma 3.3) we get that

$$\int_{\overrightarrow{\vec{k}^*-1}}^{\vec{k}^*} b\left(n^{1-\alpha} \cdot \left(x_{d-i} - \frac{k_{d-i}}{n}, x_i - \frac{t_i}{n}\right)\right) d\vec{t}^* \le b\left(n^{1-\alpha} \cdot \left(\vec{x} - \frac{\vec{k}}{n}\right)\right)$$

$$\le \int_{\vec{k}^*}^{\overrightarrow{\vec{k}^*+1}} b\left(n^{1-\alpha} \cdot \left(x_{d-i} - \frac{k_{d-i}}{n}, x_i - \frac{t_i}{n}\right)\right) d\vec{t}^*. \tag{3.17}$$

(v) Mixing up cases (iii) and (iv). I.e.,

$$\begin{cases} \text{Let } k_i \in \mathbf{Z}\colon \lceil nx_i \rceil + 1 \le k_i \le \lceil nx_i + T_i n^\alpha \rceil \\ \text{for } \textit{only some } i \in \{1,\ldots,d\}, \text{ and,} \\ \text{let } \ell_j \in \mathbf{Z}\colon \lceil nx_j - T_j n^\alpha \rceil \le \ell_j \le \lceil nx_j \rceil - 1 \\ \text{for } \textit{only some } j \in \{1,\ldots,d\}. \end{cases} \quad (3.18)$$

Here $i \ne j$. Call $\vec{k}_i^* := (k_i)$, $\vec{\ell}_j^* := (\ell_j)$, $\vec{t}_i^* = (t_i)$, $\vec{t}_j^* = (t_j)$. We see for

$$b\left(n^{1-\alpha} \cdot \left(x_{d-i} - \frac{k_{d-i}}{n}, x_i - \frac{t_i}{n}\right)\right)$$

(similar to (3.17)) that:

$$\int_{\vec{\ell}_j^*-1}^{\vec{\ell}_j^*} b\left(n^{1-\alpha} \cdot \left(x_{d-i-j} - \frac{k_{d-i-j}}{n}, x_j - \frac{t_j}{n}, x_i - \frac{t_i}{n}\right)\right) \cdot d\vec{t}_j^*$$

$$\le b\left(n^{1-\alpha} \cdot \left(x_{d-i} - \frac{k_{d-i}}{n}, x_i - \frac{t_i}{n}\right)\right) \quad (3.19)$$

$$\le \int_{\vec{\ell}_j^*}^{\vec{\ell}_j^*+1} b\left(n^{1-\alpha} \cdot \left(x_{d-i-j} - \frac{k_{d-i-j}}{n}, x_j - \frac{t_j}{n}, x_i - \frac{t_i}{n}\right)\right) \cdot d\vec{t}_j^*.$$

Then, by integrating (3.19) and using (3.16) we get:

$$\int_{\vec{k}_i^*}^{\vec{k}_i^*+1} \int_{\vec{\ell}_j^*-1}^{\vec{\ell}_j^*} b\left(n^{1-\alpha} \cdot \left(x_{d-i-j} - \frac{k_{d-i-j}}{n}, x_j - \frac{t_j}{n}, x_i - \frac{t_i}{n}\right)\right) d\vec{t}_j^* \cdot d\vec{t}_i^*$$

$$\le b\left(n^{1-\alpha} \cdot \left(\vec{x} - \frac{\vec{k}}{n}\right)\right) \quad (3.20)$$

$$\le \int_{\vec{k}_i^*-1}^{\vec{k}_i^*} \int_{\vec{\ell}_j^*}^{\vec{\ell}_j^*+1} b\left(n^{1-\alpha} \cdot \left(x_{d-i-j} - \frac{k_{d-i-j}}{n}, x_j - \frac{t_j}{n}, x_i - \frac{t_i}{n}\right)\right) d\vec{t}_j^* \cdot d\vec{t}_i^*.$$

In general let the indices $i_1, i_2, \ldots, i_\gamma \in \{1,\ldots,d\}$ with $1 \le |i_1|, |i_2|, \ldots, |i_\gamma| \le$

3. *Rates of Multivariate Neural Network Operators*                101

$d$ be such that $i_1 + i_2 + \cdots + i_\gamma = d;\ 1 \le \gamma \le d$. Therefore

$$\int_{k_1^* \pm \{0,1\}}^{k_1^* \pm \{0,1\}} \cdots \int_{k_\gamma^* \pm \{0,1\}}^{k_\gamma^* \pm \{0,1\}} b\left( n^{1-\alpha} \cdot \left( x_{(d-\sum_{\ell=1}^{\gamma} i_\ell)} - \frac{k_{(d-\sum_{\ell=1}^{\gamma} i_\ell)}}{n},\right.\right.$$

$$\left.\left. x_{i_1} - \frac{t_{i_1}}{n}, \ldots, x_{i_\gamma} - \frac{t_{i_\gamma}}{n} \right) \right) d\vec{t}_{i_1}^* \cdots d\vec{t}_{i_\gamma}^*$$

$$\le b\left( n^{1-\alpha} \cdot \left( \vec{x} - \frac{\vec{k}}{n} \right) \right) \le \int_{k_1^* \pm \{0,1\}}^{k_1^* \pm \{0,1\}} \cdots$$

$$\int_{k_\gamma^* \pm \{0,1\}}^{k_\gamma^* \pm \{0,1\}} b\left( n^{1-\alpha} \cdot \left( x_{(d-\sum_{\ell=1}^{\gamma} i_\ell)} - \frac{k_{(d-\sum_{\ell=1}^{\gamma} i_\ell)}}{n},\right.\right.$$

$$\left.\left. x_{i_1} - \frac{t_{i_1}}{n}, \ldots, x_{i_\gamma} - \frac{t_{i_\gamma}}{n} \right) \right) d\vec{t}_{i_1}^* \cdots d\vec{t}_{i_\gamma}^* \quad (3.21)$$

Double inequality (3.21) includes all possibilities indicated by the upper and lower integral bounds $\vec{k}_{i_\ell}^* \pm \vec{1}$ or $\vec{0}$, $\ell = 1, \ldots, \gamma$. Here $\vec{k}_{i_\ell}^* := (k_{i_\ell})$, $\vec{t}_{i_\ell}^* := (t_{i_\ell})$. Next denote by $\boxed{\cdot}$ any of $\lfloor \cdot \rfloor$ or $\lceil \cdot \rceil$. Then, as $n \to +\infty$, we have that

$$\frac{1}{I \cdot n^{\alpha \cdot d}} \cdot b\left( n^{1-\alpha} \cdot \left( x_1 - \frac{k_1}{n}, \ldots, x_i - \frac{\boxed{nx_i}}{n}, \ldots, x_d - \frac{k_d}{n} \right) \right) \to 0,$$

for one or more $i$: $1 \le i < d$. Furthermore for $i = 1$, or any other $i$: $1 < i \le d$ (same proof) we obtain

$$\frac{1}{I \cdot n^{\alpha \cdot d}} \cdot \sum_{k_2 = \lceil nx_2 - T_2 n^\alpha \rceil}^{[nx_2 + T_2 n^\alpha]} \cdots \sum_{k_d = \lceil nx_d - T_d n^\alpha \rceil}^{[nx_d + T_d n^\alpha]} b\left( n^{1-\alpha} \cdot \left( x_1 - \frac{\boxed{nx_1}}{n} \right),\right.$$

$$\left. n^{1-\alpha} \cdot \left( x_2 - \frac{k_2}{n} \right), \ldots, n^{1-\alpha} \cdot \left( x_d - \frac{k_d}{n} \right) \right)$$

$$\le \frac{1}{I \cdot n^{\alpha \cdot d}} \cdot b^* \cdot \sum_{k_2 = \lceil nx_2 - T_2 n^\alpha \rceil}^{[nx_2 + T_2 n^\alpha]} \cdots \sum_{k_d = \lceil nx_d - T_d n^\alpha \rceil}^{[nx_d + T_d n^\alpha]} 1$$

$$\leq \frac{b^*}{I \cdot n^{\alpha \cdot d}} \cdot \prod_{i=2}^{d}(2T_i \cdot n^\alpha + 1) = \frac{b^*}{I \cdot n^\alpha} \cdot \prod_{i=2}^{d}\left(2T_i + \frac{1}{n^\alpha}\right) \to 0,$$

as $n \to +\infty$.

The same result holds true if two or more coordinates are of the form $\boxed{nx_i}$, $i = 1, \ldots, d$ (in that case we have fewer sums involved). Obviously, we obtain the same result as the last one, when only upper or lower index sums are involved, etc. For $\vec{x} \in \mathbf{R}^d$ we get (see (3.11))

$$\frac{1}{I \cdot n^{\alpha \cdot d}} \cdot \int_{\lceil n\vec{x} \rceil + \vec{1}}^{\lceil n\vec{x} + \vec{T}n^\alpha \rceil + \vec{1}} b\left(n^{1-\alpha} \cdot \left(\vec{x} - \frac{\vec{t}}{n}\right)\right) d\vec{t} \leq S_n^1(\vec{x})$$

$$\leq \frac{1}{I \cdot n^{\alpha \cdot d}} \cdot \int_{\lceil n\vec{x} \rceil}^{\lceil n\vec{x} + \vec{T}n^\alpha \rceil} b\left(n^{1-\alpha} \cdot \left(\vec{x} - \frac{\vec{t}}{n}\right)\right) d\vec{t}.$$

That is, by the change of variable method we have that

$$\frac{1}{I} \cdot \int_{n^{1-\alpha} \cdot \left(\vec{x} - \frac{(\lceil n\vec{x} + \vec{T}n^\alpha \rceil + \vec{1})}{n}\right)}^{n^{1-\alpha} \cdot \left(\vec{x} - \frac{(\lceil n\vec{x} \rceil + \vec{1})}{n}\right)} b(\vec{t}) d\vec{t} \leq S_n^1(\vec{x})$$

$$\leq \frac{1}{I} \cdot \int_{n^{1-\alpha} \cdot \left(\vec{x} - \frac{\lceil n\vec{x} + \vec{T}n^\alpha \rceil}{n}\right)}^{n^{1-\alpha} \cdot \left(\vec{x} - \frac{\lceil n\vec{x} \rceil}{n}\right)} b(\vec{t}) d\vec{t}.$$

Therefore

$$\lim_{n \to +\infty} S_n^1(\vec{x}) = \frac{1}{I} \cdot \int_{-\vec{T}}^{\vec{0}} b(\vec{t}) d\vec{t}. \tag{3.22}$$

Also working as above (see (3.12)) we find that

$$\frac{1}{I \cdot n^{\alpha \cdot d}} \cdot \int_{\lceil n\vec{x} - \vec{T}n^\alpha \rceil - \vec{1}}^{\lceil n\vec{x} \rceil - \vec{1}} b\left(n^{1-\alpha} \cdot \left(\vec{x} - \frac{\vec{t}}{n}\right)\right) \cdot d\vec{t}$$

$$\leq S_n^2(\vec{x}) \leq \frac{1}{I \cdot n^{\alpha \cdot d}} \cdot \int_{\lceil n\vec{x} - \vec{T}n^\alpha \rceil}^{\lceil n\vec{x} \rceil} b\left(n^{1-\alpha} \cdot \left(\vec{x} - \frac{\vec{t}}{n}\right)\right) \cdot d\vec{t}.$$

And

$$\lim_{n \to +\infty} S_n^2(\vec{x}) = \frac{1}{I} \cdot \int_{\vec{0}}^{\vec{T}} b(\vec{t}) d\vec{t}. \tag{3.23}$$

## 3. Rates of Multivariate Neural Network Operators

Working analogously (use of (3.21) and the following comments) as in (3.22) and (3.23), for the case of $S_n^5(\vec{x})$ we get:

i) for any $d$-fold mixed subsum $\tilde{S}_n^j(\vec{x})$, e.g.,

$$\lim_{n\to+\infty} \tilde{S}_n^j(\vec{x}) = \frac{1}{I} \cdot \int_0^{T_1} \int_{-T_2}^0 \int_0^{T_3} \cdots \int_{-T_d}^0 b(\vec{t})d\vec{t} \qquad (3.24)$$

we get similar results for all possibilities of $0, \pm T_i$, $i = 1, \ldots, d$ as integral bounds.

ii) Also, any $(d-r)$-fold mixed subsum $\tilde{S}_n^j(\vec{x})$, $r: 1 \le r < d$, wherein the $r$ ($i$th) coordinates ($i \in \{1, \ldots, d\}$) of $b$ contain $\lceil nx_i \rceil$ instead of $k_i$ converges to 0 as $n \to +\infty$. $\qquad (3.25)$

Observe that

$$I = \int_B b(\vec{t})d\vec{t} = \int_{\vec{0}}^{\vec{T}} b(\vec{t})d\vec{t}$$

$$+ \int_{-\vec{T}}^{\vec{0}} b(\vec{t})d\vec{t} + \cdots \text{(any other mixed } d\text{-fold integral).} \qquad (3.26)$$

Finally, taking into account (3.15) along with: (3.13), (3.22), (3.23), (3.24), (3.25), and (3.26), we find that

$$\lim_{n\to+\infty} S_n(\vec{x}) = \frac{I}{I} = 1,$$

pointwise, any $\vec{x} \in \mathbf{R}^d$. ∎

The second main result follows:

**THEOREM 3.2**
Let $\vec{x} \in \mathbf{R}^d$, $f \in C^N(\mathbf{R}^d)$, $N \in \mathbf{N}$, such that all of its partial derivatives $f_{\tilde{\alpha}}$ of order $N$, $\tilde{\alpha}: |\tilde{\alpha}| = N$, are uniformly continuous or continuous and bounded. Then the following holds

$$|(\mathcal{F}_n(f))(\vec{x}) - f(\vec{x})| \le |f(\vec{x})| \cdot \left| \sum_{\vec{k}=\lceil n\vec{x}-\vec{T}n^\alpha \rceil}^{[n\vec{x}+\vec{T}n^\alpha]} \frac{1}{I \cdot n^{\alpha \cdot d}} \right.$$

$$\left. \cdot b\left(n^{1-\alpha} \cdot \left(\vec{x} - \frac{\vec{k}}{n}\right)\right) - 1 \right| + \frac{b(\vec{0})}{I} \cdot \left(\prod_{i=1}^d \left(2T_i + \frac{1}{n^\alpha}\right)\right)$$

$$\cdot \left\{ \sum_{j=1}^{N} \frac{(T^*)^j}{j! n^{j(1-\alpha)}} \cdot \left( \left( \sum_{i=1}^{d} \left| \frac{\partial}{\partial x_i} \right| \right)^j f(\vec{x}) \right) \right\}$$

$$+ \frac{(T^*)^N \cdot d^N}{N! n^{N(1-\alpha)}} \cdot \frac{b(\vec{0})}{I} \cdot \left( \prod_{i=1}^{d} \left( 2T_i + \frac{1}{n^\alpha} \right) \right)$$

$$\cdot \max_{\tilde{\alpha}:|\tilde{\alpha}|=N} \omega_1 \left( f_{\tilde{\alpha}}, \frac{T^*}{n^{1-\alpha}} \right). \tag{3.27}$$

*Inequality (3.27) is attained by constant functions. Also (3.27) gives us with rates the pointwise convergence of (cf. Lemma 3.1) $\mathcal{F}_n(f) \to f$ over $\mathbf{R}^d$, as $n \to +\infty$.*

**PROOF** Put

$$g_{\vec{k}/n}(t) := f\left( \vec{x} + t\left( \frac{\vec{k}}{n} - \vec{x} \right) \right), \quad 0 \le t \le 1.$$

Then

$$g_{\vec{k}/n}^{(j)}(t) = \left\{ \left( \sum_{i=1}^{d} \left( \frac{k_i}{n} - x_i \right) \frac{\partial}{\partial x_i} \right)^j f \right\}$$

$$\cdot \left( x_1 + t\left( \frac{k_1}{n} - x_1 \right), \ldots, x_d + t\left( \frac{k_d}{n} - x_d \right) \right)$$

and $g_{\vec{k}/n}(0) = f(\vec{x})$. By Taylor's formula we obtain

$$f\left( \frac{k_1}{n}, \ldots, \frac{k_d}{n} \right) = g_{\vec{k}/n}(1) = \sum_{j=0}^{N} \frac{g_{\vec{k}/n}^{(j)}(0)}{j!} + \mathcal{R}_N\left( \frac{\vec{k}}{n}, 0 \right),$$

where

$$\mathcal{R}_N\left( \frac{\vec{k}}{n}, 0 \right) := \int_0^1 \left( \int_0^{t_1} \cdots \left( \int_0^{t_{N-1}} (g_{\vec{k}/n}^{(N)}(t_N) \right. \right.$$

$$\left.- g^{(N)}_{\vec{k}/n}(0)\right) dt_N\right)\cdots\right) dt_1.$$

Here we denote by

$$f_{\tilde{\alpha}} := \frac{\partial^{\tilde{\alpha}} f}{\partial x^{\tilde{\alpha}}}, \quad \tilde{\alpha} := (\alpha_1,\ldots,\alpha_d), \quad \alpha_i \in \mathbf{Z}^+,$$

$i = 1,\ldots,d$ such that $|\tilde{\alpha}| := \sum_{i=1}^{d}\alpha_i = N$. Hence

$$\frac{f\left(\frac{\vec{k}}{n}\right)\cdot b\left(n^{1-\alpha}\left(\vec{x} - \frac{\vec{k}}{n}\right)\right)}{I\cdot n^{\alpha\cdot d}} = \sum_{j=0}^{N}\frac{g^{(j)}_{\vec{k}/n}(0)}{j!}$$

$$\cdot\frac{b\left(n^{1-\alpha}\left(\vec{x} - \frac{\vec{k}}{n}\right)\right)}{I\cdot n^{\alpha\cdot d}} + \frac{b\left(n^{1-\alpha}\left(\vec{x} - \frac{\vec{k}}{n}\right)\right)}{I\cdot n^{\alpha\cdot d}}\cdot\mathcal{R}_N\left(\frac{\vec{k}}{n},0\right).$$

Therefore

$$(\mathcal{F}_n(f))(\vec{x}) - f(\vec{x}) = \sum_{\vec{k}=\lceil n\vec{x}-\vec{T}n^\alpha\rceil}^{\lfloor n\vec{x}+\vec{T}n^\alpha\rfloor}\frac{f\left(\frac{\vec{k}}{n}\right)}{I\cdot n^{\alpha\cdot d}}\cdot b\left(n^{1-\alpha}\left(\vec{x} - \frac{\vec{k}}{n}\right)\right) - f(\vec{x})$$

$$= -f(\vec{x}) + \sum_{j=0}^{N}\frac{1}{j!}\cdot\left(\sum_{\vec{k}=\lceil n\vec{x}-\vec{T}n^\alpha\rceil}^{\lfloor n\vec{x}+\vec{T}n^\alpha\rfloor}g^{(j)}_{\vec{k}/n}(0)\cdot\frac{b\left(n^{1-\alpha}\left(\vec{x} - \frac{\vec{k}}{n}\right)\right)}{I\cdot n^{\alpha\cdot d}}\right) + \mathcal{R}^*,$$

where

$$\mathcal{R}^* := \sum_{\vec{k}=\lceil n\vec{x}-\vec{T}n^\alpha\rceil}^{\lfloor n\vec{x}+\vec{T}n^\alpha\rfloor}\frac{b\left(n^{1-\alpha}\left(\vec{x} - \frac{\vec{k}}{n}\right)\right)}{I\cdot n^{\alpha\cdot d}}\cdot\mathcal{R}_N\left(\frac{\vec{k}}{n},0\right).$$

Thus

$$(\mathcal{F}_n(f))(\vec{x}) - f(\vec{x}) = f(\vec{x})\cdot\left(\sum_{\vec{k}=\lceil n\vec{x}-\vec{T}n^\alpha\rceil}^{\lfloor n\vec{x}+\vec{T}n^\alpha\rfloor}\frac{1}{I\cdot n^{\alpha\cdot d}}\right.$$

$$\left.\cdot b\left(n^{1-\alpha}\left(\vec{x} - \frac{\vec{k}}{n}\right)\right) - 1\right)$$

$$+ \sum_{j=1}^{N} \frac{1}{j!} \cdot \left\{ \sum_{\vec{k}=\lceil n\vec{x}-\vec{T}n^{\alpha} \rceil}^{[n\vec{x}+\vec{T}n^{\alpha}]} g_{\vec{k}/n}^{(j)}(0) \cdot \frac{b\left(n^{1-\alpha}\left(\vec{x}-\frac{\vec{k}}{n}\right)\right)}{I \cdot n^{\alpha \cdot d}} \right\} + \mathcal{R}^{*}.$$

Consequently we find

$$|(\mathcal{F}_n(f))(\vec{x}) - f(\vec{x})| \leq |f(\vec{x})| \cdot \left| \sum_{\vec{k}=\lceil n\vec{x}-\vec{T}n^{\alpha} \rceil}^{[n\vec{x}+\vec{T}n^{\alpha}]} \frac{1}{I \cdot n^{\alpha \cdot d}} \right.$$

$$\left. \cdot b\left(n^{1-\alpha}\left(\vec{x}-\frac{\vec{k}}{n}\right)\right) - 1 \right| + \frac{b(\vec{0})}{I \cdot n^{\alpha \cdot d}}$$

$$\cdot \left\{ \sum_{j=1}^{n} \frac{1}{j!} \cdot \left( \sum_{\vec{k}=\lceil n\vec{x}-\vec{T}n^{\alpha} \rceil}^{[n\vec{x}+\vec{T}n^{\alpha}]} |g_{\vec{k}/n}^{(j)}(0)| \right) \right\} + |\mathcal{R}^{*}| =: \otimes.$$

Notice that

$$|g_{\vec{k}/n}^{(j)}(0)| \leq \left(\frac{T^{*}}{n^{1-\alpha}}\right)^{j} \cdot \left( \left( \sum_{i=1}^{d} \left| \frac{\partial}{\partial x_{i}} \right| \right)^{j} f(\vec{x}) \right).$$

Furthermore

$$\otimes \leq |f(\vec{x})| \cdot \left| \sum_{\vec{k}=\lceil n\vec{x}-\vec{T}n^{\alpha} \rceil}^{[n\vec{x}+\vec{T}n^{\alpha}]} \frac{1}{I \cdot n^{\alpha \cdot d}} \cdot b\left(n^{1-\alpha}\left(\vec{x}-\frac{\vec{k}}{n}\right)\right) - 1 \right|$$

$$+ \frac{b(\vec{0})}{I \cdot n^{\alpha \cdot d}} \cdot \left\{ \sum_{j=1}^{N} \frac{1}{j!} \cdot \left(\frac{T^{*}}{n^{1-\alpha}}\right)^{j} \cdot \left( \left( \sum_{i=1}^{d} \left| \frac{\partial}{\partial x_{i}} \right| \right)^{j} f(\vec{x}) \right) \right\}$$

$$\cdot \left( \sum_{\vec{k}=\lceil n\vec{x}-\vec{T}n^{\alpha} \rceil}^{[n\vec{x}+\vec{T}n^{\alpha}]} 1 \right) + |\mathcal{R}^{*}|.$$

## 3. Rates of Multivariate Neural Network Operators

That is, by (3.9) we obtain

$$|(\mathcal{F}_n(f))(\vec{x}) - f(\vec{x})| \leq |f(\vec{x})| \cdot \left| \sum_{\vec{k}=\lceil n\vec{x}-\vec{T}n^\alpha \rceil}^{[n\vec{x}+\vec{T}n^\alpha]} \frac{1}{I \cdot n^{\alpha \cdot d}} \right.$$

$$\left. \cdot b\left(n^{1-\alpha}\left(\vec{x} - \frac{\vec{k}}{n}\right)\right) - 1 \right| + \left(\frac{b(\vec{0})}{I} \cdot \prod_{i=1}^{d}\left(2T_i + \frac{1}{n^\alpha}\right)\right)$$

$$\cdot \left\{ \sum_{j=1}^{N} \frac{(T^*)^j}{j! n^{j(1-\alpha)}} \cdot \left(\left(\sum_{i=1}^{d}\left|\frac{\partial}{\partial x_i}\right|\right)^j f(\vec{x})\right) \right\} + |\mathcal{R}^*|. \qquad (3.28)$$

Next we need to estimate $|\mathcal{R}^*|$. For that we observe ($0 \leq t_N \leq 1$)

$$|g_{\vec{k}/n}^{(N)}(t_N) - g_{\vec{k}/n}^{(N)}(0)| = \left|\left(\sum_{i=1}^{d}\left(\frac{k_i}{n} - x_i\right)\frac{\partial}{\partial x_i}\right)^n f\left(\vec{x} + t_N\left(\frac{\vec{k}}{n} - \vec{x}\right)\right)\right.$$

$$\left. - \left(\sum_{i=1}^{d}\left(\frac{k_i}{n} - x_i\right)\frac{\partial}{\partial x_i}\right)^N f(\vec{x})\right| \leq \frac{(T^*)^N \cdot d^N}{n^{N \cdot (1-\alpha)}}$$

$$\cdot \max_{\tilde{\alpha}: |\tilde{\alpha}|=N} \omega_1\left(f_{\tilde{\alpha}}, \frac{T^*}{n^{1-\alpha}}\right).$$

Hence

$$\left|\mathcal{R}_N\left(\frac{\vec{k}}{n}, 0\right)\right| \leq \int_0^1 \left(\int_0^{t_1} \cdots \left(\int_0^{t_{N-1}} |g_{\vec{k}/n}^{(N)}(t_n) - g_{\vec{k}/n}^{(N)}(0)| dt_n\right)\right.$$

$$\cdots dt_1 \leq \frac{(T^*)^N \cdot d^N}{N! \cdot n^{N \cdot (1-\alpha)}} \cdot \max_{\tilde{\alpha}: |\tilde{\alpha}|=N} \omega_1\left(f_{\tilde{\alpha}}, \frac{T^*}{n^{1-\alpha}}\right).$$

Consequently

$$|\mathcal{R}^*| \leq \sum_{\vec{k}=\lceil n\vec{x}-\vec{T}n^\alpha \rceil}^{[n\vec{x}+\vec{T}n^\alpha]} \frac{b\left(n^{1-\alpha}\left(\vec{x} - \frac{\vec{k}}{n}\right)\right)}{I \cdot n^{\alpha \cdot d}} \cdot \left|\mathcal{R}_N\left(\frac{\vec{k}}{n}, 0\right)\right|$$

$$\leq \frac{b(\vec{0})}{I \cdot n^{\alpha \cdot d}} \cdot \sum_{\vec{k} = \lceil n\vec{x} - \vec{T}n^\alpha \rceil}^{[n\vec{x} + \vec{T}n^\alpha]} \left| \mathcal{R}_N \left( \frac{\vec{k}}{n}, 0 \right) \right|$$

$$\leq \frac{(T^*)^N \cdot d^N}{N! \cdot n^{N \cdot (1-\alpha)}} \cdot \max_{\tilde{\alpha}: |\tilde{\alpha}| = N} \omega_1 \left( f_{\tilde{\alpha}}, \frac{T^*}{n^{1-\alpha}} \right)$$

$$\cdot \frac{b(\vec{0})}{I \cdot n^{\alpha \cdot d}} \cdot \sum_{\vec{k} = \lceil n\vec{x} - \vec{T}n^\alpha \rceil}^{[n\vec{x} + \vec{T}n^\alpha]} 1.$$

That is, by (3.9) we obtain

$$|\mathcal{R}^*| \leq \frac{(T^*)^N \cdot d^N}{N! \cdot n^{N \cdot (1-\alpha)}} \cdot \frac{b(\vec{0})}{I} \cdot \left( \prod_{i=1}^d \left( 2T_i + \frac{1}{n^\alpha} \right) \right)$$

$$\cdot \max_{\tilde{\alpha}: |\tilde{\alpha}| = N} \omega_1 \left( f_{\tilde{\alpha}}, \frac{T^*}{n^{1-\alpha}} \right). \qquad (3.29)$$

At the end inserting (3.29) into (3.28) we prove (3.27). ∎

**COROLLARY 3.1**
*Here, additionally assume that $b$ is continuous on $\mathbf{R}^d$. Let $\vec{x} \in \Gamma := \prod_{i=1}^d [-\gamma_i, \gamma_i] \subset \mathbf{R}^d$, $\gamma_i > 0$ and consider*

$$n \geq \max_{i \in \{1, \ldots, d\}} (T_i + \gamma_i, T_i^{-1/\alpha}).$$

*Take $p \geq 1$. Then*

$$\|\mathcal{F}_n f - f\|_{p, \Gamma} \leq \|f\|_{\infty, \Gamma} \cdot \left\| \sum_{\vec{k} = \lceil n\vec{x} - \vec{T}n^\alpha \rceil}^{[n\vec{x} + \vec{T}n^\alpha]} \frac{1}{I \cdot n^{\alpha \cdot d}} \right.$$

$$\left. \cdot b \left( n^{1-\alpha} \left( \vec{x} - \frac{\vec{k}}{n} \right) \right) - 1 \right\|_{p, \Gamma} + \omega_1 \left( f, \frac{T^*}{n^{1-\alpha}} \right)$$

$$\cdot \frac{b(\vec{0})}{I} \cdot \prod_{i=1}^d \left( 2T_i + \frac{1}{n^\alpha} \right) \cdot 2^{d/p} \prod_{i=1}^d \gamma_i^{1/p}, \qquad (3.30)$$

## 3. Rates of Multivariate Neural Network Operators 109

attained by constant functions. From (3.30) we get the $L_p$ convergence of $\mathcal{F}_n f$ to $f$ with rates.

**PROOF** From (3.8) we have

$$|\mathcal{F}_n(f)(\vec{x}) - f(\vec{x})| \leq \|f\|_{\infty,\Gamma} \cdot \left| \sum_{\vec{k}=\lceil n\vec{x}-\vec{T}n^\alpha \rceil}^{\lfloor n\vec{x}+\vec{T}n^\alpha \rfloor} \frac{1}{I \cdot n^{\alpha \cdot d}} \right.$$

$$\left. \cdot b\left(n^{1-\alpha}\left(\vec{x} - \frac{\vec{k}}{n}\right)\right) - 1 \right| + \omega_1\left(f, \frac{T^*}{n^{1-\alpha}}\right)$$

$$\cdot \frac{b(\vec{0})}{I} \cdot \prod_{i=1}^{d}\left(2T_i + \frac{1}{n^\alpha}\right). \tag{3.31}$$

Inequality (3.30) is now established by integration of (3.31) and the properties of $L_p$-norm. From (3.9) we obtain that

$$\sum_{\vec{k}=\lceil n\vec{x}-\vec{T}n^\alpha \rceil}^{\lfloor n\vec{x}+\vec{T}n^\alpha \rfloor} \frac{1}{I \cdot n^{\alpha \cdot d}} \cdot b\left(n^{1-\alpha}\left(\vec{x} - \frac{\vec{k}}{n}\right)\right) \leq \frac{b(\vec{0})}{I} \cdot \prod_{i=1}^{d}(2T_i + 1).$$

Hence

$$\left| \sum_{\vec{k}=\lceil n\vec{x}-\vec{T}n^\alpha \rceil}^{\lfloor n\vec{x}+\vec{T}n^\alpha \rfloor} \frac{1}{I \cdot n^{\alpha \cdot d}} \cdot b\left(n^{1-\alpha}\left(\vec{x} - \frac{\vec{k}}{n}\right)\right) - 1 \right|^p$$

$$\leq \left(\frac{b(\vec{0})}{I} \cdot \prod_{i=1}^{d}(2T_i + 1) + 1\right)^p =: M > 0,$$

for all $n \in \mathbf{N}$, and all $\vec{x} \in \Gamma$. Also, from Lemma 3.1, we get

$$\lim_{n \to +\infty} \left| \sum_{\vec{k}=\lceil n\vec{x}-\vec{T}n^\alpha \rceil}^{\lfloor n\vec{x}+\vec{T}n^\alpha \rfloor} \frac{1}{I \cdot n^{\alpha \cdot d}} \cdot b\left(n^{1-\alpha}\left(\vec{x} - \frac{\vec{k}}{n}\right)\right) - 1 \right|^p = 0,$$

all $\vec{x} \in \Gamma$. Now the bounded convergence theorem implies that

$$\lim_{n\to+\infty}\left\|\sum_{\vec{k}=\lceil n\vec{x}-\vec{T}n^\alpha\rceil}^{[n\vec{x}+\vec{T}n^\alpha]} \frac{1}{I\cdot n^{\alpha\cdot d}}\cdot b\left(n^{1-\alpha}\left(\vec{x}-\frac{\vec{k}}{n}\right)\right) - 1\right\|_{p,\Gamma} = 0.$$

■

## COROLLARY 3.2

Same assumptions as in Corollary 3.1. Then

$$\|\mathcal{F}_n f - f\|_{p,\Gamma} \leq \|f\|_{\infty,\Gamma}\cdot\left\|\sum_{\vec{k}=\lceil n\vec{x}-\vec{T}n^\alpha\rceil}^{[n\vec{x}+\vec{T}n^\alpha]}\frac{1}{I\cdot n^{\alpha\cdot d}}\right.$$

$$\left.\cdot b\left(n^{1-\alpha}\left(\vec{x}-\frac{\vec{k}}{n}\right)\right) - 1\right\|_{p,\Gamma} + \frac{b(\vec{0})}{I}$$

$$\cdot\left(\prod_{i=1}^{d}\left(2T_i + \frac{1}{n^\alpha}\right)\right)$$

$$\cdot\left\{\sum_{j=1}^{N}\frac{(T^*)^j}{j!n^{j(1-\alpha)}}\cdot\left\|\left(\sum_{i=1}^{d}\left|\frac{\partial}{\partial x_i}\right|\right)^j f\right\|_{p,\Gamma}\right\}$$

$$+ \frac{(T^*)^N\cdot d^N}{N!n^{N(1-\alpha)}}\cdot\frac{b(\vec{0})}{I}\cdot\left(\prod_{i=1}^{d}\left(2T_i + \frac{1}{n^\alpha}\right)\right)$$

$$\cdot\max_{\tilde{\alpha}:|\tilde{\alpha}|=N}\omega_1\left(f_{\tilde{\alpha}},\frac{T^*}{n^{1-\alpha}}\right)\cdot 2^{d/p}\cdot\prod_{i=1}^{d}\gamma_i^{1/p}, \qquad (3.32)$$

attained by constants. Here from (3.32) we again obtain the $L_p$ convergence of $\mathcal{F}_n(f)$ to $f$ with rates.

## 3. Rates of Multivariate Neural Network Operators

**PROOF**  From (3.27) we have

$$|(\mathcal{F}_n(f))(\vec{x}) - f(\vec{x})| \leq \|f\|_{\infty,\Gamma} \cdot \Bigg| \sum_{\vec{k}=\lceil n\vec{x}-\vec{T}n^\alpha \rceil}^{[n\vec{x}+\vec{T}n^\alpha]} \frac{1}{I \cdot n^{\alpha \cdot d}}$$

$$\cdot b\left(n^{1-\alpha}\left(\vec{x}-\frac{\vec{k}}{n}\right)\right) - 1 \Bigg| + \frac{b(\vec{0})}{I} \cdot \left(\prod_{i=1}^{d}\left(2T_i + \frac{1}{n^\alpha}\right)\right)$$

$$\cdot \left\{ \sum_{j=1}^{N} \frac{(T^*)^j}{j! n^{j(1-\alpha)}} \cdot \left(\left(\sum_{i=1}^{d}\left|\frac{\partial}{\partial x_i}\right|\right)^j f(\vec{x})\right)\right\}$$

$$+ \frac{(T^*)^N \cdot d^N}{N! n^{N(1-\alpha)}} \cdot \frac{b(\vec{0})}{I} \cdot \left(\prod_{i=1}^{d}\left(2T_i + \frac{1}{n^\alpha}\right)\right)$$

$$\cdot \max_{\tilde{\alpha}:|\tilde{\alpha}|=N} \omega_1\left(f_{\tilde{\alpha}}, \frac{T^*}{n^{1-\alpha}}\right) \tag{3.33}$$

Inequality (3.32) now comes by integration of (3.33) and the properties of $L_p$-norm. ∎

---

## 3.2 The Multivariate "Squashing Operators" and their Convergence to the Unit with Rates

We use:

**DEFINITION 3.4**  Let the nonnegative function $S : \mathbf{R}^d \to \mathbf{R}$, $d \geq 1$, $S$ has compact support $\mathcal{B} := \prod_{i=1}^{d}[-T_i, T_i]$, $T_i > 0$ and is nondecreasing for each coordinate. $S$ can be continuous only on either $\prod_{i=1}^{d}(-\infty, T_i]$ or $\mathcal{B}$ and can have jump discontinuities. We call $S$ the multivariate "squashing function" (see also [62]). Suppose that

$$I^* := \int_{\mathcal{B}} S(\vec{t}) d\vec{t} > 0.$$

**EXAMPLE 3.2**

Let $\vec{S}$ as above when $d = 1$. Then

$$\hat{\vec{S}}(\vec{x}) := \hat{S}(x_1) \cdots \hat{S}(x_d), \quad \vec{x} := (x_1, \ldots, x_d) \in \mathbf{R}^d,$$

is a multivariate "squashing function."

Let $f : \mathbf{R}^d \to \mathbf{R}$ be either a uniformly continuous or a continuous and bounded function. Let $\vec{x}, \vec{x}' \in \mathcal{B}$ such that $x_{i_k} \le x'_{i_k}$ for some $i_k \in \{1, \ldots, d\}$; $k = 1, \ldots, r \le d$. Then

$$S(x_1, \ldots, x_{i_1}, \ldots, x_{i_2}, \ldots, x_{i_3}, \ldots, x_{i_k}, \ldots, x_d)$$

$$\le S(x_1, \ldots, x'_{i_1}, \ldots, x'_{i_2}, \ldots, x'_{i_3}, \ldots, x'_{i_k}, \ldots, x_d).$$

Clearly

$$\max_{\vec{x} \in \mathcal{B}} S(\vec{x}) = S(\vec{T}), \quad \vec{T} := (T_1, \ldots, T_d).$$

For $\vec{x} \in \mathbf{R}^d$ we define the multivariate "squashing operator"

$$(G_n(f))(\vec{x}) := \sum_{k_1=-n^2}^{n^2} \cdots \sum_{k_d=-n^2}^{n^2} \frac{f\left(\frac{k_1}{n}, \ldots, \frac{k_d}{n}\right)}{I^* \cdot n^{\alpha \cdot d}}$$

$$\cdot S\left(n^{1-\alpha}\left(x_1 - \frac{k_1}{n}\right), \ldots, n^{1-\alpha}\left(x_d - \frac{k_d}{n}\right)\right), \quad (3.34)$$

where $0 < \alpha < 1$ and $n \in \mathbf{N}$:

$$n \ge \max_{i \in \{1, \ldots, d\}} \{T_i + |x_i|, T_i^{-1/\alpha}\}. \quad (3.5)$$

$G_n$ is a positive linear operator. It is clear that

$$(G_n(f))(\vec{x}) = \sum_{\vec{k} = \lceil n\vec{x} - \vec{T}n^\alpha \rceil}^{[n\vec{x} + \vec{T}n^\alpha]} \frac{f\left(\frac{\vec{k}}{n}\right)}{I^* \cdot n^{\alpha \cdot d}} \cdot S\left(n^{1-\alpha}\left(\vec{x} - \frac{\vec{k}}{n}\right)\right). \quad (3.35)$$

Here we study the pointwise convergence with rates of $(G_n(f))(\vec{x}) \to f(\vec{x})$, as $n \to +\infty$, $\vec{x} \in \mathbf{R}^d$. This is given by the next result.

## 3. Rates of Multivariate Neural Network Operators

**THEOREM 3.3**

*Under the above terms and assumptions we get that*

$$|(G_n(f))(\vec{x}) - f(\vec{x})| \le |f(\vec{x})| \cdot \left| \sum_{\vec{k}=\lceil n\vec{x}-\vec{T}n^\alpha \rceil}^{\lceil n\vec{x}+\vec{T}n^\alpha \rceil} \frac{1}{I^* \cdot n^{\alpha \cdot d}} \right.$$

$$\left. \cdot S\left(n^{1-\alpha}\left(\vec{x} - \frac{\vec{k}}{n}\right)\right) - 1 \right| + \frac{S(\vec{T})}{I^*}$$

$$\cdot \prod_{i=1}^{d}\left(2T_i + \frac{1}{n^\alpha}\right) \cdot \omega_1\left(f, \frac{T^*}{n^{1-\alpha}}\right). \tag{3.36}$$

*Inequality (3.36) is attained by constant functions.*

**PROOF**    We see that

$$|(G_n(f))(\vec{x}) - f(\vec{x})| = \left| \sum_{\vec{k}=\lceil n\vec{x}-\vec{T}n^\alpha \rceil}^{\lceil n\vec{x}+\vec{T}n^\alpha \rceil} \frac{f\left(\frac{\vec{k}}{n}\right)}{I^* \cdot n^{\alpha \cdot d}} \cdot S\left(n^{1-\alpha}\left(\vec{x} - \frac{\vec{k}}{n}\right)\right) - f(\vec{x}) \right|$$

$$= \left| \sum_{\vec{k}=\lceil n\vec{x}-\vec{T}n^\alpha \rceil}^{\lceil n\vec{x}+\vec{T}n^\alpha \rceil} \frac{\left(f\left(\frac{\vec{k}}{n}\right) - f(\vec{x})\right)}{I^* \cdot n^{\alpha \cdot d}} \cdot S\left(n^{1-\alpha}\left(\vec{x} - \frac{\vec{k}}{n}\right)\right) + f(\vec{x}) \right.$$

$$\left. \cdot \sum_{\vec{k}=\lceil n\vec{x}-\vec{T}n^\alpha \rceil}^{\lceil n\vec{x}+\vec{T}n^\alpha \rceil} \frac{1}{I^* \cdot n^{\alpha \cdot d}} \cdot S\left(n^{1-\alpha}\left(\vec{x} - \frac{\vec{k}}{n}\right)\right) - f(\vec{x}) \right|$$

$$\le \sum_{\vec{k}=\lceil n\vec{x}-\vec{T}n^\alpha \rceil}^{\lceil n\vec{x}+\vec{T}n^\alpha \rceil} \frac{|f\left(\frac{\vec{k}}{n}\right) - f(\vec{x})|}{I^* \cdot n^{\alpha \cdot d}} \cdot S\left(n^{1-\alpha}\left(\vec{x} - \frac{\vec{k}}{n}\right)\right) + |f(\vec{x})|$$

$$\cdot \left| \sum_{\vec{k}=\lceil n\vec{x}-\vec{T}n^\alpha \rceil}^{\lceil n\vec{x}+\vec{T}n^\alpha \rceil} \frac{1}{I^* \cdot n^{\alpha \cdot d}} S\left(n^{1-\alpha}\left(\vec{x} - \frac{\vec{k}}{n}\right)\right) - 1 \right|$$

$$\leq \sum_{\vec{k}=\lceil n\vec{x}-\vec{T}n^\alpha \rceil}^{[n\vec{x}+\vec{T}n^\alpha]} \frac{\omega_1\left(f, \left\|\frac{\vec{k}}{n}-\vec{x}\right\|_\infty\right)}{I^* \cdot n^{\alpha \cdot d}} \cdot S\left(n^{1-\alpha}\left(\vec{x}-\frac{\vec{k}}{n}\right)\right) + |f(\vec{x})|$$

$$\cdot \left| \sum_{\vec{k}=\lceil n\vec{x}-\vec{T}n^\alpha \rceil}^{[n\vec{x}+\vec{T}n^\alpha]} \frac{1}{I^* \cdot n^{\alpha \cdot d}} \cdot S\left(n^{1-\alpha}\left(\vec{x}-\frac{\vec{k}}{n}\right)\right) - 1 \right|.$$

That is,

$$|(G_n(f))(\vec{x}) - f(\vec{x})| \leq \omega_1\left(f, \frac{T^*}{n^{1-\alpha}}\right) \cdot S(\vec{T})$$

$$\cdot \frac{1}{I^* \cdot n^{\alpha \cdot d}} \cdot \left( \sum_{\vec{k}=\lceil n\vec{x}-\vec{T}n^\alpha \rceil}^{[n\vec{x}+\vec{T}n^\alpha]} 1 \right) + |f(\vec{x})|$$

$$\cdot \left| \sum_{\vec{k}=\lceil n\vec{x}-\vec{T}n^\alpha \rceil}^{[n\vec{x}+\vec{T}n^\alpha]} \frac{1}{I^* \cdot n^{\alpha \cdot d}} \cdot S\left(n^{1-\alpha}\left(\vec{x}-\frac{\vec{k}}{n}\right)\right) - 1 \right|. \quad (3.37)$$

By (3.9) we get

$$\frac{1}{n^{\alpha \cdot d}} \cdot \left( \sum_{\vec{k}=\lceil n\vec{x}-\vec{T}n^\alpha \rceil}^{[n\vec{x}+\vec{T}n^\alpha]} 1 \right) \leq \prod_{i=1}^{d} \left(2T_i + \frac{1}{n^\alpha}\right). \quad (3.38)$$

Using (3.38) into (3.37) we obtain (3.36). ∎

We use

**LEMMA 3.4**
*It holds*

$$D_n(\vec{x}) := \sum_{\vec{k}=\lceil n\vec{x}-\vec{T}n^\alpha \rceil}^{[n\vec{x}+\vec{T}n^\alpha]} \frac{1}{I^* \cdot n^{\alpha \cdot d}} \cdot S\left(n^{1-\alpha}\left(\vec{x}-\frac{\vec{k}}{n}\right)\right) \to 1,$$

*pointwise, as* $n \to +\infty$, *where* $\vec{x} \in \mathbf{R}^d$.

## 3. Rates of Multivariate Neural Network Operators

**PROOF** Take sufficiently large $n \in \mathbf{N}$. Then there are $k_i \in \mathbf{Z}$ such that

$$k_i, k_i \pm 1 \in \left[\lceil nx_i - T_i n^\alpha \rceil, \lceil nx_i + T_i n^\alpha \rceil\right],$$

all $i = 1, \ldots, d$.

i) Consider $\vec{t} := (t_1, \ldots, t_d)$ such that

$$k_i - 1 \leq t_i \leq k_i, \quad \text{all } i = 1, \ldots, d.$$

Then

$$n^{1-\alpha}\left(x_i - \frac{(k_i - 1)}{n}\right) \geq n^{1-\alpha}\left(x_i - \frac{t_i}{n}\right) \geq n^{1-\alpha}\left(x_i - \frac{k_i}{n}\right),$$

all $i = 1, \ldots, d$. By the nondecreasingness of $S$ on $\mathcal{B}$ we obtain

$$S\left(n^{1-\alpha}\left(\vec{x} - \frac{\overrightarrow{(k-1)}}{n}\right)\right) \geq S\left(n^{1-\alpha}\left(\vec{x} - \frac{\vec{t}}{n}\right)\right)$$

$$\geq S\left(n^{1-\alpha}\left(\vec{x} - \frac{\vec{k}}{n}\right)\right).$$

Thus

$$\int_{\overrightarrow{k-1}}^{\vec{k}} S\left(n^{1-\alpha}\left(\vec{x} - \frac{\vec{t}}{n}\right)\right) d\vec{t} \geq S\left(n^{1-\alpha}\left(\vec{x} - \frac{\vec{k}}{n}\right)\right).$$

ii) Consider $\vec{t}$ such that

$$k_i \leq t_i \leq k_i + 1, \quad \text{all } i = 1, \ldots, d.$$

Then

$$n^{1-\alpha}\left(x_i - \frac{k_i}{n}\right) \geq n^{1-\alpha}\left(x_i - \frac{t_i}{n}\right) \geq n^{1-\alpha}\left(x_i - \frac{(k_i + 1)}{n}\right),$$

all $i = 1, \ldots, d$. Again we find

$$S\left(n^{1-\alpha}\left(\vec{x} - \frac{\vec{k}}{n}\right)\right) \geq S\left(n^{1-\alpha}\left(\vec{x} - \frac{\vec{t}}{n}\right)\right)$$

$$\geq S\left(n^{1-\alpha}\left(\vec{x} - \frac{\overrightarrow{(k+1)}}{n}\right)\right).$$

Therefore

$$S\left(n^{1-\alpha}\left(\vec{x} - \frac{\vec{k}}{n}\right)\right) \geq \int_{\vec{k}}^{\overrightarrow{k+1}} S\left(n^{1-\alpha}\left(\vec{x} - \frac{\vec{t}}{n}\right)\right) d\vec{t}.$$

We have proved that

$$\int_{\vec{k}}^{\overrightarrow{k+1}} S\left(n^{1-\alpha}\left(\vec{x} - \frac{\vec{t}}{n}\right)\right) d\vec{t} \leq S\left(n^{1-\alpha}\left(\vec{x} - \frac{\vec{k}}{n}\right)\right)$$

$$\leq \int_{\overrightarrow{k-1}}^{\vec{k}} S\left(n^{1-\alpha}\left(\vec{x} - \frac{\vec{t}}{n}\right)\right) d\vec{t}, \quad (3.39)$$

for any $\vec{k} := (k_1, \ldots, k_d) \in \mathbf{Z}^d$ such that

$$\lceil nx_i - T_i n^\alpha \rceil \leq k_i - 1 \leq k_i \leq k_i + 1 \leq \lfloor nx_i + T_i n^\alpha \rfloor,$$

all $i = 1, \ldots, d$.

Consequently (by (3.30)) we obtain

$$\frac{1}{I^* \cdot n^{\alpha \cdot d}} \cdot \int_{\lceil n\vec{x} - \vec{T}n^\alpha \rceil + \vec{1}}^{\lfloor n\vec{x} + \vec{T}n^\alpha \rfloor} S\left(n^{1-\alpha}\left(\vec{x} - \frac{\vec{t}}{n}\right)\right) d\vec{t} \leq D_n(\vec{x})$$

$$\leq \frac{1}{I^* \cdot n^{\alpha \cdot d}} \cdot \int_{\lceil n\vec{x} - \vec{T}n^\alpha \rceil}^{\lfloor n\vec{x} + \vec{T}n^\alpha \rfloor - \vec{1}} S\left(n^{1-\alpha}\left(\vec{x} - \frac{\vec{t}}{n}\right)\right) d\vec{t}.$$

And by the change of variable method we find

$$\frac{1}{I^*} \cdot \int_{n^{1-\alpha}\left(\vec{x} - \frac{\lfloor n\vec{x} + \vec{T}n^\alpha \rfloor}{n}\right)}^{n^{1-\alpha}\left(\vec{x} - \frac{(\lceil n\vec{x} - \vec{T}n^\alpha \rceil + 1)}{n}\right)} S(\vec{t}) d\vec{t} \leq D_n(\vec{x})$$

$$\leq \frac{1}{I^*} \cdot \int_{n^{1-\alpha}\left(\vec{x} - \frac{(\lfloor n\vec{x} + \vec{T}n^\alpha \rfloor - 1)}{n}\right)}^{n^{1-\alpha}\left(\vec{x} - \frac{\lceil n\vec{x} - \vec{T}n^\alpha \rceil}{n}\right)} S(\vec{t}) d\vec{t}.$$

## 3. Rates of Multivariate Neural Network Operators

Therefore

$$\lim_{n\to+\infty} D_n(\vec{x}) = \frac{1}{I^*} \cdot \int_{-\vec{T}}^{\vec{T}} S(\vec{t})d\vec{t} = 1,$$

pointwise, any $\vec{x} \in \mathbf{R}^d$. ∎

As a related result we give

### THEOREM 3.4

Let $\vec{x} \in \mathbf{R}^d$, $f \in C^N(\mathbf{R}^d)$, $N \in \mathbf{N}$, such that all of its partial derivatives $f_{\tilde{\alpha}}$ of order $N$, $\tilde{\alpha}:|\tilde{\alpha}| = N$, are uniformly continuous or continuous and bounded. Then it holds

$$|(G_n(f))(\vec{x}) - f(\vec{x})| \leq |f(\vec{x})| \cdot \left| \sum_{\vec{k}=\lceil n\vec{x}-\vec{T}n^\alpha \rceil}^{[n\vec{x}+\vec{T}n^\alpha]} \frac{1}{I^* \cdot n^{\alpha \cdot d}} \right.$$

$$\left. \cdot S\left(n^{1-\alpha}\left(\vec{x}-\frac{\vec{k}}{n}\right)\right) - 1 \right| + \frac{S(\vec{T})}{I^*} \cdot \left(\prod_{i=1}^d \left(2T_i + \frac{1}{n^\alpha}\right)\right)$$

$$\cdot \left\{ \sum_{j=1}^N \frac{(T^*)^j}{j! n^{j(1-\alpha)}} \cdot \left(\left(\sum_{i=1}^d \left|\frac{\partial}{\partial x_i}\right|\right)^j f(\vec{x})\right) \right\}$$

$$+ \frac{(T^*)^N \cdot d^N}{N! n^{N(1-\alpha)}} \cdot \frac{S(\vec{T})}{I^*} \cdot \left(\prod_{i=1}^d \left(2T_i + \frac{1}{n^\alpha}\right)\right)$$

$$\cdot \max_{\tilde{\alpha}:|\tilde{\alpha}|=N} \omega_1\left(f_{\tilde{\alpha}}, \frac{T^*}{n^{1-\alpha}}\right). \tag{3.40}$$

Inequality (3.40) is attained by constant functions. Also (3.40) gives us with rates the pointwise convergence of (cf. Lemma 3.4) $G_n(f) \to f$ over $\mathbf{R}^d$, as $n \to +\infty$.

**PROOF** As in the proof of Theorem 3.2 we have

$$\frac{f\left(\frac{\vec{k}}{n}\right) S\left(n^{1-\alpha}\left(\vec{x}-\frac{\vec{k}}{n}\right)\right)}{I^* \cdot n^{\alpha \cdot d}} = \sum_{j=0}^N \frac{g_{\vec{k}/n}^{(j)}(0)}{j!} \cdot \frac{S\left(n^{1-\alpha}\left(\vec{x}-\frac{\vec{k}}{n}\right)\right)}{I^* \cdot n^{\alpha \cdot d}}$$

$$+ \frac{S\left(n^{1-\alpha}\left(\vec{x} - \frac{\vec{k}}{n}\right)\right)}{I^* \cdot n^{\alpha \cdot d}} \cdot \mathcal{R}_N\left(\frac{\vec{k}}{n}, 0\right).$$

Thus

$$(G_n(f))(\vec{x}) - f(\vec{x}) = \sum_{\vec{k}=\lceil n\vec{x}-\vec{T}n^\alpha \rceil}^{[n\vec{x}+\vec{T}n^\alpha]} \frac{f\left(\frac{\vec{k}}{n}\right)}{I^* \cdot n^{\alpha \cdot d}}$$

$$\cdot S\left(n^{1-\alpha}\left(\vec{x} - \frac{\vec{k}}{n}\right)\right) - f(\vec{x}) = -f(\vec{x})$$

$$+ \sum_{j=0}^{N} \frac{1}{j!} \left( \sum_{\vec{k}=\lceil n\vec{x}-\vec{T}n^\alpha \rceil}^{[n\vec{x}+\vec{T}n^\alpha]} g_{\vec{k}/n}^{(j)}(0) \cdot \frac{S\left(n^{1-\alpha}\left(\vec{x} - \frac{\vec{k}}{n}\right)\right)}{I^* \cdot n^{\alpha \cdot d}} \right) + \tilde{\mathcal{R}}^*,$$

where

$$\tilde{\mathcal{R}}^* := \sum_{\vec{k}=\lceil n\vec{x}-\vec{T}n^\alpha \rceil}^{[n\vec{x}+\vec{T}n^\alpha]} \frac{S\left(n^{1-\alpha}\left(\vec{x} - \frac{\vec{k}}{n}\right)\right)}{I^* \cdot n^{\alpha \cdot d}} \cdot \mathcal{R}_N\left(\frac{\vec{k}}{n}, 0\right).$$

Therefore

$$|(G_n f)(\vec{x}) - f(\vec{x})|$$

$$\leq |f(\vec{x})| \cdot \left| \sum_{\vec{k}=\lceil n\vec{x}-\vec{T}n^\alpha \rceil}^{[n\vec{x}+\vec{T}n^\alpha]} \frac{1}{I^* \cdot n^{\alpha \cdot d}} \cdot S\left(n^{1-\alpha}\left(\vec{x} - \frac{\vec{k}}{n}\right)\right) - 1 \right|$$

$$+ \frac{S(\vec{T})}{I^* \cdot n^{\alpha \cdot d}} \cdot \left( \sum_{j=1}^{N} \frac{1}{j!} \cdot \left( \sum_{\vec{k}=\lceil n\vec{x}-\vec{T}n^\alpha \rceil}^{[n\vec{x}+\vec{T}n^\alpha]} |g_{\vec{k}/n}^{(j)}(0)| \right) \right) + |\tilde{\mathcal{R}}^*|.$$

As before we get

$$|(G_n(f))(\vec{x}) - f(\vec{x})| \leq |f(\vec{x})| \cdot \left| \sum_{\vec{k}=\lceil n\vec{x}-\vec{T}n^\alpha \rceil}^{[n\vec{x}+\vec{T}n^\alpha]} \frac{1}{I^* \cdot n^{\alpha \cdot d}} \right.$$

$$\cdot S\left(n^{1-\alpha}\left(\vec{x}-\frac{\vec{k}}{n}\right)\right)-1\Bigg|+\frac{S(\vec{T})}{I^*}$$

$$\cdot\left(\prod_{i=1}^{d}\left(2T_i+\frac{1}{n^\alpha}\right)\right)\cdot\left\{\sum_{j=1}^{N}\frac{(T^*)^j}{j!n^{j(1-\alpha)}}\right.$$

$$\left.\cdot\left(\left(\sum_{i=1}^{d}\left|\frac{\partial}{\partial x_i}\right|\right)^j f(\vec{x})\right)\right\}+|\tilde{\mathcal{R}}^*|.$$

As seen earlier, we obtain

$$|\tilde{\mathcal{R}}^*|\leq\frac{S(\vec{T})}{I^*\cdot n^{\alpha\cdot d}}\cdot\sum_{\vec{k}=\lceil n\vec{x}-\vec{T}n^\alpha\rceil}^{[n\vec{x}+\vec{T}n^\alpha]}\left|\mathcal{R}_N\left(\frac{\vec{k}}{n},0\right)\right|$$

$$\leq\frac{S(\vec{T})}{I^*\cdot n^{\alpha\cdot d}}\cdot\frac{(T^*)^N\cdot d^N}{N!n^{N(1-\alpha)}}\cdot\max_{\tilde{\alpha}:|\tilde{\alpha}|=N}\omega_1\left(f_{\tilde{\alpha}},\frac{T^*}{n^{1-\alpha}}\right)$$

$$\cdot\left(\sum_{\vec{k}=\lceil n\vec{x}-\vec{T}n^\alpha\rceil}^{[n\vec{x}+\vec{T}n^\alpha]}1\right).$$

And by (3.9) we finally find

$$|\tilde{\mathcal{R}}^*|\leq\frac{(T^*)^N\cdot d^N}{N!\cdot n^{N(1-\alpha)}}\cdot\frac{S(\vec{T})}{I^*}\cdot\left(\prod_{i=1}^{d}\left(2T_i+\frac{1}{n^\alpha}\right)\right)\cdot\max_{\tilde{\alpha}:|\tilde{\alpha}|=N}\omega_1\left(f_{\tilde{\alpha}},\frac{T^*}{n^{1-\alpha}}\right).$$

∎

**REMARK 3.2** Let $\vec{x}\in\mathcal{B}$, $S\in C^r(\mathcal{B})$, $r\in\mathbf{N}$, such that the partial derivative $S_{\tilde{\alpha}}(\vec{x})\geq 0$, for some $\tilde{\alpha}:|\tilde{\alpha}|=r$, over $\mathcal{B}$. Let $f\geq 0$ and $n\in\mathbf{N}$:

$$n\geq\max_{i\in\{1,\ldots,d\}}(2T_i,T_i^{-1/\alpha}),\quad 0<\alpha<1.$$

Then

$$(G_n(f))_{\tilde{\alpha}}(\vec{x}) = \sum_{\vec{k}=\lceil n\vec{x}-\vec{T}n^\alpha\rceil}^{[n\vec{x}+\vec{T}n^\alpha]} \frac{f\left(\frac{\vec{k}}{n}\right)}{I^*} \cdot n^{r-(r+d)\cdot\alpha} \cdot S_{\tilde{\alpha}}\left(n^{1-\alpha}\left(\vec{x}-\frac{\vec{k}}{n}\right)\right) \geq 0,$$

over $\mathcal{B}$. ∎

# Chapter 4

## Asymptotic Weak Convergence of Cardaliaguet–Euvrard Neural Network Operators

An $N$th order asymptotic expansion is given for the error of weak approximation of a special class of functions by the well-known Cardaliaguet–Euvrard neural network operators. This class consists of functions $f$ that are $N$ times continuously differentiable over $\mathbf{R}$, so that all $f, f', \ldots, f^{(N)}$ have the same compact support and $f^{(N)}$ is of bounded variation. This asymptotic expansion contains products of integrals of the network activation bell-shaped function $b$ and $f$. The rate of this convergence depends only on the first derivative of involved functions. This treatment is based on [25].

### 4.1 Background

We need to use

**DEFINITION 4.1** ([100], p. 206). *Let $(X, A, \mu)$ be a measure space, let $1 \le p < \infty$, and let $f$ and $(f_n)_{n=1}^{\infty}$ be functions in $\mathcal{L}_p(X, A, \mu)$. If $p > 1$, then $(f_n)$ is said to* converge *to $f$* weakly *(in $\mathcal{L}_p$) if*

$$\lim_{n \to \infty} \int_X f_n g \, d\mu = \int_X f g \, d\mu \tag{4.1}$$

*for every $g \in \mathcal{L}_{p'}$ (where $p' > 1$ such that $\frac{1}{p} + \frac{1}{p'} = 1$). If $p = 1$, then $(f_n)$*

is said to converge to $f$ weakly *(in $\mathcal{L}_1$)* if

$$\lim_{n\to\infty} \int_X f_n g \, d\mu = \int_X fg \, d\mu$$

for every bounded A-measurable function $g$ on $X$.

We also need

**DEFINITION 4.2** [62]. *A function $b: \mathbf{R} \to \mathbf{R}$ is said to be* bell-shaped *if $b$ belongs to $L_1(\mathbf{R})$ and $\int_{-\infty}^{\infty} b(x)dx \neq 0$, if it is nondecreasing on $(-\infty, a)$ and nonincreasing on $[a, +\infty)$, where $a \in \mathbf{R}$. $b(x) \geq 0$ and $b$ takes a global maximum; it is the* center *of the bell-shaped function. The function $b(x)$ may have jump discontinuities. Here $b$ can be of compact support contained in $[-T, T]$, $T > 0$, or more generally $b$ can have the property*

$$\int_{-\infty}^{\infty} |x|^r b(x) \, dx < \infty \quad \text{for all} \quad r = 0, \ldots, N-1,$$

where $N \in \mathbf{N}$ such that $N > \frac{3-2\alpha}{1-\alpha}$, $0 < \alpha < 1$. Put

$$I := \int_{-\infty}^{\infty} b(x) \, dx, \tag{4.2}$$

notice that $I > 0$.

**EXAMPLE 4.1**
1) $b(x)$ can be the *characteristic* function on $[-1, 1]$.
2) $b(x)$ can be the *hat* function on $[-1, 1]$, i.e.,

$$b(x) = \begin{cases} 1+x, & -1 \leq x \leq 0, \\ 1-x, & 0 < x \leq 1, \\ 0, & \text{elsewhere.} \end{cases}$$

3) Let $b(x) = e^{-|x|}$, $x \in \mathbf{R}$. Then

$$\int_{-\infty}^{\infty} |x|^r e^{-|x|} \, dx = 2r!, \quad \text{any } r \in \mathbf{Z}_+.$$

For $N \in \mathbf{N}$, $T^* > 0$ we would like to study the class of functions $\Phi_N := \{f \in C^N(\mathbf{R}); f, f', \ldots, f^{(N)}$ have compact supports contained in $[-T^*, T^*]$, $f^{(N)} \in BV([-T^*, T^*])\}$.

## 4. Cardaliaguet–Euvrard Neural Network Operators

**EXAMPLE 4.2**

Let
$$f^*(x) := \begin{cases} e^{-(\frac{1}{1-x^2})}, & |x| < 1 \\ 0, & \text{elsewhere.} \end{cases}$$

Then $f^* \in C^\infty(\mathbf{R})$, also $f^*$ and all of its derivatives have the same compact support $[-1, 1]$. Proving that $\Phi_N \neq \emptyset$.

Here we take $N \in \mathbf{N}$ such that $N > \frac{3-2\alpha}{1-\alpha}$, where $0 < \alpha < 1$. For $f \in \Phi_N$, any $n \in \mathbf{N}$ and $x \in \mathbf{R}$ we consider the Cardaliaguet-Euvrard neural network operators [62]

$$(F_n(f))(x) := \sum_{k=-n^2}^{n^2} \frac{f(\frac{k}{n})}{In^\alpha} b\left(n^{1-\alpha}\left(x - \frac{k}{n}\right)\right). \tag{4.3}$$

Because $f$ has compact support contained in $[-T^*, T^*]$, when $n \geq T^*$, we have that

$$(F_n(f))(x) = \sum_{\substack{k \in \mathbf{Z} \\ -T^* \leq \frac{k}{n} \leq T^*}} \frac{f(\frac{k}{n})}{In^\alpha} b\left(n^{1-\alpha}\left(x - \frac{k}{n}\right)\right). \tag{4.4}$$

In this chapter we study the weak convergence asymptotically of $F_n(f)$ to $f \in \Phi_N$, by taking $n \geq T^*$. More precisely we present an asymptotic expansion for the error $e_n(f, g)$ of the weak convergence of $F_n(f)$ to $f$ (compare with Definition 4.1), where

$$e_n(f, g) := \int_{-\infty}^{\infty} (F_n(f))(x)g(x)\,dx - \int_{-\infty}^{\infty} f(x)g(x)\,dx, \quad g \in \phi_N. \tag{4.5}$$

Here, either $g \in \Phi_N$, or more generally, $g \in A_N := \{g : \mathbf{R} \to \mathbf{R}$ such that $g, g', \ldots, g^{(N)} \in \mathcal{L}_1(\mathbf{R})$ and $g^{(N)} \in BV(\mathbf{R}) \cap C(\mathbf{R})\}$.

We observe that for all $x \in \mathbf{R}$

$$|(F_n(f))(x)| \leq \sum_{k=-n^2}^{n^2} \frac{|f(\frac{k}{n})|}{In^\alpha} b\left(n^{1-\alpha}\left(x - \frac{k}{n}\right)\right)$$

$$\leq \frac{\|f\|_\infty \|b\|_\infty}{In^\alpha} \sum_{k=-n^2}^{n^2} 1 \leq \frac{\|f\|_\infty \|b\|_\infty}{In^\alpha}(2n^2 + 1) < \infty.$$

That is
$$\|F_n(f)\|_\infty < \infty.$$

Hence
$$\left|\int_{-\infty}^\infty g(x)(F_n(f))(x)\,dx\right| \leq \left(\int_{-\infty}^\infty |g(x)|\,dx\right) \cdot \|F_n(f)\|_\infty < \infty.$$

Thus $|e_n(f,g)| < \infty$.

Also notice that in the case of $b$ having a compact support, $(F_n(f))(x)$ is not a trivial sum at least for all $x \in [-T^*, T^*]$. Rewriting we have

$$e_n(f,g) = I^* - \int_{-\infty}^\infty f(x)g(x)\,dx, \qquad (4.6)$$

where

$$I^* := \sum_{\substack{k \in \mathbf{Z} \\ -T^* \leq \frac{k}{n} \leq T^*}} \frac{f(\frac{k}{n})}{In^\alpha} \int_{-\infty}^\infty g(x)b\left(n^{1-\alpha}\left(x - \frac{k}{n}\right)\right)dx. \qquad (4.7)$$

Here, for each $n \in \mathbf{N}$, the associated neural network has the following structure: it is a three-layer feedforward network with one hidden layer. It has one input unit and one output unit. The hidden layer has $(2n^2 + 1)$ processing units. To each pair of connecting units (input to each processing unit) we assign the same weight $n^{1-\alpha}$. The threshold values $\frac{k}{n^\alpha}$ are one for each processing unit $k$. The activation function $b$ is the same for each processing unit. The weights associated with the output unit are $f\left(\frac{k}{n}\right)/In^\alpha$, one for each processing unit $k$.

The above parameters fully describe the structure of our neural network for each $n \in \mathbf{N}$.

## 4.2 General Result

We give the following

## THEOREM 4.1
For $n \geq T^*$ it holds

$$e_n(f,g) = \sum_{r=1}^{N-1} \frac{m_r \int_{-T^*}^{T^*} f(t) g^{(r)}(t)\, dt}{I r! n^{r(1-\alpha)}} + o\left(\frac{1}{n^{(1-\alpha)(N-1)-1}}\right), \quad (4.8)$$

where

$$m_r := \int_{-\infty}^{\infty} x^r b(x)\, dx \quad (4.9)$$

and $N \in \mathbf{N}$ such that

$$N > \frac{3 - 2\alpha}{1 - \alpha}.$$

**REMARK 4.1** (1) Notice that

$$\int_{-T^*}^{T^*} f g^{(r)} = (-1)^r \int_{-T^*}^{T^*} f^{(r)} g.$$

(2) From $N > \frac{3-2\alpha}{1-\alpha}$ we have that $(1-\alpha)(N-1) - 1 > 1 - \alpha$ and

$$\lim_{n \to \infty} n^{1-\alpha} \cdot o\left(\frac{1}{n^{(1-\alpha)(N-1)-1}}\right) = 0.$$

Finally, from (4.8) we derive that

$$\lim_{n \to \infty} n^{1-\alpha} e_n(f, g) = \left(\frac{\int_{-\infty}^{\infty} t b(t)\, dt}{\int_{-\infty}^{\infty} b(t)\, dt}\right) \left(\int_{-T^*}^{T^*} f(t) g'(t)\, dt\right). \quad (4.10)$$

Clearly $\lim_{n \to \infty} e_n(f, g) = 0$. Notice that the above rate of convergence depends only on the first derivative of $g$. ∎

**PROOF** (of Theorem 4.1). Here we need to find the appropriate asymptotic expansion for $I^*$ (see (4.7)). Set $y := n^{1-\alpha}\left(x - \frac{k}{n}\right)$, then

$$x = \frac{y}{n^{1-\alpha}} + \frac{k}{n} \quad \text{and} \quad dx = \frac{dy}{n^{1-\alpha}}.$$

Then

$$I^* = \sum_{\substack{k \in \mathbf{Z} \\ -T^* \leq \frac{k}{n} \leq T^*}} \frac{f(\frac{k}{n})}{In} \int_{-\infty}^{\infty} g\left(\frac{y}{n^{1-\alpha}} + \frac{k}{n}\right) b(y)\, dy.$$

Applying Taylor's theorem we obtain

$$g\left(\frac{y}{n^{1-\alpha}} + \frac{k}{n}\right) = \sum_{r=0}^{N-1} \frac{g^{(r)}(\frac{k}{n})}{r!}\left(\frac{y}{n^{1-\alpha}}\right)^r$$

$$+ \frac{\left(\frac{y}{n^{1-\alpha}}\right)^{N-1}}{(N-1)!}\int_0^1 (1-s)^{N-1} g^{(N)}\left(\frac{k}{n} + \frac{y}{n^{1-\alpha}}s\right) ds.$$

So that

$$I^* = \frac{1}{In}\sum_{\substack{k\in\mathbf{Z}\\-T^*\leq \frac{k}{n}\leq T^*}} f\left(\frac{k}{n}\right)\left[\int_{-\infty}^{\infty}\left\{\sum_{r=0}^{N-1}\frac{g^{(r)}(\frac{k}{n})}{r!}\left(\frac{y}{n^{1-\alpha}}\right)^r\right.\right.$$

$$\left.\left. + \frac{\left(\frac{y}{n^{1-\alpha}}\right)^{N-1}}{(N-1)!}\int_0^1 (1-s)^{N-1} g^{(N)}\left(\frac{k}{n}+\frac{y}{n^{1-\alpha}}s\right) ds\right\}b(y)\,dy\right].$$

Therefore

$$I^* = \frac{1}{In}\sum_{\substack{k\in\mathbf{Z}\\-T^*\leq \frac{k}{n}\leq T^*}} f\left(\frac{k}{n}\right)\cdot\left[\sum_{r=0}^{N-1}\frac{g^{(r)}(\frac{k}{n})}{r!n^{r(1-\alpha)}}m_r + \frac{\gamma_{nN}^{(k)}(g)}{(N-1)!n^{(1-\alpha)(N-1)}}\right],$$

where

$$m_r = \int_{-\infty}^{\infty} y^r b(y)\,dy$$

(notice by assumptions that $|m_r|<\infty$) and

$$\gamma_{nN}^{(k)}(g) := \int_{-\infty}^{\infty} y^{N-1}\left(\int_0^1 (1-s)^{N-1} g^{(N)}\left(\frac{k}{n}+\frac{y}{n^{1-\alpha}}s\right) ds\right)b(y)dy. \tag{4.11}$$

That is

$$I^* = \frac{1}{I}\left\{\sum_{r=0}^{N-1}\frac{m_r\mathcal{R}_{1r}}{r!n^{r(1-\alpha)}} + \frac{\mathcal{R}_2}{(N-1)!n^{((1-\alpha)(N-1)+1)}}\right\},$$

where

$$\mathcal{R}_{1r} := \frac{1}{n}\sum_{\substack{k\in\mathbf{Z}\\-T^*\leq \frac{k}{n}\leq T^*}} f\left(\frac{k}{n}\right)g^{(r)}\left(\frac{k}{n}\right) \tag{4.12}$$

and
$$\mathcal{R}_2 := \sum_{\substack{k \in \mathbf{Z} \\ -T^* \leq \frac{k}{n} \leq T^*}} f\left(\frac{k}{n}\right) \gamma_{nN}^{(k)}(g). \quad (4.13)$$

At this point notice that $\|g^{(N)}\|_\infty < \infty$. We see that

$$|\gamma_{nN}^{(k)}(g)| \leq \int_{-\infty}^{\infty} |y|^{N-1} \left( \int_0^1 (1-s)^{N-1} \left| g^{(N)}\left(\frac{k}{n} + \frac{y}{n^{1-\alpha}} s\right) \right| ds \right) b(y) dy$$

$$\leq \frac{\|g^{(N)}\|_\infty}{N} \int_{-\infty}^{\infty} |y|^{N-1} b(y) \, dy =: M_1.$$

Thus

$$|\mathcal{R}_2| \leq \sum_{\substack{k \in \mathbf{Z} \\ -T^* \leq \frac{k}{n} \leq T^*}} \left| f\left(\frac{k}{n}\right) \right| |\gamma_{nN}^{(k)}(g)| \leq M_1 \|f\|_\infty \sum_{\substack{k \in \mathbf{Z} \\ -T^* \leq \frac{k}{n} \leq T^*}} 1$$

$$\leq M_2(2T^*n + 1),$$

where
$$M_2 := M_1 \|f\|_\infty.$$

That is
$$|\mathcal{R}_2| \leq M_2(2T^*n + 1).$$

Consequently (setting $M_3 := M_2/(N-1)!$)

$$\frac{|\mathcal{R}_2|}{(N-1)! n^{(1-\alpha)(N-1)+1}} \leq \frac{M_3(2T^*n + 1)}{n^{(1-\alpha)(N-1)+1}}$$

$$= M_3 \left\{ o\left(\frac{1}{n^{(1-\alpha)(N-1)-1}}\right) + o\left(\frac{1}{n^{(1-\alpha)(N-1)}}\right) \right\}$$

$$= o\left(\frac{1}{n^{(1-\alpha)(N-1)-1}}\right).$$

That is
$$\frac{\mathcal{R}_2}{I(N-1)! n^{((1-\alpha)(N-1)+1)}} = o\left(\frac{1}{n^{(1-\alpha)(N-1)-1}}\right). \quad (4.14)$$

Because
$$N > \frac{3-2\alpha}{1-\alpha} > \frac{2-\alpha}{1-\alpha} = 1 + \frac{1}{1-\alpha}$$
we have that $(1-\alpha)(N-1) - 1 > 0$.

Next we are estimating the bounded quantity
$$\mathcal{R}_{1r} := \frac{1}{n} \sum_{\substack{k \in \mathbf{Z} \\ -T^* \le \frac{k}{n} \le T^*}} f\left(\frac{k}{n}\right) g^{(r)}\left(\frac{k}{n}\right),$$

for all $r = 0, 1, \ldots, N-1$.

Consider $G := fg^{(r)}$. If $g \in \Phi_N$, then $g, g', \ldots, g^{(N-1)}$ are of bounded variation (likewise $f, f', \ldots, f^{(N-1)}$) on $[-T^*, T^*]$. In other words, all of $f, g, f', g', \ldots, f^{(N)}, g^{(N)}$ are of bounded variation on $\mathbf{R}$. Furthermore $G, G', \ldots, G^{(N-r)} \in C(\mathbf{R})$ have compact support contained in $[-T^*, T^*]$, all are of bounded variation on $[-T^*, T^*]$ and $\mathbf{R}$, and all of them are bounded; for all $r = 0, 1, \ldots, N-1$.

If $g \in A_N$, then by $g, g', \ldots, g^{(N)} \in \mathcal{L}_1(\mathbf{R})$ and $g^{(N)} \in BV(\mathbf{R})$, we have that $g, g', \ldots, g^{(N-1)}$ are bounded, of bounded variation on $\mathbf{R}$, uniformly continuous on $\mathbf{R}$ and tend to zero as $|x| \to \infty$, also $g^{(N)}$ is bounded. Then again $G, G', \ldots, G^{(N-r)}$ are continuous on $(-T^*, T^*)$ and $G^{(N-r)} \in BV(-T^*, T^*)$, for all $r = 0, 1, \ldots, N-1$.

Then according to Lemma 4.2(a) (see supplement) we find that

$$\mathcal{R}_{1r} = \sum_{\ell=0}^{N-r} \frac{c_\ell}{n^\ell} \int_{-T^*}^{T^*} (fg^{(r)})^{(\ell)}(t)\, dt + o\left(\frac{1}{n^{N-r}}\right), \tag{4.15}$$

where
$$c_\ell = -\sum_{j=1}^{\ell} \frac{c_{\ell-j}}{(j+1)!}, \quad \ell \ge 1,\ c_0 = 1,$$

the Bernoulli numbers. The above holds true for $r = 0, 1, \ldots, N-1$.

Because $f \in \Phi_N$, we find that
$$\int_{-T^*}^{T^*} (fg^{(r)})^{(\ell)}(t)\, dt = 0,$$

for all $\ell = 1, 2, \ldots, N-r$; for all $r = 0, \ldots, N-1$. Then
$$\mathcal{R}_{1r} = \int_{-T^*}^{T^*} f(t) g^{(r)}(t)\, dt + o\left(\frac{1}{n^{N-r}}\right). \tag{4.16}$$

## 4. Cardaliaguet–Euvrard Neural Network Operators

Thus

$$\frac{1}{I}\left(\sum_{r=0}^{N-1}\frac{m_r \mathcal{R}_{1r}}{r! n^{r(1-\alpha)}}\right)$$

$$= \frac{1}{I}\sum_{r=0}^{N-1}\left(\frac{m_r}{r!}\right)\frac{1}{n^{r(1-\alpha)}}\left\{\int_{-T^*}^{T^*} f(t)g^{(r)}(t)dt + o\left(\frac{1}{n^{N-r}}\right)\right\}$$

$$= \frac{1}{I}\left\{\sum_{r=0}^{N-1}\frac{m_r \int_{-T^*}^{T^*} f(t)g^{(r)}(t)dt}{r! n^{r(1-\alpha)}} + \sum_{r=0}^{N-1} o\left(\frac{1}{n^{N-r\alpha}}\right)\right\}$$

$$= \sum_{r=0}^{N-1}\frac{m_r \int_{-T^*}^{T^*} f(t)g^{(r)}(t)dt}{I r! n^{r(1-\alpha)}} + o\left(\frac{1}{n^{N-(N-1)\alpha}}\right).$$

That is, we established that

$$\frac{1}{I}\left(\sum_{r=0}^{N-1}\frac{m_r \mathcal{R}_{1r}}{r! n^{r(1-\alpha)}}\right) = \sum_{r=0}^{N-1}\frac{m_r \int_{-T^*}^{T^*} f(t)g^{(r)}(t)dt}{I r! n^{r(1-\alpha)}} + o\left(\frac{1}{n^{N-(N-1)\alpha}}\right). \tag{4.17}$$

And from (4.12), (4.14) and (4.17) we get that

$$I^* = \sum_{r=0}^{N-1}\frac{m_r \int_{-T^*}^{T^*} f(t)g^{(r)}(t)dt}{I r! n^{r(1-\alpha)}}$$

$$+ o\left(\frac{1}{n^{N-(N-1)\alpha}}\right) + o\left(\frac{1}{n^{(1-\alpha)(N-1)-1}}\right).$$

Finally, we have

$$I^* = \sum_{r=0}^{N-1}\frac{m_r \int_{-T^*}^{T^*} f(t)g^{(r)}(t)dt}{I r! n^{r(1-\alpha)}} + o\left(\frac{1}{n^{(1-\alpha)(N-1)-1}}\right). \tag{4.18}$$

Notice that $m_0 = I$. Putting things together (use (4.6) and (4.18)) we find (4.8), establishing the claim of the theorem. ∎

## 4.3 Supplement

The following are taken from the appendix of [58].

**LEMMA 4.1**
If $g \in L_1(-\infty, \infty) \cap BV(-\infty, \infty)$, then for every $\Delta > 0$ and $a_\Delta$,

$$\sum_{k=-\infty}^{\infty} |g(a_\Delta + k\Delta)| < \infty$$

and

$$\left| \int_{-\infty}^{\infty} g(t)dt - \Delta \sum_{k=-\infty}^{\infty} g(a_\Delta + k\Delta) \right| \leq \Delta \operatorname{Var}[g, (-\infty, \infty)].$$

where $\operatorname{Var}[g, (a, b)]$ is the total variation of $g$ over $(a, b)$.

**LEMMA 4.1'**
If $g \in L_1(-\infty, \infty) \cap BV(-\infty, \infty)$, then for every $\Delta > 0$, every real $u$, and $s_k \in [(u+k)\Delta, (u+k+1)\Delta]$, holds

$$\sum_{k=-\infty}^{\infty} |g(s_k)| < \infty$$

and

$$\left| \int_{-\infty}^{\infty} g(t)dt - \Delta \sum_{k=-\infty}^{\infty} g(s_k) \right| \leq \Delta \operatorname{Var}[g, (-\infty, \infty)].$$

**PROOF** (of Lemma 4.1). We see that

$$\left| g(a_\Delta + k\Delta) - \frac{1}{\Delta} \int_{a_\Delta + k\Delta}^{a_\Delta + (k+1)\Delta} g(t)dt \right|$$

$$\leq \frac{1}{\Delta} \int_{a_\Delta + k\Delta}^{a_\Delta + (k+1)\Delta} |g(a_\Delta + k\Delta) - g(t)|dt$$

$$\leq \operatorname{Var}[g, (a_\Delta + k\Delta, a_\Delta + (k+1)\Delta)].$$

Hence

$$|g(a_\Delta + k\Delta)| \le \frac{1}{\Delta} \int_{a_\Delta + k\Delta}^{a_\Delta + (k+1)\Delta} |g| + \text{Var}[g, (a_\Delta + k\Delta, a_\Delta + (k+1)\Delta)]$$

and thus

$$\sum_{k=-\infty}^{\infty} |g(a_\Delta + k\Delta)| \le \frac{1}{\Delta} \int_{-\infty}^{\infty} |g| + \text{Var}[g, (-\infty, \infty)] < \infty.$$

From the first inequality we find that

$$\left| \int_{-\infty}^{\infty} g(t)dt - \Delta \sum_{k=-\infty}^{\infty} g(a_\Delta + k\Delta) \right|$$

$$\le \Delta \sum_{k=-\infty}^{\infty} \text{Var}[g, (a_\Delta + k\Delta, a_\Delta + (k+1)\Delta)]$$

$$= \Delta \, \text{Var}[g, (-\infty, \infty)].$$

∎

Lemma 4.1' is established like Lemma 4.1.
The above lemmas are used in the proof of

**LEMMA 4.2**
(a) If $g, g^{(1)}, \ldots, g^{(n)}$ are continuous over the finite interval $(a, b)$ and $g^{(n)} \in BV(a, b)$, then for every real $u$ as $\Delta \downarrow 0$,

$$\mathbf{R}_\Delta(g) := \Delta \sum_{a \le (u+k)\Delta \le b} g[(u+k)\Delta] = \sum_{r=0}^{n} c_r \Delta^r \int_a^b g^{(r)}(t)dt + o(\Delta^n)$$

where the $c_r$s are universal constants (independent of $g$) given by the recursion

$$c_r = -\sum_{j=1}^{r} \frac{c_{r-j}}{(j+1)!}, \quad r \ge 1, \; c_0 = 1,$$

or by the generating function

$$C(z) = \frac{z}{e^z - 1}.$$

It is easily seen that $c_1 = -1/2$, $c_2 = 1/12$, $c_3 = 0$, $c_4 = -1/720$, $c_5 = 1/180$, etc.

**PROOF** (of Lemma 4.2)

(a) Even though the interval $(a,b)$ is finite and thus the Riemann sums are finite, we write, for convenience of notation, $\int_{-\infty}^{\infty}$ and $\sum_{k=-\infty}^{\infty}$. Taylor-expanding $g$ once we find

$$\int_{-\infty}^{\infty} g(t)dt = \sum_{k=-\infty}^{\infty} \int_{(u+k)\Delta}^{(u+k+1)\Delta} dt \Big\{ g[(u+k)\Delta]$$

$$+ [t - (u+k)\Delta] \int_0^1 g'[(u+k)\Delta w + t(1-w)]dw \Big\}$$

$$= \Delta \sum_{k=-\infty}^{\infty} g[(u+k)\Delta] + \sum_{k=-\infty}^{\infty} \int_{(u+k)\Delta}^{(u+k+1)\Delta} dt \int_{(u+k)\Delta}^{t} ds\, g'(s)$$

and hence

$$\frac{1}{\Delta}\Big\{\int_{-\infty}^{\infty} g(t)dt - \mathbf{R}_\Delta(g)\Big\} = \frac{1}{\Delta} \sum_{k=-\infty}^{\infty} \int_{(u+k)\Delta}^{(u+k+1)\Delta} ds\, g'(s)[(u+k+1)\Delta - s]$$

$$= \frac{1}{2} \sum_{k=-\infty}^{\infty} g'(s_k)\Delta \to \frac{1}{2}\int_{-\infty}^{\infty} g' \text{ as } \Delta \downarrow 0$$

by Lemma 4.1', whereby the mean value theorem $s_k \in [(u+k)\Delta, (u+k+1)\Delta]$. This proves Lemma 4.2(a) for $n=1$, with $c_0 = 1$ and $c_1 = -1/2$.

Now we suppose the result is true for $1, \ldots, n$ and we will prove it is true for $n+1$. We get

$$\int_{-\infty}^{\infty} g(t)dt = \sum_{k=-\infty}^{\infty} \int_{(u+k)\Delta}^{(u+k+1)\Delta} dt \Big\{ g[(u+k)\Delta]$$

$$+ \sum_{i=1}^{n} \frac{1}{i!} g^{(i)}[(u+k)\Delta][t - (u+k)\Delta]^i$$

$$+ \frac{1}{n!}[t - (u+k\Delta)]^{n+1} \int_0^1 g^{(n+1)}[(u+k)\Delta w + t(1-w)]w^n dw \Big\}$$

## 4. Cardaliaguet–Euvrard Neural Network Operators

$$= \mathbf{R}_\Delta(g) + \sum_{i=1}^{n} \frac{1}{(i+1)!} \mathbf{R}_\Delta(g^{(i)}) \Delta^i + R_\Delta$$

where

$$R_\Delta := \sum_{k=-\infty}^{\infty} \frac{1}{n!} \int_{(u+k)\Delta}^{(u+k+1)\Delta} dt [t - (u+k)\Delta]^{n+1}$$

$$\cdot \int_0^1 dw\, w^n g^{(n+1)}[(u+k)\Delta w + t(1-w)]$$

$$= \frac{1}{n!} \sum_{k=-\infty}^{\infty} \int_{(u+k)\Delta}^{(u+k+1)\Delta} dt \int_{(u+k)\Delta}^{t} ds (t-s)^n g^{(n+1)}(s)$$

$$= \frac{1}{(n+1)!} \sum_{k=-\infty}^{\infty} \int_{(u+k)\Delta}^{(u+k+1)\Delta} ds\, g^{(n+1)}(s)[(u+k+1)\Delta - s]^{n+1}$$

$$= \frac{1}{(n+2)!} \sum_{k=-\infty}^{\infty} g^{(n+1)}(s_k) \Delta^{n+2}$$

$$= \frac{\Delta^{n+1}}{(n+2)!} \left\{ \int_{-\infty}^{\infty} g^{(n+1)}(t) dt + o(1) \right\}$$

by Lemma 4.1', where by the mean value theorem $s_k \in [(u+k)\Delta, (u+k+1)\Delta]$. We then have by Lemma 4.1 (applied to $g^{(n+1)}$)

$$R_\Delta = \frac{\Delta^{n+1}}{(n+2)!} \mathbf{R}_\Delta(g^{(n+1)}) + o(\Delta^{n+1})$$

and applying the inductive assumptions to each $g^{(i)}$, $i = 0, \ldots, n$, we find

$$\mathbf{R}_\Delta(g) = \int_{-\infty}^{\infty} g(t) dt - \sum_{i=1}^{n+1} \frac{\Delta^i}{(i+1)!} \mathbf{R}_\Delta(g^{(i)}) + o(\Delta^{n+1})$$

$$= \int_{-\infty}^{\infty} g(t) dt - \sum_{i=1}^{n+1} \frac{\Delta^i}{(i+1)!} \left\{ \sum_{p=0}^{n-i+1} c_p \Delta^p \int_{-\infty}^{\infty} g^{(i+p)}(t) dt \right.$$

$$+ o(\Delta^{n-i+1})\}$$

$$+o(\Delta^{n+1})$$

$$= \int_{-\infty}^{\infty} g(t)dt - \sum_{r=1}^{n+1}\left\{\sum_{i=1}^{r} \frac{c_{r-i}}{(i+1)!}\right\}\Delta^r \int_{-\infty}^{\infty} g^{(r)}(t)dt + o(\Delta^{n+1})$$

$$= \sum_{r=0}^{n+1} c_r \Delta^r \int_{-\infty}^{\infty} g^{(r)}(t)dt + o(\Delta^{n+1}).$$

The generating function of $\{c_r\}$ is easily obtained. ∎

# Chapter 5

## Asymptotic Weak Convergence of Squashing Neural Network Operators

An $N$th order asymptotic expansion is given for the error of weak approximation of a special class of functions by the squashing neural network operators. This class consists of functions $f$ that are $N$ times continuously differentiable over $\mathbf{R}$, so that all $f, f', \ldots, f^{(N)}$ have the same compact support and $f^{(N)}$ is of bounded variation. This asymptotic expansion contains products of integrals of the network activation squashing function $S$ and $f$. The rate of the convergence depends only on the first derivative of the involved functions. This treatment is based on [21].

## 5.1 Background

We need to use

**DEFINITION 5.1**  *Let the nonnegative function $S : \mathbf{R} \to \mathbf{R}$, $S$ has compact support $[-T, T]$, $T > 0$, and is nondecreasing there and it can be continuous only on either $(-\infty, T]$ or $[-T, T]$. $S$ can have jump discontinuities. We call $S$ the "squashing function" (see also [62]). Suppose that*

$$J := \int_{-T}^{T} S(x)dx > 0. \tag{5.1}$$

*Clearly* $\max S(x) = S(T)$, $x \in [-T, T]$.

Here we would like to consider again the class of functions $\Phi_N := \{f \in C^N(\mathbf{R}), N \in \mathbf{N}$ such that all $f, f', \ldots, f^{(N)}$ have compact supports con-

tained in $[-T^*, T^*]$, $T^* > 0$, additionally $f^{(N)} \in BV([-T^*, T^*])\}$.
Here we take $N \in \mathbf{N}$ such that

$$N > \frac{3 - 2\alpha}{1 - \alpha}, \quad \text{where } 0 < \alpha < 1.$$

For any $n \in \mathbf{N}$ and $x \in \mathbf{R}$ we consider again the squashing neural network operators (see also [62])

$$(K_n(f))(x) := \sum_{k=-n^2}^{n^2} \frac{f(\frac{k}{n})}{Jn^\alpha} S\left(n^{1-\alpha}\left(x - \frac{k}{n}\right)\right). \tag{5.2}$$

Because $f$ has compact support contained in $[-T^*, T^*]$, when $n \geq T^*$, we get that

$$(K_n(f))(x) = \sum_{\substack{k \in \mathbf{Z} \\ -T^* \leq \frac{k}{n} \leq T^*}} \frac{f(\frac{k}{n})}{Jn^\alpha} S\left(n^{1-\alpha}\left(x - \frac{k}{n}\right)\right). \tag{5.3}$$

In this chapter we study asymptotically the weak convergence of $K_n(f)$ to $f \in \Phi_N$, by taking $n \geq T^*$. More precisely we find an asymptotic expansion for the error $\gamma_n(f, g)$ of the weak convergence of $K_n(f)$ to $f$ (compare with Definition 4.1), where

$$\gamma_n(f, g) := \int_{-\infty}^{\infty} (K_n(f))(x) g(x) \, dx - \int_{-\infty}^{\infty} f(x) g(x) \, dx, \quad g \in \phi_N. \tag{5.4}$$

Here either $g \in \Phi_N$ or more generally $g \in A_N := \{g : \mathbf{R} \to \mathbf{R} \text{ such that } g, g', \ldots, g^{(N)} \in \mathcal{L}_1(\mathbf{R}) \text{ and } g^{(N)} \in BV(\mathbf{R}) \cap C(\mathbf{R})\}$.

We observe that for all $x \in \mathbf{R}$

$$|(K_n(f))(x)| \leq \sum_{k=-n^2}^{n^2} \frac{|f(\frac{k}{n})|}{Jn^\alpha} S\left(n^{1-\alpha}\left(x - \frac{k}{n}\right)\right)$$

$$\leq \frac{\|f\|_\infty S(T)}{Jn^\alpha} \sum_{k=-n^2}^{n^2} 1 \leq \frac{\|f\|_\infty S(T)}{Jn^\alpha}(2n^2 + 1) < \infty.$$

That is
$$\|K_n(f)\|_\infty < \infty.$$

So that

$$\left|\int_{-\infty}^{\infty} g(x)(K_n(f))(x)\,dx\right| \le \left(\int_{-\infty}^{\infty} |g(x)|\,dx\right) \cdot \|K_n(f)\|_\infty < \infty.$$

Thus $|\gamma_n(f,g)| < \infty$.

Also notice that $(K_n(f))(x)$ is not a trivial sum at least for all $x \in [-T^*, T^*]$. Rewriting we see that

$$\gamma_n(f,g) = \tilde{I} - \int_{-\infty}^{\infty} f(x)g(x)\,dx, \tag{5.5}$$

where

$$\tilde{I} := \sum_{\substack{k \in \mathbf{Z} \\ -T^* \le \frac{k}{n} \le T^*}} \frac{f\left(\frac{k}{n}\right)}{Jn^\alpha} \int_{-\infty}^{\infty} g(x) S\left(n^{1-\alpha}\left(x - \frac{k}{n}\right)\right) dx. \tag{5.6}$$

Here, for each $n \in \mathbf{N}$, the associated neural network has the following structure: it is a three-layer feedforward network with one hidden layer. It has one input unit and one output unit. The hidden layer has $(2n^2 + 1)$ processing units. To each pair of connecting units (input to each processing unit) we assign the same weight $n^{1-\alpha}$. The threshold values $\frac{k}{n^\alpha}$ are one for each processing unit $k$. The activation function $S$ is the same for each processing unit. The weights associated with the output unit are $f\left(\frac{k}{n}\right)/Jn^\alpha$, one for each processing unit $k$.

The above parameters fully describe the structure of our neural network for each $n \in \mathbf{N}$.

---

## 5.2 General Result

We present the following

**THEOREM 5.1**
For $n \ge T^*$ it holds that

$$\gamma_n(f,g) = \sum_{r=1}^{N-1} \frac{m_r \int_{-T^*}^{T^*} f(t) g^{(r)}(t)\,dt}{Jr!n^{r(1-\alpha)}} + o\left(\frac{1}{n^{(1-\alpha)(N-1)-1}}\right), \tag{5.7}$$

where
$$m_r := \int_{-T}^{T} x^r S(x)\, dx \tag{5.8}$$

and $N \in \mathbf{N}$ such that
$$N > \frac{3 - 2\alpha}{1 - \alpha}.$$

**REMARK 5.1** (1) Observe that
$$\int_{-T^*}^{T^*} f g^{(r)} = (-1)^r \int_{-T^*}^{T^*} f^{(r)} g.$$

(2) From $N > \frac{3-2\alpha}{1-\alpha}$ we get that $(1-\alpha)(N-1) - 1 > 1 - \alpha$ and
$$\lim_{n \to \infty} n^{1-\alpha} \cdot o\left(\frac{1}{n^{(1-\alpha)(N-1)-1}}\right) = 0.$$

Finally, from (5.7) we find that
$$\lim_{n \to \infty} n^{1-\alpha} \gamma_n(f, g) = \left(\frac{\int_{-T}^{T} t S(t)\, dt}{\int_{-T}^{T} S(t)\, dt}\right) \left(\int_{-T^*}^{T^*} f(t) g'(t)\, dt\right). \tag{5.9}$$

Clearly $\lim_{n \to \infty} \gamma_n(f, g) = 0$. Notice that the above rate of convergence depends only on the first derivative of $g$. ∎

**PROOF** (of Theorem 5.1). Here we want to find the appropriate asymptotic expansion for $\tilde{I}$ (see (5.6)). Call $y := n^{1-\alpha}\left(x - \frac{k}{n}\right)$, then
$$x = \frac{y}{n^{1-\alpha}} + \frac{k}{n} \quad \text{and} \quad dx = \frac{dy}{n^{1-\alpha}}.$$

Hence
$$\tilde{I} = \sum_{\substack{k \in \mathbf{Z} \\ -T^* \leq \frac{k}{n} \leq T^*}} \frac{f(\frac{k}{n})}{Jn} \int_{-T}^{T} g\left(\frac{y}{n^{1-\alpha}} + \frac{k}{n}\right) S(y)\, dy.$$

Applying Taylor's theorem we obtain
$$g\left(\frac{y}{n^{1-\alpha}} + \frac{k}{n}\right) = \sum_{r=0}^{N-1} \frac{g^{(r)}(\frac{k}{n})}{r!} \left(\frac{y}{n^{1-\alpha}}\right)^r$$

## 5. Squashing Neural Network Operators Studied Asymptotically

$$+ \frac{\left(\frac{y}{n^{1-\alpha}}\right)^{N-1}}{(N-1)!} \int_0^1 (1-s)^{N-1} g^{(N)}\left(\frac{k}{n} + \frac{y}{n^{1-\alpha}} s\right) ds.$$

So that

$$\tilde{I} = \frac{1}{Jn} \sum_{\substack{k \in \mathbb{Z} \\ -T^* \leq \frac{k}{n} \leq T^*}} f\left(\frac{k}{n}\right) \left[ \int_{-T}^{T} \left\{ \sum_{r=0}^{N-1} \frac{g^{(r)}\left(\frac{k}{n}\right)}{r!} \left(\frac{y}{n^{1-\alpha}}\right)^r \right. \right.$$

$$\left. \left. + \frac{\left(\frac{y}{n^{1-\alpha}}\right)^{N-1}}{(N-1)!} \int_0^1 (1-s)^{N-1} g^{(N)}\left(\frac{k}{n} + \frac{y}{n^{1-\alpha}} s\right) ds \right\} S(y)\, dy \right].$$

Therefore

$$\tilde{I} = \frac{1}{Jn} \sum_{\substack{k \in \mathbb{Z} \\ -T^* \leq \frac{k}{n} \leq T^*}} f\left(\frac{k}{n}\right) \cdot \left[ \sum_{r=0}^{N-1} \frac{g^{(r)}\left(\frac{k}{n}\right)}{r! n^{r(1-\alpha)}} m_r + \frac{\delta_{nN}^{(k)}(g)}{(N-1)! n^{(1-\alpha)(N-1)}} \right],$$

where

$$m_r = \int_{-T}^{T} y^r S(y)\, dy$$

(notice by assumption that $|m_r| < \infty$) and

$$\delta_{nN}^{(k)}(g) := \int_{-T}^{T} y^{N-1} \left( \int_0^1 (1-s)^{N-1} g^{(N)}\left(\frac{k}{n} + \frac{y}{n^{1-\alpha}} s\right) ds \right) S(y) dy. \tag{5.10}$$

I.e.

$$\tilde{I} = \frac{1}{J} \left\{ \sum_{r=0}^{N-1} \frac{m_r \mathcal{R}_{1r}}{r! n^{r(1-\alpha)}} + \frac{\mathcal{R}_2}{(N-1)! n^{((1-\alpha)(N-1)+1)}} \right\},$$

where

$$\mathcal{R}_{1r} := \frac{1}{n} \sum_{\substack{k \in \mathbb{Z} \\ -T^* \leq \frac{k}{n} \leq T^*}} f\left(\frac{k}{n}\right) g^{(r)}\left(\frac{k}{n}\right) \tag{5.11}$$

and

$$\mathcal{R}_2 := \sum_{\substack{k \in \mathbb{Z} \\ -T^* \leq \frac{k}{n} \leq T^*}} f\left(\frac{k}{n}\right) \delta_{nN}^{(k)}(g). \tag{5.12}$$

Also notice that $\|g^{(N)}\|_\infty < \infty$. We see that

$$|\delta_{nN}^{(k)}(g)| \leq \int_{-T}^{T} |y|^{N-1} \left( \int_0^1 (1-s)^{N-1} \left| g^{(N)}\left(\frac{k}{n} + \frac{y}{n^{1-\alpha}}s\right) \right| ds \right) S(y) dy$$

$$\leq \frac{\|g^{(N)}\|_\infty}{N} \int_{-T}^{T} |y|^{N-1} S(y)\, dy =: M_1.$$

Therefore

$$|\mathcal{R}_2| \leq \sum_{\substack{k \in \mathbf{Z} \\ -T^* \leq \frac{k}{n} \leq T^*}} \left| f\left(\frac{k}{n}\right) \right| |\delta_{nN}^{(k)}(g)| \leq M_1 \|f\|_\infty \sum_{\substack{k \in \mathbf{Z} \\ -T^* \leq \frac{k}{n} \leq T^*}} 1$$

$$\leq M_2(2T^*n + 1),$$

where

$$M_2 := M_1 \|f\|_\infty.$$

That is,

$$|\mathcal{R}_2| \leq M_2(2T^*n + 1).$$

Consequently (setting $M_3 := M_2/(N-1)!$)

$$\frac{|\mathcal{R}_2|}{(N-1)!n^{(1-\alpha)(N-1)+1}} \leq \frac{M_3(2T^*n + 1)}{n^{(1-\alpha)(N-1)+1}}$$

$$= M_3 \left\{ o\left(\frac{1}{n^{(1-\alpha)(N-1)-1}}\right) + o\left(\frac{1}{n^{(1-\alpha)(N-1)}}\right) \right\}$$

$$= o\left(\frac{1}{n^{(1-\alpha)(N-1)-1}}\right).$$

That is,

$$\frac{\mathcal{R}_2}{J(N-1)!n^{((1-\alpha)(N-1)+1)}} = o\left(\frac{1}{n^{(1-\alpha)(N-1)-1}}\right). \tag{5.13}$$

Because

$$N > \frac{3-2\alpha}{1-\alpha} > \frac{2-\alpha}{1-\alpha} = 1 + \frac{1}{1-\alpha}$$

## 5. Squashing Neural Network Operators Studied Asymptotically

we find that $(1-\alpha)(N-1) - 1 > 0$.

Next we are estimating the bounded quantity

$$\mathcal{R}_{1r} := \frac{1}{n} \sum_{\substack{k \in \mathbf{Z} \\ -T^* \leq \frac{k}{n} \leq T^*}} f\left(\frac{k}{n}\right) g^{(r)}\left(\frac{k}{n}\right),$$

for all $r = 0, 1, \ldots, N-1$.

Consider $G := fg^{(r)}$. If $g \in \Phi_N$, then $g, g', \ldots, g^{(N-1)}$ are of bounded variation (likewise $f, f', \ldots, f^{(N-1)}$) on $[-T^*, T^*]$. In other words, all of $f, g, f', g', \ldots, f^{(N)}, g^{(N)}$ are of bounded variation on $\mathbf{R}$. Furthermore $G, G', \ldots, G^{(N-r)} \in C(\mathbf{R})$ have compact support contained in $[-T^*, T^*]$, all are of bounded variation on $[-T^*, T^*]$ and $\mathbf{R}$, and all of them are bounded; for all $r = 0, 1, \ldots, N-1$.

If $g \in A_N$, then by $g, g', \ldots, g^{(N)} \in \mathcal{L}_1(\mathbf{R})$ and $g^{(N)} \in BV(\mathbf{R})$ we get that $g, g', \ldots, g^{(N-1)}$ are bounded, of bounded variation on $\mathbf{R}$, uniformly continuous on $\mathbf{R}$ and tend to zero as $|x| \to \infty$, also $g^{(N)}$ is bounded. Then again $G, G', \ldots, G^{(N-r)}$ are continuous on $(-T^*, T^*)$ and $G^{(N-r)} \in BV(-T^*, T^*)$, for all $r = 0, 1, \ldots, N-1$.

Then according to Lemma 4.2 we get that

$$\mathcal{R}_{1r} = \sum_{\ell=0}^{N-r} \frac{c_\ell}{n^\ell} \int_{-T^*}^{T^*} (fg^{(r)})^{(\ell)}(t)\, dt + o\left(\frac{1}{n^{N-r}}\right), \tag{5.14}$$

where

$$c_\ell = -\sum_{j=1}^{\ell} \frac{c_{\ell-j}}{(j+1)!}, \quad \ell \geq 1,\ c_0 = 1,$$

the Bernoulli numbers. The above holds true for $r = 0, 1, \ldots, N-1$.

Since $f \in \Phi_N$ we find that

$$\int_{-T^*}^{T^*} (fg^{(r)})^{(\ell)}(t)\, dt = 0,$$

for all $\ell = 1, 2, \ldots, N-r$; for all $r = 0, \ldots, N-1$. Consequently,

$$\mathcal{R}_{1r} = \int_{-T^*}^{T^*} f(t) g^{(r)}(t)\, dt + o\left(\frac{1}{n^{N-r}}\right). \tag{5.15}$$

Furthermore

$$\frac{1}{J}\left(\sum_{r=0}^{N-1}\frac{m_r \mathcal{R}_{1r}}{r!n^{r(1-\alpha)}}\right)$$

$$= \frac{1}{J}\sum_{r=0}^{N-1}\binom{m_r}{r!}\frac{1}{n^{r(1-\alpha)}}\left\{\int_{-T^*}^{T^*}f(t)g^{(r)}(t)dt + o\left(\frac{1}{n^{N-r}}\right)\right\}$$

$$= \frac{1}{J}\left\{\sum_{r=0}^{N-1}\frac{m_r\int_{-T^*}^{T^*}f(t)g^{(r)}(t)dt}{r!n^{r(1-\alpha)}} + \sum_{r=0}^{N-1}o\left(\frac{1}{n^{N-r\alpha}}\right)\right\}$$

$$= \sum_{r=0}^{N-1}\frac{m_r\int_{-T^*}^{T^*}f(t)g^{(r)}(t)dt}{Jr!n^{r(1-\alpha)}} + o\left(\frac{1}{n^{N-(N-1)\alpha}}\right).$$

In other words, we proved that

$$\frac{1}{J}\left(\sum_{r=0}^{N-1}\frac{m_r\mathcal{R}_{1r}}{r!n^{r(1-\alpha)}}\right) = \sum_{r=0}^{N-1}\frac{m_r\int_{-T^*}^{T^*}f(t)g^{(r)}(t)dt}{Jr!n^{r(1-\alpha)}} + o\left(\frac{1}{n^{N-(N-1)\alpha}}\right). \tag{5.16}$$

Finally, from (5.11), (5.13), and (5.16) we find that

$$\tilde{I} = \sum_{r=0}^{N-1}\frac{m_r\int_{-T^*}^{T^*}f(t)g^{(r)}(t)dt}{Jr!n^{r(1-\alpha)}}$$

$$+ o\left(\frac{1}{n^{N-(N-1)\alpha}}\right) + o\left(\frac{1}{n^{(1-\alpha)(N-1)-1}}\right).$$

At the end we have

$$\tilde{I} = \sum_{r=0}^{N-1}\frac{m_r\int_{-T^*}^{T^*}f(t)g^{(r)}(t)dt}{Jr!n^{r(1-\alpha)}} + o\left(\frac{1}{n^{(1-\alpha)(N-1)-1}}\right). \tag{5.17}$$

Notice that $m_0 = J$. Putting things together (use (5.5) and (5.17)) we find (5.7), proving the claim of the theorem. ∎

# Part II

# On Wavelets

# Chapter 6

## Quantitative Monotone and Probabilistic Wavelet Type Approximation

Continuous functions are approximated by wavelet type operators. These maintain monotonicity and transform continuous probability distribution functions into probability distribution functions. The degree of this approximation is estimated by establishing some Jackson type inequalities. This treatment is based on [42].

There has been great interest in the wavelet type approximation. Among others, I. Daubechies [70] and S.G. Mallat [123] have discussed orthonormal wavelets in their celebrated articles by using multiresolution analysis. Non-orthogonal but compactly supported wavelets have also been considered, such as in [68]. We are interested in the linear wavelet type approximation to some special kinds of functions.

In this chapter we consider the wavelet type approximation of continuous monotone functions and continuous probabilistic distribution functions. We construct two kinds of naturally arising positive linear wavelet type operators, $A_k(f)$ and $B_k(f)$, which maintain monotonicity and produce probabilistic distribution functions if $f$ is so. We establish Jackson type inequalities for estimating this degree of approximation and we establish that most of them are sharp.

To do this, we first deal with the problem of monotone wavelet type approximation and then with the probabilistic one and we prove that the monotone wavelet type approximation degree can be estimated by first and second moduli of smoothness of the approximated function.

## 6.1 Approximation by $A_k(f)$ Operators

Assume that $\varphi(x)$ is a bounded right-continuous function on $\mathbf{R}$ and has compact support. Define

$$\varphi_{kj}(x) := 2^{\frac{k}{2}}\varphi(2^k x - j) \quad \text{for } k, j \in Z.$$

For $f \in C(\mathbf{R})$, we define

$$A_k(f)(x) := \sum_{j=-\infty}^{\infty} \langle f, \varphi_{kj}\rangle \varphi_{kj}(x) \quad \text{for } k \in Z, \tag{6.1}$$

where

$$\langle f, \varphi_{kj}\rangle := \int_{-\infty}^{\infty} f(t)\varphi_{kj}(t)\,dt.$$

Because $\varphi(x)$ is compactly supported, for any fixed $x \in \mathbf{R}$, the summation in (6.1) only involves finite terms, so $A_k(f)(x)$ is well-defined on $\mathbf{R}$.

Since

$$\langle f, \varphi_{kj}\rangle = 2^{\frac{k}{2}} \int_{-\infty}^{\infty} f(t)\varphi(2^k t - j)\,dt$$

$$= 2^{-\frac{k}{2}} \int_{-\infty}^{\infty} f(2^{-k}u)\varphi(u - j)\,du$$

$$= 2^{-\frac{k}{2}} \langle f(2^{-k}\cdot), \varphi_{0j}\rangle,$$

we have

$$A_k(f)(x) = \sum_{j=-\infty}^{\infty} 2^{-\frac{k}{2}} \langle f(2^{-k}\cdot), \varphi_{0j}\rangle 2^{\frac{k}{2}} \varphi(2^k x - j)$$

$$= A_0(f(2^{-k}\cdot))(2^k x). \tag{6.2}$$

Let $\operatorname{supp}\varphi(x) \subseteq [-a, a]$, $0 < a < +\infty$ and $\{C_j\}_{j=-\infty}^{\infty}$ be a sequence of real numbers. Set

$$A(x) := \sum_{j=-\infty}^{\infty} C_j \varphi_{0j}(x). \tag{6.3}$$

## 6. Monotone and Probabilistic Wavelet Type Approximation

We first study some properties of $A(x)$.

**LEMMA 6.1**
If $\varphi(x)$ is a bounded right-continuous function with supp $\varphi(x) \subseteq [-a,a]$, $0 < a < +\infty$, then $A(x)$ is also a right-continuous function on $\mathbf{R}$.

**PROOF** For any $x_0 \in \mathbf{R}$, $0 \leq x - x_0 < a$, because supp $\varphi(x) \subseteq [-a,a]$, we see that

$$\varphi(x-j) = 0 \quad \text{if } j > x_0 + 2a \quad \text{or} \quad j < x_0 - a.$$

Thus

$$A(x) = \sum_{j=-\infty}^{\infty} C_j \varphi(x-j) = \sum_{\substack{j \\ x_0 - a \leq j \leq x_0 + 2a}} c_j \varphi(x-j), \quad 0 \leq x - x_0 < a.$$

Then, from the right-continuity of $\varphi(x)$, we get

$$\lim_{x \to x_0^+} A(x) = \lim_{x \to x_0^+} \sum_{\substack{j \\ x_0 - a \leq j \leq x_0 + 2a}} C_j \varphi(x-j)$$

$$= \sum_{\substack{j \\ x_0 - a \leq j \leq x_0 + 2a}} C_j \lim_{x \to x_0^+} \varphi(x-j)$$

$$= \sum_{\substack{j \\ x_0 - a \leq j \leq x_0 + 2a}} C_j \varphi(x_0 - j) = A(x_0).$$

So that $A(x)$ is right-continuous on $\mathbf{R}$. ∎

By Lemma 6.1, from (6.1) and (6.2), we have

**LEMMA 6.2**
If $\varphi(x)$ is a bounded right-continuous function with supp $\varphi(x) \subseteq [-a,a]$, $0 < a < +\infty$, then, for any $f \in C(\mathbf{R})$ and $k \in Z$, the functions $A_k(f)(x)$ defined by (6.1) are right-continuous on $\mathbf{R}$.

## LEMMA 6.3

Assume that $\varphi(x)$ is a bounded function with $\operatorname{supp} \varphi(x) \subseteq [-a, a]$, $0 < a < +\infty$ and satisfies the following conditions:

(i) $\sum_{j=-\infty}^{\infty} \varphi(x - j) \equiv 1$ on $\mathbf{R}$,

(ii) there is a number $b$ such that $\varphi(x)$ is nondecreasing if $x \leq b$ and is nonincreasing if $x \geq b$.

Then, if $\{C_j\}_{j=-\infty}^{\infty}$ is a nondecreasing sequence, the function $A(x)$ defined by (6.3) is a nondecreasing function on $\mathbf{R}$.

**PROOF** For any fixed $x \in \mathbf{R}$ and $0 < \Delta x < 1$, let $j_0$ be the integer such that
$$x - j_0 \leq b < x - j_0 + 1.$$

We see that
$$\varphi(x + \Delta x - j) - \varphi(x - j) \geq 0 \quad \text{if } j_0 + 1 \leq j < +\infty \qquad (6.4)$$

and
$$\varphi(x + \Delta x - j) - \varphi(x - j) \leq 0 \quad \text{if } -\infty < j \leq j_0 - 1, \qquad (6.5)$$

because of property (ii) of $\varphi(x)$.

Because $\{C_j\}$ is a nondecreasing sequence and $\sum_{j=-\infty}^{\infty} \varphi(x - j) \equiv 1$ on $\mathbf{R}$, from (6.4) and (6.5), we find that

$$A(x + \Delta x) - A(x) = \sum_{j=-\infty}^{\infty} C_j [\varphi(x + \Delta x - j) - \varphi(x - j)]$$

$$= \sum_{j=-\infty}^{j_0 - 1} C_j [\varphi(x + \Delta x - j) - \varphi(x - j)] + C_{j_0} [\varphi(x + \Delta x - j_0)$$

$$- \varphi(x - j_0)] + \sum_{j=j_0+1}^{\infty} C_j [\varphi(x + \Delta x - j) - \varphi(x - j)]$$

$$\geq \sum_{j=-\infty}^{j_0 - 1} C_{j_0} [\varphi(x + \Delta x - j) - \varphi(x - j)] + C_{j_0} [\varphi(x + \Delta x - j_0)$$

$$-\varphi(x-j_0)] + \sum_{j=j_0+1}^{\infty} C_{j_0}[\varphi(x+\Delta x-j) - \varphi(x-j)]$$

$$= C_{j_0} \sum_{j=-\infty}^{\infty} [\varphi(x+\Delta x-j) - \varphi(x-j)]$$

$$= C_{j_0} \left( \sum_{j=-\infty}^{\infty} \varphi(x+\Delta x-j) - \sum_{j=-\infty}^{\infty} \varphi(x-j) \right)$$

$$= 0.$$

Thus $A(x)$ is a nondecreasing function on $\mathbf{R}$. ∎

We present

**THEOREM 6.1**
Assume that $\varphi(x)$ is a bounded function with supp $\varphi(x) \subseteq [-a, a]$, $0 < a < +\infty$ and satisfies the following conditions:

(i) $\displaystyle\sum_{j=-\infty}^{\infty} \varphi(x-j) \equiv 1$ on $\mathbf{R}$;

(ii) there is a number $b$ such that $\varphi(x)$ is nondecreasing if $x \leq b$ and is nonincreasing if $x \geq b$.

Then, for $f \in C(\mathbf{R})$, if $f$ is a nondecreasing function, the linear wavelet type operators $A_k(f)$ defined by (6.1) are also nondecreasing functions on $\mathbf{R}$ and satisfy

$$|A_k(f)(x) - f(x)| \leq \omega(f, 2^{-k+1}a), \quad x \in \mathbf{R}, \ k \in \mathbf{Z}, \qquad (6.6)$$

where $\omega(f, h)$ is the modulus of continuity of $f$. Furthermore, the inequalities (6.6) are sharp.

**REMARK** (1) Here we give some examples of $\varphi(x)$ that satisfy the conditions of Theorem 6.1:

$$\varphi(x) = \begin{cases} 1, & -\frac{1}{2} \leq x < \frac{1}{2}, \\ 0, & \text{otherwise} \end{cases}$$

and
$$\varphi(x) = \begin{cases} x+1, & -1 \le x \le 0, \\ 1-x, & 0 < x \le 1, \\ 0, & \text{otherwise.} \end{cases}$$

(2) Because $\varphi(x) = 0$ for $x \in (-\infty, -a) \cup (a, +\infty)$, if $\varphi(x)$ has property (ii), then
$$\varphi(x) \ge 0, \quad x \in \mathbf{R}. \tag{6.7}$$

Thus the linear wavelet type operator $A_k(f)$ defined by (6.1) is a positive operator.

(3) In addition, the function $\varphi(x)$ in Theorem 6.1 has the property

(iii) $\int_{-\infty}^{\infty} \varphi(x)dx = 1$.

Really, from property (i) of $\varphi(x)$, we have
$$\int_{-\infty}^{\infty} \varphi(x)\,dx = \sum_{j=-\infty}^{\infty} \int_{j}^{j+1} \varphi(x)\,dx = \sum_{j=-\infty}^{\infty} \int_{0}^{1} \varphi(x+j)\,dx$$
$$= \int_{0}^{1} \sum_{j=-\infty}^{\infty} \varphi(x+j)\,dx = 1.$$

**PROOF** First let us study the monotonicity of $A_k(f)(x)$ on $\mathbf{R}$. It follows from (6.7) that
$$\langle f, \varphi_{0j} \rangle = \int_{-\infty}^{\infty} f(t)\varphi(t-j)\,dt$$
$$= \int_{-\infty}^{\infty} f(u+j)\varphi(u)\,du$$

is a nondecreasing sequence if $f$ is nondecreasing on $\mathbf{R}$. Then, by Lemma 6.3, $A_0(f)(x)$ is a nondecreasing function on $\mathbf{R}$, and furthermore, by (6.2), for any $k \in Z$, $A_k(f)(x)$ are nondecreasing too.

Now we estimate $|A_k(f)(x) - f(x)|$. From the property (i) of $\varphi(x)$, we have
$$\sum_{j=-\infty}^{\infty} \varphi(2^k x - j) \equiv 1, \tag{6.8}$$

## 6. Monotone and Probabilistic Wavelet Type Approximation

and then

$$A_k(f)(x) - f(x) = 2^{\frac{k}{2}} \sum_{j=-\infty}^{\infty} [\langle f, \varphi_{kj} \rangle - 2^{-\frac{k}{2}} f(x)] \varphi(2^k x - j). \tag{6.9}$$

From the property (iii) of $\varphi(x)$, we see

$$\int_{-\infty}^{\infty} \varphi(u-j)\,du = 1, \quad j \in Z, \tag{6.10}$$

and hence

$$\langle f, \varphi_{kj} \rangle - 2^{-\frac{k}{2}} f(x) = 2^{\frac{k}{2}} \int_{-\infty}^{\infty} f(t) \varphi(2^k t - j)\,dt - 2^{-\frac{k}{2}} f(x)$$

$$= 2^{-\frac{k}{2}} \int_{-\infty}^{\infty} f(2^{-k} u) \varphi(u-j)\,du - 2^{-\frac{k}{2}} f(x)$$

$$= 2^{-\frac{k}{2}} \int_{-\infty}^{\infty} [f(2^{-k} u) - f(x)] \varphi(u-j)\,du. \tag{6.11}$$

Because supp $\varphi \subseteq [-a, a]$, for $2^{-k}(-a+j) \le x \le 2^{-k}(a+j)$ by (6.7), (6.10), and (6.11) we get

$$|\langle f, \varphi_{kj} \rangle - 2^{-\frac{k}{2}} f(x)| = \left| 2^{-\frac{k}{2}} \int_{-a+j}^{a+j} [f(2^{-k} u) - f(x)] \varphi(u-j)\,du \right|$$

$$\le 2^{-\frac{k}{2}} \omega(f, 2^{-k+1} a) \int_{-a+j}^{a+j} \varphi(u-j)\,du$$

$$= 2^{-\frac{k}{2}} \omega(f, 2^{-k+1} a) \int_{-\infty}^{\infty} \varphi(u-j)\,du$$

$$= 2^{-\frac{k}{2}} \omega(f, 2^{-k+1} a). \tag{6.12}$$

Therefore, from (6.7)–(6.9), (6.12) and that supp $\varphi \subseteq [-a, a]$, we find

$$|A_k(f)(x) - f(x)| \le 2^{\frac{k}{2}} \sum_{\substack{j \\ 2^k x - j \in [-a, a]}} |\langle f, \varphi_{kj} \rangle - 2^{-\frac{k}{2}} f(x)| \cdot \varphi(2^k x - j)$$

$$\leq \omega(f, 2^{-k+1}a) \cdot \sum_{j=-\infty}^{\infty} \varphi(2^k x - j)$$

$$= \omega(f, 2^{-k+1}a), \quad x \in \mathbf{R}.$$

Next we are going to show that inequalities (6.6) are sharp.

Assume that there is a constant $0 < C < 1$ such that for any $f \in C(\mathbf{R})$ and $x \in \mathbf{R}$ holds

$$|A_k(f)(x) - f(x)| \leq C\omega(f, 2^{-k+1}a). \tag{6.13}$$

Take $\varphi(x) = \chi_{[-\frac{1}{2}, \frac{1}{2})}(x)$, which satisfies all the conditions in Theorem 6.1. For this $\varphi$, (6.13) becomes

$$|A_k(f)(x) - f(x)| \leq C\omega(f, 2^{-k}). \tag{6.14}$$

Define

$$g(x) = \begin{cases} 0, & x \leq -2^{-k-1}, \\ 1, & x \geq -2^{-k-1} + 2^{-k}(1-C), \\ \frac{2^k x + (1/2)}{1-C}, & -2^{-k-1} < x < -2^{-k-1} + 2^{-k}(1-C), \end{cases}$$

where $C$ is the constant in (6.14). It is easy to verify that $g(x)$ is a nondecreasing continuous function on $\mathbf{R}$ with

$$\omega(g, 2^{-k}) = 1. \tag{6.15}$$

Thus from (6.14) we have

$$|A_k(g)(x) - g(x)| \leq C. \tag{6.16}$$

On the other hand, from the definitions of $g$ and $\varphi$, we get

$$\langle g, \varphi_{kj} \rangle = 2^{\frac{k}{2}} \int_{-\infty}^{\infty} g(t) \varphi(2^k t - j) \, dt$$

$$= 2^{\frac{k}{2}} \int_{2^{-k}(j-(1/2))}^{2^{-k}(j+(1/2))} g(t) \, dt$$

$$= \begin{cases} 0, & j \leq -1 \\ 2^{-\frac{k}{2}}, & j \geq 1 \\ 2^{\frac{k}{2}} \int_{-2^{-k-1}}^{2^{-k-1}} g(t)\,dt = 2^{-\frac{k}{2}-1}(1+C), & j = 0. \end{cases}$$

Therefore

$$A_k(g)(x) = 2^{\frac{k}{2}} \sum_{j=-\infty}^{\infty} \langle g, \varphi_{kj} \rangle \varphi(2^k x - j)$$

$$= \left(\frac{1+C}{2}\right) \varphi(2^k x) + \sum_{j=1}^{\infty} \varphi(2^k x - j),$$

and then, for $\varphi(x) = \chi_{[-\frac{1}{2}, \frac{1}{2})}(x)$, we find that

$$A_k(g)(-2^{-k-1}) = \frac{1+C}{2}.$$

Because $g(-2^{-k-1}) = 0$ and $0 < C < 1$, we obtain

$$A_k(g)(-2^{-k-1}) - g(-2^{-k-1}) = \frac{1+C}{2} > C,$$

which contradicts (6.16). Thus the inequalities (6.6) are sharp and we finish the proof of Theorem 6.1. ∎

We also need

**LEMMA 6.3′**

*Assume that $\varphi(x)$ is a bounded function with* supp $\varphi \subseteq [-a, a]$, $0 < a < +\infty$ *and satisfies the conditions (i)–(ii) of Theorem 6.1. Let $f \in C(\mathbf{R})$ be nondecreasing on $\mathbf{R}$, $\lim_{x \to +\infty} f(x) = 1$ and $\lim_{x \to -\infty} f(x) = 0$. Then for any fixed $k \in Z$ we get*

$$\lim_{x \to +\infty} A_k(f)(x) = 1$$

*and*

$$\lim_{x \to -\infty} A_k(f)(x) = 0,$$

*where $A_k(f)(x)$ is defined by (6.1).*

**PROOF** Let $k \in Z$ be fixed. From Theorem 6.1, $A_k(f)(x)$ is nondecreasing and hence $\lim_{x \to +\infty} A_k(f)(x)$ and $\lim_{x \to -\infty} A_k(f)(x)$ do exist.

We only need to prove that

$$\lim_{x \to +\infty} |A_k(f)(x) - f(x)| = 0 \tag{6.17}$$

and

$$\lim_{x \to -\infty} |A_k(f)(x) - f(x)| = 0. \tag{6.18}$$

Because $\lim_{x \to +\infty} f(x) = 1$ and $\lim_{x \to -\infty} f(x) = 0$, for any $\varepsilon > 0$, there is a positive number $N$ such that for any $x_1, x_2 > N$ we have

$$|f(x_1) - f(x_2)| < \varepsilon \tag{6.19}$$

and

$$|f(-x_1) - f(-x_2)| < \varepsilon. \tag{6.20}$$

Let $x_0 > N + 2^{-k+1}a$. From (6.11) and supp $\varphi \subseteq [-a, a]$, we see that

$$\langle f, \varphi_{kj} \rangle - 2^{-\frac{k}{2}} f(x_0) = 2^{-\frac{k}{2}} \int_{-\infty}^{\infty} [f(2^{-k}u) - f(x_0)] \varphi(u - j) \, du$$

$$= 2^{-\frac{k}{2}} \int_{-a+j}^{a+j} [f(2^{-k}u) - f(x_0)] \varphi(u - j) \, du. \tag{6.21}$$

If $2^k x_0 - a \leq j \leq 2^k x_0 + a$, for $u \in [-a+j, a+j]$ we get

$$2^{-k} u \geq 2^{-k}(-a+j) \geq 2^{-k}(2^k x_0 - 2a)$$

$$= x_0 - 2^{-k+1} a > N.$$

Then from (6.19) and (6.21), noticing (6.7) and (6.10), we find

$$|\langle f, \varphi_{kj} \rangle - 2^{-\frac{k}{2}} f(x_0)| \leq 2^{-\frac{k}{2}} \varepsilon \int_{-a+j}^{a+j} \varphi(u - j) \, du$$

$$= 2^{-\frac{k}{2}} \varepsilon \int_{-\infty}^{\infty} \varphi(u - j) \, du = 2^{-\frac{k}{2}} \varepsilon,$$

for $2^k x_0 - a \leq j \leq 2^k x_0 + a$.

It follows from this that for $x_0 > N + 2^{-k+1}a$

$$|A_k(f)(x_0) - f(x_0)| \leq 2^{\frac{k}{2}} \sum_{\substack{j \\ 2^k x_0 - j \in [-a,a]}} |\langle f, \varphi_{kj}\rangle - 2^{-\frac{k}{2}} f(x_0)|\varphi(2^k x_0 - j)$$

$$\leq \varepsilon \sum_{\substack{j \\ 2^k x_0 - j \in [-a,a]}} \varphi(2^k x_0 - j)$$

$$= \varepsilon \sum_{j=-\infty}^{\infty} \varphi(2^k x_0 - j) = \varepsilon,$$

which gives (6.17). From (6.20), by using an argument similar to that above we can prove (6.18). ∎

Next we give

### THEOREM 6.2

*Assume that $\varphi(x)$ is a bounded right-continuous function with $\operatorname{supp} \varphi(x) \subseteq [-a,a]$, $0 < a < +\infty$, and satisfies the conditions (i)–(ii) of Theorem 6.1. Let $F(x)$ be a continuous distribution function on $\mathbf{R}$. Then the linear wavelet type operators $A_k(F)$ defined by (6.1) are distribution functions and satisfy*

$$|A_k(F)(x) - F(x)| \leq \omega(F, 2^{-k+1}a), \quad x \in \mathbf{R}, \ k \in \mathbf{Z}.$$

*Furthermore the above inequalities are sharp.*

**PROOF** From the assumption on $\varphi$ and Lemma 6.2, it follows that $A_k(F)$ are right-continuous on $\mathbf{R}$. Since $F(x)$ is nondecreasing and $\lim_{x \to +\infty} F(x) = 1$ and $\lim_{x \to -\infty} F(x) = 0$, from Theorem 6.1 and Lemma 6.3' we know that $A_k(F)(x)$ are nondecreasing and $\lim_{x \to +\infty} A_k(F)(x) = 1$ and $\lim_{x \to -\infty} A_k(F)(x) = 0$. Hence $A_k(F)$ are distribution functions on $\mathbf{R}$. Theorem 6.1 provides the desired estimates of $|A_k(F)(x) - F(x)|$. The example of $g(x)$ used for the sharpness of (6.6) in Theorem 6.1 is a distribution function. Therefore, these estimates are still sharp for the distribution functions. ∎

## 6.2 Approximation by $B_k(f)$ Operators

Now we define another linear wavelet type operator for $f \in C(\mathbf{R})$ and $k \in Z$:

$$B_k(f)(x) = \sum_{j=-\infty}^{\infty} f(2^{-k}j)\varphi(2^k x - j). \qquad (6.22)$$

It is easy to verify that

$$B_k(f)(x) = B_0(f(2^{-k}\cdot))(2^k x). \qquad (6.23)$$

From (6.22), (6.23) and Lemma 6.1, we get

**LEMMA 6.4**
*If $\varphi(x)$ is a bounded right-continuous function with supp $\varphi(x) \subseteq [-a, a]$, $0 < a < +\infty$, then for any $f \in C(\mathbf{R})$ and $k \in Z$, the functions $B_k(f)(x)$ defined by (6.22) are right-continuous on $\mathbf{R}$.*

We present

**THEOREM 6.3**
*Assume that $\varphi(x)$ is a bounded function with supp $\varphi(x) \subseteq [-a, a]$, $0 < a < +\infty$ and satisfies the conditions (i)–(ii) of Theorem 6.1. Then, for $f \in C(\mathbf{R})$, if $f$ is a nondecreasing function, the linear wavelet type operators $B_k(f)$ defined by (6.22) are also nondecreasing on $\mathbf{R}$ and satisfy*

$$|B_k(f)(x) - f(x)| \leq \omega(f, 2^{-k}a), \quad x \in \mathbf{R}, \ k \in \mathbf{Z}. \qquad (6.24)$$

*Furthermore, the inequalities (6.24) are sharp.*

**PROOF** Put
$$C_j := f(j), \quad j \in Z.$$

If $f$ is nondecreasing, $\{C_j\}_{j=-\infty}^{\infty}$ is a nondecreasing sequence and then, by Lemma 6.3, $B_0(f)(x)$ is a nondecreasing function on $\mathbf{R}$. It follows from (6.23) that for any $k \in Z$, $B_k(f)(x)$ is a nondecreasing function too.

Now we establish (6.24). From the property (i) of $\varphi(x)$ and supp $\varphi \subseteq [-a, a]$, we get

$$B_k(f)(x) - f(x) = \sum_{j=-\infty}^{\infty} [f(2^{-k}j) - f(x)]\varphi(2^k x - j)$$

## 6. Monotone and Probabilistic Wavelet Type Approximation

$$= \sum_{\substack{j \\ 2^k x - j \in [-a,a]}} [f(2^{-k}j) - f(x)]\varphi(2^k x - j). \quad (6.25)$$

For $2^k x - j \in [-a, a]$, we have

$$|f(2^{-k}j) - f(x)| \leq \omega(f, |2^{-k}j - x|) \leq \omega(f, 2^{-k}a). \quad (6.26)$$

Noticing (6.7) and (6.8), we obtain from (6.25) and (6.26)

$$|B_k(f)(x) - f(x)| \leq \omega(f, 2^{-k}a) \sum_{\substack{j \\ 2^k x - j \in [-a,a]}} \varphi(2^k x - j)$$

$$= \omega(f, 2^{-k}a) \sum_{j=-\infty}^{\infty} \varphi(2^k x - j)$$

$$= \omega(f, 2^{-k}a).$$

For showing the sharpness of (6.24), we take $\varphi(x) = \chi_{[-\frac{1}{2},\frac{1}{2})}(x)$ and

$$q(x) := \begin{cases} 0, & x \leq -2^{-k-1}, \\ 1, & x \geq 0, \\ 2^{k+1}x + 1, & -2^{-k-1} < x < 0. \end{cases}$$

We have $q(x) \in C(\mathbf{R})$; nondecreasing and

$$B_k(q)(x) = \sum_{j=-\infty}^{\infty} q(2^{-k}j)\varphi(2^k x - j)$$

$$= \sum_{j=0}^{\infty} \varphi(2^k x - j).$$

Thus

$$B_k(q)(-2^{-k-1}) - q(-2^{-k-1}) = B_k(q)(-2^{-k-1})$$

$$= \sum_{j=0}^{\infty} \varphi\left(-\frac{1}{2} - j\right) = 1.$$

On the other hand, we see that
$$\omega(q, 2^{-k-1}) = 1.$$

Therefore
$$B_k(q)(-2^{-k-1}) - q(-2^{-k-1}) = \omega(q, 2^{-k-1}),$$
which implies the sharpness of (6.24). ∎

**LEMMA 6.5**
*Assume that $\varphi(x)$ is a bounded function with supp $\varphi(x) \subseteq [-a, a]$, $0 < a < +\infty$ and satisfies the conditions (i)–(ii) of Theorem 6.1. Let $f \in C(\mathbf{R})$ be nondecreasing on $\mathbf{R}$, $\lim_{x \to +\infty} f(x) = 1$ and $\lim_{x \to -\infty} f(x) = 0$. Then for any fixed $k \in Z$ we have*
$$\lim_{x \to +\infty} B_k(f)(x) = 1$$
*and*
$$\lim_{x \to -\infty} B_k(f)(x) = 0,$$
*where $B_k(f)(x)$ is defined by (6.22).*

**PROOF** Let $k \in Z$ be fixed. From Theorem 6.3, $B_k(f)(x)$ is nondecreasing on $\mathbf{R}$, so that $\lim_{x \to +\infty} B_k(f)(x)$ and $\lim_{x \to -\infty} B_k(f)(x)$ do exist.

Because we have $\lim_{x \to +\infty} f(x) = 1$ and $\lim_{x \to -\infty} f(x) = 0$, for any $\varepsilon > 0$ there is a positive number $N$ such that, for any $x_1, x_2 > N$ we have
$$|f(x_1) - f(x_2)| < \varepsilon \tag{6.27}$$
and
$$|f(-x_1) - f(-x_2)| < \varepsilon. \tag{6.28}$$

Let $x_0 > N + 2^{-k}a$. If $2^k x_0 - a \leq j \leq 2^k x_0 + a$, we have
$$2^{-k} j \geq x_0 - 2^{-k} a > N.$$

Thus, from (6.27) holds
$$|B_k(f)(x_0) - f(x_0)| = \left| \sum_{j=-\infty}^{\infty} [f(2^{-k}j) - f(x_0)] \varphi(2^k x_0 - j) \right|$$

## 6. Monotone and Probabilistic Wavelet Type Approximation

$$= \left| \sum_{\substack{j \\ 2^k x_0 - j \in [-a,a]}} [f(2^{-k}j) - f(x_0)] \varphi(2^k x_0 - j) \right|$$

$$\leq \varepsilon \sum_{\substack{j \\ 2^k x_0 - j \in [-a,a]}} \varphi(2^k x_0 - j) = \varepsilon, \quad \text{if } x_0 > N + 2^{-k}a.$$

Therefore
$$\lim_{x \to +\infty} |B_k(f)(x) - f(x)| = 0,$$

which gives
$$\lim_{x \to +\infty} B_k(f)(x) = 1.$$

Similarly, by using (6.28), we can prove
$$\lim_{x \to -\infty} B_k(f)(x) = 0.$$

∎

From Theorem 6.3, Lemma 6.4, and Lemma 6.5 we give the following.

**THEOREM 6.4**
*Assume that $\varphi(x)$ is a bounded right-continuous function with $\operatorname{supp} \varphi(x) \subseteq [-a,a]$, $0 < a < +\infty$ and satisfies the conditions (i)–(ii) of Theorem 6.1. Let $F(x)$ be a continuous distribution function on $\mathbf{R}$. Then the linear wavelet type operators $B_k(F)$ defined by (6.22) are distribution functions and satisfy*

$$|B_k(F)(x) - F(x)| \leq \omega(F, 2^{-k}a), \quad x \in \mathbf{R}, \ k \in Z.$$

*Furthermore the above estimates are sharp.*

---

## 6.3 More Results on $A_k(f)$ and $B_k(f)$ Operators

In this section we discuss the approximation degree by using $w_2(f,t)$, the 2nd modulus of smoothness of $f$ on $\mathbf{R}$. First we estimate $|B_k(f)(x) - f(x)|$.

## THEOREM 6.5

Assume that $\varphi(x)$ is a bounded right-continuous function with $\operatorname{supp} \varphi(x) \subseteq [-a, a]$, $0 < a < +\infty$, and satisfies the conditions (i)–(ii) of Theorem 6.1 and also has the following property:

$$(iv) \quad \sum_{j=-\infty}^{\infty} j\varphi(x-j) = x, \quad x \in \mathbf{R}.$$

Then, for $f \in C(\mathbf{R})$, if $f$ is a nondecreasing function, the linear wavelet type operators $B_k(f)$ defined by (6.22) are also nondecreasing on $\mathbf{R}$ and satisfy

$$|B_k(f)(x) - f(x)| \leq C\omega_2(f, 2^{-k+1}a), \quad x \in \mathbf{R},\ k \in \mathbf{Z}, \qquad (6.29)$$

where $C$ is an absolute constant.

**PROOF** By Theorem 6.3, we only need to prove (6.29).

For any fixed $x_0$, let $p_{x_0}(x)$ be the best approximation of $f$ on $I := [x_0 - 2^{-k}a, x_0 + 2^{-k}a]$ from $\mathbf{P}_1$ the set of all algebraic polynomials of degree $\leq 1$. By Whitney theorem [154], we get that

$$|p_{x_0}(x) - f(x)| \leq C\omega_2(f, 2^{-k+1}a, I)$$

$$\leq C\omega_2(f, 2^{-k+1}a), \quad x \in I, \qquad (6.30)$$

where $C$ is an absolute constant, $\omega_2(f, t, I)$ is the 2nd modulus of smoothness of $f$ on $I$, and $\omega_2(f, t)$ is the 2nd modulus of smoothness of $f$ on $\mathbf{R}$.

From the conditions (i) and (iv), we have

$$p_{x_0}(x) = B_k(p_{x_0})(x), \quad x \in \mathbf{R}. \qquad (6.31)$$

Thus, from (6.31) and (6.30), we observe that

$$|B_k(f)(x_0) - f(x_0)| = |B_k(f)(x_0) - B_k(p_{x_0})(x_0) + p_{x_0}(x_0) - f(x_0)|$$

$$\leq |B_k(f - p_{x_0})(x_0)| + |p_{x_0}(x_0) - f(x_0)|$$

$$\leq |B_k(f - p_{x_0})(x_0)| + C\omega_2(f, 2^{-k+1}a). \qquad (6.32)$$

6. *Monotone and Probabilistic Wavelet Type Approximation* 161

Because supp $\varphi(x) \subseteq [-a, a]$ and $2^{-k}j \in I$ for $2^k x_0 - a \leq j \leq 2^k x_0 + a$, by (6.30), we find

$$|B_k(f - p_{x_0})(x_0)| = \left| \sum_{\substack{j \\ 2^k x_0 - j \in [-a,a]}} [f(2^{-k}j) - p_{x_0}(2^{-k}j)] \varphi(2^k x_0 - j) \right|$$

$$\leq \sum_{\substack{j \\ 2^k x_0 - j \in [-a,a]}} |f(2^{-k}j) - p_{x_0}(2^{-k}j)| \varphi(2^k x_0 - j)$$

$$\leq C\omega_2(f, 2^{-k+1}a) \sum_{\substack{j \\ 2^k x_0 - j \in [-a,a]}} \varphi(2^k x_0 - j)$$

$$= C\omega_2(f, 2^{-k+1}a).$$

Combining this with (6.32) produces

$$|B_k(f)(x_0) - f(x_0)| \leq 2C\omega_2(f, 2^{-k+1}a).$$

This finishes the proof. ∎

From Theorem 6.5, Lemma 6.4, and Lemma 6.5 we get

**THEOREM 6.6**
*Assume that $\varphi(x)$ satisfies all the assumptions of Theorem 6.5. Let $F(x)$ be a continuous distribution function on $\mathbf{R}$. Then the linear wavelet type operators $B_k(F)$ defined by (6.22) are distribution functions and satisfy*

$$|B_k(F)(x) - F(x)| \leq C\omega_2(F, 2^{-k+1}a), \quad x \in \mathbf{R}, \ k \in \mathbf{Z},$$

*where $C$ is an absolute constant.*

It is easy to confirm that

$$\varphi(x) = \begin{cases} x + 1, & -1 \leq x \leq 0, \\ 1 - x, & 0 < x \leq 1, \\ 0, & \text{otherwise}. \end{cases} \quad (6.33)$$

fulfills all the assumptions in Theorem 6.5.

For $|A_k(f)(x) - f(x)|$, we obtain

**THEOREM 6.7**
*Assume that $\varphi(x)$ satisfies all the assumptions of Theorem 6.5 and has the properties (iii) of Theorem 6.1 and*

$$(v) \int_{-\infty}^{\infty} u\varphi(u)\,du = 0.$$

*Then, for $f \in C(\mathbf{R})$, if $f$ is a nondecreasing function, the linear wavelet type operators $A_k(f)$ defined by (6.1) are nondecreasing on $\mathbf{R}$ and satisfy*

$$|A_k(f)(x) - f(x)| \leq C\omega_2(f, 2^{-k+1}a), \quad x \in \mathbf{R},\ k \in Z, \tag{6.34}$$

*where $C$ is an absolute constant.*

**PROOF** First we notice that such a $\varphi(x)$ has the property

$$(v) \int_{-\infty}^{\infty} x\varphi(x)\,dx = 0.$$

Really, from properties (i) and (iv) of $\varphi$, we observe that

$$\int_{-\infty}^{\infty} x\varphi(x)\,dx = \sum_{j=-\infty}^{\infty} \int_{j}^{j+1} x\varphi(x)\,dx$$

$$= \sum_{j=-\infty}^{\infty} \int_{0}^{1} (j+u)\varphi(j+u)\,du$$

$$= \int_{0}^{1} \sum_{j=-\infty}^{\infty} j\varphi(j+u)\,du + \int_{0}^{1} u \sum_{j=-\infty}^{\infty} \varphi(j+u)\,du$$

$$= \int_{0}^{1} -u\,du + \int_{0}^{1} u\,du = 0.$$

From the properties (i), (iii), (iv), and (v) of $\varphi$, it is easy to verify for any $p \in \mathbf{P}_1$ that

$$p(x) = A_k(p)(x).$$

Then, by using a method similar to the proof of Theorem 6.5, we can get (6.34). ∎

Finally we give

**THEOREM 6.8**
*Assume that $\varphi(x)$ satisfies all the assumptions of Theorem 6.5. Let $F(x)$ be a continuous distribution function on $\mathbf{R}$. Then, the linear wavelet type operators $A_k(F)$ defined by (6.1) are distribution functions and satisfy*

$$|A_k(F)(x) - F(x)| \leq C\omega_2(F, 2^{-k+1}a), \quad x \in \mathbf{R}, \ k \in Z,$$

*where $C$ is an absolute constant.*

# Chapter 7

## Quantitative One-Dimensional High Order Wavelet Type Approximation

High order differentiable functions of one variable are approximated by wavelet type operators. The high order of this approximation is estimated by giving some Jackson type inequalities. Sharpness of some of these inequalities over convex functions is presented nontrivially. This treatment is based on [23].

### 7.1 Estimates and Sharpness

We present the first main result

**THEOREM 7.1**
Let $f \in C^N(\mathbf{R})$, $N \geq 1$, $x \in \mathbf{R}$ and $k \in \mathbf{Z}$. Let $\varphi$ be a bounded function of compact support $\subseteq [-a, a]$, $a > 0$ such that $\sum_{j=-\infty}^{\infty} \varphi(x-j) = 1$ all $x \in \mathbf{R}$. Suppose $\varphi \geq 0$. Put

$$(B_k(f))(x) := \sum_{j=-\infty}^{\infty} f\left(\frac{j}{2^k}\right) \varphi(2^k x - j).$$

Then

$$|(B_k(f))(x) - f(x)| \leq \sum_{i=1}^{N} \frac{|f^{(i)}(x)|}{i!} \frac{a^i}{2^{ki}} + \frac{a^N}{2^{kN} N!} \omega_1\left(f^{(N)}, \frac{a}{2^k}\right), \quad (7.1)$$

which is attained by constant function.

**REMARK 7.1** i) Given that $f^{(N)}$ is continuous and bounded or uniformly continuous on $\mathbf{R}$, we have that $\omega_1\left(f^{(N)}, \frac{a}{2^k}\right) < \infty$ and $B_k f \to f$, pointwise over $\mathbf{R}$, as $k \to \infty$.

ii) Given that $f \in C_b^N(\mathbf{R})$ (i.e., $f, f', \ldots, f^{(N)}$ are continuous and bounded) we get

$$\|B_k f - f\|_\infty \leq \sum_{i=1}^{N} \frac{\|f^{(i)}\|_\infty}{i!} \frac{a^i}{2^{ki}} + \frac{a^N}{2^{kN} N!} \omega_1\left(f^{(N)}, \frac{a}{2^k}\right). \qquad (7.2)$$

That is, $B_k f \to f$, uniformly over $\mathbf{R}$, as $k \to \infty$. ∎

**PROOF** (of Theorem 7.1). Because $f \in C^N(\mathbf{R})$, $N \geq 1$ we have

$$f\left(\frac{j}{2^k}\right) = f(x) + \sum_{i=1}^{N} \frac{f^{(i)}(x)}{i!} \left(\frac{j}{2^k} - x\right)^i$$

$$+ \int_x^{j/2^k} (f^{(N)}(t) - f^{(N)}(x)) \frac{(\frac{j}{2^k} - t)^{N-1}}{(N-1)!} dt.$$

Therefore

$$f\left(\frac{j}{2^k}\right) \varphi(2^k x - j) = f(x) \varphi(2^k x - j) + \sum_{i=1}^{N} \frac{f^{(i)}(x)}{i!} \left(\frac{j}{2^k} - x\right)^i \varphi(2^k x - j)$$

$$+ \varphi(2^k x - j) \int_x^{j/2^k} (f^{(N)}(t) - f^{(N)}(x)) \frac{(\frac{j}{2^k} - t)^{N-1}}{(N-1)!} dt.$$

and

$$\sum_{j=-\infty}^{\infty} f\left(\frac{j}{2^k}\right) \varphi(2^k x - j) = f(x) \sum_{j=-\infty}^{\infty} \varphi(2^k x - j)$$

$$+ \sum_{j=-\infty}^{\infty} \sum_{i=1}^{N} \frac{f^{(i)}(x)}{i!} \left(\frac{j}{2^k} - x\right)^i \varphi(2^k x - j)$$

## 7. High Order Wavelet Type Approximation

$$+ \sum_{j=-\infty}^{\infty} \varphi(2^k x - j) \int_x^{j/2^k} (f^{(N)}(t) - f^{(N)}(x)) \frac{(\frac{j}{2^k} - t)^{N-1}}{(N-1)!} dt.$$

Hence

$$\mathcal{E}_k(x) := (B_k(f))(x) - f(x)$$

$$= \sum_{i=1}^{N} \frac{f^{(i)}(x)}{i!} \left( \sum_{j=-\infty}^{\infty} \left( \frac{j}{2^k} - x \right)^i \varphi(2^k x - j) \right) + \mathcal{R},$$

where

$$\mathcal{R} := \sum_{j=-\infty}^{\infty} \varphi(2^k x - j) \int_x^{j/2^k} (f^{(N)}(t) - f^{(N)}(x)) \frac{(\frac{j}{2^k} - t)^{N-1}}{(N-1)!} dt.$$

Thus

$$|\mathcal{E}_k(x)| \leq \sum_{i=1}^{N} \frac{|f^{(i)}(x)|}{i!} \left| \sum_{j=-\infty}^{\infty} \left( \frac{j}{2^k} - x \right)^i \varphi(2^k x - j) \right| + |\mathcal{R}|.$$

Because $\varphi$ is of compact support $\subseteq [-a, a]$ in order to have nonzero terms in the last inequality we need

$$-a \leq 2^k x - j \leq a \quad \text{i.e.,} \quad \left| x - \frac{j}{2^k} \right| \leq \frac{a}{2^k}.$$

Moreover

$$2^k x - a \leq j \leq 2^k x + a,$$

reducing

$$\sum_{j=-\infty}^{\infty} \cdots = \sum_{j=\lceil 2^k x - a \rceil}^{[2^k x + a]},$$

where $[\cdot]$, $\lceil \cdot \rceil$ denote the integral part and ceiling of the number, respectively. Therefore

$$|\mathcal{E}_k(x)| \leq \sum_{i=1}^{N} \frac{|f^{(i)}(x)|}{i!} \frac{a^i}{2^{ki}} + |\mathcal{R}|. \tag{7.3}$$

Set
$$\Gamma_j(x) := \left| \int_x^{j/2^k} (f^{(N)}(t) - f^{(N)}(x)) \frac{(\frac{j}{2^k} - t)^{N-1}}{(N-1)!} dt \right|.$$

Here we find
$$|\mathcal{R}| \leq \sum_{j=-\infty}^{\infty} \varphi(2^k x - j) \Gamma_j(x).$$

Next we estimate $\Gamma_j(x)$.

i) If $x \leq j/2^k$, then

$$\Gamma_j(x) \leq \int_x^{j/2^k} |f^{(N)}(t) - f^{(N)}(x)| \frac{(\frac{j}{2^k} - t)^{N-1}}{(N-1)!} dt$$

$$\leq \int_x^{j/2^k} \omega_1(f^{(N)}, |t-x|) \frac{(\frac{j}{2^k} - t)^{N-1}}{(N-1)!} dt$$

$$\leq \omega_1\left(f^{(N)}, \left|x - \frac{j}{2^k}\right|\right) \int_x^{j/2^k} \frac{(\frac{j}{2^k} - t)^{N-1}}{(N-1)!} dt$$

$$\leq \omega_1\left(f^{(N)}, \frac{a}{2^k}\right) \frac{(\frac{j}{2^k} - x)^N}{N!} \leq \omega_1\left(f^{(N)}, \frac{a}{2^k}\right) \frac{a^N}{N! 2^{kN}}.$$

That is, if $x \leq j/2^k$ we find that
$$\Gamma_j(x) \leq \omega_1\left(f^{(N)}, \frac{a}{2^k}\right) \frac{a^N}{2^{kN} N!}.$$

ii) If $x \geq j/2^k$, then

$$\Gamma_j(x) = \left| \int_{j/2^k}^x (f^{(N)}(t) - f^{(N)}(x)) \frac{(\frac{j}{2^k} - t)^{N-1}}{(N-1)!} dt \right|$$

$$\leq \int_{j/2^k}^x |f^{(N)}(t) - f^{(N)}(x)| \frac{(t - \frac{j}{2^k})^{N-1}}{(N-1)!} dt$$

$$\leq \int_{j/2^k}^x \omega_1(f^{(N)}, |t-x|) \frac{(t - \frac{j}{2^k})^{N-1}}{(N-1)!} dt$$

# 7. High Order Wavelet Type Approximation

$$\leq \omega_1\left(f^{(N)}, \left|x - \frac{j}{2^k}\right|\right) \int_{j/2^k}^{x} \frac{(t - \frac{j}{2^k})^{N-1}}{(N-1)!} dt$$

$$\leq \omega_1\left(f^{(N)}, \frac{a}{2^k}\right) \frac{(x - \frac{j}{2^k})^N}{N!} \leq \omega_1\left(f^{(N)}, \frac{a}{2^k}\right) \frac{a^N}{2^{kN} N!}.$$

So that, if $x \geq j/2^k$, then

$$\Gamma_j(x) \leq \omega_1\left(f^{(N)}, \frac{a}{2^k}\right) \frac{a^N}{2^{kN} N!}.$$

We proved that
$$\Gamma_j(x) \leq \Gamma^*, \quad \text{any } x \in \mathbf{R},$$

where
$$\Gamma^* := \omega_1\left(f^{(N)}, \frac{a}{2^k}\right) \frac{a^N}{2^{kN} N!}.$$

Also
$$|\mathcal{R}| \leq \left(\sum_{j=-\infty}^{\infty} \varphi(2^k x - j)\right) \Gamma^* = \Gamma^*.$$

In other words,
$$|\mathcal{R}| \leq \frac{a^N}{2^{kN} N!} \omega_1\left(f^{(N)}, \frac{a}{2^k}\right). \tag{7.4}$$

That is, from (7.3) and (7.4) we obtain

$$|\mathcal{E}_k(x)| \leq \sum_{i=1}^{N} \frac{|f^{(i)}(x)|}{i!} \frac{a^i}{2^{ki}} + \frac{a^N}{2^{kN} N!} \omega_1\left(f^{(N)}, \frac{a}{2^k}\right),$$

establishing (7.1). ∎

The second main result comes next.

### THEOREM 7.2
*Same assumptions as in Theorem 7.1. Additionally, suppose that $\varphi$ is Lebesgue measurable (then $\int_{-\infty}^{+\infty} \varphi(x) dx = 1$). Define*

$$\varphi_{kj}(x) := 2^{k/2} \varphi(2^k x - j) \quad \text{all } k, j \in \mathbf{Z},$$

$$\langle f, \varphi_{kj}\rangle := \int_{-\infty}^{\infty} f(t) \varphi_{kj}(t) dt,$$

and

$$(A_k(f))(x) := \sum_{j=-\infty}^{\infty} \langle f, \varphi_{kj}\rangle \varphi_{kj}(x)$$

$$= \sum_{j=-\infty}^{\infty} \left(\int_{-\infty}^{\infty} f\left(\frac{u}{2^k}\right)\varphi(u-j)du\right)\varphi(2^k x - j).$$

Then it holds

$$|(A_k(f))(x) - f(x)| \le \sum_{i=1}^{N} \frac{|f^{(i)}(x)|}{i!} \frac{a^i}{2^{i(k-1)}} + \frac{a^N}{N! 2^{N(k-1)}} \omega_1\left(f^{(N)}, \frac{a}{2^{k-1}}\right),$$
(7.5)

which is attained by constants.

**REMARK 7.2** i) Given that $f^{(N)}$ is continuous and bounded or uniformly continuous on $\mathbf{R}$, we find that $\omega_1(f^{(N)}, \frac{a}{2^{k-1}}) < \infty$ and $A_k f \to f$, pointwise over $\mathbf{R}$, as $k \to \infty$.

ii) Given that $f \in C_b^N(\mathbf{R})$, we have

$$\|A_k f - f\|_\infty \le \sum_{i=1}^{N} \frac{\|f^{(i)}\|_\infty}{i!} \frac{a^i}{2^{i(k-1)}} + \frac{a^N}{N! 2^{N(k-1)}} \omega_1\left(f^{(N)}, \frac{a}{2^{k-1}}\right). \quad (7.6)$$

That is, $A_k f \to f$, uniformly over $\mathbf{R}$, as $k \to \infty$. ∎

**COROLLARY 7.1**

Additionally suppose that $\varphi$ is right continuous. Here $f \in C^1(\mathbf{R})$ is a probability distribution function such that $\lambda := f'$ is the density function. Then, $k \in \mathbf{Z}$,

$$|(\mathcal{B}_k f)(x) - f(x)| \le \frac{a}{2^k}\left(\lambda(x) + \omega_1\left(\lambda, \frac{a}{2^k}\right)\right) \quad (7.7)$$

and

$$|(A_k f)(x) - f(x)| \le \frac{a}{2^{k-1}}\left(\lambda(x) + \omega_1\left(\lambda, \frac{a}{2^{k-1}}\right)\right).$$

From [42] $A_k(f)$, $B_k(f)$ are distribution functions in this case, given that $\varphi$ is bell-shaped.

## 7. High Order Wavelet Type Approximation

**PROOF** (of Theorem 7.2). By $\sum_{j=-\infty}^{\infty} \varphi(2^k x - j) = 1$ we have that

$$\int_{-\infty}^{\infty} \varphi(u-j)du = 1.$$

Notice that

$$(A_k(f))(x) - f(x) = 2^{k/2} \sum_{j=-\infty}^{\infty} [\langle f, \varphi_{k_j} \rangle - 2^{-k/2} f(x)] \varphi(2^k x - j). \quad (7.8)$$

Also we see that

$$\langle f, \varphi_{k_j} \rangle - 2^{-k/2} f(x) = 2^{k/2} \int_{-\infty}^{\infty} f(t) \varphi(2^k t - j) dt - 2^{-k/2} f(x)$$

$$= 2^{-k/2} \int_{-\infty}^{\infty} f\left(\frac{u}{2^k}\right) \varphi(u-j) du - 2^{-k/2} f(x)$$

$$= 2^{-k/2} \int_{-\infty}^{\infty} \left[ f\left(\frac{u}{2^k}\right) - f(x) \right] \varphi(u-j) du.$$

By $\operatorname{supp} \varphi \subseteq [-a, a]$ we have that $2^{-k}(-a+j) \le x \le 2^{-k}(a+j)$. Thus

$$|\langle f, \varphi_{k_j} \rangle - 2^{-k/2} f(x)| = |2^{-k/2} \Lambda|, \quad (7.9)$$

where

$$\Lambda := \int_{-a+j}^{a+j} \left[ f\left(\frac{u}{2^k}\right) - f(x) \right] \varphi(u-j) du.$$

Next we see that

$$f\left(\frac{u}{2^k}\right) - f(x) = \sum_{i=1}^{N} \frac{f^{(i)}(x)}{i!} \left(\frac{u}{2^k} - x\right)^i$$

$$+ \int_{x}^{u/2^k} (f^{(N)}(t) - f^{(N)}(x)) \frac{(\frac{u}{2^k} - t)^{N-1}}{(N-1)!} dt.$$

Hence

$$\left( f\left(\frac{u}{2^k}\right) - f(x) \right) \varphi(u-j) = \left( \sum_{i=1}^{N} \frac{f^{(i)}(x)}{i!} \left(\frac{u}{2^k} - x\right)^i \right) \varphi(u-j)$$

$$+ \varphi(u-j) \int_x^{u/2^k} (f^{(N)}(t) - f^{(N)}(x)) \frac{(\frac{u}{2^k} - t)^{N-1}}{(N-1)!} dt.$$

Now we notice that $|\frac{u}{2^k} - x| \leq \frac{a}{2^{k-1}}$ and

$$\Lambda = \sum_{i=1}^{N} \frac{f^{(i)}(x)}{i!} \int_{-a+j}^{a+j} \left(\frac{u}{2^k} - x\right)^i \varphi(u-j) du + \mathcal{R}^*, \tag{7.10}$$

where

$$\mathcal{R}^* := \left\{ \int_{-a+j}^{a+j} \varphi(u-j) \left( \int_x^{u/2^k} (f^{(N)}(t) - f^{(N)}(x)) \frac{(\frac{u}{2^k} - t)^{N-1}}{(N-1)!} dt \right) du \right\}. \tag{7.11}$$

Then

$$|\Lambda| \leq \sum_{i=1}^{N} \frac{|f^{(i)}(x)|}{i!} \frac{a^i}{2^{i(k-1)}} + |\mathcal{R}^*|. \tag{7.12}$$

Moreover

$$|\mathcal{R}^*| \leq \int_{-a+j}^{a+j} \varphi(u-j) \rho(u) du, \tag{7.13}$$

where

$$\rho(u) := \left| \int_x^{u/2^k} (f^{(N)}(t) - f^{(N)}(x)) \frac{(\frac{u}{2^k} - t)^{N-1}}{(N-1)!} dt \right|. \tag{7.14}$$

In the following we estimate $\rho(u)$: i) If $x \leq \frac{u}{2^k}$, then

$$\rho(u) \leq \int_x^{u/2^k} |f^{(N)}(t) - f^{(N)}(x)| \cdot \frac{(\frac{u}{2^k} - t)^{N-1}}{(N-1)!} dt$$

$$\leq \int_x^{u/2^k} \omega_1(f^{(N)}, |t-x|) \frac{(\frac{u}{2^k} - t)^{N-1}}{(N-1)!} dt$$

$$\leq \omega_1\left(f^{(N)}, \left|x - \frac{u}{2^k}\right|\right) \int_x^{u/2^k} \frac{(\frac{u}{2^k} - t)^{N-1}}{(N-1)!} dt$$

$$\leq \omega_1\left(f^{(N)}, \frac{a}{2^{k-1}}\right) \frac{(\frac{u}{2^k} - x)^N}{N!}$$

$$\leq \omega_1\left(f^{(N)}, \frac{a}{2^{k-1}}\right) \frac{a^N}{2^{N(k-1)} N!}.$$

## 7. High Order Wavelet Type Approximation

So if $x \le \frac{u}{2^k}$, we obtain that

$$\rho(u) \le \omega_1\left(f^{(N)}, \frac{a}{2^{k-1}}\right) \frac{a^N}{N!2^{N(k-1)}}.$$

ii) If $x \ge \frac{u}{2^k}$ then

$$\rho(u) \le \int_{u/2^k}^{x} |f^{(N)}(t) - f^{(N)}(x)| \frac{(t - \frac{u}{2^k})^{N-1}}{(N-1)!} dt$$

$$\le \omega_1\left(f^{(N)}, \frac{a}{2^{k-1}}\right) \frac{(x - \frac{u}{2^k})^N}{N!}$$

$$\le \omega_1\left(f^{(N)}, \frac{a}{2^{k-1}}\right) \frac{a^N}{N!2^{N(k-1)}}.$$

So in both cases we get that

$$\rho(u) \le \omega_1\left(f^{(N)}, \frac{a}{2^{k-1}}\right) \frac{a^N}{N!2^{N(k-1)}} =: T. \tag{7.15}$$

Consequently we have (cf. (7.13), (7.15))

$$|\mathcal{R}^*| \le \left(\int_{-a+j}^{a+j} \varphi(u - j) du\right) T = T.$$

That is

$$|\mathcal{R}^*| \le \omega_1\left(f^{(N)}, \frac{a}{2^{k-1}}\right) \frac{a^N}{N!2^{N(k-1)}}. \tag{7.16}$$

Therefore (cf. (7.12), (7.16))

$$|\Lambda| \le \sum_{i=1}^{N} \frac{|f^{(i)}(x)|}{i!} \frac{a^i}{2^{i(k-1)}} + \frac{a^N}{N!2^{N(k-1)}}\omega_1\left(f^{(N)}, \frac{a}{2^{k-1}}\right) =: \theta, \tag{7.17}$$

and

$$2^{-k/2}|\Lambda| \le 2^{-k/2}\theta.$$

So that (cf. (7.9))

$$|\langle f, \varphi_{kj}\rangle - 2^{-k/2} f(x_j)| \le 2^{-k/2}\theta. \tag{7.18}$$

Finally, from (7.8), (7.17), and (7.18) we obtain

$$|(A_k(f))(x) - f(x)| \leq 2^{k/2} \sum_{j=-\infty}^{\infty} |\langle f, \varphi_{kj}\rangle - 2^{-k/2}f(x)||\varphi(2^k x - j)$$

$$\leq 2^{k/2} \sum_{j=-\infty}^{\infty} 2^{-k/2}\theta\varphi(2^k x - j)$$

$$= \theta 2^{k/2} 2^{-k/2} \left(\sum_{j=-\infty}^{\infty} \varphi(2^k x - j)\right) = \theta,$$

establishing (7.5). ∎

Another result follows.

**THEOREM 7.3**
*Same assumptions as in Theorem 7.1. Define*

$$C_k(f)(x) := \sum_{j=-\infty}^{\infty} \gamma_{kj}(f)\varphi(2^k x - j)$$

$$:= \sum_{j=-\infty}^{\infty} \left(2^k \int_{2^{-k}j}^{2^{-k}(j+1)} f(t)dt\right) \varphi(2^k x - j).$$

*That is,*

$$\gamma_{kj}(f) := 2^k \int_{2^{-k}j}^{2^{-k}(j+1)} f(t)dt = 2^k \int_{0}^{2^{-k}} f\left(t + \frac{j}{2^k}\right) dt.$$

*Then* $(x \in \mathbf{R}, k \in \mathbf{Z})$

$$|(C_k(f))(x) - f(x)| \leq \sum_{i=1}^{N} \frac{|f^{(i)}(x)|(a+1)^i}{i!\, 2^{ki}} + \frac{(a+1)^N}{2^{kN} N!} \omega_1\left(f^{(N)}, \frac{a+1}{2^k}\right), \tag{7.19}$$

*which is attained by constants.*

## 7. High Order Wavelet Type Approximation

**REMARK 7.3** i) Given that $f^{(N)}$ is continuous and bounded or uniformly continuous we have that $\omega_1(f^{(N)}, \frac{a+1}{2^k}) < \infty$ and thus $(C_k f)(x) \to f(x)$, pointwise over $\mathbf{R}$, as $k \to \infty$.

ii) Given also that $f \in C_b^N(\mathbf{R})$, we have that

$$\|C_k f - f\|_\infty \leq \sum_{i=1}^{N} \frac{\|f^{(i)}\|_\infty}{i!} \frac{(a+1)^i}{2^{ki}} + \frac{(a+1)^N}{2^{kN} N!} \omega_1\left(f^{(N)}, \frac{a+1}{2^k}\right), \quad (7.20)$$

and hence $C_k f \to f$, uniformly on $\mathbf{R}$, as $k \to \infty$.

iii) ($N=1$ case). We obtain that

$$|(C_k f)(x) - f(x)| \leq \left(\frac{a+1}{2^k}\right)\left(|f'(x)| + \omega_1\left(f', \frac{a+1}{2^k}\right)\right). \quad (7.21)$$

**PROOF** (of Theorem 7.3). We see that

$$f\left(t + \frac{j}{2^k}\right) = f(x) + \sum_{i=1}^{N} \frac{f^{(i)}(x)}{i!}\left(t + \frac{j}{2^k} - x\right)^i$$

$$+ \int_x^{t+\frac{j}{2^k}} (f^{(N)}(\tau) - f^{(N)}(x)) \frac{(t + \frac{j}{2^k} - \tau)^{N-1}}{(N-1)!} d\tau.$$

Then

$$\gamma_{kj}(f) = 2^k \int_0^{2^{-k}} f\left(t + \frac{j}{2^k}\right) dt$$

$$= f(x) + \sum_{i=1}^{N} \frac{f^{(i)}(x)}{i!} 2^k \left(\int_0^{2^{-k}} \left(t + \frac{j}{2^k} - x\right)^i dt\right)$$

$$+ 2^k \int_0^{2^{-k}} \left(\int_x^{t+\frac{j}{2^k}} (f^{(N)}(\tau) - f^{(N)}(x)) \frac{(t + \frac{j}{2^k} - \tau)^{N-1}}{(N-1)!} d\tau\right) dt.$$

Consequently

$$(C_k(f))(x) - f(x) = \sum_{i=1}^{N} \frac{f^{(i)}(x)}{i!} \sum_{j=-\infty}^{\infty} \varphi(2^k x - j) 2^k \int_0^{2^{-k}} \left(t + \frac{j}{2^k} - x\right)^i dt + \tilde{\mathcal{R}}, \quad (7.22)$$

where

$$\tilde{\mathcal{R}} := \sum_{j=-\infty}^{\infty} \varphi(2^k x - j) 2^k \int_0^{2^{-k}} \chi(t) dt, \qquad (7.23)$$

and

$$\chi(t) := \int_x^{t+\frac{j}{2^k}} (f^{(N)}(\tau) - f^{(N)}(x)) \frac{(t + \frac{j}{2^k} - \tau)^{N-1}}{(N-1)!} d\tau. \qquad (7.24)$$

Here $0 \leq t \leq 2^{-k}$, $|x - \frac{j}{2^k}| \leq \frac{a}{2^k}$. Next we estimate $\chi(t)$:

i) If $x \leq t + \frac{j}{2^k}$, then

$$|\chi(t)| \leq \int_x^{t+\frac{j}{2^k}} \omega_1(f^{(N)}, |\tau - x|) \frac{(t + \frac{j}{2^k} - \tau)^{N-1}}{(N-1)!} d\tau \leq \text{(as earlier)}$$

$$\leq \omega_1 \left( f^{(N)}, \frac{a+1}{2^k} \right) \frac{(a+1)^N}{2^{kN} N!}.$$

ii) If $x \geq t + \frac{j}{2^k}$, then

$$|\chi(t)| \leq \int_{t+\frac{j}{2^k}}^x \omega_1(f^{(N)}, |\tau - x|) \frac{(\tau - (t + \frac{j}{2^k}))^{N-1}}{(N-1)!} d\tau \leq \text{(as earlier)}$$

$$\leq \omega_1 \left( f^{(N)}, \frac{a+1}{2^k} \right) \frac{(a+1)^N}{2^{kN} N!}.$$

So it is always true that

$$|\chi(t)| \leq \omega_1 \left( f^{(N)}, \frac{a+1}{2^k} \right) \frac{(a+1)^N}{2^{kN} N!} =: \lambda. \qquad (7.25)$$

Hence (cf. (7.25))

$$|\tilde{\mathcal{R}}| \leq \sum_{j=-\infty}^{\infty} \varphi(2^k x - j) \left( 2^k \int_0^{2^{-k}} |\chi(t)| dt \right)$$

$$\leq \sum_{j=-\infty}^{\infty} \varphi(2^k x - j) 2^k \int_0^{2^{-k}} \lambda \, dt$$

# 7. High Order Wavelet Type Approximation

$$= \lambda \left( \sum_{j=-\infty}^{\infty} \varphi(2^k x - j) \right) = \lambda 1.$$

That is

$$|\tilde{\mathcal{R}}| \leq \omega_1 \left( f^{(N)}, \frac{a+1}{2^k} \right) \frac{(a+1)^N}{2^{kN} N!}. \qquad (7.26)$$

Moreover, notice that

$$\left| 2^k \int_0^{2^{-k}} \left( t + \frac{j}{2^k} - x \right)^i dt \right| \leq 2^k \int_0^{2^{-k}} \left( |t| + \left| x - \frac{j}{2^k} \right| \right)^i dt$$

$$\leq 2^k \int_0^{2^{-k}} \left( \frac{1}{2^k} + \frac{a}{2^k} \right)^i dt = \frac{(a+1)^i}{2^{ki}}. \qquad (7.27)$$

Thus (cf. (7.27))

$$\left| \sum_{j=-\infty}^{\infty} \varphi(2^k x - j) 2^k \int_0^{2^{-k}} \left( t + \frac{j}{2^k} - x \right)^i dt \right|$$

$$\leq \left( \sum_{j=-\infty}^{\infty} \varphi(2^k x - j) \right) \left( \frac{(a+1)^i}{2^{ki}} \right) = \frac{(a+1)^i}{2^{ki}}. \qquad (7.28)$$

Eventually from (7.22), (7.26), and (7.28) we obtain that

$$|(C_k f)(x) - f(x)| \leq \sum_{i=1}^N \frac{|f^{(i)}(x)|}{i!} \frac{(a+1)^i}{2^{ki}} + \omega_1 \left( f^{(N)}, \frac{a+1}{2^k} \right) \frac{(a+1)^N}{2^{kN} N!},$$

proving (7.19). ∎

The quadrature wavelet type corresponding operator is encountered next.

**THEOREM 7.4**
*Same assumptions as in Theorem 7.1. Define* $(k, j \in \mathbf{Z}, x \in \mathbf{R})$

$$(D_k f)(x) := \sum_{j=-\infty}^{\infty} \delta_{kj}(f) \varphi(2^k x - j),$$

where
$$\delta_{kj}(f) := \sum_{r=0}^{n} w_r f\left(\frac{j}{2^k} + \frac{r}{2^k n}\right),$$

$n \in \mathbf{N}$, $w_r \geq 0$, $\sum_{r=0}^{n} w_r = 1$. Then

$$|(D_k f)(x) - f(x)| \leq \sum_{i=1}^{N} \frac{|f^{(i)}(x)|}{i!} \frac{(a+1)^i}{2^{ki}} + \frac{(a+1)^N}{2^{kN} N!} \omega_1\left(f^{(N)}, \frac{(a+1)}{2^k}\right), \tag{7.29}$$

which is attained by constants.

**REMARK 7.4** i) Given that $f^{(N)}$ is continuous and bounded or uniformly continuous we have that $\omega_1(f^{(N)}, \frac{a+1}{2^k}) < \infty$ and thus $D_k f \to f$, pointwise over $\mathbf{R}$, as $k \to \infty$.

ii) Given also that $f \in C_b^N(\mathbf{R})$ we get that

$$\|D_k f - f\|_\infty \leq \sum_{i=1}^{N} \frac{\|f^{(i)}\|_\infty}{i!} \frac{(a+1)^i}{2^{ki}} + \frac{(a+1)^N}{2^{kN} N!} \omega_1\left(f^{(N)}, \frac{(a+1)}{2^k}\right), \tag{7.30}$$

and so $D_k f \to f$, uniformly on $\mathbf{R}$, as $k \to \infty$.

iii) ($N = 1$ case). We have that

$$|(D_k f)(x) - f(x)| \leq \frac{(a+1)}{2^k}\left(|f'(x)| + \omega_1\left(f', \frac{a+1}{2^k}\right)\right). \tag{7.31}$$

**PROOF** (of Theorem 7.4). Again we observe that

$$f\left(\frac{j}{2^k} + \frac{r}{2^k n}\right) = f(x) + \sum_{i=1}^{N} \frac{f^{(i)}(x)}{i!}\left(\frac{j}{2^k} + \frac{r}{2^k n} - x\right)^i$$

$$+ \int_x^{\frac{j}{2^k} + \frac{r}{2^k n}} (f^{(N)}(t) - f^{(N)}(x)) \frac{(\frac{j}{2^k} + \frac{r}{2^k n} - t)^{N-1}}{(N-1)!} dt.$$

Hence

$$\delta_{kj}(f) = \sum_{r=0}^{N} w_r f\left(\frac{j}{2^k} + \frac{r}{2^k n}\right)$$

$$= f(x) + \sum_{i=1}^{N} \frac{f^{(i)}(x)}{i!}\left(\sum_{r=0}^{n} w_r \left(\frac{j}{2^k} + \frac{r}{2^k n} - x\right)\right)^i + \sum_{r=0}^{n} w_r \psi_{jr},$$

## 7. High Order Wavelet Type Approximation

where

$$\psi_{jr} := \int_x^{\frac{j}{2^k}+\frac{r}{2^k n}} (f^{(N)}(t) - f^{(N)}(x)) \frac{(\frac{j}{2^k}+\frac{r}{2^k n}-t)^{N-1}}{(N-1)!} dt. \quad (7.32)$$

Thus

$$(\mathcal{E}_k(f))(x) := (D_k f)(x) - f(x) = \sum_{i=1}^{N} \frac{f^{(i)}(x)}{i!} \left( \sum_{j=-\infty}^{\infty} \varphi(2^k x - j) \right) \cdot$$

$$\cdot \left( \sum_{r=0}^{n} w_r \left( \frac{j}{2^k} + \frac{r}{2^k n} - x \right)^i \right) + \hat{\mathcal{R}}, \quad (7.33)$$

where

$$\hat{\mathcal{R}} := \sum_{j=-\infty}^{\infty} \varphi(2^k x - j) \left( \sum_{r=0}^{n} w_r \psi_{jr} \right). \quad (7.34)$$

Using a similar method as in earlier theorems we obtain that

$$|\psi_{jr}| \leq \omega_1 \left( f^{(N)}, \frac{a+1}{2^k} \right) \frac{(a+1)^N}{2^{kN} N!}. \quad (7.35)$$

And

$$|\hat{\mathcal{R}}| \leq \omega_1 \left( f^{(N)}, \frac{a+1}{2^k} \right) \frac{(a+1)^N}{2^{kN} N!}. \quad (7.36)$$

Finally, from (7.33) and (7.36) we find that

$$|(\mathcal{E}_k f)(x)|$$

$$\leq \sum_{i=1}^{N} \frac{|f^{(i)}(x)|}{i!} \left\{ \left( \sum_{j=-\infty}^{\infty} \varphi(2^k x - j) \right) \left( \sum_{r=0}^{n} w_r \left| \frac{j}{2^k} + \frac{r}{2^k n} - x \right|^i \right) \right\}$$

$$+ |\hat{\mathcal{R}}| \leq \sum_{i=1}^{N} \frac{|f^{(i)}(x)|}{i!} \left\{ \left( \sum_{j=-\infty}^{\infty} \varphi(2^k x - j) \right) \left( \sum_{r=0}^{n} w_r \right) \frac{(a+1)^i}{2^{ki}} \right\}$$

$$+ |\hat{\mathcal{R}}| \leq \sum_{i=1}^{N} \frac{|f^{(i)}(x)|}{i!} \frac{(a+1)^i}{2^{ki}} + \omega_1 \left( f^{(N)}, \frac{a+1}{2^k} \right) \frac{(a+1)^N}{2^{kN} N!}$$

proving (7.29). ∎

Related results are given in

### PROPOSITION 7.1
Here $\varphi$ is as in Theorem 7.1, $f \in C(\mathbf{R})$, $x \in \mathbf{R}$, $k \in \mathbf{Z}$. Operators $C_k$, $D_k$ are as in Theorem 7.3, 7.4 respectively. Then

$$|(C_k(f))(x) - f(x)| \leq \omega_1\left(f, \frac{a+1}{2^k}\right), \qquad (7.37)$$

and

$$|(D_k(f))(x) - f(x)| \leq \omega_1\left(f, \frac{a+1}{2^k}\right). \qquad (7.38)$$

**PROOF** See that

$$(C_k(f))(x) - f(x) = \sum_{j=-\infty}^{\infty}\left(2^k \int_0^{2^{-k}} \left(f\left(t + \frac{j}{2^k}\right) - f(x)\right) dt\right) \varphi(2^k x - j).$$

Therefore

$$|(C_k(f))(x) - f(x)|$$

$$\leq \sum_{j=-\infty}^{\infty}\left(2^k \int_0^{2^{-k}} \left|f\left(t + \frac{j}{2^k}\right) - f(x)\right| dt\right) \varphi(2^k x - j)$$

$$\leq \sum_{j=-\infty}^{\infty}\left(2^k \int_0^{2^{-k}} \omega_1\left(f, \left|x - t - \frac{j}{2^k}\right|\right) dt\right) \varphi(2^k x - j)$$

$$\leq \sum_{j=-\infty}^{\infty}\left(2^k \int_0^{2^{-k}} \omega_1\left(f, |t| + \left|x - \frac{j}{2^k}\right|\right) dt\right) \varphi(2^k x - j)$$

$$\leq \sum_{j=-\infty}^{\infty}\left(2^k \int_0^{2^{-k}} \omega_1\left(f, \frac{1+a}{2^k}\right) dt\right) \varphi(2^k x - j) = \omega_1\left(f, \frac{1+a}{2^k}\right),$$

which establishes (7.37).

## 7. High Order Wavelet Type Approximation

Next we see that

$$(D_k(f))(x) - f(x) = \sum_{j=-\infty}^{\infty} \left\{ \sum_{r=0}^{n} w_r \left( f\left(\frac{j}{2^k} + \frac{r}{2^k n}\right) - f(x) \right) \right\} \varphi(2^k x - j).$$

Therefore

$$|(D_k f)(x) - f(x)| \le \sum_{j=-\infty}^{\infty} \left( \sum_{r=0}^{n} w_r \left| f\left(\frac{j}{2^k} + \frac{r}{2^k n}\right) - f(x) \right| \right) \varphi(2^k x - j)$$

$$\le \sum_{j=-\infty}^{\infty} \left( \sum_{r=0}^{n} w_r \omega_1 \left( f, \left|\frac{j}{2^k} + \frac{r}{2^k n} - x\right| \right) \right) \varphi(2^k x - j)$$

$$\le \sum_{j=-\infty}^{\infty} \left( \sum_{r=0}^{n} w_r \omega_1 \left( f, \frac{a+1}{2^k} \right) \right) \varphi(2^k x - j)$$

$$= \omega_1 \left( f, \frac{a+1}{2^k} \right),$$

proving (7.38). ∎

See that

$$(C_k(f))(x) = C_0(f(2^{-k} \cdot))(2^k x),$$

and (7.39)

$$(D_k f)(x) = D_0(f(2^{-k} \cdot))(2^k x).$$

Optimality is presented in

**PROPOSITION 7.2**
*Inequalities (7.37), (7.38) are sharp over convex $f \in C(\mathbf{R})$.*

**PROOF** Although it is enough to prove it for $k = 0$, for any other $k \ne 0$ it follows similarly.
i) We prove that

$$|(C_0(f))(x) - f(x)| \le \omega_1(f, a+1), \tag{7.40}$$

is sharp over convex $f \in C(\mathbf{R})$. Suppose that there exists $0 < c < 1$ such that
$$|(C_0(f))(x) - f(x)| \le c\omega_1(f, a+1). \tag{7.41}$$

Choose $f(x) := g(x) := (x-1)_+$ which is a convex continuous function on $\mathbf{R}$. Here, $\omega_1(g, h) = h$, any $h > 0$. Pick ($\lceil \cdot \rceil$ is the ceiling of number)
$$\beta := \left\lceil \frac{|2c - \frac{1}{2}|}{1-c} \right\rceil + 1 \quad \left( > \frac{2c - \frac{1}{2}}{1-c} \right).$$

Hence
$$\beta + \frac{1}{2} > c(\beta + 2). \tag{7.42}$$

Set
$$\varphi_1(x) := \begin{cases} 1+x, & -1 \le x \le 0, \\ 1-x, & 0 \le x \le 1, \\ 0, & \text{elsewhere}, \end{cases}$$

and
$$\varphi_3(x) := \varphi_1(x+\beta) := \begin{cases} x+\beta+1, & -\beta-1 \le x \le -\beta, \\ 1-x-\beta, & -\beta \le x \le 1-\beta, \\ 0, & \text{elsewhere}, \end{cases}$$

which is continuous and bounded on $\mathbf{R}$. See that $\varphi_3$ fulfills
1) $\sum_{j=-\infty}^{+\infty} \varphi_3(x-j) = 1$ on $\mathbf{R}$,
2) $\sum_{j=-\infty}^{\infty} j\varphi_3(x-j) = x + \beta$ on $\mathbf{R}$,
3) $\varphi_3$ is convex over $(-\infty, -1-\beta]$, and $[1-\beta, +\infty)$, also $\varphi_3$ is concave over $[-1-\beta, 1-\beta]$,
4) $\varphi_3$ is nondecreasing for $x \le -\beta$ and nonincreasing for $x \ge -\beta$, i.e., $\varphi_3 \ge 0$.

That is, $\varphi_3$ has all the properties of $\varphi$ in Theorem 7.1. Here
$$\varphi_3(x) \ne 0 \quad \text{iff} \quad -1-\beta < x < 1-\beta.$$

Therefore for $j \in \mathbf{Z}$ we have that
$$\varphi_3(1-j) \ne 0 \quad \text{iff} \quad \beta < j < \beta + 2.$$

Thus
$$\varphi_3(1-(\beta+1)) = \varphi_3(-\beta) = \varphi_1(0) = 1.$$

In other words, $\varphi_3(1-j) = 0$, all $j \in \mathbf{Z} - \{\beta+1\}$. Also
$$\ell_{0,\beta+1}(g) := \int_0^1 g(t+\beta+1)dt = \int_0^1 (t+\beta)dt = \frac{1}{2} + \beta.$$

## 7. High Order Wavelet Type Approximation

That is
$$\ell_{0,\beta+1}(g) = \beta + \frac{1}{2}. \tag{7.43}$$

Here $\ell_{0,j}(g) := \gamma_{0,j}(g)$, all $j \in \mathbf{Z}$. Therefore

$$C_0(g)(1) = \sum_{j=-\infty}^{\infty} \ell_{0,j}(g)\varphi_3(1-j) \stackrel{(7.43)}{=} \ell_{0,\beta+1}(g)\varphi_3(1-(\beta+1)) = \beta + \frac{1}{2}.$$

Also $g(1) = 0$. Thus by assumption (7.41) we obtain that (here $a = \beta + 1$, by supp $\varphi_3 \subseteq [-\beta - 1, \beta + 1]$)

$$\beta + \frac{1}{2} = |(C_0(g))(1) - g(1)| \leq c\omega_1(g, a+1)$$

$$= c(a+1)$$

$$= c(\beta + 2).$$

That is,
$$\left(\beta + \frac{1}{2}\right) \leq c(\beta + 2),$$

which contradicts (7.42).

ii) We prove that

$$|(D_0(f))(x) - f(x)| \leq \omega_1(f, a+1) \tag{7.44}$$

is sharp over convex $f \in C(\mathbf{R})$. The proof is similar to (i).

Let $0 < c < 1$ be such that

$$|(D_0 f)(x) - f(x)| \leq c\omega_1(f, a+1). \tag{7.45}$$

Choose
$$0 < \gamma := \frac{\sum_{r=0}^{n} rw_r}{n} \leq 1$$

and
$$\beta := \left\lceil \frac{|2c - \gamma|}{1-c} \right\rceil + 1 > \frac{2c - \gamma}{1-c}.$$

Hence
$$\beta + \gamma > c(\beta + 2). \tag{7.46}$$

Observe that ($g$ as in (i))

$$\delta_{0,\beta+1}(g) = \sum_{r=0}^{n} w_r g\left(\beta + 1 + \frac{r}{n}\right) = \sum_{r=0}^{n} w_r \left(\beta + \frac{r}{n}\right) = \beta + \gamma.$$

That is
$$\delta_{0,\beta+1}(g) = \beta + \gamma. \tag{7.47}$$

Therefore (cf. (7.47))

$$(D_0(g))(1) = \sum_{j=-\infty}^{\infty} \delta_{0,j}(g)\varphi_3(1-j) = \delta_{0,\beta+1}(g)\varphi_3(1-(\beta+1))$$

$$= (\beta + \gamma)\varphi_3(-\beta) = \beta + \gamma.$$

That is $(D_0(g))(1) = \beta + \gamma$. Eventually (cf. (7.45))

$$\beta + \gamma = |(D_0(g))(1) - g(1)| \le c(\beta + 2),$$

contradicting (7.46). ∎

# Chapter 8

## More on Shape and Probability Preserving One-Dimensional Wavelet Type Operators

Here we study the wavelet type operators $C_k, D_k$ of Chapter 7 again, in more depth. Shape and probabilistic maintaining properties of these operators are presented, as well as their optimal approximation to the unit operator. Associated Jackson type inequalities are given. Here we follow the basic treatment of author [20]. More precisely, our setting is as follows:

Let $f \in C^N(\mathbf{R})$, $N \geq 0$, $x \in \mathbf{R}$, $k, j \in \mathbf{Z}$. Let $\varphi$ be a bounded function of compact support $\subseteq [-a, a]$, $a > 0$ such that $\sum_{j=-\infty}^{\infty} \varphi(x-j) = 1$ all $x \in \mathbf{R}$. Also assume there exists $b \in \mathbf{R}$ such that $\varphi$ is nondecreasing for $x \leq b$ and $\varphi$ is nonincreasing for $x \geq b$ (that is, $\varphi \geq 0$).

Define

$$C_k(f)(x) := \sum_{j=-\infty}^{\infty} \gamma_{kj}(f)\varphi(2^k x - j)$$

$$:= \sum_{j=-\infty}^{\infty} \left( 2^k \int_{2^{-k}j}^{2^{-k}(j+1)} f(t)dt \right) \varphi(2^k x - j), \tag{8.1}$$

i.e.,

$$\gamma_{kj}(f) := 2^k \int_{2^{-k}j}^{2^{-k}(j+1)} f(t)dt = 2^k \int_0^{2^{-k}} f\left(t + \frac{j}{2^k}\right) dt. \tag{8.2}$$

Also define

$$D_k(f)(x) := \sum_{j=-\infty}^{\infty} \delta_{kj}(f)\varphi(2^k x - j), \tag{8.3}$$

185

where

$$\delta_{kj}(f) := \sum_{r=0}^{n} w_r f\left(\frac{j}{2^k} + \frac{r}{2^k n}\right), \quad n \in \mathbf{N}, w_r \geq 0, \sum_{r=0}^{n} w_r = 1. \quad (8.4)$$

We observe that

$$(C_k(f))(x) = (C_0(f(2^{-k}\cdot)))(2^k x)$$

$$(D_k(f))(x) = (D_0(f(2^{-k}\cdot)))(2^k x). \quad (8.5)$$

## 8.1 Estimates, Maintenance, and Sharpness

Let $\{c_j\}_{j=-\infty}^{\infty} \in \mathbf{R}$, set

$$A(x) := \sum_{j=-\infty}^{\infty} c_j \varphi(x - j). \quad (8.6)$$

Based on Lemma 6.1 we have

**LEMMA 8.1**
If $\varphi(x)$ is a bounded right-continuous function with supp $\varphi(x) \subseteq [-a, a]$, $a > 0$, then for any $f \in C(\mathbf{R})$, the functions $(C_k f)(x)$ and $(D_k f)(x)$ are right-continuous on $\mathbf{R}$, and $k \in \mathbf{Z}$.

Based on Lemma 6.3 we find

**LEMMA 8.2**
Let $\varphi$ as assumed. Let $f \in C(\mathbf{R})$ which is nondecreasing. Then $(C_k f)(x)$, $(D_k f)(x)$ are nondecreasing, any $k \in \mathbf{Z}$.

**PROOF** Let $j_1 \leq j_2$ and $f \in C(\mathbf{R})$ be nondecreasing.
i) Then

$$\gamma_{kj_1}(f) = 2^k \int_0^{2^{-k}} f\left(t + \frac{j_1}{2^k}\right) dt \leq 2^k \int_0^{2^{-k}} f\left(t + \frac{j_2}{2^k}\right) dt = \gamma_{kj_2}(f).$$

That is, $\gamma_{kj}(f)$ is nondecreasing in $j$. Thus $(C_0(f))(x)$ (Lemma 6.3), is nondecreasing in $x \in \mathbf{R}$. Consequently, by (8.5), $(C_k(f))(x)$ is nondecreasing on $\mathbf{R}$.

ii) We obtain that

$$\delta_{kj_1}(f) = \sum_{r=0}^{n} w_r f\left(\frac{j_1}{2^k} + \frac{r}{2^k n}\right) \leq \sum_{r=0}^{n} w_r f\left(\frac{j_2}{2^k} + \frac{r}{2^k n}\right) = \delta_{kj_2}(f).$$

That is, $\delta_{kj}(f)$ is nondecreasing in $j$. Then $(D_0(f))(x)$ (Lemma 6.3) is nondecreasing in $x \in \mathbf{R}$. Hence by (8.5), $(D_k(f))(x)$ is nondecreasing on $\mathbf{R}$. ∎

We also use

**LEMMA 8.3**
Assume that $\varphi(x)$ is a bounded function with $\operatorname{supp} \varphi \subseteq [-a, a]$, $a > 0$ and fulfills

i) $\sum_{j=-\infty}^{\infty} \varphi(x - j) = 1$ on $\mathbf{R}$,

ii) there exists $b \in \mathbf{R}$ such that $\varphi$ is nondecreasing for $x \leq b$ and $\varphi$ is nonincreasing for $x \geq b$.

Let $f \in C(\mathbf{R})$ be nondecreasing on $\mathbf{R}$, $\lim_{x \to +\infty} f(x) = 1$, $\lim_{x \to -\infty} f(x) = 0$, $k \in \mathbf{Z}$. Then it holds

$$\lim_{x \to +\infty} C_k(f)(x) = \lim_{x \to +\infty} D_k(f)(x) = 1,$$

and

$$\lim_{x \to -\infty} C_k(f)(x) = \lim_{x \to -\infty} D_k(f)(x) = 0.$$

**PROOF** Here $C_k(f)(x)$, $D_k(f)(x)$ are nondecreasing between 0 and 1. Then,

$$\lim_{x \to \pm\infty} C_k(f)(x), \quad \lim_{x \to \pm\infty} D_k(f)(x)$$

do exist. We need to prove that

$$\lim_{x \to \pm\infty} \begin{cases} |C_k(f)(x) - f(x)| \\ |D_k(f)(x) - f(x)| \end{cases} = 0.$$

Because $\lim_{x \to +\infty} f(x) = 1$ and $\lim_{x \to -\infty} f(x) = 0$, for any $\varepsilon > 0$, $\exists N > 0$ real such that for any $x_1, x_2 > N$ we have

$$|f(x_1) - f(x_2)| < \varepsilon$$

and

$$|f(-x_1) - f(-x_2)| < \varepsilon.$$

Let $x_0 > N + 2^{-k}(a+1)$. We have that

$$C_k(f)(x_0) - f(x_0) = \sum_{j=-\infty}^{\infty} \left( 2^k \int_0^{2^{-k}} \left( f\left(t + \frac{j}{2^k}\right) - f(x_0) \right) dt \right) \varphi(2^k x_0 - j).$$

Because

$$\left| t + \frac{j}{2^k} - x_0 \right| \leq \frac{a+1}{2^k},$$

we get that $t + \frac{j}{2^k} \geq x_0 - \frac{a+1}{2^k} > N$. Thus

$$|C_k(f)(x_0) - f(x_0)| \leq \sum_{j=-\infty}^{+\infty} \varepsilon \varphi(2^k x_0 - j) = \varepsilon,$$

if $x_0 > N + 2^{-k}(a+1)$. That is

$$\lim_{x \to +\infty} |C_k(f)(x) - f(x)| = 0.$$

Also notice that

$$D_k(f)(x_0) - f(x_0) = \sum_{j=-\infty}^{+\infty} \left( \sum_{r=0}^{n} w_r \left( f\left(\frac{j}{2^k} + \frac{r}{2^k n}\right) - f(x_0) \right) \right) \varphi(2^k x_0 - j).$$

Because

$$\left| \frac{r}{2^k n} + \frac{j}{2^k} - x_0 \right| \leq \frac{a+1}{2^k}$$

we find that

$$\left( \frac{r}{2^k n} + \frac{j}{2^k} \right) \geq x_0 - \frac{(a+1)}{2^k} > N.$$

Therefore

$$|D_k(f)(x_0) - f(x_0)| \leq \sum_{j=-\infty}^{+\infty} \left( \sum_{r=0}^{n} w_r \varepsilon \right) \varphi(2^k x_0 - j) = \varepsilon,$$

if $x_0 > N + 2^{-k}(a+1)$. That is, $\lim_{x \to +\infty} |D_k(f)(x) - f(x)| = 0$.
We have proved that

$$\lim_{x \to +\infty} C_k(f)(x) = \lim_{x \to +\infty} D_k(f)(x) = 1.$$

Similarly, we establish that

$$\lim_{x \to -\infty} C_k(f)(x) = \lim_{x \to -\infty} D_k(f)(x) = 0,$$

(use of $|f(-x_1) - f(-x_2)| < \varepsilon$). ∎

Next comes the first main result.

### THEOREM 8.1
*Let $\varphi$ as in Lemma 8.3. Additionally suppose that $\varphi$ is right continuous. Let $f \in C(\mathbf{R})$ be a probability distribution function. Then, for any $k \in \mathbf{Z}$, we have that $C_k(f)$, $D_k(f)$ are probability distribution functions.*

**PROOF** By Lemmas 8.1–8.3. ∎

### PROPOSITION 8.1
*Let $f$ be a probability distribution function in $C^1(\mathbf{R})$. Set $\lambda := f'$. Then*

$$|C_k(f)(x) - f(x)| \leq \left(\frac{a+1}{2^k}\right)\left(\lambda(x) + \omega_1\left(f', \frac{a+1}{2^k}\right)\right), \quad (8.7)$$

*and*

$$|D_k(f)(x) - f(x)| \leq \left(\frac{a+1}{2^k}\right)\left(\lambda(x) + \omega_1\left(f', \frac{a+1}{2^k}\right)\right). \quad (8.8)$$

**PROOF** By (7.21) and (7.31). ∎

We need to mention

### PROPOSITION 7.1
*Let $f \in C(\mathbf{R})$, $x \in \mathbf{R}$, $k \in \mathbf{Z}$. Let $\varphi$ be a bounded function of compact support $\subseteq [-a, a]$, $a > 0$ such that $\sum_{j=-\infty}^{\infty} \varphi(x - j) = 1$ on $\mathbf{R}$. Also there*

exists $b \in \mathbf{R}$ such that $\varphi$ is nondecreasing for $x \leq b$ and $\varphi$ is nonincreasing for $x \geq b$. Then it holds

$$|(C_k(f))(x) - f(x)| \leq \omega_1\left(f, \frac{a+1}{2^k}\right), \tag{8.9}$$

and

$$|(D_k(f))(x) - f(x)| \leq \omega_1\left(f, \frac{a+1}{2^k}\right). \tag{8.10}$$

Optimality is presented in

**THEOREM 8.2**
*Inequalities (8.9) and (8.10) are sharp (in fact attained).*

**PROOF** i) Let

$$q(x) := \begin{cases} 1, & x \geq 0, \\ 0, & x \leq -\frac{1}{2^{k+1}}, \\ 2^{k+1} \cdot x + 1, & -\frac{1}{2^{k+1}} \leq x \leq 0. \end{cases}$$

Consider

$$\varphi(x) := \chi_{[-\frac{1}{2},\frac{1}{2})}(x),$$

the characteristic function on $[-\frac{1}{2}, \frac{1}{2})$. That is, here $a = \frac{1}{2}$. We see that

$$\omega_1\left(q, \frac{a+1}{2^k}\right) = \omega_1\left(q, \frac{3}{2^{k+1}}\right) = \omega_1\left(q, \frac{1}{2^{k+1}}\right) = 1.$$

Also we obtain

$$\gamma_{k(-1)}(q) = 2^k \int_{-\frac{1}{2^k}}^{0} q(t)dt = 2^k \int_{-\frac{1}{2^{k+1}}}^{0} q(t)dt = \frac{1}{4},$$

i.e., $\gamma_{k(-1)}(q) = \frac{1}{4}$. Moreover

$$\gamma_{k(-2)}(q) = 2^k \int_{-\frac{2}{2^k}}^{-\frac{1}{2^k}} q(t)dt = 0,$$

## 8. More on One-Dimensional Wavelet Type Operators

and $\gamma_{kj}(q) = 0$, all $j \leq -2$, and $\gamma_{kj}(q) = 1$, all $j \geq 0$. Thus

$$C_k(q)(x) = \frac{1}{4}\varphi(2^k x + 1) + \sum_{j=0}^{+\infty} \varphi(2^k x - j).$$

Next we calculate

$$C_k(q)\left(-\frac{1}{2^{k+1}}\right) = \frac{1}{4}\varphi\left(\frac{1}{2}\right) + \sum_{j=0}^{+\infty} \varphi\left(-\frac{1}{2} - j\right) = 1 + \sum_{j=1}^{+\infty} \varphi\left(-\frac{1}{2} - j\right) = 1.$$

That is,

$$C_k(q)\left(-\frac{1}{2^{k+1}}\right) = 1, \quad \text{and} \quad q\left(-\frac{1}{2^{k+1}}\right) = 0.$$

Therefore

$$\left|C_k(q)\left(-\frac{1}{2^{k+1}}\right) - q\left(-\frac{1}{2^{k+1}}\right)\right| = \omega_1\left(q, \frac{3}{2^{k+1}}\right) = 1,$$

i.e., inequality (8.9) is attained.

ii) It is treated similarly to (i). Notice that $\delta_{kj}(q) = 1$, all $j \geq 0$, and $\delta_{kj}(q) = 0$, all $j \leq -2$. See that

$$\varphi\left(2^k\left(-\frac{1}{2^{k+1}}\right) - (-1)\right) = \varphi\left(\frac{1}{2}\right) = 0.$$

Moreover

$$(D_k q)\left(-\frac{1}{2^{k+1}}\right) - q\left(-\frac{1}{2^{k+1}}\right) = \sum_{j=-\infty}^{+\infty} \delta_{kj}(q)\varphi\left(2^k\left(-\frac{1}{2^{k+1}}\right) - j\right) - 0$$

$$= \sum_{j=0}^{+\infty} 1\varphi\left(-\frac{1}{2} - j\right) = 1.$$

Then

$$\left|(D_k q)\left(-\frac{1}{2^{k+1}}\right) - q\left(-\frac{1}{2^{k+1}}\right)\right| = 1 = \omega_1\left(q, \frac{3}{2^{k+1}}\right),$$

i.e., inequality (8.10) is attained. ∎

**REMARK 8.1** Because $q$, in the last proof, is a continuous distribution function, inequalities (8.9) and (8.10) are also sharp over continuous probability distribution functions.

We use

**LEMMA 8.4**
(Anastassiou and Yu [43]). *Assume that $\varphi(x)$ is a bounded continuous function on $\mathbf{R}$, supp $\varphi(x) \subseteq [-a, a]$, $a > 0$, and fulfills*

i) $\sum_{j=-\infty}^{\infty} \varphi(x-j)$ and $\sum_{j=-\infty}^{\infty} j\varphi(x-j)$ are linear functions on $\mathbf{R}$.

ii) there exist real numbers $b_1$ and $b_2$, $b_1 \leq b_2$ such that $\varphi(x)$ is convex on $(-\infty, b_1]$ and $[b_2, +\infty)$ respectively, and $\varphi(x)$ is concave on $[b_1, b_2]$.

*Then, if $\{C_j\}$ is a convex sequence, i.e., $\{C_j - C_{j-1}\}$ is nondecreasing, the function $A(x)$ defined by (8.6) is convex on $\mathbf{R}$.*

As a related result we present

**PROPOSITION 8.2**
*Let $f \in C(\mathbf{R})$, which is convex. Let $\varphi$ as in Lemma 8.4. Then $C_k(f)(x)$, $D_k(f)(x)$ are convex, any $k \in \mathbf{Z}$.*

**PROOF** By (8.5) it is enough to prove it for $k = 0$.
i) Here
$$C_0(f)(x) = \sum_{j=-\infty}^{\infty} \ell_{0j} \varphi(x-j),$$
where
$$\ell_{0j} := \int_0^1 f(t+j)\,dt.$$

Let $0 \leq \lambda \leq 1$, by convexity of $f$ we have that
$$f(t + \lambda j_1 + (1-\lambda) j_2) \leq \lambda f(t + j_1) + (1-\lambda) f(t + j_2).$$

Then
$$\int_0^1 f(t + \lambda j_1 + (1-\lambda) j_2)\,dt \leq \lambda \int_0^1 f(t + j_1)\,dt$$

## 8. More on One-Dimensional Wavelet Type Operators

$$+ (1-\lambda) \int_0^1 f(t+j_2)dt,$$

i.e.,
$$\ell_{0(\lambda j_1 + (1-\lambda)j_2)} \leq \lambda \ell_{0j_1} + (1-\lambda)\ell_{0j_2}.$$

That is, $\ell_{0j}$ is convex in $j \in \mathbf{Z}$. By Lemma 8.4 we find that $(C_0 f)$ is convex.

ii) Here
$$(D_0 f)(x) = \sum_{j=-\infty}^{\infty} \delta_{0j}(f)\varphi(x-j),$$

where
$$\delta_{0j}(f) := \sum_{r=0}^{n} w_r f\left(j + \frac{r}{n}\right).$$

Let $0 \leq \lambda \leq 1$, by convexity of $f$ we obtain that
$$f\left(\lambda j_1 + (1-\lambda)j_2 + \frac{r}{n}\right) \leq \lambda f\left(j_1 + \frac{r}{n}\right) + (1-\lambda)f\left(j_2 + \frac{r}{n}\right).$$

Summing up over $r$ we get
$$\delta_{0,\lambda j_1 + (1-\lambda)j_2}(f) = \sum_{r=0}^{n} w_r f\left(\lambda j_1 + (1-\lambda)j_2 + \frac{r}{n}\right)$$
$$\leq \lambda \sum_{r=0}^{n} w_r f\left(j_1 + \frac{r}{n}\right)$$
$$+ (1-\lambda) \sum_{r=0}^{n} w_r f\left(j_2 + \frac{r}{n}\right) = \lambda \delta_{0j_1} + (1-\lambda)\delta_{0j_2}.$$

That is, $\delta_{0,j}(f)$ is convex in $j \in \mathbf{Z}$. Hence by Lemma 8.4 we prove that $(D_0 f)(x)$ is convex. ∎

Define
$$\Delta^3 c_j := \sum_{i=0}^{3} (-1)^{3-i} \binom{3}{i} c_{j-3+i} = -c_{j-3} + 3c_{j-2} - 3c_{j-1} + c_j, \text{ any } j \in \mathbf{Z}.$$
(8.11)

For 3th convexity we need to prove that the sequence $\{c_j\}_{j=-\infty}^{+\infty}$ satisfies
$$\Delta^3 c_j \geq 0. \tag{8.12}$$

**LEMMA 8.5**
(Anastassiou and Yu [43]). *Assume that $\varphi(x)$ is a bounded continuous function on $\mathbf{R}$, $\operatorname{supp} \varphi(x) \subseteq [-a, a]$, $a > 0$ and fulfills*

i) $\sum_{j=-\infty}^{\infty} \varphi(x-j)$, $\sum_{j=-\infty}^{\infty} j\varphi(x-j)$ *and* $\sum_{j=-\infty}^{\infty} j^2 \varphi(x-j)$ *are quadratic functions on* $\mathbf{R}$,

ii) *there exist real numbers $b_1, b_2$, and $b_3$, $b_1 \leq b_2 \leq b_3$ such that $\varphi(x)$ is 3th convex on $(-\infty, b_1]$ and $[b_2, b_3]$ respectively, and $\varphi(x)$ is 3th concave on $[b_1, b_2]$ and $[b_3, +\infty)$ respectively.*

*Then, if $\{c_j\}$ is a 3th convex sequence, the function $A(x)$ defined by (8.6) is 3th convex on $\mathbf{R}$.*

**EXAMPLE 8.1**
Of $\varphi$ as in Lemma 8.5 ([43])

$$\varphi_2(x) = \begin{cases} 0, & x \leq -\frac{3}{2}, \\ \frac{1}{2}(\frac{3}{2}+x)^2, & -\frac{3}{2} < x \leq -\frac{1}{2}, \\ 1 + x - (x+\frac{1}{2})^2, & -\frac{1}{2} < x < \frac{1}{2}, \\ \frac{1}{2}(\frac{3}{2}-x)^2, & \frac{1}{2} \leq x \leq \frac{3}{2} \\ 0, & x > \frac{3}{2}. \end{cases}$$

Here $b_1 = -1$, $b_2 = 0$, $b_3 = 1$.

The related result comes next.

**PROPOSITION 8.3**
*Let $\varphi$ as in Lemma 8.5. Let $f \in C(\mathbf{R})$ which is 3th-convex. Then $C_k(f)(x)$ and $D_k(f)(x)$ are 3th-convex, any $k \in \mathbf{Z}$.*

**PROOF** Enough to prove it for $k = 0$, then by (8.5) we obtain it for all $k \in \mathbf{Z}$.
i) By assumption we have

$$0 \leq \Delta_h^3 f(x) = \sum_{i=0}^{3}(-1)^{3-i}\binom{3}{i}f(x+ih)$$

$$= -f(x) + 3f(x+h) - 3f(x+2h) + f(x+3h). \quad (8.13)$$

Notice that (8.12) is equivalent to

$$-c_j + 3c_{j+1} - 3c_{j+2} + c_{j+3} \geq 0, \quad (8.14)$$

# 8. More on One-Dimensional Wavelet Type Operators

all $j \in \mathbf{Z}$. But (8.13) implies that

$$-f(t+j) + 3f(t+j+1) - 3f(t+j+2) + f(t+j+3) \geq 0. \quad (8.15)$$

Here

$$\gamma_{0j}(f) = \int_0^1 f(t+j)dt.$$

By integrating (8.15) over $[0,1]$ we obtain that

$$-\gamma_{0j}(f) + 3\gamma_{0,(j+1)}(f) - 3\gamma_{0,(j+2)}(f) + \gamma_{0,(j+3)}(f) \geq 0,$$

any $j \in \mathbf{Z}$. That is proving that $\{\gamma_{0j}(f)\}_{j=-\infty}^{\infty}$ is 3th-convex. Hence by Lemma 8.5 we find that $C_0(f)(x)$ is 3th-convex.

ii) Here

$$\delta_{0j}(f) = \sum_{r=0}^{n} w_r f\left(j + \frac{r}{n}\right).$$

By assumption on $f$ and as before we find that

$$-\delta_{0j}(f) + 3\delta_{0,(j+1)}(f) - 3\delta_{0,(j+2)}(f) + \delta_{0,(j+3)}(f) \geq 0, \quad \text{all } j \in \mathbf{Z}.$$

That is establishing that $\{\delta_{0j}(f)\}_{j=-\infty}^{\infty}$ is 3th-convex. Hence by Lemma 8.5 we get that $D_0(f)(x)$ is 3th-convex. ∎

We use

**LEMMA 8.6**
Let $a \leq b$, $a, b \in \mathbf{R}$. Let $\ell$ be the maximum number of integers contained in $[a, b]$. Then $\ell \leq b - a + 1$.

**PROOF** Let $k_1, \ldots, k_\ell \in \mathbf{Z}$ be all the integers contained in $[a, b]$. Without loss of generality we suppose that $k_1 < k_2 < \cdots < k_\ell$. Here $k_{i+1} - k_i = 1$, $i = 1, \ldots, \ell - 1$; $0 \leq k_1 - a \leq 1$, $0 \leq b - k_\ell \leq 1$. Thus

$$b - a = (k_1 - a) + (b - k_\ell) + \sum_{i=1}^{\ell-1}(k_{i+1} - k_i)$$

$$= (k_1 - a) + (b - k_\ell) + (\ell - 1) \geq \ell - 1.$$

Obviously $\ell \leq b - a + 1$. ∎

Notice that for $y \geq x$ we obtain

$$\sum_{\mathbf{Z} \ni j=x}^{y} 1 = \sum_{j=\lceil x \rceil}^{[y]} 1 = [y] - \lceil x \rceil + 1 \leq y - x + 1.$$

That is,

$$\sum_{\mathbf{Z} \ni j=x}^{y} 1 \leq y - x + 1. \qquad (8.16)$$

As a complement we give

**PROPOSITION 8.4**
Let $\varphi$ as in Lemma 8.5 such that $\sum_{j=-\infty}^{\infty} \varphi(x-j) = 1$. Let $f \in C(\mathbf{R})$. Then it holds

$$|C_k(f)(x) - f(x)| \leq c\omega_1\left(f, \frac{a+1}{2^k}\right), \qquad (8.17)$$

and

$$|D_k(f)(x) - f(x)| \leq c\omega_1\left(f, \frac{a+1}{2^k}\right), \qquad (8.18)$$

where $c$ depends only on $\varphi$.

**PROOF**  Here $|\varphi(x)| \leq M$, where $M > 0$.
i) Notice that

$$C_k(f)(x) - f(x) = \sum_{j=-\infty}^{\infty} \left(2^k \int_0^{2^{-k}} \left(f\left(t + \frac{j}{2^k}\right) - f(x)\right) dt\right) \varphi(2^k x - j).$$

So that

$$|C_k(f)(x) - f(x)| \leq \omega_1\left(f, \frac{a+1}{2^k}\right) \sum_{j=\lceil 2^k x - a \rceil}^{[2^k x + a]} |\varphi(2^k x - j)|$$

$$\overset{(8.16)}{\leq} M(2a+1)\omega_1\left(f, \frac{a+1}{2^k}\right).$$

The last proves (8.17).

ii) Next we see that

$$|(D_k f)(x) - f(x)|$$

$$= \left| \sum_{j=-\infty}^{\infty} \left( \sum_{r=0}^{n} w_r \left( f\left( \frac{j}{2^k} + \frac{r}{2^k n} \right) - f(x) \right) \right) \varphi(2^k x - j) \right|$$

$$\leq \omega_1 \left( f, \frac{a+1}{2^k} \right) \sum_{j=\lceil 2^k x - a \rceil}^{\lfloor 2^k x + a \rfloor} |\varphi(2^k x - j)| \stackrel{(8.16)}{\leq} M(2a+1)\omega_1 \left( f, \frac{a+1}{2^k} \right).$$

The above establishes (8.18). ∎

Coconvexity is studied finally.

**PROPOSITION 8.5**
Assume that $\varphi(x)$ is a bounded continuous function on $\mathbf{R}$, supp $\varphi(x) \subseteq [-a, a]$, $a > 0$ and satisfies:

(i) $\sum_{j=-\infty}^{\infty} \varphi(x-j)$ and $\sum_{j=-\infty}^{\infty} j\varphi(x-j)$ are linear functions on $\mathbf{R}$,

(ii) there exist real numbers $b_1$ and $b_2$, $b_1 \leq b_2$ such that $\varphi(x)$ is convex on $(-\infty, b_1]$ and $[b_2, +\infty)$ respectively, and $\varphi(x)$ is concave on $[b_1, b_2]$. Let $f(x) \in C(\mathbf{R})$ be concave on $(x_0, +\infty)$. Then $C_0(f)(x)$, $D_0(f)(x)$ are concave on $(x_0 + a, +\infty)$.

That is, $C_k(f)(x)$ and $D_k(f)(x)$ are concave on $(x_0 + \frac{a}{2^k}, +\infty)$, any $k \in \mathbf{Z}$.

**PROOF** Let $x \in (x_0 + a, +\infty)$.
i) Recall

$$C_0(f)(x) = \sum_{j=-\infty}^{\infty} \left( \int_0^1 f(t+j)dt \right) \varphi(x-j)$$

$$= \sum_{x-a \leq j \leq x+a} \left( \int_0^1 f(t+j)dt \right) \varphi(x-j)$$

$$= \sum_{x_0 < j} \left( \int_0^1 f(t+j)dt \right) \varphi(x-j).$$

Let $j_0$ be the smallest integer such that $j_0 > x_0$, and put

$$C_j := \int_0^1 f(t+j)dt, \quad j = j_0, j_0+1, \ldots.$$

Let $0 \leq \lambda \leq 1$, by $f$-concavity on $(x_0, +\infty)$ we have that

$$f(t + \lambda j_1 + (1-\lambda)j_2) \geq \lambda f(t+j_1) + (1-\lambda)f(t+j_2).$$

Integrating the last inequality over $[0,1]$ we get that

$$C_{\lambda j_1 + (1-\lambda)j_2} \geq \lambda C_{j_1} + (1-\lambda)C_{j_2}.$$

In other words, $\{C_j\}$ is a concave sequence for $j = j_0, j_0+1, \ldots$. Set

$$C_j := 2C_{j+1} - C_{j+2}, \quad j = j_0-1, j_0-2, \ldots,.$$

Hence $\{C_j\}_{j=-\infty}^{\infty}$ is a concave sequence.
Moreover, notice that

$$(C_0 f)(x) = \sum_{j=-\infty}^{\infty} C_j \varphi(x-j),$$

for any $x \in (x_0 + a, +\infty)$.
By Lemma 8.4, the function $\sum_{j=-\infty}^{\infty} C_j \varphi(x-j)$ is concave on $\mathbf{R}$. That is, $(C_0 f)(x)$ is concave on $(x_0 + a, +\infty)$.
ii) Recall

$$(D_0 f)(x) = \sum_{j=-\infty}^{\infty} \left( \sum_{r=0}^{n} w_r f\left(j + \frac{r}{n}\right) \right) \varphi(x-j)$$

$$= \sum_{x-a \leq j \leq x+a} \left( \sum_{r=0}^{n} w_r f\left(j + \frac{r}{n}\right) \right) \varphi(x-j)$$

$$= \sum_{x_0 < j} \left( \sum_{r=0}^{n} w_r f\left(j + \frac{r}{n}\right) \right) \varphi(x-j).$$

# 8. More on One-Dimensional Wavelet Type Operators

Let $j_0$ be the smallest integer such that $j_0 > x_0$ and put

$$C_j := \left(\sum_{r=0}^{n} w_r f\left(j + \frac{r}{n}\right)\right), \quad j = j_0, j_0 + 1, \ldots.$$

Again by concavity of $f(x)$ over $(x_0, +\infty)$, we have that

$$C_{\lambda j_1 + (1-\lambda) j_2} \geq \lambda C_{j_1} + (1 - \lambda) C_{j_2},$$

any $0 \leq \lambda \leq 1$. In other words, $C_j$ is concave for $j = j_0, j_0 + 1, \ldots$. Put

$$C_j := 2C_{j+1} - C_{j+2}, \quad j = j_0 - 1, j_0 - 2, \ldots.$$

That is, $\{C_j\}_{j=-\infty}^{\infty}$ is a concave sequence.

We see that

$$(D_0 f)(x) = \sum_{j=-\infty}^{\infty} C_j \varphi(x - j),$$

for any $x \in (x_0 + a, +\infty)$. Again $\sum_{j=-\infty}^{\infty} C_j \varphi(x-j)$ is concave on $\mathbf{R}$. Then $(D_0 f)(x)$ is concave on $(x_0 + a, +\infty)$. ∎

Setting facts together we get

### THEOREM 8.3
Assume that $\varphi(x)$ is a bounded continuous function on $\mathbf{R}$ with supp $\varphi(x) \subseteq [-a, a]$, $a > 0$ that fulfills:

i) $\sum_{j=-\infty}^{\infty} \varphi(x - j) = 1$ on $\mathbf{R}$; $\sum_{j=-\infty}^{\infty} j\varphi(x - j)$ is linear on $\mathbf{R}$,

ii) there exists $b_0 \in \mathbf{R}$ such that $\varphi$ is nondecreasing for $x \leq b_0$ and $\varphi$ is nonincreasing for $x \geq b_0$,

iii) there exist $b_1, b_2 \in \mathbf{R}$, $b_1 \leq b_2$, such that $\varphi$ is convex on $(-\infty, b_1]$ and $[b_2, +\infty)$ respectively, and $\varphi$ is concave on $[b_1, b_2]$.

Let $F$ be a continuous distribution function on $\mathbf{R}$ that is concave on $(x_0, +\infty)$. Then $(C_k F)(x)$, $(D_k F)(x)$ are distributions functions on $\mathbf{R}$, which are concave on $(x_0 + \frac{a}{2^k}, +\infty)$, and fulfill

$$|(C_k F)(x) - F(x)| \leq \omega_1\left(F, \frac{a+1}{2^k}\right), \tag{8.19}$$

and
$$|(D_kF)(x) - F(x)| \leq \omega_1\left(F, \frac{a+1}{2^k}\right), \tag{8.20}$$
any $x \in \mathbf{R}$, any $k \in \mathbf{Z}$.

# Chapter 9

## Quantitative Multidimensional High Order Wavelet Type Approximation

High order differentiable functions in several variables are approximated by multidimensional wavelet type operators. The high order of this approximation is estimated by giving some multivariate Jackson type inequalities. Sharpness of some of these inequalities over continuous functions is presented nontrivially. This treatment is based on [29].

All the presented inequalities involve the multivariate first modulus of continuity $\omega_1$ of $f^{(N)}$, $N \geq 0$, where $f \in C^N(\mathbf{R}^r)$, $r \geq 1$ and a multivariate scale function of compact support $\varphi$, which is left without any assumption about its orthogonality.

We need to mention the following multivariate first modulus of continuity $\omega_1$: Let $f \in C(\mathbf{R}^r)$ which is bounded or uniformly continuous, we define ($h > 0$)

$$\omega_1(f, h) = \sup_{\substack{\text{all } x_i, x'_i \in \mathbf{R} \\ |x_i - x'_i| \leq h, \text{ for } i=1,\ldots,r}} |f(x_1, \ldots, x_r) - f(x'_1, \ldots, x'_r)|.$$

## 9.1 Estimates and Sharpness

Next comes the first main result.

**THEOREM 9.1**
Let $f \in C^N(\mathbf{R}^r)$, $N$ and $r \geq 1$; $\vec{x} \in \mathbf{R}^r$ and $k \in \mathbf{Z}$. Let $\varphi \geq 0$ be a bounded function on $\mathbf{R}^r$ of compact support $\subseteq \prod_{i=1}^{r}[-a_i, a_i]$, $0 < a_i < +\infty$,

$a := \max(a_1, \ldots, a_r)$. Suppose that

$$\sum_{j_1=-\infty}^{\infty} \cdots \sum_{j_r=-\infty}^{\infty} \varphi(x_1 - j_1, \ldots, x_r - j_r) = 1, \quad \text{all } \vec{x} := (x_1, \ldots, x_r) \in \mathbf{R}^r, \tag{9.1}$$

in short

$$\sum_{\vec{j}=-\infty}^{\infty} \varphi(\vec{x} - \vec{j}) = 1, \quad \text{all } \vec{x} \in \mathbf{R}^r, \tag{9.1'}$$

where $\vec{j} := (j_1, \ldots, j_r)$. Put

$$B_k(f)(x_1, \ldots, x_r) := \sum_{j_1=-\infty}^{\infty} \cdots \sum_{j_r=-\infty}^{\infty} f\left(\frac{j_1}{2^k}, \ldots, \frac{j_r}{2^k}\right);$$

$$\varphi(2^k x_1 - j_1, \ldots, 2^k x_r - j_r), \quad \text{any } k \in \mathbf{Z}, \tag{9.2}$$

all $(x_1, \ldots, x_r) \in \mathbf{R}^r$; in short

$$B_k(f)(\vec{x}) = \sum_{\vec{j}=-\infty}^{\infty} f\left(\frac{\vec{j}}{2^k}\right) \varphi(2^k \vec{x} - \vec{j}), \quad \text{any } k \in \mathbf{Z}, \text{ all } \vec{x} \in \mathbf{R}^r. \tag{9.2'}$$

Here we further assume that all of the partial derivatives of $f$ of order $N$, denoted by

$$f_{\tilde{\alpha}} := \frac{\partial^{\tilde{\alpha}} f}{\partial x^{\tilde{\alpha}}} \left(\tilde{\alpha} := (\alpha_1, \ldots, \alpha_r), \alpha_i \in \mathbf{Z}^+, i = 1, \ldots, r : |\tilde{\alpha}| = \sum_{i=1}^{r} \alpha_i = N\right),$$

are continuous and bounded or uniformly continuous on $\mathbf{R}^r$.

Then it holds

$$|(B_k(f))(\vec{x}) - f(\vec{x})| \le \sum_{j=1}^{N} \frac{a^j}{j! 2^{kj}} \left(\left(\sum_{i=1}^{r} \left|\frac{\partial}{\partial x_i}\right|\right)^j f(\vec{x})\right)$$

$$+ \frac{a^N r^N}{N! 2^{kN}} \max_{\tilde{\alpha}: |\tilde{\alpha}|=N} \omega_1\left(f_{\tilde{\alpha}}, \frac{a}{2^k}\right), \quad \text{any } k \in \mathbf{Z}, \tag{9.3}$$

which is attained by constant functions.

## 9. Quantitative Multidimensional Approximation

**REMARK 9.1** i) Clearly here $B_k f \to f$ pointwise over $\mathbf{R}^r$, as $k \to \infty$.
ii) Given that $f \in C_b^N(\mathbf{R}^r)$ (i.e., all $f$ and its partial derivatives up to order $N$ are continuous and bounded) we get

$$\|B_k f - f\|_\infty \le \sum_{j=1}^N \frac{a^j}{j! 2^{kj}} \left( \left( \sum_{i=1}^r \left\| \frac{\partial}{\partial x_i} \right\|_\infty \right)^j f \right)$$

$$+ \frac{a^N r^N}{N! 2^{kN}} \max_{\tilde{\alpha}: |\tilde{\alpha}| = N} \omega_1 \left( f_{\tilde{\alpha}}, \frac{a}{2^k} \right), \quad \text{any } k \in \mathbf{Z}. \quad (9.4)$$

That is $B_k f \to f$, uniformly over $\mathbf{R}^r$, as $k \to \infty$.
iii) When $N = 1$ from (9.3) we find that

$$|(B_k f)(\vec{x}) - f(\vec{x})| \le \frac{a}{2^k} \left\{ \left( \sum_{i=1}^r \left| \frac{\partial f(\vec{x})}{\partial x_i} \right| \right) \right.$$

$$\left. + r \max_{i \in \{1,\ldots,r\}} \omega_1 \left( \frac{\partial f}{\partial x_i}, \frac{a}{2^k} \right) \right\}, \quad \text{any } k \in \mathbf{Z}. \quad (9.5)$$

■

**PROOF** (of Theorem 9.1). Put

$$g_{\frac{\vec{j}}{2^k}}(t) := f\left( \vec{x} + t \left( \frac{\vec{j}}{2^k} - \vec{x} \right) \right), \quad \text{all } 0 \le t \le 1.$$

Then for $j = 1, 2, \ldots, N$ we see that

$$g_{\frac{\vec{j}}{2^k}}^{(j)}(t) = \left\{ \left( \sum_{i=1}^r \left( \frac{j_i}{2^k} - x_i \right) \frac{\partial}{\partial x_i} \right)^j f \right\}$$

$$\cdot \left( x_1 + t \left( \frac{j_1}{2^k} - x_1 \right), \ldots, x_r + t \left( \frac{j_r}{2^k} - x_r \right) \right)$$

and

$$g_{\frac{\vec{j}}{2^k}}(0) = f(\vec{x}).$$

By Taylor's formula we have

$$f\left(\frac{\vec{j}}{2^k}\right) = g_{\frac{\vec{j}}{2^k}}(1) = \sum_{j=0}^{N} \frac{g_{\frac{\vec{j}}{2^k}}^{(j)}(0)}{j!} + \mathcal{R}_N\left(\frac{\vec{j}}{2^k}, 0\right),$$

where

$$\mathcal{R}_N\left(\frac{\vec{j}}{2^k}, 0\right) := \int_0^1 \left(\int_0^{t_1} \cdots \left(\int_0^{t_{N-1}} \left(g_{\frac{\vec{j}}{2^k}}^{(N)}(t_N) - g_{\frac{\vec{j}}{2^k}}^{(N)}(0)\right) dt_N\right) \cdots\right) dt_1.$$

Then

$$f\left(\frac{\vec{j}}{2^k}\right)\varphi(2^k\vec{x} - \vec{j}) = \sum_{j=0}^{N} \frac{g_{\frac{\vec{j}}{2^k}}^{(j)}(0)}{j!}\varphi(2^k\vec{x} - \vec{j}) + \varphi(2^k\vec{x} - \vec{j})\mathcal{R}_n\left(\frac{\vec{j}}{2^k}, 0\right).$$

We get that

$$(B_k f)(\vec{x}) - f(\vec{x}) = \sum_{j=1}^{N} \sum_{\vec{j}=-\infty}^{\infty} \frac{g_{\frac{\vec{j}}{2^k}}^{(j)}(0)}{j!} \cdot \varphi(2^k\vec{x} - \vec{j}) + \mathcal{R},$$

where

$$\mathcal{R} := \sum_{\vec{j}=-\infty}^{\infty} \varphi(2^k\vec{x} - \vec{j})\mathcal{R}_N\left(\frac{\vec{j}}{2^k}, 0\right).$$

Notice that

$$\sum_{\vec{j}=-\infty}^{\infty} \cdots = \sum_{\vec{j}=\lceil 2^k\vec{x}-\vec{a}\rceil}^{\lfloor 2^k\vec{x}+\vec{a}\rfloor},$$

where $\vec{a} = (a_1, \ldots, a_r)$ and $\lfloor \cdot \rfloor$, $\lceil \cdot \rceil$ denote the integral part and ceiling of the number, respectively.

Also it holds that

$$\left|x_i - \frac{j_i}{2^k}\right| \le \frac{a_i}{2^k}, \quad i = 1, \ldots, r.$$

## 9. Quantitative Multidimensional Approximation

Moreover, we notice that

$$|g^{(j)}_{\frac{\vec{j}}{2^k}}(0)| \le \left(\frac{a}{2^k}\right)^j \left(\left(\sum_{i=1}^{r}\left|\frac{\partial}{\partial x_i}\right|\right)^j f(\vec{x})\right).$$

Thus

$$\left|\sum_{j=1}^{N}\sum_{\vec{j}=-\infty}^{\infty}\frac{g^{(j)}_{\frac{\vec{j}}{2^k}}(0)}{j!}\varphi(2^k\vec{x}-\vec{j})\right|$$

$$\le \sum_{j=1}^{N}\frac{a^j}{j!2^{kj}}\left(\left(\sum_{i=1}^{r}\left|\frac{\partial}{\partial x_i}\right|\right)^j f(\vec{x})\right)\sum_{\vec{j}=-\infty}^{\infty}\varphi(2^k\vec{x}-\vec{j})$$

$$= \sum_{j=1}^{N}\frac{a^j}{j!2^{kj}}\left(\left(\sum_{i=1}^{r}\left|\frac{\partial}{\partial x_i}\right|\right)^j f(\vec{x})\right). \qquad (9.6)$$

From (9.6) we obtain

$$|B_k(f)(\vec{x}) - f(\vec{x})| \le \sum_{j=1}^{N}\frac{a^j}{j!2^{kj}}\left(\left(\sum_{i=1}^{r}\left|\frac{\partial}{\partial x_i}\right|\right)^j f(\vec{x})\right) + |\mathcal{R}|. \qquad (9.7)$$

But we notice that

$$\mathcal{R} = \sum_{\vec{j}=\lceil 2^k\vec{x}-\vec{a}\rceil}^{[2^k\vec{x}+\vec{a}]}\varphi(2^k\vec{x}-\vec{j})\mathcal{R}_N\left(\frac{\vec{j}}{2^k},0\right).$$

Next we estimate $|\mathcal{R}|$: For that we see $(0 \le t_N \le 1)$

$$|g^{(N)}_{\frac{\vec{j}}{2^k}}(t_N) - g^{(N)}_{\frac{\vec{j}}{2^k}}(0)| = \left|\left(\sum_{i=1}^{r}\left(\frac{j_i}{2^k}-x_i\right)\frac{\partial}{\partial x_i}\right)^N f\left(\vec{x}+t_N\left(\frac{\vec{j}}{2^k}-\vec{x}\right)\right)\right.$$

$$\left. - \left(\sum_{i=1}^{r}\left(\frac{j_i}{2^k}-x_i\right)\frac{\partial}{\partial x_i}\right)^N f(\vec{x})\right| \le \frac{a^N r^N}{2^{kN}}\max_{\vec{\alpha}:|\vec{\alpha}|=N}\omega_1\left(f_{\vec{\alpha}},\frac{a}{2^k}\right).$$

Therefore

$$\left|\mathcal{R}_N\left(\frac{\vec{j}}{2^k},0\right)\right| \leq \int_0^1 \left(\int_0^{t_1} \cdots \left(\int_0^{t_{N-1}} |g_{\frac{\vec{j}}{2^k}}^{(N)}(t_N) - g_{\frac{\vec{j}}{2^k}}^{(N)}(0)|dt_N\right)\cdots dt_1\right)$$

$$\leq \frac{a^N r^N}{N! 2^{kN}} \max_{\tilde{\alpha}:|\tilde{\alpha}|=N} \omega_1\left(f_{\tilde{\alpha}}, \frac{a}{2^k}\right) =: \gamma.$$

That is

$$|\mathcal{R}| \leq \left(\sum_{\vec{j}=-\infty}^{\infty} \varphi(2^k \vec{x} - \vec{j})\right) \gamma = \gamma. \tag{9.8}$$

Finally, from (9.7) and (9.8) we obtain (9.3). ∎

A second main result follows.

**THEOREM 9.2**
*Here we consider $f$ and $\varphi$ as in Theorem 9.1 and use the same notations. Additionally, suppose that $\varphi$ is Lebesgue measurable. Define*

$$A_k(f)(\vec{x}) := \sum_{\vec{j}=-\infty}^{\infty} \alpha_{k\vec{j}}(f)\varphi(2^k \vec{x} - \vec{j}), \tag{9.9}$$

*where*

$$\alpha_{k\vec{j}}(f) := \int_{\mathbf{R}^r} f\left(\frac{\vec{u}}{2^k}\right) \varphi(\vec{u} - \vec{j}) d\vec{u}, \quad k \in \mathbf{Z}. \tag{9.10}$$

*Then for any $k \in \mathbf{Z}$ we get*

$$|A_k(f)(\vec{x}) - f(\vec{x})| \leq \sum_{j=1}^{N} \frac{a^j}{j! 2^{(k-1)j}} \left(\left(\sum_{i=1}^{r} \left|\frac{\partial}{\partial x_i}\right|\right)^j f(\vec{x})\right)$$

$$+ \frac{a^N r^N}{N! 2^{(k-1)N}} \max_{\tilde{\alpha}:|\tilde{\alpha}|=N} \omega_1\left(f_{\tilde{\alpha}}, \frac{a}{2^{k-1}}\right), \tag{9.11}$$

*which is attained by constant functions.*

**REMARK 9.2** i) Clearly here $A_k f \to f$ pointwise on $\mathbf{R}^r$, as $k \to \infty$.

ii) Given that $f \in C_b^N(\mathbf{R}^r)$ we find

$$\|A_k f - f\|_\infty \le \sum_{j=1}^{N} \frac{a^j}{j! 2^{(k-1)j}} \left( \left( \sum_{i=1}^{r} \left\| \frac{\partial}{\partial x_i} \right\|_\infty \right)^j f \right)$$

$$+ \frac{a^N r^N}{N! 2^{(k-1)N}} \max_{\tilde{\alpha}: |\tilde{\alpha}| = N} \omega_1 \left( f_{\tilde{\alpha}}, \frac{a}{2^{k-1}} \right), \text{ any } k \in \mathbf{Z}. (9.12)$$

That is $A_k f \to f$, uniformly on $\mathbf{R}^r$, as $k \to \infty$.

iii) When $N = 1$ from (9.11) we have that

$$|(A_k f)(\vec{x}) - f(\vec{x})| \le \frac{a}{2^{k-1}} \left\{ \left( \sum_{i=1}^{r} \left| \frac{\partial f(\vec{x})}{\partial x_i} \right| \right) \right. \quad (9.13)$$

$$\left. + r \max_{i \in \{1, \dots, r\}} \omega_1 \left( \frac{\partial f}{\partial x_i}, \frac{a}{2^{k-1}} \right) \right\}, \text{ any } k \in \mathbf{Z}.$$

**PROOF** (of Theorem 9.2). Here, from $\sum_{\vec{j}=-\infty}^{\infty} \varphi(\vec{x} - \vec{j}) = 1$, any $\vec{x} \in \mathbf{R}^r$, we have that

$$\int_{-\infty}^{\infty} \int_{-\infty}^{\infty} \cdots \int_{-\infty}^{\infty} \varphi(x_1, \dots, x_r) dx_1 \dots dx_r$$

$$= \sum_{j_1=-\infty}^{\infty} \int_{j_1}^{j_1+1} \cdots \left( \sum_{j_r=-\infty}^{\infty} \int_{j_r}^{j_r+1} \varphi(x_1, \dots, x_r) dx_r \right) \dots dx_1$$

$$= \sum_{j_1=-\infty}^{\infty} \int_{0}^{1} \cdots \left( \sum_{j_r=-\infty}^{\infty} \int_{0}^{1} \varphi(x_1 + j_1, \dots, x_r + j_r) dx_r \right) \dots dx_1$$

$$= \int_{0}^{1} \cdots \left( \int_{0}^{1} \sum_{j_1=-\infty}^{\infty} \cdots \sum_{j_r=-\infty}^{\infty} \varphi(x_1 + j_1, \dots, x_r + j_r) dx_r \right) \dots dx_1$$

$$= \int_{\vec{0}}^{\vec{1}} \left( \sum_{\vec{j}=-\infty}^{\infty} \varphi(\vec{x} + \vec{j}) \right) d\vec{x} = \int_{\vec{0}}^{\vec{1}} 1 d\vec{x} = 1.$$

In other words,
$$\int_{\mathbf{R}^r} \varphi(\vec{x}) d\vec{x} = 1,$$
and in particular
$$\int_{\mathbf{R}^r} \varphi(\vec{x} - \vec{j}) d\vec{x} = 1, \qquad (9.14)$$
any $\vec{j} := (j_1, \ldots, j_r) \in \mathbf{Z}^r$. We see that
$$A_k(f)(\vec{x}) - f(\vec{x}) = \sum_{\vec{j}=-\infty}^{\infty} (\alpha_{k\vec{j}}(f) - f(\vec{x})) \varphi(2^k \vec{x} - \vec{j}), \qquad (9.15)$$
where
$$\alpha_{k\vec{j}}(f) - f(\vec{x}) = \int_{\mathbf{R}^r} \left( f\left(\frac{\vec{u}}{2^k}\right) - f(\vec{x}) \right) \varphi(\vec{u} - \vec{j}) d\vec{u}.$$

Put
$$g_{\frac{\vec{u}}{2^k}}(t) := f\left( \vec{x} + t \left( \frac{\vec{u}}{2^k} - \vec{x} \right) \right), \quad 0 \le t \le 1.$$

Then $(j = 1, \ldots, N)$
$$g_{\frac{\vec{u}}{2^k}}^{(j)}(t) = \left\{ \left( \sum_{i=1}^{r} \left( \frac{u_i}{2^k} - x_i \right) \frac{\partial}{\partial x_i} \right)^j f \right\}$$
$$\left( x_1 + t\left(\frac{u_1}{2^k} - x_1\right), \ldots, x_r + t\left(\frac{u_r}{2^k} - x_r\right) \right)$$

and
$$g_{\frac{\vec{u}}{2^k}}(0) = f(\vec{x}).$$

By Taylor's formula we have that
$$f\left(\frac{\vec{u}}{2^k}\right) = g_{\frac{\vec{u}}{2^k}}(1) = \sum_{j=0}^{N} \frac{g_{\frac{\vec{u}}{2^k}}^{(j)}(0)}{j!} + \mathcal{R}_N\left(\frac{\vec{u}}{2^k}, 0\right),$$

where

$$\mathcal{R}_N\left(\frac{\vec{u}}{2^k}, 0\right)$$
$$:= \int_0^1 \left(\int_0^{t_1} \cdots \left(\int_0^{t_{N-1}} \left(g_{\frac{\vec{u}}{2^k}}^{(N)}(t_N) - g_{\frac{\vec{u}}{2^k}}^{(N)}(0)\right) dt_N\right) \cdots\right) dt_1.$$

Therefore

$$\left(f\left(\frac{\vec{u}}{2^k}\right) - f(\vec{x})\right)\varphi(\vec{u}-\vec{j}) = \sum_{j=1}^N \frac{g_{\frac{\vec{u}}{2^k}}^{(j)}(0)}{j!} \varphi(\vec{u}-\vec{j}) + \mathcal{R},$$

where

$$\mathcal{R} := \mathcal{R}_N\left(\frac{\vec{u}}{2^k}, 0\right)\varphi(\vec{u}-\vec{j}).$$

Here we have that

$$\left|\frac{u_i}{2^k} - x_i\right| \le \frac{a_i}{2^{k-1}}, \quad i=1,\ldots,r.$$

Furthermore, we obtain

$$|g_{\frac{\vec{u}}{2^k}}^{(j)}(0)| \le \left(\frac{a}{2^{k-1}}\right)^j \left(\left(\sum_{i=1}^r \left|\frac{\partial}{\partial x_i}\right|\right)^j f(\vec{x})\right).$$

Then

$$|\alpha_{k\vec{j}}(f) - f(\vec{x})| \le \sum_{j=1}^N \frac{1}{j!} \int_{\mathbf{R}^r} |g_{\frac{\vec{u}}{2^k}}^{(j)}(0)|\varphi(\vec{u}-\vec{j})d\vec{u} + \mathcal{R}^* =: (*),$$

where

$$\mathcal{R}^* := \int_{\mathbf{R}^r} \left|\mathcal{R}_N\left(\frac{\vec{u}}{2^k},0\right)\right| \varphi(\vec{u}-\vec{j})d\vec{u}.$$

In other words,

$$(*) \le \sum_{j=1}^N \frac{1}{j!} \frac{a^j}{2^{(k-1)j}} \left(\left(\sum_{i=1}^r \left|\frac{\partial}{\partial x_i}\right|\right)^j f(\vec{x})\right) \cdot \int_{\mathbf{R}^r} \varphi(\vec{u}-\vec{j})d\vec{u} + \mathcal{R}^*.$$

That is

$$|\alpha_{k\vec{j}}(f) - f(\vec{x})| \overset{(9.14)}{\leq} \sum_{j=1}^{N} \frac{a^j}{j! 2^{(k-1)j}} \left( \left( \sum_{i=1}^{r} \left| \frac{\partial}{\partial x_i} \right| \right)^j f(\vec{x}) \right) + \mathcal{R}^*. \quad (9.16)$$

Let $0 \leq t_N \leq 1$, then

$$\left| g_{\frac{\vec{u}}{2^k}}^{(N)}(t_N) - g_{\frac{\vec{u}}{2^k}}^{(N)}(0) \right| = \left| \left( \sum_{i=1}^{r} \left( \frac{u_i}{2^k} - x_i \right) \frac{\partial}{\partial x_i} \right)^N f\left( \vec{x} + t_N \left( \frac{\vec{u}}{2^k} - \vec{x} \right) \right) \right.$$

$$\left. - \left( \sum_{i=1}^{r} \left( \frac{u_i}{2^k} - x_i \right) \frac{\partial}{\partial x_i} \right)^N f(\vec{x}) \right| \leq \frac{a^N r^N}{2^{(k-1)N}} \max_{\tilde{\alpha}: |\tilde{\alpha}|=N} \omega_1 \left( f_{\tilde{\alpha}}, \frac{a}{2^{k-1}} \right).$$

Thus

$$\left| \mathcal{R}_N \left( \frac{\vec{u}}{2^k}, 0 \right) \right| \leq \int_0^1 \left( \int_0^{t_1} \cdots \left( \cdots \int_0^{t_{N-1}} |g_{\frac{\vec{u}}{2^k}}^{(N)}(t_N) - g_{\frac{\vec{u}}{2^k}}^{(N)}(0)| dt_N \right) \cdots dt_1 \right)$$

$$\leq \frac{a^N r^N}{N! 2^{(k-1)N}} \max_{\tilde{\alpha}: |\tilde{\alpha}|=N} \omega_1 \left( f_{\tilde{\alpha}}, \frac{a}{2^{k-1}} \right) =: \chi.$$

That is,

$$\left| \mathcal{R}_N \left( \frac{\vec{u}}{2^k}, 0 \right) \right| \leq \chi.$$

Consequently

$$\mathcal{R}^* \leq \chi \cdot \int_{\mathbf{R}^r} \varphi(\vec{u} - \vec{j}) d\vec{u} \overset{(9.14)}{=} \chi.$$

That is,

$$\mathcal{R}^* \leq \chi. \quad (9.17)$$

From (9.16) and (9.17) we get that

$$|\alpha_{k\vec{j}}(f) - f(\vec{x})| \leq \sum_{j=1}^{N} \frac{a^j}{j! 2^{(k-1)j}} \left( \left( \sum_{i=1}^{r} \left| \frac{\partial}{\partial x_i} \right| \right)^j f(\vec{x}) \right)$$

$$+ \frac{a^N r^N}{N! 2^{(k-1)N}} \max_{\tilde{\alpha}: |\tilde{\alpha}|=N} \omega_1 \left( f_{\tilde{\alpha}}, \frac{a}{2^{k-1}} \right) =: \rho. \quad (9.18)$$

## 9. Quantitative Multidimensional Approximation

Finally, from (9.15) we obtain that

$$|A_k(f)(\vec{x}) - f(\vec{x})| \leq \sum_{\vec{j}=-\infty}^{\infty} |\alpha_{k\vec{j}}(f) - f(\vec{x})| \varphi(2^k\vec{x} - \vec{j})$$

$$\stackrel{(9.18)}{\leq} \rho \sum_{\vec{j}=-\infty}^{\infty} \varphi(2^k\vec{x} - \vec{j}) = \rho,$$

proving (9.11). ∎

The next result follows.

**THEOREM 9.3**
*Here are some assumptions and notations as in Theorem 9.1. Put*

$$\gamma_{k\vec{j}}(f) := 2^{rk} \int_{2^{-k}\vec{j}}^{2^{-k}(\vec{j}+\vec{1})} f(\vec{t}) d\vec{t} = 2^{rk} \int_{\vec{0}}^{2^{-k}} f\left(\vec{t} + \frac{\vec{j}}{2^k}\right) d\vec{t}, \quad (9.19)$$

*and*

$$C_k(f)(\vec{x}) := \sum_{\vec{j}=-\infty}^{\infty} \gamma_{k\vec{j}}(f) \varphi(2^k\vec{x} - \vec{j}), \quad (9.20)$$

*all $\vec{x} \in \mathbf{R}^r$ and $k \in \mathbf{Z}$. Then it holds*

$$|C_k(f)(\vec{x}) - f(\vec{x})| \leq \sum_{j=1}^{N} \frac{(a+1)^j}{j! 2^{kj}} \left(\left(\sum_{i=1}^{r} \left|\frac{\partial}{\partial x_i}\right|\right)^j f(\vec{x})\right)$$

$$+ \frac{(a+1)^N r^N}{N! 2^{kN}} \max_{\tilde{\alpha}: |\tilde{\alpha}|=N} \omega_1\left(f_{\tilde{\alpha}}, \frac{a+1}{2^k}\right), \quad (9.21)$$

*which is attained by constant functions.*

**REMARK 9.3** i) Clearly here $C_k f \to f$ pointwise on $\mathbf{R}^r$, as $k \to \infty$.
ii) Given that $f \in C_b^N(\mathbf{R}^r)$ we get

$$\|C_k f - f\|_\infty \leq \sum_{j=1}^{N} \frac{(a+1)^j}{j! 2^{kj}} \left(\left(\sum_{i=1}^{r} \left\|\frac{\partial}{\partial x_i}\right\|_\infty\right)^j f\right)$$

$$+ \frac{(a+1)^N r^N}{N! 2^{kN}} \max_{\tilde{\alpha}: |\tilde{\alpha}|=N} \omega_1\left(f_{\tilde{\alpha}}, \frac{a+1}{2^k}\right), \quad (9.22)$$

any $k \in \mathbf{Z}$. That is $C_k f \to f$, uniformly on $\mathbf{R}^r$, as $k \to \infty$.

iii) When $N = 1$ from (9.21) we have that

$$|(C_k f)(\vec{x}) - f(\vec{x})| \leq \left(\frac{a+1}{2^k}\right) \left\{ \left( \sum_{i=1}^{r} \left| \frac{\partial f(\vec{x})}{\partial x_i} \right| \right) \right.$$

$$\left. + r \max_{i \in \{1, \ldots, r\}} \omega_1\left(\frac{\partial f}{\partial x_i}, \frac{a+1}{2^k}\right) \right\}, \quad (9.23)$$

any $k \in \mathbf{Z}$. ∎

**PROOF** (of Theorem 9.3). We observe that

$$C_k(f)(\vec{x}) - f(\vec{x}) = \sum_{\vec{j}=-\infty}^{\infty} (\gamma_{k\vec{j}}(f) - f(\vec{x})) \varphi(2^k \vec{x} - \vec{j}).$$

Put

$$g_{\vec{t} + \frac{\vec{j}}{2^k}}(\tau) := f\left(\vec{x} + \tau \left(\vec{t} + \frac{\vec{j}}{2^k} - \vec{x}\right)\right), \quad 0 \leq \tau \leq 1.$$

Therefore

$$g^{(j)}_{\vec{t} + \frac{\vec{j}}{2^k}}(\tau) = \left\{ \left( \sum_{i=1}^{r} \left(t_i + \frac{j_i}{2^k} - x_i\right) \frac{\partial}{\partial x_i} \right)^j f \right\}$$

$$\cdot \left( x_1 + \tau \left(t_1 + \frac{j_1}{2^k} - x_1\right), \ldots, x_r + \tau \left(t_r + \frac{j_r}{2^k} - x_r\right) \right)$$

and

$$g_{\vec{t} + \frac{\vec{j}}{2^k}}(0) = f(\vec{x}).$$

By Taylor's formula we have

$$f\left(\vec{t} + \frac{\vec{j}}{2^k}\right) = g_{\vec{t} + \frac{\vec{j}}{2^k}}(1) = \sum_{j=0}^{N} \frac{g^{(j)}_{\vec{t} + \frac{\vec{j}}{2^k}}(0)}{j!} + \mathcal{R}_N\left(\vec{t} + \frac{\vec{j}}{2^k}, 0\right),$$

where

$$\mathcal{R}_N\left(\vec{t}+\frac{\vec{j}}{2^k},0\right) := \int_0^1 \left(\int_0^{t_1} \cdots \left(\int_0^{t_{N-1}} \left(g^{(N)}_{\vec{t}+\frac{\vec{j}}{2^k}}(t_N) - g^{(N)}_{\vec{t}+\frac{\vec{j}}{2^k}}(0)\right) dt_N\right) \cdots\right) dt_1.$$

Then

$$\gamma_{k\vec{j}}(f) - f(\vec{x}) = \sum_{j=1}^{N} 2^{rk} \frac{\int_{\vec{0}}^{2^{-\vec{k}}} g^{(j)}_{\vec{t}+\frac{\vec{j}}{2^k}}(0) d\vec{t}}{j!} + 2^{rk}\int_{\vec{0}}^{2^{-\vec{k}}} \mathcal{R}_N\left(\vec{t}+\frac{\vec{j}}{2^k},0\right) d\vec{t}.$$

Here $0 \le t_i \le 2^{-k}$, and

$$\left|x_i - \frac{j_i}{2^k}\right| \le \frac{a_i}{2^k}, \quad i=1,\ldots,r.$$

Moreover, we have $(j=1,\ldots,N)$

$$|g^{(j)}_{\vec{t}+\frac{\vec{j}}{2^k}}(0)| \le \left(\frac{a+1}{2^k}\right)^j \left(\left(\sum_{i=1}^{r}\left|\frac{\partial}{\partial x_i}\right|\right)^j f(\vec{x})\right).$$

Then

$$\sum_{j=1}^{N} \frac{2^{rk}}{j!} \int_{\vec{0}}^{2^{-\vec{k}}} |g^{(j)}_{\vec{t}+\frac{\vec{j}}{2^k}}(0)| d\vec{t} \le \sum_{j=1}^{N} \frac{(a+1)^j}{j! 2^{kj}} \left(\left(\sum_{i=1}^{r}\left|\frac{\partial}{\partial x_i}\right|\right)^j f(\vec{x})\right). \quad (9.24)$$

Let $0 \le \tau_N \le 1$, then

$$|g^{(N)}_{\vec{t}+\frac{\vec{j}}{2^k}}(\tau_N) - g^{(N)}_{\vec{t}+\frac{\vec{j}}{2^k}}(0)|$$

$$= \left|\left(\sum_{i=1}^{r}\left(t_i + \frac{j_i}{2^k} - x_i\right)\frac{\partial}{\partial x_i}\right)^N f\left(\vec{x} + \tau_N\left(\vec{t}+\frac{\vec{j}}{2^k}-\vec{x}\right)\right)\right.$$

$$-\left(\sum_{i=1}^{r}\left(t_i + \frac{j_i}{2^k} - x_i\right)\frac{\partial}{\partial x_i}\right)^N f(\vec{x})\Bigg|$$

$$\leq \frac{(a+1)^N r^N}{2^{kN}} \max_{\tilde{\alpha}:|\tilde{\alpha}|=N} \omega_1\left(f_{\tilde{\alpha}}, \frac{a+1}{2^k}\right).$$

And

$$\left|\mathcal{R}_N\left(\vec{t} + \frac{\vec{j}}{2^k}, 0\right)\right|$$

$$\leq \int_0^1 \left(\int_0^{t_1} \cdots \left(\int_0^{t_{N-1}} |g_{\vec{t}+\frac{\vec{j}}{2^k}}^{(N)}(t_N) - g_{\vec{t}+\frac{\vec{j}}{2^k}}^{(N)}(0)|dt_N\right)\cdots dt_1\right.$$

$$\leq \frac{(a+1)^N r^N}{N! 2^{kN}} \max_{\tilde{\alpha}:|\tilde{\alpha}|=N} \omega_1\left(f_{\tilde{\alpha}}, \frac{a+1}{2^k}\right). \tag{9.25}$$

Consequently (by (9.24) and (9.25))

$$|\gamma_{k\vec{j}}(f) - f(\vec{x})| \leq \sum_{j=1}^{N} \frac{(a+1)^j}{j! 2^{kj}} \left(\left(\sum_{i=1}^{r}\left|\frac{\partial}{\partial x_i}\right|\right)^j f(\vec{x})\right)$$

$$+ \frac{(a+1)^N r^N}{N! 2^{kN}} \max_{\tilde{\alpha}:|\tilde{\alpha}|=N} \omega_1\left(f_{\tilde{\alpha}}, \frac{a+1}{2^k}\right) =: \lambda. \tag{9.26}$$

Eventually we get that

$$|C_k(f)(\vec{x}) - f(\vec{x})| \overset{(9.26)}{\leq} \lambda \sum_{\vec{j}=-\infty}^{\infty} \varphi(2^k \vec{x} - \vec{j}) = \lambda,$$

proving (9.21). ∎

The quadrature wavelet type multivariate operator is studied next.

**THEOREM 9.4**
*Same assumptions and notations as in Theorem 9.1. Put $(k \in \mathbf{Z}, \vec{j} \in \mathbf{Z}^r,$*

## 9. Quantitative Multidimensional Approximation

$\vec{x} \in \mathbf{R}^r$)

$$(D_k f)(\vec{x}) := \sum_{\vec{j}=-\infty}^{\infty} \delta_{k\vec{j}}(f) \varphi(2^k \vec{x} - \vec{j}), \quad (9.27)$$

where

$$\delta_{k\vec{j}}(f) := \sum_{\vec{\ell}=\vec{0}}^{\vec{n}} w_{\vec{\ell}} f\left(\frac{\vec{j}}{2^k} + \frac{\vec{\ell}}{2^k \vec{n}}\right),$$

$$\vec{\ell} \in \mathbf{Z}_+^r, \quad \vec{n} \in \mathbf{N}^r, \quad w_{\vec{\ell}} \geq 0, \quad \sum_{\vec{\ell}=\vec{0}}^{\vec{n}} w_{\vec{\ell}} = 1.$$

That is,

$$\delta_{k,j_1,\ldots,j_r}(f) = \sum_{\ell_1=0}^{n_1} \sum_{\ell_2=0}^{n_2} \cdots \sum_{\ell_r=0}^{n_r} w_{\ell_1,\ldots,\ell_r}$$

$$\cdot f\left(\frac{j_1}{2^k} + \frac{\ell_1}{2^k n_1}, \frac{j_2}{2^k} + \frac{\ell_2}{2^k n_2}, \ldots, \frac{j_r}{2^k} + \frac{\ell_r}{2^k n_r}\right),$$

$$w_{\ell_1,\ell_2,\ldots,\ell_r} \geq 0, \quad \sum_{\ell_1=0}^{n_1} \cdots \sum_{\ell_r=0}^{n_r} w_{\ell_1,\ldots,\ell_r} = 1.\Bigg)$$

Then it holds

$$|(D_k f)(\vec{x}) - f(\vec{x})| \leq \sum_{j=1}^{N} \frac{(a+1)^j}{j! 2^{kj}} \left(\left(\sum_{i=1}^{r} \left|\frac{\partial}{\partial x_i}\right|\right)^j f(\vec{x})\right)$$

$$+ \frac{(a+1)^N r^N}{N! 2^{kN}} \max_{\tilde{\alpha}: |\tilde{\alpha}|=N} \omega_1\left(f_{\tilde{\alpha}}, \frac{a+1}{2^k}\right), \quad (9.28)$$

which is attained by constant functions.

**REMARK 9.4** i) Clearly here $D_k f \to f$ pointwise on $\mathbf{R}^r$, as $k \to \infty$.
ii) Given that $f \in C_b^N(\mathbf{R}^r)$ we find

$$\|D_k f - f\|_\infty \leq \sum_{j=1}^{N} \frac{(a+1)^j}{j! 2^{kj}} \left(\left(\sum_{i=1}^{r} \left\|\frac{\partial}{\partial x_i}\right\|_\infty\right)^j f\right)$$

$$+ \frac{(a+1)^N r^N}{N! 2^{kN}} \max_{\tilde{\alpha}:|\tilde{\alpha}|=N} \omega_1 \left(f_{\tilde{\alpha}}, \frac{a+1}{2^k}\right), \quad \text{any } k \in \mathbf{Z}. \quad (9.29)$$

That is, $D_k f \to f$, uniformly on $\mathbf{R}^r$, as $k \to \infty$.

iii) When $N = 1$ from (9.28) we have that

$$|(D_k f)(\vec{x}) - f(\vec{x})| \leq \left(\frac{a+1}{2^k}\right) \left\{ \left(\sum_{i=1}^{r} \left|\frac{\partial f(\vec{x})}{\partial x_i}\right|\right) \right.$$

$$\left. + r \max_{i \in \{1,\ldots,r\}} \omega_1 \left(\frac{\partial f}{\partial x_i}, \frac{a+1}{2^k}\right)\right\}, \quad \text{any } k \in \mathbf{Z}. \quad (9.30)$$

∎

**PROOF** (of Theorem 9.4). We see that

$$(D_k f)(\vec{x}) - f(\vec{x}) = \sum_{\vec{j}=-\infty}^{\infty} (\delta_{k\vec{j}}(f) - f(\vec{x})) \varphi(2^k \vec{x} - \vec{j}),$$

where

$$\delta_{k\vec{j}}(f) - f(\vec{x}) = \sum_{\vec{\ell}=\vec{0}}^{\vec{n}} w_{\vec{\ell}} \left( f\left(\frac{\vec{j}}{2^k} + \frac{\vec{\ell}}{2^k \vec{n}}\right) - f(\vec{x}) \right).$$

Put

$$g_{\frac{\vec{j}}{2^k} + \frac{\vec{\ell}}{2^k \vec{n}}}(\tau) := f\left(\vec{x} + \tau\left(\frac{\vec{j}}{2^k} + \frac{\vec{\ell}}{2^k \vec{n}} - \vec{x}\right)\right), \quad 0 \leq \tau \leq 1.$$

Therefore $(j = 1, \ldots, N)$

$$g^{(j)}_{\frac{\vec{j}}{2^k} + \frac{\vec{\ell}}{2^k \vec{n}}}(\tau) = \left\{ \left(\sum_{i=1}^{r} \left(\frac{j_i}{2^k} + \frac{\ell_i}{2^k n_i} - x_i\right) \frac{\partial}{\partial x_i}\right)^j f \right\}$$

$$\cdot \left(x_1 + \tau\left(\frac{j_1}{2^k} + \frac{\ell_1}{2^k n_1} - x_1\right), \ldots, x_r + \tau\left(\frac{j_r}{2^k} + \frac{\ell_r}{2^k n_r} - x_r\right)\right)$$

and

$$g_{\frac{\vec{j}}{2^k} + \frac{\vec{\ell}}{2^k \vec{n}}}(0) = f(\vec{x}).$$

## 9. Quantitative Multidimensional Approximation

By Taylor's formula we have

$$f\left(\frac{\vec{j}}{2^k} + \frac{\vec{\ell}}{2^k \vec{n}}\right) - f(\vec{x}) = g_{\frac{\vec{j}}{2^k} + \frac{\vec{\ell}}{2^k \vec{n}}}(1) - f(\vec{x})$$

$$= \sum_{j=1}^{N} \frac{g_{\frac{\vec{j}}{2^k} + \frac{\vec{\ell}}{2^k \vec{n}}}^{(j)}(0)}{j!} + \mathcal{R}_N\left(\frac{\vec{j}}{2^k} + \frac{\vec{\ell}}{2^k \vec{n}}, 0\right),$$

where

$$\mathcal{R}_N\left(\frac{\vec{j}}{2^k} + \frac{\vec{\ell}}{2^k \vec{n}}, 0\right) := \int_0^1 \left(\int_0^{t_1}\right.$$

$$\cdots \left(\int_0^{t_{N-1}} \left(g_{\frac{\vec{j}}{2^k} + \frac{\vec{\ell}}{2^k \vec{n}}}^{(N)}(t_N) - g_{\frac{\vec{j}}{2^k} + \frac{\vec{\ell}}{2^k \vec{n}}}^{(N)}(0)\right) dt_N\right) \cdots \right) dt_1.$$

Consequently

$$\delta_{k\vec{j}}(f) - f(\vec{x}) = \sum_{j=1}^{N} \sum_{\vec{\ell}=\vec{0}}^{\vec{n}} w_{\vec{\ell}} \frac{g_{\frac{\vec{j}}{2^k} + \frac{\vec{\ell}}{2^k \vec{n}}}^{(j)}(0)}{j!} + \sum_{\vec{\ell}=\vec{0}}^{\vec{n}} w_{\vec{\ell}} \mathcal{R}_N\left(\frac{\vec{j}}{2^k} + \frac{\vec{\ell}}{2^k \vec{n}}, 0\right).$$

Here it holds that

$$\left|x_i - \frac{j_i}{2^k}\right| \leq \frac{a_i}{2^k}, \quad \frac{\ell_i}{n_i} \leq 1; \quad i = 1, \ldots, r$$

and reminds us that $a := \max(a_1, \ldots, a_r)$.

Moreover, we have ($j = 1, \ldots, N$)

$$\left|g_{\frac{\vec{j}}{2^k} + \frac{\vec{\ell}}{2^k \vec{n}}}^{(j)}(0)\right| \leq \frac{(a+1)^j}{2^{kj}} \left(\left(\sum_{i=1}^{r} \left|\frac{\partial}{\partial x_i}\right|\right)^j f(\vec{x})\right).$$

Then

$$\sum_{j=1}^{N} \sum_{\vec{\ell}=\vec{0}}^{\vec{w}} w_{\vec{\ell}} \frac{\left|g_{\frac{\vec{j}}{2^k} + \frac{\vec{\ell}}{2^k \vec{n}}}^{(j)}(0)\right|}{j!} \leq \sum_{j=1}^{N} \frac{(a+1)^j}{j! 2^{kj}} \left(\left(\sum_{i=1}^{r} \left|\frac{\partial}{\partial x_i}\right|\right)^j f(\vec{x})\right). \quad (9.31)$$

Let $0 \leq \tau_N \leq 1$, then

$$\left| g^{(N)}_{\frac{\vec{j}}{2^k} + \frac{\vec{\ell}}{2^k \vec{n}}}(\tau_N) - g^{(N)}_{\frac{\vec{j}}{2^k} + \frac{\vec{\ell}}{2^k \vec{n}}}(0) \right|$$

$$= \left| \left( \sum_{i=1}^{r} \left( \frac{j_i}{2^k} + \frac{\ell_i}{2^k n_i} - x_i \right) \frac{\partial}{\partial x_i} \right)^N f\left( \vec{x} + \tau_N \left( \frac{\vec{j}}{2^k} + \frac{\vec{\ell}}{2^k \vec{n}} - \vec{x} \right) \right) \right.$$

$$\left. - \left( \sum_{i=1}^{r} \left( \frac{j_i}{2^k} + \frac{\ell_i}{2^k n_i} - x_i \right) \frac{\partial}{\partial x_i} \right)^N f(\vec{x}) \right|$$

$$\leq \frac{(a+1)^N r^N}{2^{kN}} \max_{\tilde{\alpha}:|\tilde{\alpha}|=N} \omega_1\left( f_{\tilde{\alpha}}, \frac{a+1}{2^k} \right).$$

And

$$\left| \mathcal{R}_N\left( \frac{\vec{j}}{2^k} + \frac{\vec{\ell}}{2^k \vec{n}}, 0 \right) \right| \leq \frac{(a+1)^N r^N}{N! 2^{kN}} \max_{\tilde{\alpha}:|\tilde{\alpha}|=N} \omega_1\left( f_{\tilde{\alpha}}, \frac{a+1}{2^k} \right) =: \theta.$$

That is,

$$\sum_{\vec{\ell}=\vec{0}}^{\vec{n}} w_{\vec{\ell}} \left| \mathcal{R}_N\left( \frac{\vec{j}}{2^k} + \frac{\vec{\ell}}{2^k \vec{n}}, 0 \right) \right| \leq \theta. \tag{9.32}$$

Setting things together we find (by (9.31) and (9.32))

$$|\delta_{k\vec{j}}(f) - f(\vec{x})| \leq \sum_{j=1}^{N} \frac{(a+1)^j}{j! 2^{kj}} \left( \left( \sum_{i=1}^{r} \left| \frac{\partial}{\partial x_i} \right| \right)^j f(\vec{x}) \right)$$

$$+ \frac{(a+1)^N r^N}{N! 2^{kN}} \max_{\tilde{\alpha}:|\tilde{\alpha}|=N} \omega_1\left( f_{\tilde{\alpha}}, \frac{a+1}{2^k} \right) =: \rho \tag{9.33}$$

Eventually we observe that

$$|(D_k f)(\vec{x}) - f(\vec{x})| \stackrel{(9.33)}{\leq} \rho \sum_{\vec{j}=-\infty}^{\infty} \varphi(2^k \vec{x} - \vec{j}) = \rho,$$

establishing (9.28). ∎

# 9. Quantitative Multidimensional Approximation

**Note.** We see that

$$A_k(f)(\vec{x}) = A_0(f(2^{-k}\cdot))(2^k\vec{x}),$$

$$B_k(f)(\vec{x}) = B_0(f(2^{-i}\cdot))(2^k\vec{x}), \tag{9.34}$$

$$C_k(f)(\vec{x}) = C_0(f(2^{-k}\cdot))(2^k\vec{x}),$$

and

$$D_k(f)(\vec{x}) = D_0(f(2^{-k}\cdot))(2^k\vec{x}),$$

all $x \in \mathbf{Z}$ and $\vec{x} \in \mathbf{R}^r$.

In [41] and [45] it was proved that

$$|(B_k f)(\vec{x}) - f(\vec{x})| \leq \omega_1\left(f, \frac{a}{2^k}\right), \tag{9.35}$$

where $f \in C(\mathbf{R}^r)$, $a := \max(a_1, \ldots, a_r)$, any $k \in \mathbf{Z}$, all $\vec{x} \in \mathbf{R}^r$. Also (9.35) was proved to be sharp.

In the following we treat the other operators similarly.

## PROPOSITION 9.1

Let $f \in C(\mathbf{R}^r)$, $r \geq 1$, $a := \max(a_1, \ldots, a_r)$, $k \in \mathbf{Z}$, $\vec{x} \in \mathbf{R}^r$.

*i)* Under the notations and assumptions of Theorem 9.2 we get

$$|A_k(f)(\vec{x}) - f(\vec{x})| \leq \omega_1\left(f, \frac{a}{2^{k-1}}\right). \tag{9.36}$$

*ii)* Under the notations and assumptions of Theorems 9.1, 9.3 we have

$$|C_k(f)(\vec{x}) - f(\vec{x})| \leq \omega_1\left(f, \frac{a+1}{2^k}\right). \tag{9.37}$$

*iii)* Under the notations and assumptions of Theorems 9.1, 9.4 it holds

$$|D_k(f)(\vec{x}) - f(\vec{x})| \leq \omega_1\left(f, \frac{a+1}{2^k}\right). \tag{9.38}$$

**PROOF** i) We observe that

$$|\alpha_{k\vec{j}}(f) - f(\vec{x})| = \left|\int_{\mathbf{R}^r} \left(f\left(\frac{\vec{t}}{2^k}\right) - f(\vec{x})\right) \varphi(\vec{t} - \vec{j}) d\vec{t}\right|$$

$$\leq \int_{\mathbf{R}^r} \omega_1\left(f, \left|\frac{\vec{t}}{2^k} - \vec{x}\right|\right) \varphi(\vec{t} - \vec{j}) d\vec{t}$$

$$\leq \omega_1\left(f, \frac{a}{2^{k-1}}\right) \int_{\mathbf{R}^r} \varphi(\vec{t} - \vec{j}) d\vec{t} = \omega_1\left(f, \frac{a}{2^{k-1}}\right) 1.$$

In other words,
$$|\alpha_{k\vec{j}}(f) - f(\vec{x})| \leq \omega_1\left(f, \frac{a}{2^{k-1}}\right).$$

And
$$|(A_k f)(\vec{x}) - f(\vec{x})| \leq \sum_{\vec{j}=-\infty}^{\infty} |\alpha_{k\vec{j}}(f) - f(\vec{x})| \varphi(2^k \vec{x} - \vec{j})$$

$$\leq \omega_1\left(f, \frac{a}{2^{k-1}}\right) \sum_{\vec{j}=-\infty}^{\infty} \varphi(2^k \vec{x} - \vec{j}) = \omega_1\left(f, \frac{a}{2^{k-1}}\right).$$

That is (9.36) is proved.

ii) Here we observe that

$$|\gamma_{k\vec{j}}(f)(\vec{x}) - f(\vec{x})| = \left|2^{rk} \int_{\vec{0}}^{2^{-k}} f\left(\vec{t} + \frac{\vec{j}}{2^k}\right) d\vec{t} - f(\vec{x})\right|$$

$$\leq 2^{rk} \int_{\vec{0}}^{2^{-k}} \left|f\left(\vec{t} + \frac{\vec{j}}{2^k}\right) - f(\vec{x})\right| d\vec{t} \leq \omega_1\left(f, \frac{a+1}{2^k}\right).$$

Hence
$$|C_k(f)(\vec{x}) - f(\vec{x})| \leq \sum_{\vec{j}=-\infty}^{\infty} |\gamma_{k\vec{j}}(f) - f(\vec{x})| \varphi(2^k \vec{x} - \vec{j})$$

$$\leq \omega_1\left(f, \frac{a+1}{2^k}\right) \sum_{\vec{j}=-\infty}^{\infty} \varphi(2^k \vec{x} - \vec{j}) = \omega_1\left(f, \frac{a+1}{2^k}\right),$$

proving (9.37).

iii) We see that

$$|\delta_{k\vec{j}}(f) - f(\vec{x})| = \left|\sum_{\vec{\ell}=\vec{0}}^{\vec{n}} w_{\vec{\ell}}\left(f\left(\frac{\vec{j}}{2^k} + \frac{\vec{\ell}}{2^k\vec{n}}\right) - f(\vec{x})\right)\right|$$

$$\leq \sum_{\vec{\ell}=\vec{0}}^{\vec{n}} w_{\vec{\ell}}\left|f\left(\frac{\vec{j}}{2^k} + \frac{\vec{\ell}}{2^k\vec{n}}\right) - f(\vec{x})\right| \leq \omega_1\left(f, \frac{a+1}{2^k}\right).$$

Therefore

$$|D_k(f)(\vec{x}) - f(\vec{x})| \leq \sum_{\vec{j}=-\infty}^{\infty} |\delta_{k\vec{j}}(f) - f(\vec{x})|\varphi(2^k\vec{x} - \vec{j})$$

$$\leq \omega_1\left(f, \frac{a+1}{2^k}\right) \sum_{\vec{j}=-\infty}^{\infty} \varphi(2^k\vec{x} - \vec{j}) = \omega_1\left(f, \frac{a+1}{2^k}\right),$$

establishing (9.38). ∎

Optimality of the last inequalities is established next.

**PROPOSITION 9.2**
*Inequalities (9.36), (9.37), (9.38) are sharp!*

**PROOF**  i) Here we want to prove that

$$|A_k(f)(\vec{x}) - f(\vec{x})| \leq \omega_1\left(f, \frac{a}{2^{k-1}}\right)$$

is sharp, where $a := \max(a_1, \ldots, a_r)$. Also

$$A_k(f)(\vec{x}) := \sum_{\vec{j}=-\infty}^{\infty} \alpha_{k\vec{j}}(f)\varphi(2^k\vec{x} - \vec{j}),$$

where

$$\alpha_{k\vec{j}}(f) := \int_{\mathbf{R}^r} f\left(\frac{\vec{u}}{2^k}\right)\varphi(\vec{u} - \vec{j})d\vec{u}.$$

Suppose that there exists $0 < c^r < 1$ (which implies $0 < c < 1$) such that

$$|A_k(f)(\vec{x}) - f(\vec{x})| \leq c^r \omega_1\left(f, \frac{a}{2^{k-1}}\right). \tag{9.39}$$

Here, pick as

$$\varphi(\vec{x}) := \prod_{i=1}^{r} \chi_{[-\frac{1}{2}, \frac{1}{2})}(x_i),$$

where $\vec{x} := (x_1, \ldots, x_r)$ and $\chi_{[-\frac{1}{2}, \frac{1}{2})}$ stands for the characteristic function on $[-\frac{1}{2}, \frac{1}{2})$.

One can easily see that this $\varphi$ fulfills all assumptions of Proposition 9.1. Also take

$$f(x_1, \ldots, x_r) := \prod_{i=1}^{r} g(x_i),$$

where

$$g(x_i) := \begin{cases} 0, & x_i \leq -2^{-k-1}, \\ 1, & x_i \geq -2^{-k-1} + 2^{-k}(1-c), \\ \frac{2^k x_i + (\frac{1}{2})}{1-c}, & -2^{-k-1} < x_i < -2^{-k-1} + 2^{-k}(1-c). \end{cases}$$

Here $a = \frac{1}{2}$ and

$$\omega_1\left(f, \frac{a}{2^{k-1}}\right) = \omega_1\left(f, \frac{1}{2^k}\right).$$

Notice that $(0 < \frac{1-c}{2^k} < \frac{1}{2^k})$

$$\omega_1\left(f, \frac{1}{2^k}\right) = \omega_1\left(\prod_{i=1}^{r} g(x_i), \frac{1}{2^k}\right)$$

$$= \sup_{\substack{\text{all } x_i', x_i'' \in \mathbf{R} \\ |x_i' - x_i''| \leq \frac{1}{2^k},\, i=1,\ldots,r}} \left|\prod_{i=1}^{r} g(x_i') - \prod_{i=1}^{r} g(x_i'')\right| = 1.$$

In other words,

$$\omega_1\left(f, \frac{1}{2^k}\right) = 1.$$

That is (by (9.39))

$$|A_k(f)(\vec{x}) - f(\vec{x})| \leq c^r, \quad \text{all } \vec{x} \in \mathbf{R}^r. \tag{9.40}$$

## 9. Quantitative Multidimensional Approximation

Here we notice that

$$\alpha_{k\vec{j}}(f) = \int_{\mathbf{R}^r} \prod_{i=1}^{r} g\left(\frac{u_i}{2^k}\right) \prod_{i=1}^{r} \chi_{[-\frac{1}{2},\frac{1}{2})}(u_i - j_i) \prod_{i=1}^{r} du_i$$

$$= \prod_{i=1}^{r} \int_{\mathbf{R}} g\left(\frac{u_i}{2^k}\right) \chi_{[-\frac{1}{2},\frac{1}{2})}(u_i - j_i) du_i$$

$$= \prod_{i=1}^{r} 2^k \int_{-\frac{1}{2}+j_i}^{\frac{1}{2}+j_i} g\left(\frac{u_i}{2^k}\right) d\frac{u_i}{2^k} = 2^{kr} \prod_{i=1}^{r} \int_{2^{-k}(-\frac{1}{2}+j_i)}^{2^{-k}(\frac{1}{2}+j_i)} g(v) dv.$$

That is,

$$\alpha_{k\vec{j}}(f) = 2^{kr} \left( \prod_{i=1}^{r} \int_{-2^{-k-1}+2^{-k}j_i}^{2^{-k-1}+2^{-k}j_i} g(v) dv \right).$$

We get that $\alpha_{k\vec{j}}(f) = 0$, if any or some $j_i \leq -1$ from

$$\vec{j} = (j_1, \ldots, j_i, \ldots, j_r).$$

If $j_i \geq 1$, then

$$\int_{-2^{-k-1}+2^{-k}j_i}^{2^{-k-1}+2^{-k}j_i} g(v) dv = 2^{-k}.$$

If $j_i = 0$, then

$$\int_{-2^{-k-1}}^{2^{-k-1}} g(v) dv = 2^{-(k+1)}(1+c).$$

Therefore

$$\alpha_{k\vec{j}}(f) = 2^{m_{\vec{j}}-r}(1+c)^{(r-m_{\vec{j}})},$$

when $j_i \geq 0$ for all $i = 1, 2, \ldots, r$, where $m_{\vec{j}} \in \{0, 1, 2, \ldots, r\}$.
Consequently

$$A_k(f)(\vec{x}) = \sum_{\vec{j}=\vec{0}}^{\infty} \alpha_{k\vec{j}}(f) \left( \prod_{i=1}^{r} \chi_{[-\frac{1}{2},\frac{1}{2})}(2^k x_i - j_i) \right)$$

(i.e., all $j_i \geq 0$, $i = 1, 2, \ldots, r$). Set $x_i := -2^{-k-1}$, all $i = 1, \ldots, r$, then

$$\chi_{[-\frac{1}{2},\frac{1}{2})}(2^k x_i - j_i) = \chi_{[-\frac{1}{2},\frac{1}{2})}\left(-\frac{1}{2} - j_i\right) = 0,$$

if either $j_i \geq 1$ or $j_i \leq -1$. But

$$\chi_{[-\frac{1}{2},\frac{1}{2})}(2^k x_i - j_i) = 1, \quad \text{if } j_i = 0.$$

Therefore

$$\prod_{i=1}^{r} \chi_{[-\frac{1}{2},\frac{1}{2})}(2^k x_i - j_i) = \prod_{i=1}^{r} \chi_{[-\frac{1}{2},\frac{1}{2})}\left(-\frac{1}{2} - j_i\right) = 1$$

if all $j_i = 0$, $i = 1, \ldots, r$; otherwise it is zero. And ($\vec{i}$ is the unit vector in $\mathbf{R}^r$)

$$A_k(f)(-2^{-k-1}\vec{i}) = \alpha_{k\vec{0}}(f) = \left(\frac{1+c}{2}\right)^r.$$

Also we find that

$$f(-2^{-k-1}\vec{i}) = 0.$$

Then

$$|A_k(f)(-2^{-k-1}\vec{i}) - f(-2^{-k-1}\vec{i})| = \left(\frac{1+c}{2}\right)^r > c^r,$$

is absurd (see (9.40)). This proves that (9.36) is sharp.

ii) Here we establish that

$$|C_k(f)(\vec{x}) - f(\vec{x})| \leq \omega_1\left(f, \frac{a+1}{2^k}\right)$$

is sharp.

Let

$$q(x_i) := \begin{cases} 1, & x_i \geq 0, \\ 0, & x_i \leq -\frac{1}{2^{k+1}}, \\ 2^{k+1}x_i + 1, & -\frac{1}{2^{k+1}} \leq x_i \leq 0, \end{cases}$$

and take

$$f(\vec{x}) := \prod_{i=1}^{r} q(x_i).$$

Consider

$$\varphi(\vec{x}) := \prod_{i=1}^{r} \chi_{[-\frac{1}{2},\frac{1}{2})}(x_i),$$

i.e., $a = \frac{1}{2}$. Here (notice $\frac{1}{2^{k+1}} < \frac{3}{2^{k+1}}$ and the shape of $q$)

$$\omega_1\left(f, \frac{a+1}{2^k}\right) = \omega_1\left(f, \frac{3}{2^{k+1}}\right)$$

## 9. Quantitative Multidimensional Approximation

$$= \sup_{\substack{\text{all } x'_i, x''_i \in \mathbf{R}, \\ |x'_i - x''_i| \le \frac{3}{2^{k+1}} \text{ all } i=1,\ldots,r}} \left| \prod_{i=1}^{r} q(x'_i) - \prod_{i=1}^{r} q(x''_i) \right|$$

$$= \omega_1\left(f, \frac{1}{2^{k+1}}\right) = 1.$$

That is
$$\omega_1\left(f, \frac{3}{2^{k+1}}\right) = 1.$$

We get that

$$\gamma_{k(-\vec{1})}(f) = 2^{rk} \int_{-2^{-k}}^{\vec{0}} f(\vec{t})d\vec{t} = 2^{rk} \left( \prod_{i=1}^{r} \int_{-2^{-k}}^{0} q(t_i)dt_i \right)$$

$$= 2^{rk} \left( \prod_{i=1}^{r} \int_{-2^{-(k+1)}}^{0} q(t_i)dt_i \right) = 4^{-r}.$$

That is,
$$\gamma_{k(-\vec{1})}(f) = \frac{1}{4^r}.$$

Let $\vec{j} = (j_1, \ldots, j_i, \ldots, j_r)$ such that some $j_i \le -2$, then

$$\int_{j_i/2^k}^{(j_i+1)/2^k} q(t_i)dt_i = 0$$

and
$$\gamma_{k\vec{j}}(f) = 2^{rk} \prod_{i=1}^{r} \left( \int_{\frac{j_i}{2^k}}^{\frac{j_i+1}{2^k}} q(t_i)dt_i \right) = 0.$$

That is
$$\gamma_{k\vec{j}}(f) = 0, \quad \text{if any } j_i \le -2.$$

Let all $j_i \ge 0$, then

$$\gamma_{k\vec{j}}(f) = 2^{rk} \prod_{i=1}^{r} \int_{\frac{j_i}{2^k}}^{\frac{j_i+1}{2^k}} q(t_i)dt_i = 2^{rk} \prod_{i=1}^{r} \left(\frac{1}{2^k}\right) = 1.$$

That is $\gamma_{k\vec{j}}(f) = 1$, when all $j_i \geq 0$; $\vec{j} = (j_1, \ldots, j_i, \ldots, j_r)$. We evaluate ($\vec{i}$ is the unit vector in $\mathbf{R}^r$)

$$C_k(f)\left(-\frac{1}{2^{k+1}}\vec{i}\right) = \sum_{\vec{j}=-\infty}^{\infty} \gamma_{k\vec{j}}(f)\varphi\left(-\frac{1}{2}\vec{i} - \vec{j}\right)$$

$$= \sum_{\vec{j}=-\infty}^{\infty} \gamma_{k\vec{j}}(f)\prod_{i=1}^{r} \chi_{[-\frac{1}{2},\frac{1}{2})}\left(-\frac{1}{2} - j_i\right)$$

$$= \sum_{\vec{j}=-\infty}^{\vec{0}} \gamma_{k\vec{j}}(f)\prod_{i=1}^{r} \chi_{[-\frac{1}{2},\frac{1}{2})}\left(-\frac{1}{2} - j_i\right)$$

(i.e., all $j_i \leq 0$)

$$= \gamma_{k\vec{0}}(f)\prod_{i=1}^{r} \chi_{[-\frac{1}{2},\frac{1}{2})}\left(-\frac{1}{2}\right) = 1.$$

That is,

$$C_k(f)\left(-\frac{1}{2^{k+1}}\vec{i}\right) = 1,$$

and $f(-\frac{1}{2^{k+1}}\vec{i}) = 0$. At the end we see that

$$\left|C_k(f)\left(-\frac{1}{2^{k+1}}\vec{i}\right) - f\left(-\frac{1}{2^{k+1}}\vec{i}\right)\right| = 1 = \omega_1\left(f, \frac{3}{2^{k+1}}\right).$$

In other words, inequality (9.37) is attained, that is sharp.

iii) Here we prove that

$$|D_k(f)(\vec{x}) - f(\vec{x})| \leq \omega_1\left(f, \frac{a+1}{2^k}\right)$$

is sharp. We take $q$, $f$, $\varphi$ as in (ii) of this proof. Again

$$\omega_1\left(f, \frac{a+1}{2^k}\right) = \omega_1\left(f, \frac{3}{2^{k+1}}\right) = 1.$$

We notice that $\delta_{k\vec{j}}(f) = 1$, when all $j_i \geq 0$; $\vec{j} = (j_1, \ldots, j_i, \ldots, j_r)$. Also we see that

$$\delta_{k\vec{j}}(f) = 0, \text{ if some } j_i \leq -2,$$

## 9. Quantitative Multidimensional Approximation

the last is true since $q$ is increasing, etc. Furthermore

$$(D_k f)\left(-\frac{1}{2^{k+1}}\vec{i}\right) = \sum_{\vec{j}=-\infty}^{\infty} \delta_{k\vec{j}}(f)\varphi\left(-\frac{1}{2}\vec{i}-\vec{j}\right)$$

$$= \sum_{\vec{j}=-\infty}^{\vec{0}} \delta_{k\vec{j}}(f) \prod_{i=1}^{r} \chi_{[-\frac{1}{2},\frac{1}{2})}\left(-\frac{1}{2}-j_i\right)$$

$$= \delta_{k\vec{0}}(f) \prod_{i=1}^{r} \chi_{[-\frac{1}{2},\frac{1}{2})}\left(-\frac{1}{2}\right) = 1.$$

That is

$$(D_k f)\left(-\frac{1}{2^{k+1}}\vec{i}\right) = 1,$$

and

$$f\left(-\frac{1}{2^{k+1}}\vec{i}\right) = 0.$$

Finally, we see that

$$\left|(D_k f)\left(-\frac{1}{2^{k+1}}\vec{i}\right) - f\left(-\frac{1}{2^{k+1}}\vec{i}\right)\right| = 1 = \omega_1\left(f, \frac{3}{2^{k+1}}\right),$$

proving (9.38) attained, thus it is sharp. ∎

# Chapter 10

## Rate of Convergence of Probabilistic Discrete Wavelet Approximation

Let $\Phi(x) := \begin{cases} 1, & x \geq 0 \\ 0, & x < 0 \end{cases}$ and consider

$$\mathcal{F} := \Big\{ F \text{ distribution functions on } \mathbf{R} :$$

$$\int_{-\infty}^{+\infty} |F(x) - \Phi(x)|^p \, dx < +\infty, \quad 1 \leq p < +\infty \Big\} \quad (10.1)$$

For $F \in \mathcal{F}$, there exist linear wavelet type operators $L_k(F, x)$ that are also distribution functions where the defining mother wavelet is $\Phi(x)$. These approximate $F(x)$ in the $L_p(\mathbf{R})$-norm. The degree of this approximation is given by establishing an $L_p$-Jackson type inequality. This treatment is based on [44].

Basically, we present here another type of wavelet approximation to probabilistic distribution functions. This is motivated by the following very important theorem of Analysis [66, p. 221]. Let $E$ be a locally compact Hausdorff space. It is proved that the discrete measures of the form

$$\sum_{i=1}^{n} a_i \delta_{x_i}$$

where $a_i \geq 0$, $x_i \in E$, $\delta_{x_i}$ is the unit (Dirac) measure, are dense in the weak*-topology in $M^+(E)$, the set of all positive Radon measures on $E$. In this chapter we study the following form of probabilistic discrete wavelets:

$$\sum_{j=-\infty}^{\infty} a_j \Phi(x-j) \qquad (10.2)$$

where $\Phi(x)$ is the distribution function (10.1) associated with the unit measure at $x = 0$.

Here we show that for any $F \in \mathcal{F}$ there exist linear operators $L_k(F, x)$ that are distribution functions of the form (10.2) and converge to $F(x)$ in $L_p(\mathbf{R})$-norm with the approximation errors bounded by $2\omega(F, 2^{-k})_p$, where $\omega(F, t)_p$ is the $L_p$-modulus of continuity of $F$. Since $\omega(\Phi, t)_p < +\infty$, by $L^p$-triangle inequality one can easily see that for $F \in \mathcal{F}$ we have $\omega(F, t)_p < +\infty$. One should notice that any distribution function $F \notin L_p(\mathbf{R})$.

## 10.1 Approximation by $L_k(f, x)$ Operators

For any distribution function $F \in \mathcal{F}$, we define

$$L_0(F, x) := \sum_{j=-\infty}^{\infty} (C_j - C_{j-1}) \Phi(x-j) \qquad (10.3)$$

where

$$C_j := \int_j^{j+1} F(t)\, dt, \quad j \in Z \qquad (10.4)$$

and

$$L_k(F, x) := L_0(F(2^{-k} \cdot), 2^k x), \quad k \in Z. \qquad (10.5)$$

We observe that

$$L_k(F, x) = \sum_{j=-\infty}^{\infty} (C_{k,j} - C_{k,j-1}) \Phi(2^k x - j) \qquad (10.6)$$

where

$$C_{k,j} := 2^k \cdot \int_{2^{-k} j}^{2^{-k}(j+1)} F(t)\, dt, \quad j \in Z,\ k \in Z. \qquad (10.7)$$

In the next lemma we show that $L_k(F, x)$ are well defined.

# 10. Probabilistic Discrete Wavelet Approximation

**LEMMA 10.1**
Let $F(x)$ be a distribution function on $\mathbf{R}$. Then $L_k(F, x)$ are well defined by (10.3) and (10.5), and are also distribution functions.

**PROOF** Since $F(x)$ is a distribution function, $F(x)$ is nondecreasing on $\mathbf{R}$ and hence we have

$$C_{k,j} - C_{k,j-1} \geq 0 \tag{10.8}$$

by (10.7). Furthermore, because $\lim_{x \to +\infty} F(x) = 1$ and $\lim_{x \to -\infty} F(x) = 0$, we see that

$$\lim_{j \to +\infty} C_{k,j} = \lim_{j \to +\infty} 2^k \int_{2^{-k}j}^{2^{-k}(j+1)} F(t)\,dt$$

$$= \lim_{j \to +\infty} \int_0^1 F(2^{-k}(t+j))\,dt$$

$$= \int_0^1 \lim_{j \to +\infty} F(2^{-k}(t+j))\,dt$$

$$= 1, \tag{10.9}$$

and

$$\lim_{j \to -\infty} C_{k,j} = \int_0^1 \lim_{j \to -\infty} F(2^{-k}(t+j))\,dt = 0. \tag{10.10}$$

Here we also used the fact that $F(x) \geq 0$ is nondecreasing for taking the limit under the integration.

The relations (10.8–10.10) prove that $L_k(F, x)$ is a probabilistic mixture of point mass laws and so it is a distribution function as desired. ∎

We present

**THEOREM 10.1**
Let $F(x)$ be a distribution function in $\mathcal{F}$ and $1 \leq p < +\infty$. Then $L_k(F, x)$ defined by (10.3) and (10.5) are distribution functions such that

$$\|L_k(F, x) - F(x)\|_p \leq 2\omega(F, 2^{-k})_p. \tag{10.11}$$

**PROOF** By Lemma 10.1, we only have to establish (10.11).

For any fixed $x$, let $j_0$ be the integer such that $j_0 \le x < j_0 + 1$. From the definition of $\Phi(x)$, we have

$$\Phi(x-j) = 0, \quad j \ge j_0 + 1$$

and

$$\Phi(x-j) = 1, \quad j \le j_0.$$

Thus, from these and (10.10), we get

$$L_0(F, x) = \sum_{j=-\infty}^{j_0} (C_j - C_{j-1})\Phi(x-j) + \sum_{j=j_0+1}^{+\infty} (C_j - C_{j-1})\Phi(x-j)$$

$$= \sum_{j=-\infty}^{j_0} (C_j - C_{j-1}) = C_{j_0} - \lim_{j \to -\infty} C_j = C_{j_0}.$$

Therefore

$$\|L_0(F, x) - F(x)\|_p = \left\| \int_{j_0}^{j_0+1} F(t)\, dt - F(x) \right\|_p$$

$$\le \left\| \int_{j_0}^{x} [F(x) - F(t)]\, dt \right\|_p + \left\| \int_{x}^{j_0+1} [F(t) - F(x)]\, dt \right\|_p.$$

Because $F(x)$ is nondecreasing and noticing $j_0 \le x < j_0 + 1$, we obtain

$$\left\| \int_{j_0}^{x} [F(x) - F(t)]\, dt \right\|_p \le \left\| \int_{x-1}^{x} [F(x) - F(t)]\, dt \right\|_p$$

and

$$\left\| \int_{x}^{j_0+1} [F(t) - F(x)]\, dt \right\|_p \le \left\| \int_{x}^{x+1} [F(t) - F(x)]\, dt \right\|_p.$$

And then

$$\|L_0(F, x) - F(x)\|_p \le \left\| \int_{x-1}^{x} [F(x) - F(t)]\, dt \right\|_p$$

$$+ \left\| \int_{x}^{x+1} [F(t) - F(x)]\, dt \right\|_p.$$

## 10. Probabilistic Discrete Wavelet Approximation

$$\leq \int_0^1 \|F(x) - F(x-t)\|_p \, dt + \int_0^1 \|F(x+t) - F(x)\|_p \, dt$$

$$\leq 2\omega(F, 1)_p. \tag{10.12}$$

So that we have established (10.11) for $k = 0$.

Next we observe that

$$\omega(F(2^{-k}\cdot), 1)_p = \sup_{0 \leq t \leq 1} \left( \int_{-\infty}^{\infty} |F(2^{-k}(x+t)) - F(2^{-k}x)|^p \, dx \right)^{1/p}$$

$$= \sup_{0 \leq t \leq 1} 2^{k/p} \left( \int_{-\infty}^{\infty} |F(u + 2^{-k}t) - F(u)|^p \, du \right)^{1/p}$$

$$= 2^{k/p} \omega(F, 2^{-k})_p. \tag{10.13}$$

We also have

$$\|L_k(F, x) - F(x)\|_p = \|L_0(F(2^{-k}\cdot), 2^k x) - F(2^{-k} \cdot 2^k x)\|_p$$

$$= 2^{-k/p} \|L_0(F(2^{-k}\cdot), x) - F(2^{-k}x)\|_p. \tag{10.14}$$

For a distribution function $F \in \mathcal{F}$, it is easy to see that $F(2^{-k}\cdot)$ is also a distribution function in $\mathcal{F}$. Then, from (10.12), (10.13), and (10.14), we find

$$\|L_k(F, x) - F(x)\|_p = 2^{-k/p} \|L_0(F(2^{-k}\cdot), x) - F(2^{-k}x)\|_p$$

$$\leq 2^{-k/p} \cdot 2\omega(F(2^{-k}\cdot), 1)_p$$

$$\leq 2^{-k/p} \cdot 2 \cdot 2^{k/p} \omega(F, 2^{-k})_p$$

$$= 2\omega(F, 2^{-k})_p$$

which produces (10.11). This ends the proof of the theorem. ∎

# Chapter 11

## Asymptotic Nonorthogonal Wavelet Approximation for Deterministic Signals

An $n$th order asymptotic expansion is given for the $\mathcal{L}_2$-error in an nonorthogonal (in general) wavelet approximation at resolution $2^{-k}$ of deterministic signals $f$. These signals over the whole real line are supposed to have $n$ continuous derivatives of bounded variation. The engaged nonorthogonal (in general) scale function $\varphi$ fulfills the partition of unity property and it is of compact support. The asymptotic expansion has only even terms of products of integrals involving $\varphi$ with integrals of squares of (the first $\lceil \frac{n}{2} \rceil - 1$ only) derivatives of $f$. This treatment is based on [32].

## 11.1 Introduction

Multiresolution signal decomposition and wavelet orthonormal bases of $L_2(\mathbf{R})$ have received a lot of attention in huge numbers of mathematical, signal, and image processing articles and books, e.g., see [59], [67], [71], [123], and [130]. However, the nonorthogonal case is less discussed.

Here we only mention that a multiresolution decomposition of $L_2(\mathbf{R})$ is a sequence $\{V_k\}_{k=-\infty}^{+\infty}$ of closed subspaces of $L_2(\mathbf{R})$ such that for all $k,j \in \mathbf{Z}$:

(1)   $V_k \subset V_{k+1}$,

(2)   $\displaystyle\bigcup_{k=-\infty}^{+\infty} V_k$ is dense in $L_2(\mathbf{R})$,

(3)   $\displaystyle\bigcap_{k=-\infty}^{+\infty} V_k = \emptyset$,

(4)  $f(t) \in V_k$ iff $f(2t) \in V_{k+1}$,

(5)  $f(t) \in V_k \Leftrightarrow f(t - 2^{-k} \cdot j) \in V_k$.

In the orthogonal case the approximation of $f \in L_2(\mathbf{R})$ at resolution $2^{-k}$ is the orthogonal projection $\hat{f}_k$ of $f$ on $V_k$. This is calculated by the use of a wavelet orthonormal basis $\{\varphi_{k,j}(t) := 2^{k/2} \cdot \varphi(2^k t - j)\}_{j=-\infty}^{+\infty}$ for $V_k$, which is generated by the *scale* function $\varphi \in L_2(\mathbf{R})$. A simple example of an orthogonal scale function of compact support and having the partition of unity property is

$$\varphi_0(x) = \begin{cases} 1, & -\frac{1}{2} \leq x < \frac{1}{2}, \\ 0, & \text{otherwise.} \end{cases}$$

To many applications such as image compression or edge detection, the orthogonal environment is not enough, because some natural constraints cannot be achieved. For example, it is impossible to make use of finite impulse response linear phase filters. To overcome such difficulties in general, we perform a different kind of $\mathcal{L}_2$-approximation. Namely, here the approximants of $f \in L_2(\mathbf{R})$ at resolution $2^{-k}$ will be the operators

$$(B_k f)(x) := \sum_{j=-\infty}^{+\infty} f\left(\frac{j}{2^k}\right) \cdot \varphi(2^k x - j),$$

where $\varphi$ is a Lebesgue measurable bounded function with compact support having the partition of unity property, and *not* generating (in general) an orthonormal basis $\varphi_{k,j}$ for $V_k$.

A well-known example of a *nonorthogonal* $\varphi$ as described above, is the $m \geq 2$th order cardinal $B$-spline $N_m$ defined through convolution

$$N_m(x) := (N_{m-1} * N_1)(x) = \int_0^1 N_{m-1}(x - t)dt,$$

where $N_1$ is the characteristic function of the interval $[0, 1)$.

We get that

(i)  $N_m(x) = \dfrac{1}{(m-1)!} \cdot \sum_{\tau=0}^{m}(-1)^\tau \cdot \dbinom{m}{\tau} \cdot (x - \tau)_+^{m-1}$, $m \geq 2$,

(ii)  support $N_m = [0, m]$,

(iii)  $N_m(x) > 0$, for $0 < x < m$,

(iv) $\sum_{j=-\infty}^{+\infty} N_m(x-j) = 1$ all $x \in \mathbf{R}$, and

(v) $N_m$ is bounded and Lebesgue measurable.

For a detailed discussion and properties of $N_m$ see [67], pp. 85–86. So we establish in Theorem 11.1 that $(B_k f)$ converges to $f$ as $k \to +\infty$ in $\mathcal{L}_2$ norm, under mild appropriate assumptions on $f$. In fact, we present an asymptotic expansion for $e_k(f) := \|B_k f - f\|_2$, which is the $L_2$ approximation error at resolution $2^{-k}$. This is an $n$th order asymptotic expansion having a certain degree of smoothness (the existing $n$ derivatives of the deterministic signal $f$).

The exact rate of convergence and asymptotic constant are determined and their dependence on $f$ and on $\varphi$ are given. For related work about deterministic signals $f$ in the orthogonal case, see [58], [59], [123], Theorem 3, and [130].

## 11.2 Asymptotic Wavelet Approximation at Resolution $2^{-k}$

We present the main result:

**THEOREM 11.1**
*Let $f$ be such that $f, f', f'', \ldots, f^{(n)}$ are functions in $L_1(\mathbf{R})$ and $f^{(n)} \in BV(\mathbf{R}) \cap C(\mathbf{R})$, $n \geq 2$. Let $\varphi$ be a Lebesque measurable bounded function with support $\varphi \subseteq [-a, a]$, $0 < a < +\infty$, such that*

$$\sum_{j=-\infty}^{+\infty} \varphi(x-j) = 1, \quad \forall x \in \mathbf{R}. \tag{11.1}$$

Put

$$(B_k f)(x) := \sum_{j=-\infty}^{+\infty} f\left(\frac{j}{2^k}\right) \cdot \varphi(2^k x - j) \tag{11.2}$$

and

$$e_k^2(f) := \int_{-\infty}^{+\infty} (f(x) - (B_k f)(x))^2 \cdot dx, \quad k \in \mathbf{Z} \tag{11.3}$$

Set

$$\beta := \begin{cases} [2a], & \text{if } 2a \neq \text{integer,} \\ 2a - 1, & \text{if } 2a \text{ is an integer,} \end{cases} \tag{11.4}$$

where $[\cdot]$ is the integral part function. Put

$$\ell := \begin{cases} \frac{n-1}{2}, & n \text{ is odd,} \\ \frac{n-2}{2}, & n \text{ is even} \end{cases} = \lceil \frac{n}{2} \rceil - 1, \tag{11.5}$$

where $\lceil \cdot \rceil$ is the ceiling of the number. Call

$$\rho_\gamma(f) := \frac{(-1)^\gamma}{(2\gamma)!} \cdot \int_{-\infty}^{+\infty} (f^{(\gamma)}(t))^2 dt, \quad \gamma = 1, \ldots, \ell. \tag{11.6}$$

Set $(\gamma = 1, \ldots, \ell)$

$$\lambda_\gamma(\varphi) := \int_{-\infty}^{+\infty} \varphi(u) \cdot \left\{ -2 \cdot u^{2 \cdot \gamma} \right.$$

$$\left. + \sum_{q=1}^{\beta} q^{2 \cdot \gamma} \cdot (\varphi(u+q) + \varphi(u-q)) \right\} \cdot du. \tag{11.7}$$

Then, for $k \in \mathbf{N}$, we have that

$$e_k^2(f) = \sum_{\gamma=1}^{\ell^*} \frac{\rho_\gamma(f) \cdot \lambda_\gamma(\varphi)}{4^{k \cdot \gamma}} + o\left(\frac{1}{2^{k \cdot (n-2)}}\right), \tag{11.8}$$

where

$$\ell^* := \begin{cases} \ell, & \text{if } n \text{ is even,} \\ \ell - 1, & \text{if } n \text{ is odd.} \end{cases}$$

**REMARK 11.1** We see that $e_k(f) \to 0$ as $k \to +\infty$, i.e., for $n \geq 2$ we get that $B_k(f) \to f$ as $k \to +\infty$ in $\mathcal{L}_2$-norm.

Notice that the odd powers of $2^{-k}$ are missing in the asymptotic expansion. In particular, when $n = 2$ we obtain that $e_k(f) = o(1)$, and when $n = 3$ we find that $e_k^2(f) = o(2^{-k})$.

Next we denote by

$$\lambda_1(\varphi) := \int_{-\infty}^{+\infty} \varphi(u) \cdot \left\{ -2 \cdot u^2 + \sum_{q=1}^{\beta} q^2 \cdot (\varphi(u+q) + \varphi(u-q)) \right\} \cdot du$$

## 11. Nonorthogonal Wavelet Approximation for Deterministic Signals

and when $n = 4$ we observe that

$$e_k^2(f) = \frac{-\lambda_1(\varphi) \cdot \int_{-\infty}^{+\infty} (f'(t))^2 dt}{2^{2k+1}} + o\left(\frac{1}{2^{2k}}\right).$$

Furthermore, in the case of $n \geq 4$, when $\lambda_1(\varphi) \neq 0$, we get that

$$2^{2k} e_k^2(f) \to \frac{-\lambda_1(\varphi) \cdot \int_{-\infty}^{+\infty} (f'(t))^2 \cdot dt}{2},$$

as $k \to +\infty$, and this rate cannot be improved by additional differentiability of $f$ greater than first order. ∎

**PROOF** (of Theorem 11.1). Here we would like to prove for $e_k^2(f)$ the asymptotic expansion (11.8). This is done in the following paragraphs.

(§1) We start by unfolding $e_k^2(f)$ as far as possible. Notice that

$$e_k^2(f) = \int_{-\infty}^{+\infty} f^2(x) dx - 2 \int_{-\infty}^{\infty} f(x) \cdot (B_k f)(x) \cdot dx$$

$$+ \int_{-\infty}^{+\infty} (B_k f)^2(x) \cdot dx, \quad k \in \mathbf{Z}. \tag{11.9}$$

First we would like to prove that $e_k(f) < +\infty$. Because $f, f', f'', \ldots, f^{(n)} \in L_1(\mathbf{R})$ and $f^{(n)} \in BV(\mathbf{R})$, $(n \geq 2)$, by Remark 1 of [58] we have that $f, f', \ldots, f^{(n-1)}$ are all bounded, of bounded variation, and uniformly continuous on the real line and tend to zero as $|x| \to +\infty$. Also, $f^{(n)}$ is bounded. Furthermore $f^2 \in L_1(\mathbf{R})$, (i.e., $\|f\|_2 < +\infty$) along with $f^2 \in BV(\mathbf{R})$. Hence from Lemma 1 of [58] we find that

$$\sum_{j=-\infty}^{+\infty} \left| f\left(\frac{j}{2^k}\right) \right|^r < +\infty, \quad \text{any } k \in \mathbf{Z}, \tag{11.10}$$

for $r = 1, 2$ (for the case $r = 1$ notice that $|f| \in L_1(\mathbf{R}) \cap BV(\mathbf{R})$).
Next we need to prove that

$$\xi_1 := \sum_{i=-\infty}^{+\infty} \sum_{j=-\infty}^{+\infty} \left| f\left(\frac{i}{2^k}\right) \right| \cdot \left| f\left(\frac{j}{2^k}\right) \right| \cdot \int_{-\infty}^{+\infty} |\varphi(2^k x - i)|$$

$$\cdot |\varphi(2^k x - j)| \cdot dx < +\infty. \tag{11.11}$$

By a change of variable we obtain

$$\xi_1 = \frac{1}{2^k} \cdot \sum_{i=-\infty}^{+\infty} \sum_{j=-\infty}^{+\infty} \left|f\left(\frac{i}{2^k}\right)\right| \cdot \left|f\left(\frac{j}{2^k}\right)\right| \cdot \psi(i-j), \qquad (11.12)$$

where

$$\psi(q) := \int_{-\infty}^{+\infty} |\varphi(y)| \cdot |\varphi(y+q)| \cdot dy \geq 0, \qquad (11.13)$$

and $\psi(q) < +\infty$, $q \in \mathbf{Z}$. Thus we need to prove that

$$\xi_2 := \sum_{i=-\infty}^{+\infty} \sum_{j=-\infty}^{+\infty} \left|f\left(\frac{i}{2^k}\right)\right| \cdot \left|f\left(\frac{j}{2^k}\right)\right| \cdot \psi(i-j) < +\infty. \qquad (11.14)$$

However, (cf. (11.50))

$$\xi_2 = \left(\sum_{i=-\infty}^{+\infty} \left(f\left(\frac{i}{2^k}\right)\right)^2\right) \cdot \psi(0) + \sum_{q=1}^{\beta} \left\{ \sum_{j=-\infty}^{+\infty} \left|f\left(\frac{j+q}{2^k}\right)\right| \cdot \right.$$

$$\left. \cdot \left|f\left(\frac{j}{2^k}\right)\right| \cdot \psi(q) + \sum_{i=-\infty}^{+\infty} \left|f\left(\frac{i}{2^k}\right)\right| \cdot \left|f\left(\frac{i+q}{2^k}\right)\right| \cdot \psi(-q) \right\}$$

$$= \left(\sum_{i=-\infty}^{+\infty} \left(f\left(\frac{i}{2^k}\right)\right)^2\right) \cdot \psi(0) + \sum_{q=1}^{\beta} \left\{ \sum_{i=-\infty}^{+\infty} \left|f\left(\frac{i}{2^k}\right)\right| \right.$$

$$\left. \cdot \left|f\left(\frac{i+q}{2^k}\right)\right| \cdot (\psi(q) + \psi(-q)) \right\} =: \xi_3. \qquad (11.15)$$

That is, we want to prove $\xi_3 < +\infty$ ($\xi_1 < +\infty$ iff $\xi_3 < +\infty$). But from (11.10), (11.13) and earlier comments on $f$ we have that

$$\sum_{i=-\infty}^{+\infty} \left|f\left(\frac{i}{2^k}\right)\right| \cdot \left|f\left(\frac{i+q}{2^k}\right)\right| \cdot (\psi(q) + \psi(-q))$$

$$\leq \|f\|_\infty \cdot \left(\sum_{i=-\infty}^{+\infty} \left|f\left(\frac{i}{2^k}\right)\right|\right) \cdot (\psi(q) + \psi(-q)) < +\infty.$$

## 11. Nonorthogonal Wavelet Approximation for Deterministic Signals

Therefore, again from (11.10), (11.13), and $\beta$ being finite, we find that $\xi_3 < +\infty$, and equivalently that $\xi_1 < +\infty$. Hence from (11.11) we get that

$$\sum_{i=-\infty}^{+\infty} \sum_{j=-\infty}^{+\infty} f\left(\frac{i}{2^k}\right) \cdot f\left(\frac{j}{2^k}\right) \cdot \int_{-\infty}^{+\infty} \varphi(2^k x - i) \cdot \varphi(2^k x - j) \, dx$$

$$= \int_{-\infty}^{+\infty} (B_k f)^2(x) \cdot dx = \|B_k f\|_2^2 < +\infty, \qquad (11.16)$$

in particular,

$$\|B_k f\|_2 < +\infty. \qquad (11.17)$$

And by

$$\|f - B_k f\|_2 \le \|f\|_2 + \|B_k f\|_2 < +\infty,$$

we obtain that

$$e_k(f) < +\infty, \quad k \in \mathbf{Z}. \qquad (11.18)$$

In the following we need to establish that

$$\theta_1 := \sum_{j=-\infty}^{+\infty} \left| f\left(\frac{j}{2^k}\right) \right| \cdot \int_{-\infty}^{+\infty} |f(x)| \cdot |\varphi(2^k x - j)| \cdot dx < +\infty. \qquad (11.19)$$

Notice that

$$\theta_1 = \frac{1}{2^k} \cdot \sum_{j=-\infty}^{+\infty} \left| f\left(\frac{j}{2^k}\right) \right| \cdot \int_{-a}^{a} \left| f\left(\frac{y+j}{2^k}\right) \right| \cdot |\varphi(y)| \cdot dy$$

$$\le \frac{1}{2^k} \cdot \left( \sum_{j=-\infty}^{+\infty} \left| f\left(\frac{j}{2^k}\right) \right| \right) \cdot \|f\|_\infty \cdot \int_{-a}^{a} |\varphi(y)| dy < +\infty. \qquad (11.20)$$

R.H.S. (11.20) is finite by (11.10) and $f, \varphi$ are bounded functions. Thus $\theta_1 < +\infty$. That is,

$$\int_{-\infty}^{+\infty} f(x) \cdot (B_k f)(x) \cdot dx = \sum_{j=-\infty}^{+\infty} f\left(\frac{j}{2^k}\right) \cdot \int_{-\infty}^{+\infty} f(x) \cdot \varphi(2^k x - j) \cdot dx.$$

$$(11.21)$$

We have proved now that (cf. (11.9), (11.16), and (11.21))

$$e_k^2(f) = \int_{-\infty}^{+\infty} f^2(x)dx + \sum_{i=-\infty}^{+\infty} \sum_{j=-\infty}^{+\infty} f\left(\frac{i}{2^k}\right) \cdot f\left(\frac{j}{2^k}\right)$$

$$\cdot \int_{-\infty}^{+\infty} \varphi(2^k x - i) \cdot \varphi(2^k x - j) \cdot dx$$

$$-2 \cdot \sum_{j=-\infty}^{+\infty} f\left(\frac{j}{2^k}\right) \cdot \int_{-\infty}^{+\infty} f(x) \cdot \varphi(2^k x - j) \cdot dx. \quad (11.22)$$

(§2) Here we obtain the appropriate asymptotic expansion for

$$\int_{-\infty}^{+\infty} f(x) \cdot (B_k f)(x) \cdot dx = \sum_{j=-\infty}^{+\infty} f\left(\frac{j}{2^k}\right) \cdot \int_{-\infty}^{+\infty} f(x) \cdot \varphi(2^k x - j) dx$$

$$= \frac{1}{2^k} \cdot \sum_{j=-\infty}^{+\infty} f\left(\frac{j}{2^k}\right) \cdot \int_{-\infty}^{+\infty} f\left(\frac{y+j}{2^k}\right) \cdot \varphi(y) \cdot dy$$

$$= \frac{1}{2^k} \cdot \sum_{j=-\infty}^{+\infty} f\left(\frac{j}{2^k}\right) \cdot \int_{-a}^{a} f\left(\frac{y+j}{2^k}\right) \cdot \varphi(y) \cdot dy. \quad (11.23)$$

However, (by Taylor's theorem)

$$f\left(\frac{y+j}{2^k}\right) = \sum_{r=0}^{n-1} \frac{f^{(r)}\left(\frac{j}{2^k}\right)}{r!} \cdot \left(\frac{y}{2^k}\right)^r + \frac{\left(\frac{y}{2^k}\right)^{n-1}}{(n-1)!}$$

$$\cdot \int_0^1 (1-s)^{n-1} \cdot f^{(n)}\left(\frac{j}{2^k} + \frac{y}{2^k} \cdot s\right) \cdot ds, \quad k \in \mathbf{Z}. \quad (11.24)$$

So that

$$\int_{-\infty}^{+\infty} f(x) \cdot (B_k f)(x) \cdot dx = \frac{1}{2^k} \cdot \sum_{j=-\infty}^{+\infty} f\left(\frac{j}{2^k}\right)$$

$$\cdot \int_{-a}^{a} \left\{ \sum_{r=0}^{n-1} \frac{f^{(r)}\left(\frac{j}{2^k}\right)}{r!} \cdot \frac{y^r}{2^{kr}} + \frac{y^{n-1}}{2^{k(n-1)} \cdot (n-1)!} \right.$$

$$\left. \cdot \int_0^1 (1-s)^{n-1} \cdot f^{(n)}\left(\frac{j+y\cdot s}{2^k}\right) \cdot ds \right\} \cdot \varphi(y) \cdot dy$$

$$= \frac{1}{2^k} \cdot \sum_{j=-\infty}^{+\infty} f\left(\frac{j}{2^k}\right) \cdot \left\{ \sum_{r=0}^{n-1} \frac{1}{2^{kr} \cdot r!} \cdot f^{(r)}\left(\frac{j}{2^k}\right) \cdot m_r \right.$$

$$\left. + \frac{1}{2^{k\cdot(n-1)} \cdot (n-1)!} \cdot \gamma_{kn}^{(j)}(f) \right\}. \tag{11.25}$$

Here

$$m_r := \int_{-a}^{a} y^r \cdot \varphi(y) \cdot dy, \quad r = 0, \ldots, n-1, \tag{11.26}$$

and

$$\gamma_{kn}^{(j)}(f) := \int_{-a}^{a} y^{n-1} \cdot \left( \int_0^1 (1-s)^{n-1} \cdot f^{(n)}\left(\frac{j+y\cdot s}{2^k}\right) \cdot ds \right) \cdot \varphi(y) \cdot dy, \tag{11.27}$$

with $j \in \mathbf{Z}$. Therefore

$$\int_{-\infty}^{+\infty} f(x) \cdot (B_k f)(x) \cdot dx$$

$$= \frac{1}{2^k} \cdot \left\{ \sum_{r=0}^{n-1} \frac{m_r}{2^{kr} \cdot r!} \cdot \mathcal{R}_1 + \frac{\mathcal{R}_2}{2^{k\cdot(n-1)} \cdot (n-1)!} \right\}, \tag{11.28}$$

where

$$\mathcal{R}_1 := \sum_{j=-\infty}^{+\infty} f\left(\frac{j}{2^k}\right) \cdot f^{(r)}\left(\frac{j}{2^k}\right), \tag{11.29}$$

and

$$\mathcal{R}_2 := \sum_{j=-\infty}^{+\infty} f\left(\frac{j}{2^k}\right) \cdot \gamma_{kn}^{(j)}(f). \tag{11.30}$$

Equality (11.28) is valid given that $\mathcal{R}_1$ and $\mathcal{R}_2$ converge, a claim that we establish next. Notice that for $0 \le r \le n-1$: $g := f \cdot f^{(r)}$, $g'$,

$g'', \ldots, g^{(n-r)} \in L_1(\mathbf{R})$ and $g^{(n-r)} \in BV(\mathbf{R})$. From [58] and Lemma 2(b) we get that

$$\frac{1}{2^k} \cdot \sum_{j=-\infty}^{+\infty} f\left(\frac{j}{2^k}\right) \cdot f^{(r)}\left(\frac{j}{2^k}\right)$$

$$= \int_{-\infty}^{+\infty} f(x) \cdot f^{(r)}(x) \cdot dx + o\left(\frac{1}{2^{(n-r) \cdot k}}\right), \quad (11.31)$$

where $0 \le r \le n-1$. Again from [58], now Lemma 3 (or from [59] and Lemma 2) we obtain

$$\int_{-\infty}^{\infty} f \cdot f^{(r)} \cdot dx = \begin{cases} 0, & r \text{ is odd} \\ (-1)^m \cdot \int_{-\infty}^{+\infty} (f^{(m)}(t))^2 \cdot dt, & r = 2m. \end{cases} \quad (11.32)$$

Hence

$$\frac{1}{2^k} \cdot \sum_{j=-\infty}^{+\infty} f\left(\frac{j}{2^k}\right) \cdot f^{(r)}\left(\frac{j}{2^k}\right)$$

$$= = \begin{cases} o\left(\frac{1}{2^{(n-r) \cdot k}}\right), & r = \text{odd}, \\ (-1)^m \cdot \int_{-\infty}^{+\infty} (f^{(m)}(t))^2 \cdot dt + o\left(\frac{1}{2^{(n-2m) \cdot k}}\right), & r = 2m, \end{cases} \quad (11.33)$$

where $0 \le r \le n-1$. That is, we have proved that $\mathcal{R}_1$ converges.

Next we estimate

$$\frac{1}{2^k} \cdot \mathcal{R}_2 = \frac{1}{2^k} \cdot \sum_{j=-\infty}^{+\infty} f\left(\frac{j}{2^k}\right)$$

$$\cdot \left(\int_{-a}^{a} y^{n-1} \cdot \left(\int_0^1 (1-s)^{n-1} \cdot f^{(n)}\left(\frac{j+y \cdot s}{2^k}\right) \cdot ds\right) \cdot \varphi(y) \cdot dy\right).$$

From [58] and Remark 1 we get that $M := \|f^{(n)}\|_\infty < +\infty$. And so we obtain

$$\left| \int_0^1 (1-s)^{n-1} \cdot f^{(n)} \left( \frac{j+y \cdot s}{2^k} \right) \cdot ds \right| \leq M \cdot \int_0^1 (1-s)^{n-1} \cdot ds = \frac{M}{n}.$$

Therefore,

$$\left| \frac{1}{2^k} \cdot \mathcal{R}_2 \right| \leq \left( \frac{1}{2^k} \cdot \sum_{j=-\infty}^{+\infty} \left| f \left( \frac{j}{2^k} \right) \right| \right) \cdot \frac{M \cdot \theta}{n},$$

where

$$\theta := \int_{-a}^{a} |y|^{n-1} \cdot |\varphi(y)| \cdot dy < +\infty. \qquad (11.34)$$

Because $|f| \in L_1(\mathbf{R}) \cap BV(\mathbf{R})$, by Lemma 1 of [58] we find that

$$\frac{1}{2^k} \cdot \sum_{j=-\infty}^{+\infty} \left| f \left( \frac{j}{2^k} \right) \right| = \int_{-\infty}^{+\infty} |f(t)| \cdot dt + o(1). \qquad (11.35)$$

Hence

$$\left| \frac{1}{2^k} \cdot \mathcal{R}_2 \right| \leq \left( \int_{-\infty}^{+\infty} |f(t)| \cdot dt + o(1) \right) \cdot \frac{M \cdot \theta}{n} < +\infty.$$

That is, $\mathcal{R}_2$ converges too.

Put

$$K := M \cdot \theta < +\infty, \qquad (11.36)$$

and

$$L := \int_{-\infty}^{+\infty} |f(t)| dt < +\infty. \qquad (11.37)$$

Thus

$$\left| \frac{\mathcal{R}_2}{2^k \cdot 2^{k \cdot (n-1)} \cdot (n-1)!} \right| \leq (L + o(1)) \cdot \frac{K}{2^{k \cdot (n-1)} \cdot n!}$$

$$= \frac{L \cdot K}{2^{k \cdot (n-1)} \cdot n!} + \frac{K \cdot o(1)}{2^{k \cdot (n-1)} \cdot n!}$$

$$= o \left( \frac{1}{2^{k \cdot (n-2)}} \right) + o \left( \frac{1}{2^{k \cdot (n-1)}} \right)$$

$$= o \left( \frac{1}{2^{k \cdot (n-2)}} \right) + o \left( \frac{1}{2^{k \cdot (n-2)}} \right)$$

$$= o\left(\frac{1}{2^{k \cdot (n-2)}}\right), \quad k \in \mathbf{N}, \; n \geq 2.$$

So we have proved that

$$\frac{\mathcal{R}_2}{2^k \cdot 2^{k \cdot (n-1)} \cdot (n-1)!} = o\left(\frac{1}{2^{k \cdot (n-2)}}\right), \quad k \in \mathbf{N}, \; n \geq 2. \qquad (11.38)$$

Merging (11.31) and (11.38) into (11.28) we obtain

$$\int_{-\infty}^{+\infty} f(x) \cdot (B_k f)(x) \cdot dx = \sum_{r=0}^{n-1} \frac{m_r}{2^{kr} \cdot r!} \cdot \left\{ \int_{-\infty}^{+\infty} f(x) \cdot f^{(r)}(x) \cdot dx \right.$$

$$\left. + o\left(\frac{1}{2^{(n-r) \cdot k}}\right) \right\} + o\left(\frac{1}{2^{k \cdot (n-2)}}\right), (11.39)$$

where $k \in \mathbf{N}, \; n \geq 2$.

Next we work on (11.39). In the case where $n$ is odd, then $n - 1 = 2\ell$ is even. So we find (by (11.32))

$$\int_{-\infty}^{+\infty} f(x) \cdot (B_k f)(x) \cdot dx = \left\{ \sum_{\gamma=0}^{\ell} \frac{m_{2\gamma}}{2^{k \cdot 2\gamma} \cdot (2\gamma)!} \right.$$

$$\left. \cdot \left\{ (-1)^\gamma \cdot \int_{-\infty}^{+\infty} (f^{(\gamma)}(t))^2 \cdot dt + o\left(\frac{1}{2^{(n-2\gamma) \cdot k}}\right) \right\} \right\}$$

$$+ \left\{ \sum_{\gamma=0}^{\ell-1} \frac{m_{2\gamma+1}}{2^{k \cdot (2\gamma+1)} \cdot (2\gamma+1)!} \cdot o\left(\frac{1}{2^{(n-2\gamma-1) \cdot k}}\right) \right\} + o\left(\frac{1}{2^{k(n-2)}}\right)$$

$$= \sum_{\gamma=0}^{\ell} \frac{m_{2\gamma}}{(2\gamma)!} \cdot \frac{1}{4^{k \cdot \gamma}} \cdot (-1)^\gamma \cdot \int_{-\infty}^{+\infty} (f^{(\gamma)}(t))^2 \cdot dt$$

$$+ \sum_{\gamma=0}^{\ell} \frac{m_{2\gamma}}{(2\gamma)!} \cdot \frac{1}{2^{k \cdot 2\gamma}} \cdot o\left(\frac{1}{2^{(n-2\gamma) \cdot k}}\right)$$

$$+ \sum_{\gamma=0}^{\ell-1} \frac{m_{2\gamma+1}}{(2\gamma+1)!} \cdot \frac{1}{2^{k \cdot (2\gamma+1)}} \cdot o\left(\frac{1}{2^{(n-2\gamma-1) \cdot k}}\right) + o\left(\frac{1}{2^{k(n-2)}}\right)$$

$$= \sum_{\gamma=0}^{\ell} \frac{m_{2\gamma}}{(2\gamma)!} \cdot \frac{1}{4^{k\gamma}} \cdot (-1)^{\gamma} \cdot \int_{-\infty}^{+\infty} (f^{(\gamma)}(t))^2 \cdot dt$$

$$+ \left( \sum_{\gamma=0}^{\ell} \frac{m_{2\gamma}}{(2\gamma)!} \right) \cdot o\left( \frac{1}{2^{k \cdot n}} \right)$$

$$+ \left( \sum_{\gamma=0}^{\ell-1} \frac{m_{2\gamma+1}}{(2\gamma+1)!} \right) \cdot o\left( \frac{1}{2^{k \cdot n}} \right) + o\left( \frac{1}{2^{k \cdot (n-2)}} \right)$$

$$= \sum_{\gamma=0}^{\ell} \frac{m_{2\gamma}}{(2\gamma)!} \cdot \frac{1}{4^{k\gamma}} \cdot (-1)^{\gamma} \cdot \int_{-\infty}^{+\infty} (f^{(\gamma)}(t))^2 \cdot dt$$

$$+ o\left( \frac{1}{2^{k \cdot n}} \right) + o\left( \frac{1}{2^{k \cdot (n-2)}} \right)$$

$$= \sum_{\gamma=0}^{\ell} \frac{m_{2\gamma}}{(2\gamma)!} \cdot \frac{1}{4^{k\gamma}} \cdot (-1)^{\gamma} \cdot \int_{-\infty}^{+\infty} (f^{(\gamma)}(t))^2 \cdot dt + o\left( \frac{1}{2^{k \cdot (n-2)}} \right).$$

Hence, when $n$ is odd $(n - 1 = 2\ell)$ we have established that

$$\int_{-\infty}^{+\infty} f(x) \cdot (B_k f)(x) \cdot dx = \sum_{\gamma=0}^{\ell} \frac{m_{2\gamma}}{(2\gamma)!} \cdot \frac{1}{4^{k \cdot \gamma}} \cdot (-1)^{\gamma}$$

$$\cdot \int_{-\infty}^{+\infty} (f^{(\gamma)}(t))^2 \cdot dt + o\left( \frac{1}{2^{k \cdot (n-2)}} \right), \quad n \geq 2, k \in \mathbf{N}. \quad (11.40)$$

We continue work on (11.39). In the case where $n$ is even, then $n - 2 =: 2 \cdot \ell$ is even. So we get (by (11.32))

$$\int_{-\infty}^{+\infty} f(x) \cdot (B_k f)(x) \cdot dx = \left\{ \sum_{\gamma=0}^{\ell} \frac{m_{2\gamma}}{2^{k \cdot 2\gamma} \cdot (2\gamma)!} \right.$$

$$\left. \cdot \left\{ (-1)^{\gamma} \cdot \int_{-\infty}^{+\infty} (f^{(\gamma)}(t))^2 \cdot dt + o\left( \frac{1}{2^{(n-2\gamma) \cdot k}} \right) \right\} \right\}$$

$$+ \left\{ \sum_{\gamma=0}^{\ell} \frac{m_{2\gamma+1}}{2^{k \cdot (2\gamma+1)} \cdot (2\gamma+1)!} \cdot o\left(\frac{1}{2^{(n-2\gamma-1) \cdot k}}\right) \right\} + o\left(\frac{1}{2^{k(n-2)}}\right)$$

$$= \sum_{\gamma=0}^{\ell} \frac{m_{2\gamma}}{(2\gamma)!} \cdot \frac{1}{4^{k \cdot \gamma}} \cdot (-1)^{\gamma} \cdot \int_{-\infty}^{+\infty} (f^{(\gamma)}(t))^2 \cdot dt$$

$$+ \left( \sum_{\gamma=0}^{\ell} \frac{m_{2\gamma}}{(2\gamma)!} \right) \cdot o\left(\frac{1}{2^{k \cdot n}}\right)$$

$$+ \left( \sum_{\gamma=0}^{\ell} \frac{m_{2\gamma+1}}{(2\gamma+1)!} \right) \cdot o\left(\frac{1}{2^{k \cdot n}}\right) + o\left(\frac{1}{2^{k \cdot (n-2)}}\right).$$

Thus when $n$ is even ($n - 2 =: 2 \cdot \ell$) we have proved that

$$\int_{-\infty}^{+\infty} f(x) \cdot (B_k f)(x) \cdot dx = \sum_{\gamma=0}^{\ell} \frac{m_{2\gamma}}{(2\gamma)!} \cdot \frac{1}{4^{k \cdot \gamma}} \cdot (-1)^{\gamma}$$

$$\cdot \int_{-\infty}^{+\infty} (f^{(\gamma)}(t))^2 \cdot dt + o\left(\frac{1}{2^{k \cdot (n-2)}}\right), \quad n \geq 2, k \in \mathbf{N}. \quad (11.41)$$

Letting $\ell$ as in (11.5) we finally arrive at the asymptotic expansion for

$$\int_{-\infty}^{+\infty} f(x) \cdot (B_k f)(x) \cdot dx = \sum_{\gamma=0}^{\ell} \frac{m_{2\gamma}}{(2\gamma)!} \cdot \frac{1}{4^{k \cdot \gamma}} \cdot (-1)^{\gamma}$$

$$\cdot \int_{-\infty}^{+\infty} (f^{(\gamma)}(t))^2 \cdot dt + o\left(\frac{1}{2^{k \cdot (n-2)}}\right), \quad (11.42)$$

where $k \in \mathbf{N}$ and $n \geq 2$. Here

$$m_{2\gamma} := \int_{-a}^{a} y^{2\gamma} \cdot \varphi(y) \cdot dy, \quad \gamma = 0, 1, \ldots, \ell. \quad (11.43)$$

It is interesting to notice from (11.1) that

$$m_0 := \int_{-\infty}^{+\infty} \varphi(y) \cdot dy = 1. \quad (11.44)$$

Actually, from (11.1) we get that

$$\sum_{j=-\infty}^{+\infty} \varphi(x+j) = 1, \quad \forall x \in \mathbf{R}. \tag{11.1*}$$

Furthermore,

$$\int_{-\infty}^{+\infty} \varphi(x) \cdot dx = \sum_{j=-\infty}^{+\infty} \int_{j}^{j+1} \varphi(x) \cdot dx = \sum_{j=-\infty}^{+\infty} \int_{0}^{1} \varphi(x+j) dx$$

$$= \int_{0}^{1} \left( \sum_{j=-\infty}^{+\infty} \varphi(x+j) \right) \cdot dx \stackrel{(11.1^*)}{=} \int_{0}^{1} 1 \cdot dx = 1,$$

proving (11.44).

(§3) Next we would like to obtain a suitable asymptotic expansion for

$$\int_{-\infty}^{+\infty} (B_k f)^2(x) \cdot dx = \sum_{i=-\infty}^{+\infty} \sum_{j=-\infty}^{+\infty} f\left(\frac{i}{2^k}\right) f\left(\frac{j}{2^k}\right)$$

$$\cdot \int_{-\infty}^{+\infty} \varphi(2^k x - i) \cdot \varphi(2^k x - j) \cdot dx. \tag{11.45}$$

Equality (11.45) makes sense due to (11.16). By a change of variable we get that

$$\int_{-\infty}^{+\infty} (B_k f)^2(x) \cdot dx = \frac{1}{2^k} \cdot \sum_{i=-\infty}^{+\infty} \sum_{j=-\infty}^{+\infty} f\left(\frac{i}{2^k}\right) \cdot f\left(\frac{j}{2^k}\right) \cdot \sigma_{i-j}, \tag{11.46}$$

where

$$\sigma_{i-j} := \int_{-\infty}^{+\infty} \varphi(u) \cdot \varphi(u+i-j) \cdot du, \tag{11.47}$$

exists for any $i, j \in \mathbf{Z}$.

Because the support of $\varphi(u)$ is in $[-a, a]$, the support of $\varphi(u+i-j)$ is in $[-a+j-i, a+j-i]$. So $\varphi(u), \varphi(u+i-j)$ have no common support iff $|j-i| \geq 2 \cdot a$. That is, $\varphi(u), \varphi(u+i-j)$ have common support iff $|j-i| \leq \beta$, where $\beta$ is defined by (11.4). Thus $\sigma_{i-j} \neq 0$ iff $|j - i| \leq \beta$. Hence

$$\sum_{i=-\infty}^{+\infty} \sum_{j=-\infty}^{+\infty} = \sum_{i=j} + \sum \sum_{i-j=\pm 1} + \cdots + \sum \sum_{i-j=\pm \beta}, \qquad (11.48)$$

and

$$\sum \sum_{i-j=\pm q} = \sum \sum_{i=j+q \ j} + \sum \sum_{i \ j=i+q}, \qquad (11.49)$$

for all $1 \leq q \leq \beta$.

Finally, we obtain that

$$\sum_{i=-\infty}^{+\infty} \sum_{j=-\infty}^{+\infty} = \sum_{i=j=-\infty}^{+\infty} \sum + \sum_{q=1}^{\beta} \left( \sum_{i=j+q} \sum_{j=-\infty}^{+\infty} + \sum_{i=-\infty}^{+\infty} \sum_{j=i+q} \right). \qquad (11.50)$$

Therefore

$$\int_{-\infty}^{+\infty} (B_k f)^2(x) \cdot dx = \left( \frac{1}{2^k} \cdot \sum_{i=-\infty}^{+\infty} f^2 \left( \frac{i}{2^k} \right) \right) \cdot \sigma_0$$

$$+ \sum_{q=1}^{\beta} \left\{ \frac{1}{2^k} \cdot \left( \sum_{j=-\infty}^{+\infty} f \left( \frac{j+q}{2^k} \right) \cdot f \left( \frac{j}{2^k} \right) \right) \cdot \sigma_q \right.$$

$$\left. + \frac{1}{2^k} \cdot \left( \sum_{i=-\infty}^{+\infty} f \left( \frac{i}{2^k} \right) \cdot f \left( \frac{i+q}{2^k} \right) \right) \cdot \sigma_{-q} \right\},$$

where

$$\sigma_0 = \int_{-\infty}^{+\infty} \varphi^2(u) \cdot du. \qquad (11.51)$$

That is, we get that

$$\int_{-\infty}^{+\infty} (B_k f)^2(x) \cdot dx = \left( \frac{1}{2^k} \cdot \sum_{i=-\infty}^{+\infty} f^2 \left( \frac{i}{2^k} \right) \right) \cdot \sigma_0$$

$$+ \sum_{q=1}^{\beta} \left\{ \frac{1}{2^k} \cdot \left( \sum_{i=-\infty}^{+\infty} f \left( \frac{i}{2^k} \right) \cdot f \left( \frac{i+q}{2^k} \right) \right) \cdot (\sigma_q + \sigma_{-q}) \right\}. \qquad (11.52)$$

## 11. Nonorthogonal Wavelet Approximation for Deterministic Signals 251

Notice that $f^2, (f^2)', \ldots, (f^2)^{(n)} \in L_1(\mathbf{R})$, all continuous and $(f^2)^{(n)} \in BV(\mathbf{R})$. Then from [58], Lemma 2(b) we obtain

$$\frac{1}{2^k} \cdot \sum_{i=-\infty}^{+\infty} f^2\left(\frac{i}{2^k}\right) = \int_{-\infty}^{+\infty} f^2(t)dt + o\left(\frac{1}{2^{k \cdot n}}\right). \qquad (11.53)$$

By Taylor's theorem we observe that

$$f\left(\frac{i+q}{2^k}\right) = \sum_{r=0}^{n-1} \frac{f^{(r)}\left(\frac{i}{2^k}\right)}{r!} \cdot \left(\frac{q}{2^k}\right)^r + \frac{\left(\frac{q}{2^k}\right)^{n-1}}{(n-1)!}$$

$$\cdot \int_0^1 (1-s)^{n-1} \cdot f^{(n)}\left(\frac{i+q \cdot s}{2^k}\right) \cdot ds. \qquad (11.54)$$

Consequently

$$\frac{1}{2^k} \cdot \sum_{i=-\infty}^{+\infty} f\left(\frac{i}{2^k}\right) \cdot f\left(\frac{i+q}{2^k}\right)$$

$$= \frac{1}{2^k} \cdot \sum_{i=-\infty}^{+\infty} \left\{ \left( \sum_{r=0}^{n-1} \frac{q^r}{2^{k \cdot r} \cdot r!} \cdot f\left(\frac{i}{2^k}\right) \cdot f^{(r)}\left(\frac{i}{2^k}\right) \right) \right.$$

$$+ \frac{q^{n-1}}{2^{k \cdot (n-1)} \cdot (n-1)!} \cdot \left( f\left(\frac{i}{2^k}\right) \cdot \int_0^1 (1-s)^{n-1} \cdot f^{(n)}\left(\frac{i+q \cdot s}{2^k}\right) ds \right) \right\}$$

$$= \sum_{r=0}^{n-1} \frac{q^r}{2^{k \cdot r} \cdot r!} \cdot E_1 + \frac{q^{n-1}}{2^{k \cdot (n-1)} \cdot (n-1)!} \cdot E_2. \qquad (11.55)$$

Here, as before (11.33)

$$E_1 := \frac{1}{2^k} \cdot \sum_{i=-\infty}^{+\infty} f\left(\frac{i}{2^k}\right) \cdot f^{(r)}\left(\frac{i}{2^k}\right)$$

$$= \begin{cases} o\left(\dfrac{1}{2^{(n-r)\cdot k}}\right), & r \text{ is odd,} \\ (-1)^m \cdot \displaystyle\int_{-\infty}^{+\infty} (f^{(m)}(t))^2 \cdot dt + o\left(\dfrac{1}{2^{(n-2m)\cdot k}}\right), & r = 2m, \end{cases} \quad (11.56)$$

where $0 \leq r \leq n-1$. That is, $E_1$ converges. Also

$$E_2 := \frac{1}{2^k} \cdot \sum_{i=-\infty}^{+\infty} f\left(\frac{i}{2^k}\right) \cdot \int_0^1 (1-s)^{n-1} \cdot f^{(n)}\left(\frac{i+q\cdot s}{2^k}\right) \cdot ds. \quad (11.57)$$

Still we need to prove that $E_2$ converges so that (11.55) is valid. Again $M := \|f^{(n)}\|_\infty < +\infty$. Therefore

$$\left|\int_0^1 (1-s)^{n-1} \cdot f^{(n)}\left(\frac{i+q\cdot s}{2^k}\right) \cdot ds\right| \leq \frac{M}{n},$$

and

$$|E_2| \leq \left(\frac{1}{2^k} \cdot \sum_{i=-\infty}^{+\infty} \left|f\left(\frac{i}{2^k}\right)\right|\right) \cdot \frac{M}{n} \quad \text{(by (11.35))}$$

$$= \left(\int_{-\infty}^{+\infty} |f(t)| \cdot dt + o(1)\right) \cdot \frac{M}{n} < +\infty.$$

In other words, $E_2$ converges. Furthermore (by (11.37) $L < +\infty$) we obtain

$$\frac{q^{n-1}}{2^{k\cdot(n-1)} \cdot (n-1)!} \cdot |E_2| \leq \frac{M \cdot q^{n-1}}{n! \cdot 2^{k\cdot(n-1)}} \cdot (L + o(1))$$

$$= \frac{\tau_1}{2^{k\cdot(n-1)}} + \frac{\tau_2}{2^{k\cdot(n-1)}} \cdot o(1)$$

$$\left(\tau_1 := \frac{M\cdot L\cdot q^{n-1}}{n!} < +\infty, \ \tau_2 := \frac{M\cdot q^{n-1}}{n!} < +\infty, k \in \mathbf{N}, n \geq 2\right)$$

$$= o\left(\frac{1}{2^{k\cdot(n-2)}}\right) + o\left(\frac{1}{2^{k\cdot(n-1)}}\right) = o\left(\frac{1}{2^{k\cdot(n-2)}}\right).$$

## 11. Nonorthogonal Wavelet Approximation for Deterministic Signals 253

That is, we have proved that

$$\frac{q^{n-1}}{2^{k \cdot (n-1)} \cdot (n-1)!} \cdot E_2 = o\left(\frac{1}{2^{k \cdot (n-2)}}\right), \tag{11.58}$$

where $k \in \mathbf{N}$, $n \geq 2$. Next we estimate

$$\sum_{r=0}^{n-1} \frac{q^r}{2^{k \cdot r} \cdot r!} \cdot E_1 \stackrel{(11.56)}{=} \sum_{\substack{r=0 \\ (r \text{ odd}=1)}}^{n-1} \frac{q^r}{2^{kr} \cdot r!} \cdot o\left(\frac{1}{2^{(n-r) \cdot k}}\right)$$

$$+ \sum_{\substack{r=0 \\ (r \text{ even}=0) \\ r=2 \cdot m}}^{n-1} \frac{q^r}{2^{k \cdot r} \cdot r!} \cdot \left((-1)^m \cdot \int_{-\infty}^{+\infty} (f^{(m)}(t))^2 \cdot dt + o\left(\frac{1}{2^{(n-2m) \cdot k}}\right)\right)$$

$$= \left(\sum_{r \text{ odd}=1}^{n-1} \frac{q^r}{r!}\right) \cdot o\left(\frac{1}{2^{k \cdot n}}\right) + \sum_{\substack{r=0 \\ (r \text{ even}=0) \\ r=2 \cdot m}}^{n-1} \frac{q^r}{r!} \cdot \frac{1}{2^{k \cdot r}}$$

$$\cdot (-1)^m \cdot \int_{-\infty}^{+\infty} (f^{(m)}(t))^2 \cdot dt + \left(\sum_{\substack{r=0 \\ (r \text{ even}=0) \\ r=2 \cdot m}}^{n-1} \frac{q^r}{r!}\right) \cdot o\left(\frac{1}{2^{k \cdot n}}\right)$$

$$= \left(\sum_{\substack{r=0 \\ (r \text{ even}=0) \\ r=2 \cdot m}}^{n-1}\right) \frac{q^r}{r!} \cdot \frac{1}{2^{k \cdot r}} \cdot (-1)^m \cdot \int_{-\infty}^{+\infty} (f^{(m)}(t))^2 \cdot dt + o\left(\frac{1}{2^{k \cdot n}}\right).$$

Following Definition (11.5) we obtain

$$\sum_{r=0}^{n-1} \frac{q^r}{2^{k \cdot r} \cdot r!} \cdot E_1 = \sum_{\gamma=0}^{\ell} \frac{q^{2 \cdot \gamma}}{(2 \cdot \gamma)!} \cdot \frac{1}{4^{k \cdot \gamma}} \cdot (-1)^\gamma$$

$$\cdot \int_{-\infty}^{+\infty} (f^{(\gamma)}(t))^2 \cdot dt + o\left(\frac{1}{2^{k \cdot n}}\right), \tag{11.59}$$

where $k \in \mathbf{N}$, $n \geq 2$. Hence (see (11.55), (11.58), and (11.59))

$$\frac{1}{2^k} \cdot \sum_{i=-\infty}^{+\infty} f\left(\frac{i}{2^k}\right) \cdot f\left(\frac{i+q}{2^k}\right) = \sum_{\gamma=0}^{\ell} \frac{q^{2\cdot\gamma}}{(2\cdot\gamma)!} \cdot \frac{1}{4^{k\cdot\gamma}} \cdot (-1)^{\gamma}$$

$$\cdot \int_{-\infty}^{+\infty} (f^{(\gamma)}(t))^2 \cdot dt + o\left(\frac{1}{2^{k\cdot(n-2)}}\right), \ k \in \mathbf{N}, \ n \geq 2. \quad (11.60)$$

Next from (11.52), (11.53), and (11.60) we observe that ($n \geq 2$, $k \in \mathbf{N}$)

$$\int_{-\infty}^{+\infty} (B_k f)^2(x) \cdot dx = \left(\int_{-\infty}^{+\infty} f^2(t) dt + o\left(\frac{1}{2^{k\cdot n}}\right)\right) \cdot \sigma_0$$

$$+ \sum_{q=1}^{\beta} \left\{ \left(\sum_{\gamma=0}^{\ell} \frac{q^{2\cdot\gamma}}{(2\cdot\gamma)!} \cdot \frac{1}{4^{k\cdot\gamma}} \cdot (-1)^{\gamma} \right. \right.$$

$$\left. \left. \cdot \int_{-\infty}^{+\infty} (f^{(\gamma)}(t))^2 \cdot dt + o\left(\frac{1}{2^{k\cdot(n-2)}}\right) \right) \cdot (\sigma_q + \sigma_{-q}) \right\}$$

$$= \left(\int_{-\infty}^{+\infty} f^2(t) \cdot dt\right) \cdot \sigma_0 + \sum_{q=1}^{\beta} \left\{ \left\{ \sum_{\gamma=0}^{\ell} \frac{q^{2\cdot\gamma}}{(2\cdot\gamma)!} \cdot \frac{1}{4^{k\cdot\gamma}} \cdot (-1)^{\gamma} \right. \right.$$

$$\left. \left. \cdot \int_{-\infty}^{+\infty} (f^{(\gamma)}(t))^2 \cdot dt \right\} \cdot (\sigma_q + \sigma_{-q}) \right\} + o\left(\frac{1}{2^{k\cdot(n-2)}}\right).$$

Hence ($n \geq 2$, $k \in \mathbf{N}$)

$$\int_{-\infty}^{+\infty} (B_k f)^2(x) \cdot dx = \left\{ \sum_{\gamma=0}^{\ell} \left( \frac{(-1)^{\gamma} \cdot \int_{-\infty}^{+\infty} (f^{(\gamma)}(t))^2 \cdot dt}{(2\cdot\gamma)! \cdot 4^{k\cdot\gamma}} \right) \right.$$

$$\left. \cdot \left( \sum_{q=1}^{\beta} q^{2\gamma} \cdot (\sigma_q + \sigma_{-q}) \right) \right\}$$

$$+ \left(\int_{-\infty}^{+\infty} f^2(t) \cdot dt\right) \cdot \sigma_0 + o\left(\frac{1}{2^{k\cdot(n-2)}}\right).$$

Furthermore

$$\int_{-\infty}^{+\infty} (B_k f)^2(x) \cdot dx = \left\{ \sum_{\gamma=1}^{\ell} \left( \frac{(-1)^\gamma \cdot \int_{-\infty}^{+\infty} (f^{(\gamma)}(t))^2 \cdot dt}{(2 \cdot \gamma)! \cdot 4^{k \cdot \gamma}} \right) \right.$$

$$\left. \cdot \left( \sum_{q=1}^{\beta} q^{2\gamma} \cdot (\sigma_q + \sigma_{-q}) \right) \right\}$$

$$+ \left( \int_{-\infty}^{+\infty} (f(t))^2 \cdot dt \right) \cdot \left( \sigma_0 + \sum_{q=1}^{\beta} (\sigma_q + \sigma_{-q}) \right)$$

$$+ o\left( \frac{1}{2^{k \cdot (n-2)}} \right), \quad n \geq 2, \ k \in \mathbf{N}. \tag{11.61}$$

Thus

$$\sigma_0 + \sum_{q=1}^{\beta} (\sigma_q + \sigma_{-q}) = \int_{-\infty}^{+\infty} \varphi(u) \cdot \varphi(u) \cdot du$$

$$+ \sum_{q=1}^{\beta} \left( \int_{-\infty}^{+\infty} \varphi(u) \cdot \varphi(u+q) \cdot du + \int_{-\infty}^{+\infty} \varphi(u) \cdot \varphi(u-q) \cdot du \right)$$

$$= \int_{-\infty}^{+\infty} \varphi(u) \cdot \left\{ \varphi(u) + \sum_{q=1}^{\beta} \varphi(u+q) + \sum_{q=1}^{\beta} \varphi(u-q) \right\} \cdot du$$

$$= \int_{-\infty}^{+\infty} \varphi(u) \cdot \left\{ \sum_{q=-\beta}^{\beta} \varphi(u-q) \right\} \cdot du$$

$$= \int_{-\infty}^{+\infty} \varphi(u) \cdot \left( \sum_{q=-\infty}^{+\infty} \varphi(u-q) \right) \cdot du$$

$$\stackrel{(11.1)}{=} \int_{-\infty}^{\infty} \varphi(u) \cdot 1 \cdot du = 1, \quad \text{by (11.44)}.$$

That is,
$$\sigma_0 + \sum_{q=1}^{\beta}(\sigma_q + \sigma_{-q}) = 1. \tag{11.62}$$

Consequently ($k \in \mathbf{N}$, $n \geq 2$)

$$\int_{-\infty}^{+\infty} (B_k f)^2(x) \cdot dx = \left\{ \sum_{\gamma=1}^{\ell} \left( \frac{(-1)^\gamma \cdot \int_{-\infty}^{+\infty} (f^{(\gamma)}(t))^2 \cdot dt}{(2\cdot\gamma)! \cdot 4^{k\cdot\gamma}} \right) \right.$$

$$\left. \cdot \left( \sum_{q=1}^{\beta} q^{2\cdot\gamma} \cdot (\sigma_q + \sigma_{-q}) \right) \right\}$$

$$+ \int_{-\infty}^{+\infty} f^2(t) \cdot dt + o\left(\frac{1}{2^{k\cdot(n-2)}}\right). \tag{11.63}$$

And finally,

(§4) Setting things together: we get that (see (11.22), (11.42), (11.63); $k \in \mathbf{N}$, $n \geq 2$)

$$e_k^2(f) = \int_{-\infty}^{+\infty} f^2(t)dt + \left\{ \sum_{\gamma=1}^{\ell} \left( \frac{(-1)^\gamma \cdot \int_{-\infty}^{+\infty} (f^{(\gamma)}(t))^2 \cdot dt}{(2\cdot\gamma)! \cdot 4^{k\cdot\gamma}} \right) \right.$$

$$\left. \cdot \left( \sum_{q=1}^{\beta} q^{2\cdot\gamma} \cdot (\sigma_q + \sigma_{-q}) \right) \right\} + \int_{-\infty}^{+\infty} f^2(t) \cdot dt$$

$$+ o\left(\frac{1}{2^{k\cdot(n-2)}}\right) - 2 \cdot \sum_{\gamma=0}^{\ell} \frac{m_{2\cdot\gamma}}{(2\cdot\gamma)!} \cdot \frac{1}{4^{k\cdot\gamma}}$$

$$\cdot (-1)^\gamma \cdot \int_{-\infty}^{+\infty} (f^{(\gamma)}(t))^2 \cdot dt - 2 \cdot o\left(\frac{1}{2^{k\cdot(n-2)}}\right)$$

$$= \left\{ \sum_{\gamma=1}^{\ell} \left( \frac{(-1)^\gamma \cdot \int_{-\infty}^{+\infty} (f^{(\gamma)}(t))^2 \cdot dt}{(2\cdot\gamma)! \cdot 4^{k\cdot\gamma}} \right) \cdot \left( \sum_{q=1}^{\beta} q^{2\cdot\gamma} \cdot (\sigma_q + \sigma_{-q}) \right) \right\}$$

$$-2 \cdot \sum_{\gamma=1}^{\ell} \frac{m_{2\cdot\gamma}}{(2\cdot\gamma)!} \cdot \frac{1}{4^{k\cdot\gamma}} \cdot (-1)^{\gamma} \cdot \int_{-\infty}^{+\infty} (f^{(\gamma)}(t))^2 \cdot dt$$

$$+o\left(\frac{1}{2^{k\cdot(n-2)}}\right) \stackrel{(11.6)}{=} \left\{ \sum_{\gamma=1}^{\ell} \frac{\rho_\gamma(f)}{4^{k\cdot\gamma}} \cdot \left( \sum_{q=1}^{\beta} q^{2\cdot\gamma} \cdot (\sigma_q + \sigma_{-q}) \right) \right\}$$

$$-2 \cdot \sum_{\gamma=1}^{\ell} \rho_\gamma(f) \cdot \frac{m_{2\cdot\gamma}}{4^{k\cdot\gamma}} + o\left(\frac{1}{2^{k\cdot(n-2)}}\right).$$

Furthermore ($k \in \mathbf{N}$, $n \geq 2$)

$$e_k^2(f) = \sum_{\gamma=1}^{\ell} \sum_{q=1}^{\beta} \left( \frac{\rho_\gamma(f)}{4^{k\cdot\gamma}} \right) \cdot q^{2\cdot\gamma} \cdot (\sigma_q + \sigma_{-q})$$

$$+ \sum_{\gamma=1}^{\ell} (-2 \cdot m_{2\cdot\gamma}) \cdot \frac{\rho_\gamma(f)}{4^{k\cdot\gamma}} + o\left( \frac{1}{2^{k\cdot(n-2)}} \right)$$

$$= \sum_{\gamma=1}^{\ell} \left\{ \left( \sum_{q=1}^{\beta} \left( \frac{\rho_\gamma(f)}{4^{k\cdot\gamma}} \right) \cdot q^{2\cdot\gamma} \cdot (\sigma_q + \sigma_{-q}) \right) - 2 \cdot m_{2\cdot\gamma} \cdot \frac{\rho_\gamma(f)}{4^{k\cdot\gamma}} \right\}$$

$$+ o\left(\frac{1}{2^{k\cdot(n-2)}}\right) = \sum_{\gamma=1}^{\ell} \left\{ \frac{\rho_\gamma(f)}{4^{k\cdot\gamma}} \cdot \left( \left( \sum_{q=1}^{\beta} q^{2\cdot\gamma} \cdot (\sigma_q + \sigma_{-q}) \right) - 2 \cdot m_{2\cdot\gamma} \right) \right\}$$

$$+ o\left(\frac{1}{2^{k\cdot(n-2)}}\right). \tag{11.64}$$

However

$$\sum_{q=1}^{\beta} q^{2\cdot\gamma} \cdot (\sigma_q + \sigma_{-q}) - 2 \cdot m_{2\cdot\gamma} \stackrel{\{(11.26),(11.43),(11.47)\}}{=} \sum_{q=1}^{\beta} q^{2\cdot\gamma}$$

$$\cdot \left( \int_{-\infty}^{+\infty} \varphi(u) \cdot \varphi(u+q) \cdot du + \int_{-\infty}^{+\infty} \varphi(u) \cdot \varphi(u-q) \cdot du \right)$$

$$-2 \cdot \int_{-\infty}^{+\infty} u^{2\cdot\gamma} \cdot \varphi(u) \cdot du = -2 \cdot \int_{-\infty}^{+\infty} u^{2\cdot\gamma} \cdot \varphi(u) \cdot du$$

$$+ \sum_{q=1}^{\beta} q^{2\cdot\gamma} \cdot \left( \int_{-\infty}^{+\infty} \varphi(u) \cdot (\varphi(u+q) + \varphi(u-q)) \cdot du \right)$$

$$= -2 \cdot \int_{-\infty}^{+\infty} u^{2\cdot\gamma} \cdot \varphi(u) \cdot du + \int_{-\infty}^{+\infty} \varphi(u)$$

$$\cdot \left( \sum_{q=1}^{\beta} q^{2\cdot\gamma} \cdot (\varphi(u+q) + \varphi(u-q)) \right) \cdot du$$

$$= \int_{-\infty}^{+\infty} \varphi(u) \cdot \left\{ -2 \cdot u^{2\cdot\gamma} + \sum_{q=1}^{\beta} q^{2\cdot\gamma} \cdot (\varphi(u+q) \right.$$

$$\left. + \varphi(u-q)) \right\} du \stackrel{(11.7)}{=} \lambda_\gamma(\varphi) < +\infty,$$

all $\gamma = 1, \ldots, \ell$. That is,

$$\sum_{q=1}^{\beta} q^{2\cdot\gamma} \cdot (\sigma_q + \sigma_{-q}) - 2 \cdot m_{2\cdot\gamma} = \lambda_\gamma(\varphi), \qquad (11.65)$$

all $\gamma = 1, \ldots, \ell$. Also, from (11.64) we obtain that ($k \in \mathbf{N}$, $n \geq 2$)

$$e_k^2(f) = \sum_{\gamma=1}^{\ell} \left\{ \frac{\rho_\gamma(f)}{4^{k\cdot\gamma}} \cdot \left[ \left( \sum_{q=1}^{\beta} q^{2\cdot\gamma} \cdot (\sigma_q + \sigma_{-q}) \right) - 2 \cdot m_{2\cdot\gamma} \right] \right\}$$

$$+ o\left( \frac{1}{2^{k\cdot(n-2)}} \right). \qquad (11.66)$$

Finally, it is obvious that (11.65) and (11.66) imply (11.8). ■

# Chapter 12

# Wavelet Type Differentiated Shift-Invariant Integral Operators

In approximating $f^{(i)}$, when $f \in C^{(i)}(\mathbf{R})$, $i$ fixed, by $L_k^{(i)}(f)$, where $L_k$ are linear approximation operators, it is interesting to investigate which characteristics of $L_k^{(i)}(f)$ are similar to the main features of $f^{(i)}$. Also it is important to prove sufficient conditions for suitable $L_k$, so that $L_k^{(i)}(f)$ best fit and resemble $f^{(i)}$ as much as possible.

All the operators $L_k$ encountered here are of integral form, defined through convolution type integrations on $\mathbf{R}$, that involve other basic positive linear operators $l_k$ and a general scaling function $\varphi$. The derivatives of the above operators are studied thoroughly in terms of: shift invariance, preservation of high order global smoothness, convergence to the derivatives of function with rates, and preservation of probabilistic properties. Here we follow the basic work [34].

## 12.1 General Results

Let $X := C_U(\mathbf{R})$ be the space of uniformly continuous real-valued functions on $\mathbf{R}$ and $C(\mathbf{R})$ the space of continuous functions from $\mathbf{R}$ into itself. For any $f \in X$ we find $\omega_p(f;\delta) < +\infty$, $p \in \mathbf{N}$, $\delta > 0$, where $\omega_p$ is the $p$th modulus of smoothness with respect to the supremum norm defined as follows:

$$\omega_p(f;\delta) = \sup_{0 < h \leq \delta} \|\Delta_h^p f\|_\infty,$$

where

$$\Delta_h^p f(t) := \sum_{i=0}^{p} (-1)^{p-i} \binom{p}{i} f(t + ih)$$

is the forward finite difference. Let $\{l_k\}_{k \in \mathbf{Z}}$ be a sequence of positive linear operators from $C(\mathbf{R})$ into itself such that

$$l_k(f)(x) = l_0(f(2^{-k}\cdot))(x), \quad x \in \mathbf{R}, \ f \in C(\mathbf{R}). \tag{12.1}$$

Suppose that

$$l_0 : C^{(i)}(\mathbf{R}) \to C^{(i)}(\mathbf{R}), \quad i \in \mathbf{N} \text{ fixed}, \tag{12.2}$$

where $C^{(i)}(\mathbf{R})$ is the space of $i$-times continuously differentiable functions. Notice that if $f \in C^{(i)}(\mathbf{R})$ then

$$\frac{\partial^{(i)}(l_0(f))}{\partial x^i}(x-u)$$

is continuous in $x, u$.

In this chapter we study only $f \in C^{(i)}(\mathbf{R})$ such that $f^{(i)} \in C_U(\mathbf{R})$. Let $\varphi$ be a continuous real-valued function of compact support $\subseteq [-a,a]$, $\varphi \geq 0$ and

$$\int_{-\infty}^{+\infty} \varphi(x-u)du = 1, \quad x \in \mathbf{R}. \tag{12.3}$$

It is transparent that

$$\int_{-\infty}^{+\infty} \varphi(u)du = 1. \tag{12.4}$$

**EXAMPLE 12.1**
The hat function

$$\varphi(x) = \begin{cases} 1-x, & 0 \leq x \leq 1 \\ 1+x, & -1 \leq x \leq 0 \\ 0, & \text{elsewhere.} \end{cases}$$

Let $\{L_k\}_{k \in \mathbf{Z}}$ be the sequence of linear operators acting on $C^{(i)}(\mathbf{R})$ and defined as follows

$$L_0(f)(x) := \int_{-\infty}^{+\infty} l_0(f)(u)\varphi(x-u)du, \tag{12.5}$$

and

$$L_k(f)(x) := L_0(f(2^{-k}\cdot))(2^k x), \quad x \in \mathbf{R}, \ k \in \mathbf{Z}. \tag{12.6}$$

## 12. Differentiated Shift-Invariant Integral Operators

That is
$$L_k(f)(x) = \int_{-\infty}^{+\infty} l_k(f)(u)\varphi(2^k x - u)du. \qquad (12.7)$$

Notice that
$$L_0(f)(x) = \int_{-\infty}^{+\infty} l_0(f)(x-u)\varphi(u)du = \int_{-a}^{a} l_0(f)(x-u)\varphi(u)du. \qquad (12.8)$$

By Leibnitz rule and by (12.5)–(12.8) we find (see [26, relation (12)])

$$L_k^{(i)}(f)(x) = 2^{ki} \int_{-\infty}^{+\infty} l_k^{(i)}(f)(u)\varphi(2^k x - u)du. \qquad (12.9)$$

**ASSUMPTION**

For fixed $a > 0$, $f \in C^{(i)}(\mathbf{R})$, ($i \in \mathbf{N}$ fixed) such that $f^{(i)} \in X$, let us suppose that

$$\sup_{\substack{u,y \in \mathbf{R} \\ |u-y| \le \alpha}} |l_0^{(i)}(f)(u) - f^{(i)}(y)| \le \omega_1\left(f^{(i)}; \frac{ma+n}{2^r}\right), \qquad (12.10)$$

where $m \in \mathbf{N}$, $n \in \mathbf{Z}_+$, $r \in \mathbf{Z}$.

We use

**DEFINITION 12.1** Let $f_\alpha(\cdot) = f(\cdot + \alpha)$, $\alpha \in \mathbf{R}$ and $\phi$ be an operator. If $\phi(f_\alpha) = (\phi(f))_\alpha$ then $\phi$ is called a shift invariant operator.

We present

**THEOREM 12.1**
For any $f \in C^{(i)}(\mathbf{R})$ such that $f^{(i)} \in X$, $i \in \mathbf{N}$ fixed, suppose that

$$|\Delta_h^p l_0^{(i)}(f)(x-u)| \le \omega_p(f^{(i)}; h), \quad \forall\, x, u \in \mathbf{R},\ h > 0, \qquad (12.11)$$

where $p \in \mathbf{N}$ is fixed and $\Delta_h^p$ is the pth forward finite difference operator. Then

$$\omega_p(L_k^{(i)}(f); \delta) \le \omega_p(f^{(i)}; \delta), \quad \forall\, \delta > 0,\ k \in \mathbf{Z}. \qquad (12.12)$$

**PROOF** We see that

$$|\Delta_h^p L_0^{(i)}(f)(x)| = \left|\sum_{j=0}^{p}(-1)^{p-j}\binom{p}{j}L_0^{(i)}(f)(x+jh)\right|$$

$$= \left|\sum_{j=0}^{p}(-1)^{p-j}\binom{p}{j}\int_{-\infty}^{+\infty}l_0^{(i)}(f)(x+jh-u)\varphi(u)du\right|$$

$$= \left|\int_{-\infty}^{+\infty}\Delta_h^p l_0^{(i)}(f)(x-u)\varphi(u)du\right|$$

$$\leq \int_{-\infty}^{+\infty}|\Delta_h^p l_0^{(i)}(f)(x-u)|\varphi(u)du$$

$$\leq \omega_p(f^{(i)};\delta), \quad \forall\, h \in [0,\delta],\ x \in \mathbf{R}.$$

By (12.1), (12.6), (12.9) we obtain

$$|\Delta_h^p L_k^{(i)}(f)(x)| = |2^{ki}\Delta_{2^k h}^p L_0^{(i)}(f(2^{-k}\cdot))(2^k x)|$$

$$= |\Delta_{2^k h}^p L_0^{(i)}[2^{ki}f(2^{-k}\cdot)](2^k x)|$$

$$\leq \omega_p([2^{ki}f(2^{-k}\cdot)]^{(i)};2^k h) = \omega_p(f^{(i)};h).$$

∎

It is known that (see [26, Theorem 3]):

**THEOREM 12.2**
For $f \in C^{(i)}(\mathbf{R})$ such that $f^{(i)} \in X$, under the assumption (12.10), it holds

$$|L_k^{(i)}(f)(x) - f^{(i)}(x)| \leq \omega_1\left(f^{(i)}; \frac{ma+n}{2^{k+r}}\right),$$

where $i, m \in \mathbf{N}$, $n \in \mathbf{Z}_+$, $k, r \in \mathbf{Z}$.

## 12. Differentiated Shift-Invariant Integral Operators

Next we define the operators (see also [35])

$$L_{k,j}(f)(x) = \int_{-\infty}^{+\infty} l_k(f)(2^k x - ju)\varphi(u)du, \quad k \in \mathbf{Z},\ j \in \mathbf{N}. \qquad (12.13)$$

Notice that

$$L_{k,j}(f)(x) = L_{0,j}(f(2^{-k}\cdot))(2^k x), \qquad (12.14)$$

i.e., $L_{k,1} = L_k$.

We give

**THEOREM 12.3**
*Suppose that*

$$l_0^{(i)}(f(2^{-k}\cdot + \alpha))(2^k u) = l_0^{(i)}(f(2^{-k}\cdot))(2^k(u + \alpha)), \qquad (12.15)$$

*for all $k \in \mathbf{Z}$, $\alpha \in \mathbf{R}$ fixed, $u \in \mathbf{R}$, $f \in C^{(i)}(\mathbf{R})$, $i \in \mathbf{N}$ be given. Then $L_{k,j}^{(i)}$ are shift-invariant operators, for all $k \in \mathbf{Z}$, $j \in \mathbf{N}$.*

**PROOF**   We see that

$$L_{k,j}^{(i)}(f(\cdot + \alpha))(x) = L_{k,j}^{(i)}(f_\alpha)(x) = 2^{ki} L_{0,j}^{(i)}(f_\alpha(2^{-k}\cdot))(2^k x)$$

$$= 2^{ki} \int_{-\infty}^{+\infty} l_0^{(i)}(f(2^{-k}\cdot + \alpha))(2^k x - ju)\varphi(u)du$$

$$= 2^{ki} \int_{-\infty}^{+\infty} l_0^{(i)}(f(2^{-k}\cdot + \alpha))(2^k(x - 2^{-k}ju))\varphi(u)du$$

$$= 2^{ki} \int_{-\infty}^{+\infty} l_0^{(i)}(f(2^{-k}\cdot))(2^k(x - 2^{-k}ju + \alpha))\varphi(u)du$$

$$= 2^{ki} \int_{-\infty}^{+\infty} l_0^{(i)}(f(2^{-k}\cdot))(2^k(x + \alpha) - ju)\varphi(u)du$$

$$= 2^{ki} L_{0,j}^{(i)}(f(2^{-k}\cdot))(2^k(x + \alpha)) = L_{k,j}^{(i)}(f)(x + \alpha).$$

∎

In the following we study the global smoothness preservation property of $L_{k,j}^{(i)}$.

**THEOREM 12.4**
Suppose that (12.11) holds, for $f \in C^{(i)}(\mathbf{R})$, $f^{(i)} \in X$, $i \in \mathbf{N}$ fixed. Then it holds

$$\omega_p(L_{k,j}^{(i)}(f);\delta) \leq \omega_p(f^{(i)};\delta), \quad k \in \mathbf{Z}, \ j \in \mathbf{N}, \ \delta > 0. \qquad (12.16)$$

**PROOF** We observe that

$$|\Delta_h^p L_{0,j}^{(i)}(f)(x)| = \left|\sum_{q=0}^{p}(-1)^{p-q}\binom{p}{q}L_{0,j}^{(i)}(f)(x+qh)\right|$$

$$= \left|\sum_{q=0}^{p}(-1)^{p-q}\binom{p}{q}\int_{-\infty}^{+\infty}l_0^{(i)}(f)(x-ju+qh)\varphi(u)du\right|$$

$$\leq \int_{-\infty}^{+\infty}|\Delta_h^p l_0^{(i)}(f)(x-ju)|\varphi(u)du \leq \omega_p(f^{(i)};h).$$

But by (12.14) we obtain that

$$|\Delta_h^p L_{k,j}^{(i)}(f)(x)| = |2^{ki}\Delta_{2^k h}^p L_{0,j}^{(i)}(f(2^{-k}\cdot))(2^k x)|$$

$$\leq \omega_p([2^{ki}f(2^{-k}\cdot)]^{(i)};2^k h) = \omega_p(f^{(i)};h),$$

which establishes the theorem. ∎

Convergence with rates of $L_{k,j}^{(i)}$ operators follows.

**THEOREM 12.5**
Let $f \in C^{(i)}(\mathbf{R})$, $f^{(i)} \in X$, $i \in \mathbf{N}$ fixed. Suppose that (12.10) holds. Then

$$|L_{k,j}^{(i)}(f)(x) - f^{(i)}(x)| \leq \omega_1\left(f^{(i)};\frac{mja+n}{2^{k+r}}\right), \qquad (12.17)$$

where $m, j \in \mathbf{N}$, $n \in \mathbf{Z}_+$, $k, r \in \mathbf{Z}$.

## 12. Differentiated Shift-Invariant Integral Operators

**PROOF** From $\operatorname{supp}\varphi \subseteq [-a, a]$ and (12.3) we find that

$$|L_{k,j}^{(i)}(f)(x) - f^{(i)}(x)| \stackrel{(12.14)}{=} |2^{ki} L_{0,j}^{(i)}(f(2^{-k}\cdot))(2^k x) - f^{(i)}(2^{-k}(2^k x))|$$

$$\stackrel{(12.13)}{=} \left| \int_{-\infty}^{+\infty} [l_0^{(i)}(2^{ki} f(2^{-k}\cdot))(2^k x - ju) - f^{(i)}(2^{-k}(2^k x))]\varphi(u) du \right|$$

$$\leq \sup\{|l_0^{(i)}(2^{ki} f(2^{-k}\cdot))(2^k x - ju) - f^{(i)}(2^{-k}(2^k x))|; u \in [-a, a]\} =: \otimes.$$

Clearly $g := f(2^{-k}\cdot) \in C^{(i)}(\mathbf{R})$ and $g^{(i)} = 2^{-ki} f^{(i)}(2^{-k}\cdot) \in X$. Thus

$$\otimes = \sup\{|l_0^{(i)}(2^{ki} g)(2^k x - ju) - (2^{ki} g)^{(i)}(2^k x)|; u \in [-a, a]\}$$

$$\leq \sup\{|l_0^{(i)}(2^{ki} g)(v) - (2^{ki} g)^{(i)}(w)|; |v - w| \leq ja\}$$

$$\stackrel{(12.10)}{\leq} \omega_1\left((2^{ki} g)^{(i)}; \frac{mja + n}{2^r}\right)$$

$$= \omega_1\left(f^{(i)}(2^{-k}\cdot); \frac{mja + n}{2^r}\right) = \omega_1\left(f^{(i)}; \frac{mja + n}{2^{r+k}}\right),$$

which proves (12.17). ∎

**REMARK** For $j = 1$ we get the previous Theorem 12.2.

Next by using the idea that produces the generalized Jackson's operators in the classical approximation theory, let us consider the operators (see also [35])

$$I_{0,q}(f)(x) := -\int_{-\infty}^{+\infty} \sum_{j=1}^{q} (-1)^j \binom{q}{j} l_0(f)(x - ju)\varphi(u) du, \qquad (12.18)$$

and

$$I_{k,q}(f)(x) := -\int_{-\infty}^{+\infty} \sum_{j=1}^{q} (-1)^j \binom{q}{j} l_k(f)(2^k x - ju)\varphi(u) du, \qquad (12.19)$$

where $k \in \mathbf{Z}$, $q \in \mathbf{N}$.

Notice that by (12.1), (12.18), (12.19) we have that

$$I_{k,q}(f)(x) = I_{0,q}(f(2^{-k}\cdot))(2^k x), \tag{12.20}$$

$$I_{0,q}(f)(x) = -\sum_{j=1}^{q}(-1)^j \binom{q}{j} L_{0,j}(f)(x), \tag{12.21}$$

and

$$I_{k,q}(f)(x) = -\sum_{j=1}^{q}(-1)^j \binom{q}{j} L_{k,j}(f)(x). \tag{12.22}$$

We present

**THEOREM 12.6**
Let $f \in C^{(i)}(\mathbf{R})$, $i \in \mathbf{N}$ fixed, and suppose that (12.15) holds. Then $I_{k,q}^{(i)}$ is a shift invariant operator.

**PROOF** It is immediate by (12.21), (12.22) and by Theorem 12.3. ∎

Global smoothness of operators $I_{k,q}^{(i)}$ comes next.

**THEOREM 12.7**
Suppose that (12.11) is true, for $f \in C^{(i)}(\mathbf{R})$, $f^{(i)} \in X$, $i \in \mathbf{N}$ fixed. Then

$$\omega_p(I_{k,q}^{(i)}(f); \delta) \le (2^q - 1)\omega_p(f^{(i)}; \delta), \quad \delta > 0, \ p, q \in \mathbf{N}, \ k \in \mathbf{Z}. \tag{12.23}$$

**PROOF** Apply Theorem 12.4 and notice that $\sum_{j=1}^{q} \binom{q}{j} = 2^q - 1$, which establishes (12.23). ∎

We use that

$$-\sum_{j=1}^{q}(-1)^j \binom{q}{j} = 1. \tag{12.24}$$

In what follows we study the convergence to $f^{(i)}$ of $I_{k,q}^{(i)}$ operators.

**THEOREM 12.8**
For $f \in C^{(i)}(\mathbf{R})$ such that $f^{(i)} \in X$, under the assumption (12.10) it

## 12. Differentiated Shift-Invariant Integral Operators

follows that

$$|I_{k,q}^{(i)}(f)(x) - f^{(i)}(x)| \le (2^q - 1)\omega_1\left(f^{(i)}; \frac{mqa+n}{2^{k+r}}\right), \qquad (12.25)$$

where $k, r \in \mathbf{Z}$, $q, m \in \mathbf{N}$, $n \in \mathbf{Z}_+$.

**PROOF** We observe that

$$|I_{k,q}^{(i)}(f)(x) - f^{(i)}(x)| \stackrel{(12.24)}{=} \left|\sum_{j=1}^{q}(-1)^j \binom{q}{j} L_{k,j}^{(i)}(f)(x) - \sum_{j=1}^{q}(-1)^j \binom{q}{j} f^{(i)}(x)\right|$$

$$= \left|\sum_{j=1}^{q}(-1)^j \binom{q}{j}[L_{k,j}^{(i)}(f)(x) - f^{(i)}(x)]\right|$$

$$\le \sum_{j=1}^{q}\binom{q}{j}|L_{k,j}^{(i)}(f)(x) - f^{(i)}(x)|$$

$$\stackrel{(12.17)}{\le} \sum_{j=1}^{q}\binom{q}{j}\omega_1\left(f^{(i)}; \frac{mja+n}{2^{k+r}}\right)$$

$$\le (2^q - 1)\omega_1\left(f^{(i)}; \frac{mqa+n}{2^{k+r}}\right),$$

which proves (12.25). ∎

About the sharpness of the above global smoothness inequalities (12.12), (12.16), (12.23) follows the

### THEOREM 12.9
Let $i \in \mathbf{N}$ be fixed and $g_i(x) := \frac{x^{i+1}}{(i+1)!}$. Suppose that for all $u, x, y \in \mathbf{R}$ one has

$$l_0^{(i)}(g_i(2^{-k}\cdot))(2^k x - u) - l_0^{(i)}(g_i(2^{-k}\cdot))(2^k y - u) = (x-y)2^{-ki}. \qquad (12.26)$$

Then (12.12) is attained by $f := g_i$, that is sharp!

**PROOF** By (12.26), the case $p = 1$ in (12.12) was established in [26, Theorem 2]. Also, when $p > 1$, (12.12) is trivially attained. ∎

### THEOREM 12.10
Let $i \in \mathbf{N}$ be fixed and $g_i(x) := \frac{x^{i+1}}{(i+1)!}$. Suppose that for all $u, x, y \in \mathbf{R}$ one has

$$l_0^{(i)}(g_i(2^{-k}\cdot))(2^k x - ju) - l_0^{(i)}(g_i(2^{-k}\cdot))(2^k y - ju) = (x-y)2^{-ki}. \quad (12.27)$$

Then for all $x, y \in \mathbf{R}$ we get

$$L_{k,j}^{(i)}(g_i)(x) - L_{k,j}^{(i)}(g_i)(y) = x - y, \quad (12.28)$$

that is

$$\omega_1(L_{k,j}^{(i)}(g_i); \delta) = \omega_1(g_i^{(i)}; \delta), \quad \delta > 0, \quad (12.29)$$

proving (12.16) sharp when $p = 1$. Furthermore we have

$$I_{k,q}^{(i)}(g_i)(x) - I_{k,q}^{(i)}(g_i)(y) = x - y, \quad (12.30)$$

and

$$\omega_1(I_{k,q}^{(i)}(g_i); \delta) = \omega_1(g_i^{(i)}; \delta), \quad (12.31)$$

that is doing better in some cases than inequality (12.23).

Also both inequalities (12.16) and (12.23) are trivially attained by $g_i$, when $p, q \in \mathbf{N} \setminus \{1\}$, respectively.

**PROOF** We see that

$$L_{k,j}^{(i)}(f)(x) = 2^{ki} \int_{-\infty}^{+\infty} l_0^{(i)}(f(2^{-k}\cdot))(2^k x - ju)\varphi(u)du$$

and for $f = g_i$, by using (12.27) too

$$L_{k,j}^{(i)}(g_i)(x) - L_{k,j}^{(i)}(g_i)(y) = 2^{ki} \int_{-\infty}^{+\infty} (x-y)2^{-ki}\varphi(u)du$$

$$= x - y = g_i^{(i)}(x) - g_i^{(i)}(y).$$

Therefore (12.28) and (12.29) follow.

Furthermore we obtain that

$$I_{k,q}^{(i)}(g_i)(x) - I_{k,q}^{(i)}(g_i)(y) = -\sum_{j=1}^{q}(-1)^j\binom{q}{j}[L_{k,j}^{(i)}(g_i)(x) - L_{k,j}^{(i)}(g_i)(y)]$$

$$= \left[-\sum_{j=1}^{q}(-1)^j\binom{q}{j}\right](x-y) = x - y,$$

which proves (12.30) and (12.31). ■

## 12.2 Examples

Next we present examples of operators $L_k$ that fulfill the assumptions of all the results in Section 12.1. The basic function $\varphi$ here will be as in (12.3), (12.4). However, only for the operators of $A_k$-type, $\varphi$ will be additionally assumed an even function. These operators appeared also in [26] and [35].

For each $k \in \mathbf{Z}$, we define

(i)
$$A_k(f)(x) := \int_{-\infty}^{+\infty} r_k^f(u)\varphi(2^k x - u)du, \qquad (12.32)$$

where
$$l_k(f)(u) = r_k^f(u) = 2^k \int_{-\infty}^{+\infty} f(t)\varphi(2^k t - u)dt \qquad (12.33)$$

is continuous in $u \in \mathbf{R}$.

(ii)
$$B_k(f)(x) := \int_{-\infty}^{+\infty} f\left(\frac{u}{2^k}\right)\varphi(2^k x - u)du, \qquad (12.34)$$

where
$$l_k(f)(u) = f\left(\frac{u}{2^k}\right) \qquad (12.35)$$

is continuous in $u \in \mathbf{R}$.

(iii)
$$C_k(f)(x) := \int_{-\infty}^{+\infty} c_k^f(u)\varphi(2^k x - u)du, \qquad (12.36)$$

where
$$l_k(f)(u) = c_k^f(u) := 2^k \int_{2^{-k}u}^{2^{-k}(u+1)} f(t)dt \qquad (12.37)$$

is continuous in $u \in \mathbf{R}$.
(iv)
$$\Gamma_k(f)(x) := \int_{-\infty}^{+\infty} \gamma_k^f(u)\varphi(2^k x - u)du, \qquad (12.38)$$

where
$$l_k(f)(u) = \gamma_k^f(u) := \sum_{j=0}^{n} w_j f\left(\frac{u}{2^k} + \frac{j}{2^k n}\right), \qquad (12.39)$$

is continuous in $u \in \mathbf{R}$, $w_j \geq 0$, $\sum_{j=1}^{n} w_j = 1$.

Also we define the $j \in \mathbf{N}$ version of the above operators:

$$A_{k,j}(f)(x) := \int_{-\infty}^{+\infty} r_k^f(2^k x - ju)\varphi(u)du, \qquad (12.40)$$

$$B_{k,j}(f)(x) := \int_{-\infty}^{+\infty} f\left(\frac{2^k x - ju}{2^k}\right)\varphi(u)du, \qquad (12.41)$$

$$C_{k,j}(f)(x) := \int_{-\infty}^{+\infty} c_k^f(2^k x - ju)\varphi(u)du, \qquad (12.42)$$

$$\Gamma_{k,j}(f)(x) := \int_{-\infty}^{+\infty} \gamma_k^f(2^k x - ju)\varphi(u)du. \qquad (12.43)$$

Notice that $A_k = A_{k,1}$, $B_k = B_{k,1}$, $C_k = C_{k,1}$, and $\Gamma_k = \Gamma_{k,1}$, where $l_k(f)$, $l_0(f)$ as above satisfy (12.1) and (12.2).

Also, note that for all $l_k$ given by (12.33), (12.35), (12.37), (12.39) we see that
$$l_k^{(i)}(f) = 2^{-ki} l_k(f^{(i)}), \qquad (12.44)$$

and (12.6) is satisfied by the corresponding $L_k$ operators.

Define also for $k \in \mathbf{Z}$, $q \in \mathbf{N}$, $x \in \mathbf{R}$

$$I_{k,q}^A(f)(x) := -\sum_{j=1}^{q}(-1)^j \binom{q}{j} A_{k,j}(f)(x), \qquad (12.45)$$

$$I_{k,q}^B(f)(x) := -\sum_{j=1}^{q}(-1)^j \binom{q}{j} B_{k,j}(f)(x), \qquad (12.46)$$

$$I_{k,q}^C(f)(x) := -\sum_{j=1}^{q}(-1)^j \binom{q}{j} C_{k,j}(f)(x), \qquad (12.47)$$

## 12. Differentiated Shift-Invariant Integral Operators

$$I_{k,q}^{\Gamma}(f)(x) := -\sum_{j=1}^{q}(-1)^j \binom{q}{j}\Gamma_{k,j}(f)(x). \tag{12.48}$$

Clearly the above four operators satisfy (12.20)-(12.22).

By (12.44) it easily follows that if we denote by $L$ any among the operators given by (12.32), (12.34), (12.36), (12.38), (12.40)-(12.43), (12.45)-(12.48), then

$$L^{(i)}(f)(x) = L(f^{(i)})(x), \quad f \in C^{(i)}(\mathbf{R}),$$

a property not in general true [26].

Therefore, $A_k$, $B_k$, $C_k$, $\Gamma_k$ satisfy (12.12) in Theorem 12.1 as a consequence of [35, Proposition 2]; $A_{k,j}$, $B_{k,j}$, $C_{k,j}$, $\Gamma_{k,j}$ satisfy Theorem 12.3, Theorem 12.4, as consequences of [35, Proposition 2, Proposition 3]. Also $A_{k,j}$, $B_{k,j}$, $C_{k,j}$, $\Gamma_{k,j}$ satisfy Theorem 12.5 with $(m = 1, n = 0, r = -1)$, $(m = 1, n = 0, r = 0)$, $(m = 1, n = 1, r = 0)$, $(m = 1, n = 1, r = 0)$, respectively, as a direct consequence of [35, Proposition 6]; as a consequence of [35, Remark 5]. The operators $I_{k,q}^{A}$, $I_{k,q}^{B}$, $I_{k,q}^{C}$, $I_{k,q}^{\Gamma}$ satisfy Theorem 12.6, Theorem 12.7, and Theorem 12.8 (with $m = 1$, $n = 0$, $r = -1$ for $I_{k,q}^{A}$, $m = 1$, $n = 0$, $r = 0$ for $I_{k,q}^{B}$, $m = 1$, $n = 1$, $r = 0$ for $I_{k,q}^{C}$ and $I_{k,q}^{\Gamma}$); as a consequence of [35, Theorem 9]. Finally, $A_{k,j}^{(i)}$, $B_{k,j}^{(i)}$, $C_{k,j}^{(i)}$, $\Gamma_{k,j}^{(i)}$ preserve continuous probability density functions.

# Part III

# On Partial Differential Equations

# Chapter 13

## A Discrete Kac's Formula and Optimal Quantitative Approximation in the Solution of Heat Equation

In this chapter we get a closed form solution of the Dirichlet problem of the discretized heat equation with potential. Sharp quantitative estimates of the difference between actual and approximate solutions are presented in terms of the first and second moduli of continuity of some first and second order partial derivatives of the exact solution. This is done probabilistically by using a suitable random walk. This treatment is based on [31].

Let $\Omega$ be the open unit cube in $\mathbf{R}^l$, $l \geq 1$ and $\dot{\Omega} := \Omega \times I$ be an "interval" in space-time $\dot{\mathbf{R}}^l = \mathbf{R}^l \times \mathbf{R}$, where $I := (0, T)$, $T > 0$. Denote by $\dot{\Delta} = \frac{1}{2}\Delta - \partial_t$ the heat operator, where $\Delta$ stands for the Laplacian operator in $\mathbf{R}^l$. Given a continuous function $\lambda \geq 0$ on $\dot{\Omega}$, consider the Dirichlet problem in $\dot{\Omega}$:

$$\dot{\Delta} u(\dot{x}) - \lambda(\dot{x}) u(\dot{x}) = -f(\dot{x}), \quad \dot{x} \in \dot{\Omega},$$

$$\lim_{\dot{x} \to \dot{y}} u(\dot{x}) = \varphi(\dot{y}), \quad \forall \, \dot{y} \in \partial \dot{\Omega} - \{\dot{x} = (x, t) : t = T\}.$$

It is a known fact [103] that the solution $u$ of this equation is also expressed by the formula

$$u(\dot{x}) = E_{\dot{x}} \left\{ \int_0^{\tau_{\dot{\Omega}}} M_s \cdot f(\dot{x}(s)) \, ds + M_{\tau_{\dot{\Omega}}} \cdot \varphi(\dot{x}(\tau_{\dot{\Omega}})) \right\},$$

where

$$M_t = \exp\left\{ -\int_0^t \lambda(\dot{x}(s)) \, ds \right\}, \quad t \geq 0$$

and $\{\dot{x}(t,w), t < \tau_{\dot{\Omega}}\}$ is a space-time Brownian motion in domain $\dot{\Omega}$. We also study the discrete Dirichlet problem of getting $u_h$ such that

$$\dot{\Delta}_h u_h(\dot{x}) - \lambda(\dot{x}) u_h(\dot{x}) = -f(\dot{x}), \quad \forall \dot{x} \in \dot{\Omega}_{h,T},$$

$$u_h(\dot{x}) = \varphi(\dot{x}), \quad \forall \dot{x} \in \dot{\Omega}_{h,T} - \{\dot{x} = (x,t) : t = T\},$$

where $\dot{\Delta}_h$ is the discrete heat operator

$$\dot{\Delta}_h u_h(\dot{x}) = \frac{1}{2} h^{-2} \left( \sum_{k=1}^{l} u_h(x \pm h e_k, t - k(h)) - 2l u_h(x,t) \right),$$

$$\forall \dot{x} = (x,t) \in \dot{\Omega}_{h,T}.$$

Here $k(h) = \frac{h^2}{l}$ and $\{e_k\}$ is the natural basis in $\mathbf{R}^l$. Using the space-time random walk $\{\dot{x}(k,w), k < \tau_{\dot{\Omega}_{h,T}}\}$ in $\dot{\Omega}_{h,T}$ we present a discrete analog of Kac's formula for the solution $u_h$

$$u_h(\dot{x}) = E_{\dot{x}} \left\{ k \cdot \sum_{n=0}^{\tau_{\dot{\Omega}_{h,T}}-1} M_n \cdot f(\dot{x}(n)) + M_{\tau_{\dot{\Omega}_{h,T}}-1} \cdot \varphi(\dot{x}(\tau_{\dot{\Omega}_{h,T}})) \right\},$$

where

$$M_n = \prod_{s=0}^{n} (1 + k \cdot \lambda(\dot{x}(s)))^{-1}, \quad n = 0, 1, 2, \ldots$$

and $k = h^2/l$.

Setting together all the above we are able to give the next sharp inequality

(i) $\quad \|u - u_h\|_{\overline{\dot{\Omega}_{h,T}}} \leq \min\left\{T, \frac{1}{4}\right\} \cdot \left\{ \frac{1}{4} \cdot \sum_{i=1}^{l} \omega_{2,i}(h, \partial_{x_i}^2 u; \overline{\dot{\Omega}}_{h,T}) \right.$

$$\left. + \frac{1}{2} \cdot \sum_{i=1}^{l} \omega_1(k, \partial_{x_i}^2 u; \overline{\dot{\Omega}}_{h,T}) + \omega_1(k, \partial_t u; \overline{\dot{\Omega}}_{h,T}) \right\},$$

where $\|\cdot\|_{\overline{\dot{\Omega}_{h,T}}}$ is the supremum norm in $\overline{\dot{\Omega}}_{h,T}$.

Here $\omega_{2,i}$ is the second modulus of continuity of the second single partial of $u$ with respect to $x_i$, $i = 1, \ldots, l$; while $\omega_1$ stands for the first modulus of continuity of the indicated function with respect to the variable $t$.

## 13.1 Auxiliary Results

Dirichlet problem for the "heat operator: Let $R^l$ be the Euclidean space, and $\Delta = \partial_{x_1}^2 + \cdots + \partial_{x_l}^2$ be the Laplacian. We denote $\dot{R}^j = \{\dot{x} = (x,t) : x \in R^l, t \in R\}$ and let $\dot{\Delta} = \frac{1}{2}\Delta - \partial_t$ be the "heat" operator (i.e., the parabolic Laplacian). Let $\dot{\Omega} \subset \dot{R}^l$ be an open subset with nonempty boundary $\partial\dot{\Omega}$. The Dirichlet problem in $\dot{\Omega}$ consists of finding a function $u$ on $\dot{\Omega}$ such that for given functions $f$, defined in $\dot{\Omega}$, and $\varphi$, defined on $\partial\dot{\Omega}$ it holds

$$\dot{\Delta}u(\dot{x}) = -f(\dot{x}), \quad \dot{x} \in \dot{\Omega}, \tag{13.1}$$

and

$$\lim_{\dot{x}\to\dot{y}} u(\dot{x}) = \varphi(\dot{y}), \quad \forall \dot{y} \in \partial\dot{\Omega}. \tag{13.2}$$

It is well known [79] that if $\dot{\Omega}$ has compact closure and regular boundary (for example, if $\dot{\Omega}$ is a ball, or a convex polyhedron that is situated above its horizontal face) and $f$ is a Hölder function and $\varphi$ is a continuous function, then the problem (13.1–13.2) has unique solution $u(\dot{x})$. This solution can be represented in the form

$$u(\dot{x}) = G_{\dot{\Omega}}f(\dot{x}) + \Pi_{\dot{\Omega}}\varphi(\dot{x}), \tag{13.3}$$

where

$$G_{\dot{\Omega}}f(\dot{x}) = \int_{\dot{\Omega}} g_{\dot{\Omega}}(\dot{x},\dot{y})f(\dot{y})\,d\dot{y}, \tag{13.4}$$

and

$$\Pi_{\dot{\Omega}}\varphi(\dot{x}) = \int_{\partial\dot{\Omega}} \varphi(\dot{y})\Pi_{\dot{\Omega}}(\dot{x},d\dot{y}), \tag{13.5}$$

are the "parabolic" Green potential of the function $f$ and the "parabolic" harmonic function (i.e., a parabolic function) in $\dot{\Omega}$ with boundary values $\varphi$, respectively. In (13.4) and (13.5), $g_{\dot{\Omega}}(\dot{x},\dot{y})$ and $\Pi_{\dot{\Omega}}(\dot{x},d\dot{y})$ are the "parabolic" Green function and the "parabolic" harmonic measure (i.e., the parabolic measure) of $\dot{\Omega}$, respectively.

Let $\dot{x}_0 = (x_0, t_0)$ be a point of $\dot{R}^l$, and let $\{\dot{x}(\cdot, \omega), x(0) = x_0\}$ be a Brownian motion in $R^l$ starting from $x_0$. The process

$$\{\dot{x}(t,\omega), t \in R^+\} := \{(x(t,\omega), t_0 - t), t \in R^+\}$$

with state space $\dot{R}^l$ is called a space-time Brownian motion starting from $\dot{x}_0$. In this definition the space-time Brownian motion moves downward in

$\dot R^l$, that is, in the direction of decreasing ordinate values. Let $\tau_{\dot\Omega}$ be the first exit time of the process $\dot x(\cdot,\omega)$ of an open set $\dot\Omega \subset \dot R^l$. We notice two special cases:

1) if $\dot\Omega = \Omega \times R^l$ is a cylinder with the base $\Omega \subset R^l$, then $\tau_{\dot\Omega} = \tau_\Omega$, where $\tau_\Omega$ is the first exit time of Brownian motion $x(\cdot,\omega)$ of $\Omega$,

2) if $\dot\Omega = \Omega \times I$, $I = (a,b)$, is an "interval" with the base $\Omega \subset R^l$, then $\tau_{\dot\Omega} = \min\{\tau_\Omega, \tau_I\}$ where $\tau_I \le b-a$ is the first exit time of the uniform motion $t \to t_0 - t$ of $I$. Therefore, in particular, $\tau_{\dot\Omega} \le b-a$.

As usual we denote by $E_{\dot x} F(\omega)$ (resp., $E_x F(\omega)$) the mathematical expectation corresponding to the process $\{\dot x(\cdot,\omega), \dot x(0) = \dot x_0\}$ (resp., $\{x(\cdot,\omega), x(0) = x\}$). We get

$$G_{\dot\Omega} f(\dot x) = E_{\dot x}\left(\int_0^{\tau_{\dot\Omega}} f(\dot x(s))\, ds\right) \tag{13.6}$$

and

$$\Pi_{\dot\Omega}\varphi(\dot x) = E_{\dot x}[\varphi(\dot x(\tau_{\dot\Omega})); \tau_{\dot\Omega} < \infty]. \tag{13.7}$$

In particular, if $\dot\Omega = \Omega \times (0,T)$ is an "interval" and $\varphi(x,t) = 0$ for each $x \in \partial\Omega$ and $t \ge 0$, then setting $\varphi(x,0) := \varphi(x)$ we obtain

$$u(x,t) = E_x\{\varphi(x(t)); t < \tau_\Omega\}, \tag{13.8}$$

which gives the solution $u(x,t)$ of the Cauchy problem in $\Omega$

$$\left.\begin{array}{ll}\partial_t u(x,t) = \tfrac{1}{2}\Delta u(x,t), & x \in \Omega,\ t > 0, \\ u(x,0) = \varphi(x), & x \in \Omega, \\ u(x,t) = 0, & x \in \partial\Omega,\ t > 0.\end{array}\right\} \tag{13.9}$$

The next fact we mention here is the famous Kac's formula [103], see also [102], [80], [49]. This formula, in particular, shows how to get solutions of a more general class of Dirichlet problems by using Brownian motion.

Given a continuous function $\lambda(\dot x) \ge 0$, define a random process

$$M_t := \exp\left\{-\int_0^t \lambda(\dot x(s))\, ds\right\}, \quad t \ge 0,$$

and put

$$u(\dot x) := E_{\dot x}\left\{\int_0^{\tau_{\dot\Omega}} M_s \cdot f(\dot x(s))\, ds + M_{\tau_{\dot\Omega}} \cdot \varphi(\dot x(\tau_{\dot\Omega}))\right\}, \tag{13.10}$$

where $\tau_{\dot\Omega}$ is the first exit time of the space-time Brownian motion $\dot x(\cdot,\omega)$, $\dot x(0,\omega) = \dot x$, of an open regular set $\dot\Omega \subset \dot R^l$. Then, according to [103], the function $u(\dot x)$, $\dot x \in \dot\Omega$, gives a solution to the following Dirichlet problem

$$\left.\begin{aligned}\dot\Delta u(\dot x) - \lambda(\dot x)u(\dot x) &= -f(\dot x), & \dot x &\in \dot\Omega \\ \lim_{\dot x \to \dot y} u(\dot x) &= \varphi(\dot y), & \dot y &\in \partial\dot\Omega.\end{aligned}\right\} \quad (13.11)$$

In the next section we consider a discrete analog of the Kac's formula and then its application on a uniform grid.

## 13.2 The Dirichlet Problem: Discrete Case

Let $\dot Z^l$ be the $(l+1)$-dimensional integer-valued lattice. This lattice consists of points (vectors) of type $\dot x = x_1 e_1 + \cdots + x_l e_l + te$, where $e_1, \ldots, e_l, e$ comprises the orthogonal basis of $R^{l+1}$, and the coordinates $x_1, \ldots, x_l, t$ are arbitrary integers. Decreasing the $t$-coordinate by one unit and increasing or decreasing each one of the $x$-coordinates by one unit and leaving the other $x$-coordinates unchanged, we get the $2l$ neighboring lattice points to $\dot x$. Let $B$ be a subset of points of the lattice $\dot Z^l$. We call a point $\dot x \notin B$ a boundary point for the set $B$ if $\dot x$ is a neighboring point for at least one point of $B$. The set of all boundary points of the set $B$ is called the boundary of $B$, denoted by $\partial B$.

Let $f$ be a function defined at the points of the lattice $\dot Z^l$. We put

$$\dot P f(\dot x) := \frac{1}{2l}\sum_{k=1}^{l} f(\dot x \pm e_k - e).$$

We call $\dot P$ the averaging operator. The linear operator $\dot P - E$, where $E$ is the unit operator, is the discrete analog of the "parabolic" Laplacian $\dot\Delta$. Really for a sufficiently smooth function $f(\dot x)$ specified over all the space $\dot R^l$,

$$\dot\Delta f(\dot x) = \lim_{h \to 0} \frac{1}{2} h^{-2} \left( \sum_{k=1}^{l} f\left(\dot x \pm he_k - \frac{h^2}{l}e\right) - 2lf(\dot x) \right).$$

So that the "parabolic" Laplacian is found by passing to the limit from the scaled operator $\dot P - E$ as the lattice is infinitely partioned.

Let $\dot\Omega \subset \dot Z^l$ be a finite subset and $\lambda(\dot x)$ be a nonnegative function defined on $\dot\Omega$. The $\lambda$-Dirichlet problem in $\dot\Omega$ consists in finding a function $u(\dot x)$,

$\dot{x} \in \dot{\Omega} \cup \partial\dot{\Omega}$ such that for given functions $f$ defined in $\dot{\Omega}$ and $\varphi$ defined in $\partial\dot{\Omega}$ we have

$$(\dot{P} - E)u(\dot{x}) - \lambda(\dot{x})u(\dot{x}) = -f(\dot{x}), \quad \dot{x} \in \dot{\Omega}, \tag{13.12}$$

$$u(\dot{x}) = \varphi(\dot{x}), \quad \dot{x} \in \partial\dot{\Omega}. \tag{13.13}$$

First we notice that if $u_1$ and $u_2$ are two solutions of problems (13.12)–(13.13) then $u_1 \equiv u_2$. This fact follows immediately from the next minimum principle.

**THEOREM 13.1**
Let $u$ be a function on $\dot{\Omega} \cup \partial\dot{\Omega}$ such that

$$(\dot{P} - E)u(\dot{x}) - \lambda(\dot{x})u(\dot{x}) \leq 0$$

for any $\dot{x} \in \dot{\Omega}$. If $u(\dot{x}) \geq 0$ for any $\dot{x} \in \partial\dot{\Omega}$, then $u(\dot{x}) \geq 0$ for any $\dot{x} \in \dot{\Omega} \cup \partial\dot{\Omega}$.

**PROOF** Let $u_*(\dot{x})$ be a function that is defined on $\dot{\Omega} \cup \partial\dot{\Omega}$, strictly positive and satisfies the equation

$$(\dot{P} - E)u_*(\dot{x}) - \lambda(\dot{x})u_*(\dot{x}) = 0, \quad \dot{x} \in \dot{\Omega}.$$

The existence of such a function will be shown later (see Theorem 13.3). Set $\dot{P}_\lambda := (1 + \lambda)^{-1}\dot{P}$ and define the following average operator

$$P^*\varphi(\dot{x}) := \frac{1}{u_*(\dot{x})} \dot{P}_\lambda u_* \cdot \varphi(\dot{x}).$$

It is obvious that $P^*1(x) = 1$. Consider now the following function

$$\bar{u}(\dot{x}) := u(\dot{x})/u_*(\dot{x}).$$

Then we have $\bar{u}(\dot{x}) \geq 0$ for each $\dot{x} \in \partial\dot{\Omega}$, and furthermore: for any $\dot{x} \in \dot{\Omega}$

$$P^*\bar{u}(\dot{x}) = \frac{1}{u_*(\dot{x})} \cdot \frac{1}{1 + \lambda(\dot{x})} \cdot \dot{P}u(\dot{x})$$

$$\leq \frac{1}{u_*(\dot{x})} \cdot \frac{1}{1 + \lambda(\dot{x})} \cdot (1 + \lambda(\dot{x})) \cdot u(\dot{x})$$

## 13. A Discrete Kac's Formula

$$= \frac{u(\dot{x})}{u_*(\dot{x})} = \overline{u}(\dot{x}).$$

Hence we have

$$P^*\overline{u}(\dot{x}) \leq \overline{u}(\dot{x}), \quad \dot{x} \in \dot{\Omega},$$

$$\overline{u}(\dot{x}) \geq 0, \quad \dot{x} \in \partial\dot{\Omega}.$$

Assume now that there exists $\dot{x} \in \dot{\Omega}$ such that

$$\overline{u}(\dot{x}_0) = \min\{\overline{u}(\dot{x}) : \dot{x} \in \dot{\Omega} \cup \partial\dot{\Omega}\}.$$

Then we have
$$\overline{u}(\dot{x}_0) \geq P^*\overline{u}(\dot{x}_0) \geq \overline{u}(\dot{x}_0).$$

That is, $\overline{u}(\dot{x}_1) = \overline{u}(\dot{x}_0)$ for each point $\dot{x}_1$ neighboring $\dot{x}_0$. If one of these points belongs to $\partial\dot{\Omega}$, then the proof is finished, if not, then we repeat our reasoning again. It is clear that after a finite number of steps we meet the boundary $\partial\dot{\Omega}$. Hence, we finally derive that

$$\overline{u}(\dot{x}) \geq \overline{u}(\dot{x}_0) = \overline{u}(\dot{x}_1) = \cdots = \overline{u}(\dot{x}_n) \geq 0$$

for some $\dot{x}_n \in \partial\dot{\Omega}$. Because $u_*(\dot{x}) > 0$ for all $\dot{x} \in \dot{\Omega} \cup \partial\dot{\Omega}$, we obtain that $u(\dot{x}) \geq 0$ for each $\dot{x} \in \dot{\Omega} \cup \partial\dot{\Omega}$. The proof is now complete. ∎

Next our remark concerns the decomposition $u = u_1 + u_2$ of the solution $u$ of problems (13.12)–(13.13), where

$$(\dot{P} - E)u_1(\dot{x}) - \lambda(\dot{x})u_1(\dot{x}) = -f(\dot{x}), \quad \dot{x} \in \dot{\Omega}, \qquad (13.14)$$

$$u_1(\dot{x}) = 0, \quad \dot{x} \in \partial\dot{\Omega}, \qquad (13.15)$$

$$(\dot{P} - E)u_2(\dot{x}) - \lambda(\dot{x})u_2(\dot{x}) = 0, \quad \dot{x} \in \dot{\Omega}, \qquad (13.16)$$

and
$$u_2(\dot{x}) = \varphi(\dot{x}), \quad \dot{x} \in \partial\dot{\Omega}. \qquad (13.17)$$

This decomposition is the discrete analog of the decomposition (13.3). Following the same reasonings as in §13.1, we present the probabilistic representation of the "discrete" $\lambda$-Green potential $u_1 := G_{\dot{\Omega}}^{\lambda} f$ and of the "discrete" $\lambda$-parabolic function $u_2 := \prod_{\dot{\Omega}}^{\lambda} \varphi$ in (13.14)–(13.17). In this part of our presentation, we use tools from the monographs [163] and [81].

Let $\{x(n), P\}$ be a simple random walk on the lattice $\mathbf{Z}^l$, i.e., a stochastic process with independent identically distributed increments $\Delta x(n) := x(n+1) - x(n)$, $n = 0, 1, \ldots$, and such that

$$P(\Delta x(1) = x) = \begin{cases} \frac{1}{2l}, & \text{if } x = \pm e_k, \\ 0, & \text{otherwise.} \end{cases}$$

The process

$$\{\dot{x}(n), n = 0, 1, \ldots\} := \{(x(n), n_0 - n), n = 0, 1, \ldots\}$$

with state space $\dot{\mathbf{Z}}^l$ is called a space-time random walk starting from $\dot{x}(0) = (x(0), n_0)$. In this definition, space-time random walk moves downward on $\dot{\mathbf{Z}}^l$, that is, in the direction of decreasing ordinate values. It is easy to see that for each bounded function $f$

$$E\{f(\dot{x}(1)); \dot{x}(0) = \dot{x}\} = \dot{P} f(\dot{x}),$$

and more generally

$$E\{f(\dot{x}(n)); \dot{x}(0) = \dot{x}\} = \dot{P}^n f(\dot{x}).$$

For each $k = 0, 1, \ldots$ we define the next random variable

$$M_k := \prod_{s=0}^{k} [1 + \lambda(\dot{x}(s))]^{-1},$$

we put also $M_{-1} = 1$.

Let $f$ be a bounded function defined at the points of a lattice $\dot{\mathbf{Z}}^l$. Define the following linear operator

$$G^\lambda f(\dot{x}) := E_{\dot{x}} \left( \sum_{k=0}^{\infty} M_k(f(\dot{x}(k))) \right),$$

where here, and in what follows, we use $E_{\dot{x}} \Phi$, instead of $E\{\Phi; \dot{x}(0) = \dot{x}\}$, just to simplify our notations.

First we notice that if the function $f$ is supported on a finite subset $\dot{\Omega} \subset \dot{\mathbf{Z}}^l$, then the function $G^\lambda f$ takes finite values. Really, we have

$$|G^\lambda f(\dot{x})| \leq \sup_{\dot{z} \in \dot{\Omega}} |f(\dot{z})| E_{\dot{x}} \left( \sum_{\dot{x}(k) \in \dot{\Omega}}^{\infty} 1 \right).$$

## 13. A Discrete Kac's Formula

Next we notice that $\sum_{\dot{x}(k)\in\dot{\Omega}} 1$ is exactly the time that particle $\dot{x}(\cdot,\omega)$ spends in the set $\dot{\Omega}$. This time is a bounded random variable, since the particle moves downward in the direction of decreasing ordinate values. We notice also, if $\lambda_0 := \inf\{\lambda(\dot{x}): \dot{x} \in \dot{\mathbf{Z}}^l\} > 0$, then the previous statement holds true without any assumption on supp $f$. Really,

$$|G^\lambda f(\dot{x})| \leq \sup_{\dot{z}\in\dot{\mathbf{Z}}^l} |f(\dot{z})| \cdot \sum_{k=0}^{\infty}[1+\lambda_0]^{-k} < \infty.$$

It follows about the problem (13.12)–(13.13).

### THEOREM 13.2
*Let the function $f$ be supported on a finite subset $\dot{\Omega} \subset \dot{\mathbf{Z}}^l$. Then the function $u = G^\lambda f$ fulfills the following equation*

$$(\dot{P} - E)u(\dot{x}) - \lambda(\dot{x})u(\dot{x}) = -f(\dot{x}), \quad \dot{x} \in \dot{\mathbf{Z}}^l.$$

**PROOF** For each $k = 0, 1, \ldots$ denote

$$u(k, \dot{x}) := E_{\dot{x}}(M_k f(\dot{x}(k)))$$

and let $F_k$ be the $\sigma$-algebra generated by random variables $\{\dot{x}(0), \dot{x}(1), \ldots, \dot{x}(k)\}$.

We see that

$$u(k, \dot{x}) := E_{\dot{x}}(E_{\dot{x}}(M_k f(\dot{x}(k)) \mid F_1))$$

$$= [1+\lambda(\dot{x})]^{-1} E_{\dot{x}}\left(E_{\dot{x}}\left(\prod_{s=1}^{k}[1+\lambda(\dot{x}(s))]^{-1} f(\dot{x}(k)) \mid F_1\right)\right)$$

$$= [1+\lambda(\dot{x})]^{-1} E_{\dot{x}}\left(E_{\dot{x}(1)}\left(\prod_{s=0}^{k-1}[1+\lambda(\dot{x}(s))]^{-1} f(\dot{x}(k-1)) \mid F_1\right)\right)$$

$$= [1+\lambda(\dot{x})]^{-1} E_{\dot{x}}(u(k-1, \dot{x}(1))$$

$$= [1+\lambda(\dot{x})]^{-1} \dot{P}_u(k-1, \cdot)(\dot{x}).$$

Now we use this recurrence to obtain the desired result

$$G^\lambda f(\dot{x}) = \sum_{k=0}^{\infty} u(k, \dot{x})$$

$$= u(0, \dot{x}) + [1 + \lambda(\dot{x})]^{-1} \dot{P} \sum_{k=1}^{\infty} u(k-1, \cdot)(\dot{x})$$

$$= [1 + \lambda(\dot{x})]^{-1} f(\dot{x}) + [1 + \lambda(\dot{x})]^{-1} \dot{P} \sum_{k=0}^{\infty} u(k, \cdot)(\dot{x})$$

$$= [1 + \lambda(\dot{x})]^{-1} \{f(\dot{x}) + \dot{P} G^\lambda f(\dot{x})\}.$$

Therefore, we finally get

$$[1 + \lambda(\dot{x})] G^\lambda f(\dot{x}) = f(\dot{x}) + \dot{P} G^\lambda f(\dot{x}).$$

The obtained relation ends the proof of the theorem. ∎

For a finite set $\dot{\Omega} \subset \dot{\mathbf{Z}}^l$, we denote by $\tau_{\dot{\Omega}}$ the first exit time of the particle $\dot{x}(\cdot, \omega)$ of $\dot{\Omega}$. For a function $\varphi$ defined on the set $\dot{\mathbf{Z}}^l \setminus \dot{\Omega}$, we define the following linear operator

$$\Pi_{\dot{\Omega}}^\lambda \varphi(\dot{x}) := E_{\dot{x}} \{ M_{\tau_{\dot{\Omega}} - 1} \varphi(\dot{x}(\tau_{\dot{\Omega}})) \}.$$

According to our agreement, $M_{-1} = 1$ and consequently $\Pi_{\dot{\Omega}}^\lambda \varphi(\dot{x}) = \varphi(\dot{x})$ for each $\dot{x} \in \dot{\mathbf{Z}}^l \setminus \dot{\Omega}$ and, in particular, for each $\dot{x} \in \partial \dot{\Omega}$.

**THEOREM 13.3**
 The function $u(\dot{x}) := \Pi_{\dot{\Omega}}^\lambda \varphi(\dot{x})$ provides the unique solution of problem (13.16)–(13.17).

**PROOF** As it was mentioned above, the function $u$ satisfies the condition (13.17). Thus we have to verify condition (13.16). Let $\dot{x} \in \dot{\Omega}$, then for a particle starting from the point $\dot{x}$ we have $\tau_{\dot{\Omega}} \geq 1$. In what follows we suppose that $\dot{x} \in \dot{\Omega}$. We see that

$$\Pi_{\dot{\Omega}}^\lambda \varphi(\dot{x}) = \sum_{k=0}^{\infty} E_{\dot{x}} \{ M_{k-1} \varphi(\dot{x}(k)), \tau_{\dot{\Omega}} = k \}$$

## 13. A Discrete Kac's Formula

$$= \sum_{k=1}^{\infty} E_{\dot{x}}\{M_{k-1}\varphi(\dot{x}(k)), \tau_{\dot{\Omega}} = k\}.$$

For each $k = 0, 1, \ldots$ put

$$u(\dot{x}, k) := E_{\dot{x}}\{M_{k-1}\varphi(\dot{x}(k)), \tau_{\dot{\Omega}} = k\}.$$

Using the Markov property of the process $\dot{x}(\cdot, \omega)$ we have for $k \geq 1$ that

$$u(\dot{x}, k) = E_{\dot{x}}\{M_{k-1}\varphi(\dot{x}(k))1_{\tau_{\dot{\Omega}}=k}\}$$

$$= E_{\dot{x}}\left(E_{\dot{x}}\{M_{k-1}\varphi(\dot{x}(k))1_{\tau_{\dot{\Omega}}=k} \mid F_1\}\right)$$

$$= E_{\dot{x}}\left(E_{\dot{x}}\left\{[1 + \lambda(\dot{x}(0))]^{-1} \prod_{s=0}^{k-2}[1 + \lambda(\dot{x}(s+1))]^{-1}\varphi(\dot{x}(k-1+1))1_{\tau_{\dot{\Omega}}-1=k-1} \mid F_1\right\}\right)$$

$$= [1 + \lambda(\dot{x})]^{-1} E_{\dot{x}}\left(E_{\dot{x}(1)}\{M_{k-2}\varphi(\dot{x}(k-1))1_{\tau_{\dot{\Omega}}=k-1}\}\right)$$

$$= [1 + \lambda(\dot{x})]^{-1} E_{\dot{x}}(u(\dot{x}(1), k-1))$$

$$= [1 + \lambda(\dot{x})]^{-1} \dot{P} u(\cdot, k-1)(\dot{x}).$$

Now we use this recurrence relation to obtain the final results. Really, for $\dot{x} \in \dot{\Omega}$ we get

$$\Pi_{\dot{\Omega}}^{\lambda}\varphi(\dot{x}) = \sum_{k=1}^{\infty} u(\dot{x}, k)$$

$$= \sum_{k=1}^{\infty}[1 + \lambda(\dot{x})]^{-1}\dot{P}u(\cdot, k-1)(\dot{x})$$

$$= [1 + \lambda(\dot{x})]^{-1}\dot{P}\sum_{k=1}^{\infty} u(\cdot, k-1)(\dot{x})$$

$$= [1+\lambda(\dot{x})]^{-1}\dot{P}\sum_{k=0}^{\infty}u(\cdot,k)(\dot{x})$$

$$= [1+\lambda(\dot{x})]^{-1}\dot{P}\Pi_{\dot{\Omega}}^{\lambda}\varphi(\dot{x}).$$

Consequently, we have established that for each $x \in \dot{\Omega}$, the following relation is valid

$$\Pi_{\dot{\Omega}}^{\lambda}\varphi(\dot{x}) = [1+\lambda(\dot{x})]^{-1}\dot{P}\Pi_{\dot{\Omega}}^{\lambda}\varphi(\dot{x}).$$

Clearly, from this last relation the statement of the theorem is implied. ∎

We again fix a finite set $\dot{\Omega} \subset \dot{\mathbf{Z}}^l$ and we introduce the following linear operator:

$$G_{\dot{\Omega}}^{\lambda}f(\dot{x}) := E_{\dot{x}}\left(\sum_{k=0}^{\tau_{\dot{\Omega}}-1}M_k f(\dot{x}(k))\right)$$

(of course, we put $\sum_{k\in\phi} = 0$). Hence we observe that $G_{\dot{\Omega}}^{\lambda}f(\dot{x}) = 0$ for each $\dot{x} \notin \dot{\Omega}$ and, in particular for $\dot{x} \in \partial\dot{\Omega}$.

The following result concerns the relation between operators $G^{\lambda}$, $G_{\dot{\Omega}}^{\lambda}$ and $\Pi_{\dot{\Omega}}^{\lambda}$. This relation is a discrete analog of the known Dynkin's formula for Markov processes [80].

**THEOREM 13.4**
*The following relation holds valid*

$$G^{\lambda}f(\dot{x}) - \Pi_{\dot{\Omega}}^{\lambda}G^{\lambda}f(\dot{x}) = G_{\dot{\Omega}}^{\lambda}f(\dot{x}), \quad \dot{x} \in \dot{\Omega}.$$

**PROOF** We see that

$$G^{\lambda}f(\dot{x}) = E_{\dot{x}}\left(\sum_{k=0}^{\infty}M_k f(\dot{x}(k))\right)$$

$$= E_{\dot{x}}\left(\sum_{k=0}^{\tau_{\dot{\Omega}}-1}M_k f(\dot{x}(k))\right) + E_{\dot{x}}\left(\sum_{k=\tau_{\dot{\Omega}}}^{\infty}M_k f(\dot{x}(k))\right)$$

$$=: G_{\dot{\Omega}}^{\lambda}f(\dot{x}) + I.$$

## 13. A Discrete Kac's Formula

Denote by $F_{\tau_{\hat{\Omega}}}$ the $\sigma$-algebra of events $A$ such that $A \cap \{\tau_{\hat{\Omega}} \leq k\} \subset F_k$ for each $k \geq 1$, see [163], [157]. Using now the strong Markov property of the process $\dot{x}(\cdot, \omega)$ we investigate the function $I$. We observe that

$$I = E_{\dot{x}}\left(\sum_{k=\tau_{\hat{\Omega}}}^{\infty} M_k f(\dot{x}(k))\right) = E_{\dot{x}}\left(\sum_{k=0}^{\infty} M_{k+\tau_{\hat{\Omega}}} f(\dot{x}(k+\tau_{\hat{\Omega}}))\right)$$

$$= E_{\dot{x}}\left(E_{\dot{x}}\left(\sum_{k=0}^{\infty} \prod_{s=0}^{k+\tau_{\hat{\Omega}}} [1+\lambda(\dot{x}(s))]^{-1} f(\dot{x}(k+\tau_{\hat{\Omega}})) \mid F_{\tau_{\hat{\Omega}}}\right)\right)$$

$$= E_{\dot{x}}\left(E_{\dot{x}}\left(\sum_{k=0}^{\infty} \prod_{s=0}^{\tau_{\hat{\Omega}}-1} [1+\lambda(\dot{x}(s))]^{-1}\right.\right.$$

$$\left.\left. \times \prod_{s=\tau_{\hat{\Omega}}}^{k+\tau_{\hat{\Omega}}} [1+\lambda(\dot{x}(s))]^{-1} f(\dot{x}(k+\tau_{\hat{\Omega}})) \mid F_{\tau_{\hat{\Omega}}}\right)\right)$$

$$= E_{\dot{x}}\left(\prod_{s=0}^{\tau_{\hat{\Omega}}-1} [1+\lambda(\dot{x}(s))]^{-1}\right.$$

$$\left. \times E_{\dot{x}}\left\{\sum_{k=0}^{\infty} \prod_{s=0}^{k} [1+\lambda(\dot{x}(s+\tau_{\hat{\Omega}}))]^{-1} f(\dot{x}(k+\tau_{\hat{\Omega}})) \mid F_{\tau_{\hat{\Omega}}}\right\}\right)$$

$$= E_{\dot{x}}\left(M_{\tau_{\hat{\Omega}}-1} E_{\dot{x}(\tau_{\hat{\Omega}})} \left\{\sum_{k=0}^{\infty} M_k f(\dot{x}(k))\right\}\right)$$

$$= E_{\dot{x}}(M_{\tau_{\hat{\Omega}}-1} G^{\lambda}(\dot{x}(\tau_{\hat{\Omega}}))) = \Pi_{\hat{\Omega}}^{\lambda} G^{\lambda} f(\dot{x}).$$

Therefore, the proof of the theorem is now completed. ∎

### THEOREM 13.5

*The function $u(\dot{x}) := G_{\hat{\Omega}}^{\lambda} f(\dot{x})$ provides the unique solution of the problem (13.14)–(13.15).*

**PROOF** It follows immediately from Theorems 13.2, 13.3, and 13.4. ∎

Next we put together the obtained results.

### THEOREM 13.6
Let $\dot{\Omega} \subset \dot{\mathbf{Z}}^l$ be a finite set. Then the problem (13.12)–(13.13) has a unique solution $u$, which can be represented by the formula

$$u(\dot{x}) = E_{\dot{x}} \left\{ \sum_{k=0}^{\tau_{\dot{\Omega}}-1} M_k f(\dot{x}(k)) + M_{\tau_{\dot{\Omega}}-1} \varphi(\dot{x}(\tau_{\dot{\Omega}})) \right\},$$

where $M_k = \Pi_{s=0}^{k}[1 + \lambda(\dot{x}(s))]^{-1}$, $k = 0, 1, \ldots$, $M_{-1} = 1$, and $\tau_{\dot{\Omega}}$ is the first exit time of the process $\dot{x}(\cdot, \omega)$ from the set $\dot{\Omega}$.

As an important special case of Theorem 13.6, we mention the "discrete" Cauchy problem.

For a function $u(\dot{x}) = u(t, x)$ specified on the lattice $\dot{\mathbf{Z}}^l = \mathbf{Z}^1 \times \mathbf{Z}^l$ we define the following operators

$$\partial u(t, x) := u(t+1, x) - u(t, x)$$

and

$$Pu(t, x) := \frac{1}{2l} \sum_{k=1}^{l} u(t, x \pm e_k).$$

### THEOREM 13.7
Let $\Omega$ be a subset of $\mathbf{Z}^l$ and $\varphi$ be a bounded function defined on $\Omega$. The "discrete" Cauchy problem with equations

$$\partial u(t, x) = (P - E)u(t, x) - \lambda(x)u(t, x), \quad x \in \Omega, \quad (13.18)$$

$$u(0, x) = \varphi(x), \quad x \in \Omega, \quad (13.19)$$

and

$$u(t, x) = 0, \quad x \in \partial\Omega, \quad t > 0 \quad (13.20)$$

has a unique solution, which can be expressed in the following form

$$u(t, x) = E_x(M_{t-1}\varphi(x(t)), t < \tau_\Omega),$$

## 13. A Discrete Kac's Formula

where $x(\cdot,\omega)$ is the simple random walk in $\mathbf{Z}^l$ and $\tau_\Omega$ is the first exit time of $x(\cdot,\omega)$ from the set $\Omega$ (see (13.8)).

Next we come back to the general situation. For a set $\dot\Omega \subset \dot{\mathbf{Z}}^l$ and a function $u$ defined on $\dot\Omega$ we put

$$\|u\|_{\dot\Omega} := \sup\{|u(\dot x)| : \dot x \in \dot\Omega\}.$$

An auxiliary result follows.

**THEOREM 13.8**
*For every function $u$ on $\dot\Omega \cup \partial\dot\Omega$, the next inequality holds valid*

$$\|u\|_{\dot\Omega \cup \partial\dot\Omega} \leq C(\dot\Omega, \lambda)\|(\dot P - E)u - \lambda u\|_{\dot\Omega} + \|u\|_{\partial\dot\Omega}. \qquad (13.21)$$

*where*

$$C(\dot\Omega, \lambda) := \|G_{\dot\Omega}^\lambda 1\|_{\dot\Omega}.$$

**PROOF** Define the following functions

$$f(\dot x) := -(\dot P - E)u(\dot x) + \lambda(\dot x)u(\dot x), \quad \dot x \in \dot\Omega,$$

and

$$\varphi(\dot x) := u(\dot x), \quad \dot x \in \partial\dot\Omega.$$

Then we see that

(i) $(\dot P - E)u(\dot x) - \lambda(\dot x)u(\dot x) = -f(\dot x), \quad x \in \dot\Omega$; and
(ii) $u(\dot x) = \varphi(\dot x), \quad \dot x \in \partial\dot\Omega.$

From Theorems 13.3 and 13.5 we obtain that

$$u(\dot x) = G_{\dot\Omega}^\lambda f(\dot x) + \Pi_{\dot\Omega}^\lambda \varphi(\dot x).$$

By this identity we immediately get that

$$\|u\|_{\dot\Omega \cup \partial\dot\Omega} \leq \|G_{\dot\Omega}^\lambda 1\|_{\dot\Omega} + \|f\|_{\dot\Omega} + \|\Pi_{\dot\Omega}^\lambda \varphi\|_{\dot\Omega \cup \partial\dot\Omega}$$

$$\leq C(\dot\Omega, \lambda)\|(\dot P - E)u - \lambda u\|_{\dot\Omega} + \|\varphi\|_{\partial\dot\Omega}$$

$$= C(\dot\Omega, \lambda)\|(\dot P - E)u - \lambda u\|_{\dot\Omega} + \|u\|_{\partial\dot\Omega}.$$

The proof is done.  ∎

The next result is useful in further considerations.

**THEOREM 13.9**
Let $\dot\Omega = \Omega \times I$ be an interval, where

$$\Omega = \{(x_1,\ldots,x_l) \in \mathbf{Z}^l : 1 \le x_i \le N,\ i = 1,\ldots,l\},$$

$$I = \{(1,2,\ldots,T)\}, \quad T \in \mathbf{N},$$

and let also $\lambda_* := \max\{\lambda(\dot x) : \dot x \in \dot\Omega\}$. Then it holds

$$\frac{1}{2}e^{-T\lambda_*} \min\left\{T, \frac{2}{\pi^2}(N+1)^2\right\} \le C(\dot\Omega, \lambda) \le \min\left\{T, \frac{l}{4}(N+1)^2\right\}.$$

**PROOF** We denote $c(\dot\Omega) := \max\{E_{\dot x}(\tau_{\dot\Omega}) : \dot x \in \dot\Omega\}$ and prove the following inequality

$$e^{-T\lambda_*}c(\dot\Omega) \le C(\dot\Omega, \lambda) \le c(\dot\Omega). \tag{*}$$

Actually, we see that

$$C(\dot\Omega, \lambda) = \max_{\dot x \in \dot\Omega} E_{\dot x}\left\{\sum_{k=0}^{\tau_{\dot\Omega}-1} M_k\right\} \le \max_{\dot x \in \dot\Omega} E_{\dot x}\tau_{\dot\Omega} = c(\dot\Omega).$$

On the other hand, if we set $(1+\lambda_*)^{-1} := q$, then we have

$$C(\dot\Omega), \lambda) \ge \max_{\dot x \in \dot\Omega} E_{\dot x}\{q + \cdots + q^{\tau_{\dot\Omega}}\}$$

$$\ge \max_{\dot x \in \dot\Omega} E_{\dot x}\{\tau_{\dot\Omega} q^{\tau_{\dot\Omega}}\}$$

$$= \max_{\dot x \in \dot\Omega} E_{\dot x}\{\tau_{\dot\Omega}(1+\lambda_*)^{-\tau_{\dot\Omega}}\}$$

$$\ge \max_{\dot x \in \dot\Omega} E_{\dot x}\{\tau_{\dot\Omega}(1+\lambda_*)^{-T}\}$$

$$\ge e^{-T\lambda_*} \max_{\dot x \in \dot\Omega} E_{\dot x}(\tau_{\dot\Omega}) = e^{-T\lambda_*}c(\dot\Omega),$$

where on the last step of estimates we used the elementary inequalities $\log(1+\lambda_*) \leq \lambda_*$ and $\tau_{\dot\Omega} \leq T$. So that the inequality $(*)$ is proved.

At the next step we establish the following inequality

$$\frac{1}{2}\min\{T, c(\Omega)\} \leq c(\dot\Omega) \leq \min(T, c(\Omega)\}, \qquad (**)$$

where $c(\Omega) := \max_{x \in \Omega} E_x(\tau_\Omega)$ and $\tau_\Omega$ is the first exist time of simple random walk $x(\cdot, \omega)$ from $\Omega$. We see for a process $\dot x(\cdot, \omega)$ with $\dot x(0) = \dot x$, where $\dot x = (x, t)$, that we have $\tau_{\dot\Omega} = \min\{\tau_\Omega, t\}$. It impies the inequality

$$E_{\dot x}(\tau_{\dot\Omega}) \leq \min\{t, E_x(\tau_\Omega)\},$$

from which the right-hand side of $(**)$ follows. To establish the left-hand side of $(**)$, we take the functions $u(x) := E_x(\tau_\Omega)$ and $\bar u(\dot x) := tu(x)$. It is clear that $\bar u(\dot x) = 0$ for any $\dot x \in \partial\dot\Omega$. Applying to function $\bar u$, inequality (13.21) for $\lambda \equiv 0$ we obtain

$$\|\bar u\|_{\dot\Omega} \leq c(\dot\Omega)\|(\dot P - E)\bar u\|_{\dot\Omega}. \qquad (***)$$

Next we notice that $\|\bar u\|_{\dot\Omega} = T \cdot \|u\|_\Omega = T \cdot c(\Omega)$. To calculate $(\dot P - E)\bar u$ we observe that $\tau_\Omega$ coincides with the first exit time from cylinder $\dot\Omega_c := \Omega \times (-\infty, \infty)$. So the function

$$u_c(\dot x) := u(x) = E_x(\tau_\Omega) = E_{\dot x}(\tau_{\dot\Omega_c})$$

fulfills the equation

$$(\dot P - E)u_c(\dot x) = -1, \quad \dot x \in \dot\Omega_c.$$

Using the facts mentioned above, we obtain

$$(\dot P - E)\bar u(\dot x) = \dot P\bar u(\dot x) - \bar u(\dot x) = (t-1)\dot P u_c(\dot x) - tu_c(\dot x)$$

$$= (t-1)(u_c(\dot x) - 1) - tu_c(\dot x) = -(u_c(\dot x) + (t-1)).$$

Therefore, we finally find

$$\|(\dot P - E)\bar u\|_{\dot\Omega} \leq (\|u\|_\Omega + T) = (c(\Omega) + T). \qquad (****)$$

Estimates $(***)$ and $(****)$ imply that

$$Tc(\Omega) \leq c(\dot\Omega)(T + c(\Omega)),$$

and consequently

$$c(\dot\Omega) \geq \frac{T \cdot c(\Omega)}{T + c(\Omega)} \geq \frac{1}{2}\min\{T, c(\Omega)\}.$$

Hence the left-hand side of $(**)$ follows. Next it remains to find upper and lower bounds of the constant $c(\Omega) = \max_{x \in \Omega} E_x(\tau_\Omega)$. Let $\Omega_1 := \{x \in \mathbf{Z}^l : 1 \leq x_l \leq N\}$ be a slab. Because $\Omega \subset \Omega_1$, we have $\tau_\Omega \leq \tau_{\Omega_1}$, and consequently

$$E_x(\tau_\Omega) \leq E_x(\tau_{\Omega_1}).$$

Now we notice that the function

$$\overline{u}_1(\dot x) := E_{\dot x}(\tau_{\Omega_1})$$

satisfies the equation $(\dot P - E)\overline{u}_1 = -1$ in a space-time slab $\dot\Omega_1 := \Omega_1 \times (-\infty, \infty)$ and has zero boundary values. Take the function $n(\dot x) := lx_1((N+1)-x_1)$. It is easy to see that this function satisfies the equation $(\dot P - E)n = -1$ in $\dot\Omega_1$ and has zero boundary values. Hence by unicity $\overline{u}_1(\dot x) = n(\dot x)$ for all $\dot x \in \dot\Omega_1$, and therefore

$$c(\Omega) \leq \max\{\overline{u}_1(\dot x) : \dot x \in \dot\Omega_1\} = \max\{n(\dot x) : \dot x \in \dot\Omega_1\} \leq l\left(\frac{N+1}{2}\right)^2,$$

which gives the upper bound of $c(\Omega)$. To find the lower bound of $c(\Omega)$ we use a similar method. More precisely, we consider the function $u(s) := \sin\frac{\pi s}{N+1}$ and denote

$$\overline{u}(\dot x) := \prod_{k=1}^{l} u(x_k), \quad x \in \Omega.$$

It is obvious that $\overline{u}(\dot x) = 0$ for every $\dot x \in \partial\dot\Omega_c$. Let $P_i$, $i = 1, \ldots, l$ be the one-dimensional average operator

$$P_i\overline{u}(\dot x) := \frac{1}{2}\{\overline{u}(\dot x + e_i) + \overline{u}(\dot x - e_i)\}.$$

It is obvious that

$$(\dot P - E)\overline{u}(\dot x) = \frac{1}{l}\sum_{i=1}^{l}(P_i - E)\overline{u}(\dot x)$$

$$= \frac{1}{l}\sum_{i=1}^{l}(P_i - E)u(x_i)\prod_{k\neq i}u(x_k).$$

Next we compute $(P_i - E)u(x_i)$ for $1 \leq x_i \leq N$ and for $1 \leq i \leq l$. We see that
$$(P_i - E)u(x_i) = -2\sin^2\frac{\pi}{2(N+1)}u(x_i),$$
and thus we obtain that for $\dot{x}\in\dot{\Omega}_c$
$$(\dot{P} - E)\overline{u}(\dot{x}) = -2\sin^2\frac{\pi}{2(N+1)}\cdot \overline{u}(\dot{x}).$$

To find the lower bounds of $c(\Omega) = c(\dot{\Omega}_c)$ it remains to apply the inequality (13.21) for $\lambda \equiv 0$,
$$0 < \|\overline{u}\|_{\dot{\Omega}_c} \leq c(\Omega)\cdot 2\sin^2\frac{\pi}{2(N+1)}\|\overline{u}\|_{\dot{\Omega}_c}$$
$$\leq c(\Omega)\frac{\pi^2}{2(N+1)^2}\|\overline{u}\|_{\dot{\Omega}_c}.$$

Hence we finally get
$$c(\Omega) \geq \frac{2}{\pi^2}(N+1)^2,$$
which gives the lower bound of $c(\Omega)$.

Inequalities (*) and (**) along with the upper and lower bounds of $c(\Omega)$ produce the desired result. ∎

---

## 13.3 About Approximation on the Grid

We study the following Dirichlet problem on the interval $\dot{\Omega} = \{\dot{x}\in\dot{R}^l : 0 < x_i < 1,\ 0 < t < \infty,\ i = 1,\ldots,l\}$

$$\left.\begin{array}{l}\dot{\Delta}u(\dot{x}) - \lambda(\dot{x})u(\dot{x}) = -f(\dot{x}),\quad \dot{x}\in\dot{\Omega},\\ u(\dot{x}) = \varphi(\dot{x}),\hspace{3.5cm}\dot{x}\in\partial\dot{\Omega},\end{array}\right\} \quad (13.22)$$

with $\varphi$ and $\lambda \geq 0$ continuous functions and $f$, a bounded locally Hölder function. In the next we restrict our treatment to problem (13.22) for which $u \in C^{(2)}(\overline{\dot\Omega})$.

Let $h := 1/n$ with $n \in \mathbf{N}$ and $k := h^2/l$. An approximate solution $u_h$, defined on the grid $\overline{\dot\Omega}_h = \dot\Omega_h \cup \partial\dot\Omega_h$, where

$$\overline{\dot\Omega}_h := \{\dot x = (x_1, \ldots, x_l, t) \in \dot R^l : x_i = k_i h, t = jk,$$

$$k_i = 0, \ldots, n, \ i = 1, \ldots, l, \ j = 0, 1, \ldots\},$$

and

$$\dot\Omega_h := \overline{\dot\Omega}_h \cap \dot\Omega, \quad \partial\dot\Omega_h := \overline{\dot\Omega}_h \setminus \dot\Omega_h,$$

is gotten as the solution of the discrete counterpart to (13.22)

$$\left.\begin{array}{ll}\dot\Delta_h u_h(\dot x) - \lambda(\dot x) u_h(\dot x) = -f(\dot x), & \dot x \in \dot\Omega_h, \\ u_h(\dot x) = \varphi(\dot x), & \dot x \in \partial\dot\Omega_h,\end{array}\right\} \quad (13.23)$$

Here the "discrete" parabolic Laplacian $\dot\Delta_h$ is given by

$$\dot\Delta_h u(\dot x) := \frac{1}{2} h^{-2} \left( \sum_{k=1}^{l} u(\dot x \pm he_k - k \cdot e) - 2lu(\dot x) \right),$$

where $k = k(h) := h^2/l$.

We denote by $\dot{\mathbf{Z}}_h^l := \{\dot x \in \dot R^l : x = zh, \ t = nk, \ z \in \mathbf{Z}^l, \ n \in \mathbf{Z}^1\}$, and we take the space-time random walk on $\dot{\mathbf{Z}}_h^l$

$$\{\dot x(n), n = 0, 1, \ldots\} = \{(x(n), t - nk), n = 0, 1, \ldots\}$$

starting from point $\dot x = (x, t)$, where $\{x(n), n = 0, 1, \ldots\}$ is a simple random walk on the $h$-lattice $h\mathbf{Z}^l$ starting from point $x(0) = x$. Corresponding to $\{\dot x(n), \ n = 0, 1, \ldots\}$ values, functions and operators are attached to the index $h$. So, for example, the average operator $\dot P_h$ takes the following form

$$\dot P_h u(\dot x) = E_{\dot x}\{u(\dot x(1))\} = \frac{1}{2l} \sum_{k=1}^{l} u(\dot x \pm he_k - ke).$$

It is clear that with these notations, the discrete parabolic Laplacian $\dot\Delta_h$ takes the following form, $\dot\Delta_h = lh^{-2}(\dot P_h - E)$. Now applying Theorem 13.6

## 13. A Discrete Kac's Formula

we observe that the problem (13.23) has a unique solution $u_h$, which can be represented by the form

$$u_h(\dot{x}) = E_{\dot{x}} \left\{ k \sum_{n=0}^{\tau_{\dot{\Omega}_h}-1} M_n f(\dot{x}(n)) + M_{\tau_{\dot{\Omega}_h}-1} \varphi(\dot{x}(\tau_{\dot{\Omega}_h})) \right\},$$

where $k = h^2/l$ and

$$M_n = \prod_{s=0}^{n}[1 + k\lambda(\dot{x}(s))]^{-1} \sim \exp\left\{-k \cdot \sum_{s=0}^{n} \lambda(\dot{x}(s))\right\}, \quad k \downarrow 0.$$

It is implied by Donsker's extension of the De Moivre-Laplace limit theorem (see [78], [102, 1.10]) that, as $h \downarrow 0$,

$$u_h(\dot{x}) \to u(\dot{x}) := E_{\dot{x}} \left\{ \int_0^{\tau_{\dot{\Omega}}} \mathcal{M}_s f(\dot{x}(s))\, ds + \mathcal{M}_{\tau_{\dot{\Omega}}} \varphi(\dot{x}(\tau_{\dot{\Omega}})) \right\},$$

where $\dot{x}(s)$ is the space-time Brownian motion starting from point $\dot{x}(0) = \dot{x}$, $\tau_{\dot{\Omega}}$ is the first exit time of $\dot{x}(s)$ from domain $\dot{\Omega}$ and

$$\mathcal{M}_s := \exp\left\{-\int_0^s \lambda(\dot{x}(p))\, dp\right\}, \quad s > 0.$$

According to Section 13.1, $u(\dot{x})$ is exactly the solution of the Dirichlet problem (13.22) (the Kac's formula).

To estimate the rate of convergence $u_h \to u$ as $h \downarrow 0$, we use the representation

$$u_h(\dot{x}) = k G_{\dot{\Omega}_h}^{k\lambda} f(\dot{x}) + \Pi_{\dot{\Omega}_h} \varphi(\dot{x}) \tag{13.24}$$

(see Theorems 13.3 and 13.5) as well as the important inequality (13.21), which now takes the following form

$$\|u\|_{\dot{\Omega}_{h,T}} \leq \min\left\{T, \frac{1}{4}\right\} \|\dot{\Delta}_h u - \lambda u\|_{\dot{\Omega}_{h,T}} + \|u\|_{\partial \dot{\Omega}_{h,T}}, \tag{13.25}$$

where $\dot{\Omega}_{h,T} := \{\dot{x} \in \dot{\Omega}_h : t \leq T\}$. Really, by application of (13.21) and Theorem 13.9 we obtain that

$$\|u\|_{\dot{\Omega}_{h,T}} \leq C(\dot{\Omega}_{h,T}, k\lambda) \|(\dot{P}_h - E)u - k\lambda u\|_{\dot{\Omega}_{h,T}} + \|u\|_{\partial \dot{\Omega}_{h,T}}$$

$$\leq \min\left\{\frac{T}{k}, \frac{l}{4}n^2\right\} k\|\dot{\Delta}_h u - \lambda u\|_{\dot{\Omega}_{h,T}} + \|u\|_{\partial\dot{\Omega}_{h,T}}$$

$$= \min\left\{T, \frac{1}{4}\right\} \|\dot{\Delta}_h u - \lambda u\|_{\dot{\Omega}_{h,T}} + \|u\|_{\partial\dot{\Omega}_{h,T}}.$$

In the following we apply inequality (13.25) to $u_h - u$, where $u_h$ and $u$ are the solutions of the problems (13.23) and (13.22), respectively, and we get the following error estimate

$$\|u_h - u\|_{\dot{\Omega}_{h,T}} \leq \min\left\{T, \frac{1}{4}\right\} \|\dot{\Delta}_h(u_h - u) - \lambda(u_h - u)\|_{\dot{\Omega}_{h,T}}$$

$$= \min\left\{T, \frac{1}{4}\right\} \|(\dot{\Delta}_h u_h - \lambda u_h) - (\dot{\Delta}_h u - \lambda u)\|_{\dot{\Omega}_{h,T}}$$

$$= \min\left\{T, \frac{1}{4}\right\} \|-f - (\dot{\Delta}_h u - \lambda u)\|_{\dot{\Omega}_{h,T}}$$

$$= \min\left\{T, \frac{1}{4}\right\} \|(\dot{\Delta} u - \lambda u) - (\dot{\Delta}_h u - \lambda u)\|_{\dot{\Omega}_{h,T}}$$

$$= \min\left\{T, \frac{1}{4}\right\} \|\dot{\Delta} u - \dot{\Delta}_h u\|_{\dot{\Omega}_{h,T}}.$$

That is, we have established

$$\|u_h - u\|_{\overline{\dot{\Omega}}_{h,T}} \leq \min\left\{T, \frac{1}{4}\right\} \|\dot{\Delta} u - \dot{\Delta}_h u\|_{\dot{\Omega}_{h,T}}. \tag{13.26}$$

Next the rate of convergence of $u_h \to u$ as $h \downarrow 0$ can be measured by the partial moduli of continuity $\omega_1$ and $\omega_{2,i}$, $i = 1, \ldots, l$, where

$$\omega_1(\delta, f; \overline{\dot{\Omega}}) := \sup\{|f(\dot{x} + \theta e) - f(\dot{x})| : \dot{x}, \dot{x} + \theta e \in \overline{\dot{\Omega}}, |\theta| < \delta\},$$

$$\omega_{2,i}(\delta, f; \overline{\dot{\Omega}}) := \sup\{|f(\dot{x} + \theta e_i) - 2f(\dot{x}) + f(\dot{x} - \theta e_i)| : \dot{x}, \dot{x} \pm \theta e_i \in \overline{\dot{\Omega}}, |\theta| < \delta\}.$$

Namely we have

## THEOREM 13.10

*Assume that the solution $u$ of problem (13.22) is in $C^{(2)}(\overline{\Omega})$, then for the solution $u_h$ of the problem (13.23) the following inequality holds valid*

$$\|u_h - u\|_{\overline{\Omega}_{h,T}} \leq \min\left\{T, \frac{1}{4}\right\}\left[\frac{1}{4}\sum_{i=1}^{l}\omega_{2,i}(h, \partial_{x_i}^2 u; \overline{\Omega}_{h,T})\right.$$

$$\left. + \frac{1}{2}\sum_{i=1}^{l}\omega_1(k, \partial_{x_i}^2 u; \overline{\Omega}_{h,T}) + \omega_1(k, \partial_t u; \overline{\Omega}_{h,T})\right].(13.27)$$

**PROOF** For $u \in C^{(2)}(\overline{\Omega})$ and $k = h^2/l$ we see that

$$\dot{\Delta}_h u(x,t) = \frac{1}{2}h^{-2}\sum_{i=1}^{l}[u(x+he_i, t-k) + u(x-he_i, t-k) - 2u(x,t)]$$

$$= \frac{1}{2}h^{-2}\left\{\sum_{i=1}^{l}[u(x+he_i, t-k) + u(x-he_i, t-k) - 2u(x, t-k)]\right\}$$

$$- \frac{1}{k}[u(x,t) - u(x,t-k)] := \sum_{i=1}^{l}\Delta_{h,i}u(x, t-k) - \partial_{k,t}u(x,t).$$

By appropriate Taylor expansion, we find

$$\left|\Delta_{h,i}u(x, t-k) - \frac{1}{2}\partial_{x_i}^2 u(x, t-k)\right| \leq \frac{1}{4}\omega_{2,i}(h, \partial_{x_i}^2 u; \overline{\Omega}_{h,T}),$$

and

$$|\partial_{k,t}u(x,t) - \partial_t u(x,t)| \leq \omega_1(k, \partial_t u; \overline{\Omega}_{h,T}).$$

Hence we get the following estimate

$$|\dot{\Delta}_h u(x,t) - \dot{\Delta}u(x,t)| \leq \frac{1}{4}\sum_{i=1}^{l}\omega_{2,i}(h, \partial_{x_i}^2 u; \overline{\Omega}_{h,T})$$

$$+ \frac{1}{2}\sum_{i=1}^{l}\omega_1(k, \partial_{x_i}^2 u; \overline{\Omega}_{h,T}) + \omega_1(k, \partial_t u; \overline{\Omega}_{h,T}).$$

At the end we apply this estimate to (13.26) to find the desired result. ∎

**REMARK 13.1** The estimate (13.27) is sharp, i.e., there exists a function $u$ such that
$$\liminf_{h \to 0} \|u - u_h\|_{\dot{\Omega}_{h,T}} / R(h, u) > 0, \qquad (13.28)$$

where $R(h, u)$ is the right-hand side of the inequality (13.27). Actually, pick $u(\dot{x}) := x_1^4$ and compute both sides of the inequality (13.27). We find that
$$\dot{\Delta} u(\dot{x}) := 6x_1^2,$$

and
$$\dot{\Delta}_h u(\dot{x}) := 6x_1^2 + h^2.$$

Hence
$$\dot{\Delta} u(\dot{x}) - \dot{\Delta}_h u(\dot{x}) = -h^2.$$

Next we apply (13.24) to the function $u - u_h$, which takes zero values on the boundary $\partial \dot{\Omega}_h$, we find

$$u(\dot{x}) - u_h(\dot{x}) = -k G_{\dot{\Omega}_h}^{k\lambda} [(\dot{\Delta}_h u - \lambda u) - (\dot{\Delta}_h u_h - \lambda u_h)](\dot{x})$$

$$= -k G_{\dot{\Omega}_h}^{k\lambda} [(\dot{\Delta}_h u - \lambda u) - (\dot{\Delta} u - \lambda u)](\dot{x})$$

$$= -k G_{\dot{\Omega}_h}^{k\lambda} (\dot{\Delta}_h u - \dot{\Delta} u)(\dot{x}) = -k h^2 G_{\dot{\Omega}_h}^{k\lambda} 1(\dot{x}). \qquad (13.29)$$

Hence by (13.29) and Theorem 13.9 we obtain

$$\|u - u_h\|_{\overline{\dot{\Omega}}_{h,T}} \geq kh^2 \frac{1}{2} e^{-\frac{T}{k} \cdot k\lambda_*} \min \left\{ \frac{T}{k}, \frac{2}{\pi^2} h^{-2} \right\}$$

$$= \frac{1}{2} h^2 e^{-T\lambda_*} \min \left\{ T, \frac{2}{\pi^2 l} \right\}. \qquad (13.30)$$

Moreover,

$$R(h, u) = \min \left\{ T, \frac{1}{4} \right\} \cdot \frac{1}{4} \sum_{i=1}^{l} \omega_{2,i}(h, \partial_{x_i}^2 u; \overline{\dot{\Omega}}_{h,T}) \qquad (13.31)$$

$$= \min \left\{ T, \frac{1}{4} \right\} \cdot \frac{1}{4} \omega_{2,1}(h, \partial_{x_1}^2 u; \overline{\dot{\Omega}}_{h,T}) = \min \left\{ T, \frac{1}{4} \right\} 6h^2.$$

## 13.4 On Sharpness of the Error Estimate

As was mentioned in Remark 13.1, the error estimate (13.27) is sharp, i.e., there exists a function $u$ such that

$$\|u - u_h\|_{\overline{\Omega}_{h,T}} \asymp R(h,u) \quad \text{as } h \downarrow 0,$$

where $R(h,u)$ is the right-hand side of the inequality (13.27).

The fact that (13.27) is sharp with regard to the rate of convergence is next proved in connection with general Lipschitz classes, determined by an abstract modulus of continuity, i.e., by a function $\omega(t)$, continuous on $[0, +\infty)$ such that

$$0 = \omega(0) < \omega(s) \leq \omega(s+t) \leq \omega(s) + \omega(t), \quad s, t > 0.$$

Here we follow the same technique that was used in [52], [53], and [82]. These were papers devoted to the elliptic and parabolic Dirichlet problems with $\lambda(\dot{x}) \equiv 0$ in dimensions $l = 1, 2$, and $l = 1$, respectively. Our reasoning is based on the following variant of the uniform boundedness principle [76]. For a Banach space $(X, \|\cdot\|)$, let $X^*$ be the set of sublinear bounded functionals on $X$. We obtain

**THEOREM 13.11**

*Assume that for given $\{T_n\}_{n \in \mathbf{N}} \subset X^*$ and $\{S_\delta\}_{\delta > 0} \subset X^*$ there are $\{g_n\}_{n \in \mathbf{N}} \subset X$ such that*

$$\|g_n\| \leq c_1, \quad n = 1, 2, \ldots \tag{13.32}$$

$$\liminf_{n \to \infty} |T_n g_n| > 0 \tag{13.33}$$

$$|S_\delta g_n| \leq c_2 \min\left\{1, \frac{\sigma(\delta)}{\varphi_n}\right\}, \quad n = 1, 2, \ldots, \tag{13.34}$$

*where $\sigma(\delta)$ is a strictly positive function on $(0, \infty)$ and $\{\varphi_n\}_{n \in \mathbf{N}}$ is a strictly decreasing real sequence with $\lim_{n \to \infty} \varphi_n = 0$. Then for each modulus of* continuity $\omega$ as above, obeying

$$\lim_{t \to 0} \frac{\omega(t)}{t} = \infty, \qquad (13.35)$$

there exists an element $u_\omega \in X$ such that

$$|S_\delta u_\omega| \le c_\omega \cdot \omega(\sigma(\delta)), \quad 0 < \delta < 1, \qquad (13.36)$$

$$\liminf_{n \to \infty} |T_n u_\omega|/\omega(\varphi_n) > 0. \qquad (13.37)$$

Next comes the optimal result.

**THEOREM 13.12**
For every modulus of continuity $\omega$ there exists a function $u_\omega \in C^{(2)}(\overline{\dot{\Omega}})$, such that

$$R(\delta, u_\omega) \le c_\omega \omega(\delta^2), \quad 0 < \delta < 1 \qquad (13.38)$$

and

$$\liminf_{h \to 0} \|u_\omega - u_{\omega,h}\|_{\overline{\dot{\Omega}_{h,T}}} / \omega(h^2) > 0. \qquad (13.39)$$

**PROOF**   To apply Theorem 13.11 we put

$$X := C^{(2)}(\overline{\dot{\Omega}}),$$

$$T_n u := \|u - u_h\|_{\overline{\dot{\Omega}_{h,T}}}, \quad h = 1/n,$$

$$S_\delta u := R(\delta, u), \quad 0 < \delta < 1$$

and

$$g_n(\dot{x}) := n^{-2} \sum_{i=1}^{l} \sin^2 \pi n x_i, \quad \dot{x} = (x, t), \ x = (x_1, \ldots, x_l) \in \Omega.$$

Then (13.32) is satisfied with $c_1 = \pi^2 l$. We see that $g_n(\dot{x}) = g_{n,h}(\dot{x})$ for $\dot{x} \in \partial \dot{\Omega}_{h,T}$, $h = 1/n$, and $\dot{\Delta} g_n(\dot{x}) = \pi^2 l$ and $\dot{\Delta}_h g_n(\dot{x}) = 0$ for all $\dot{x} \in \dot{\Omega}_{h,T}$. Then (13.24) implies

$$T_n g_n = \|g_n - g_{n,h}\|_{\dot{\Omega}_{h,T}}$$

## 13. A Discrete Kac's Formula

$$= k\|G^{k\lambda}_{\dot\Omega_{h,T}}[\dot\Delta_h(g_n - g_{n,h}) - \lambda(g_n - g_{n,h})]\|_{\dot\Omega_{h,T}}$$

$$= k\|G^{k\lambda}_{\dot\Omega_{h,T}}[(\dot\Delta_h g_n - \lambda g_n) - (\dot\Delta_h g_{n,h} - \lambda g_{n,h})]\|_{\dot\Omega_{h,T}}$$

$$= k\|G^{k\lambda}_{\dot\Omega_{h,T}}[(\dot\Delta_h g_n - \lambda g_n) - (\dot\Delta g_n - \lambda g_n)]\|_{\dot\Omega_{h,T}}$$

$$= k\|G^{k\lambda}_{\dot\Omega_{h,T}}(\dot\Delta_h g_n - \dot\Delta g_n)\|_{\dot\Omega_{h,T}}$$

$$= k\pi^2 l \|G^{k\lambda}_{\dot\Omega_{h,T}} 1\|_{\dot\Omega_{h,T}}$$

$$\geq k\pi^2 l \frac{1}{2} \cdot e^{-\frac{T}{k}\cdot k\lambda_*} \min\left\{\frac{T}{k}, \frac{2}{\pi^2} h^{-2}\right\}$$

$$= e^{-T\lambda_*} \min\left\{1, \frac{\pi^2 lT}{2}\right\}.$$

The last inequality comes from Theorem 13.9, and hence condition (13.33) is fulfilled. To verify the condition (13.34) we see that

$$S_\delta g_n \leq \frac{1}{2}\pi^2 l$$

and

$$S_\delta g_n \leq \frac{\delta^2}{16} \sum_{i=1}^{l} \|\partial^4_{x_i} g_n\|_{\dot\Omega_{h,T}} \leq \frac{\delta^2 n^2 \pi^4 l}{2}.$$

These upper bounds on $S_\delta g_n$ produce (13.34) with $\sigma(\delta) := \pi^2 \delta^2$ and $\varphi_n := n^{-2}$. Therefore we are able to apply Theorem 13.11 and hence (13.38)–(13.39) are proved. ∎

# Part IV
# On Semigroups

# Chapter 14

## Quantitative Asymptotic Expansions of the Probabilistic Representation Formulae for $(C_0)$ m-Parameter Operator Semigroups

In this chapter we study quantitative asymptotic expansions for the probabilistic representation formulae of discrete type and asymptotic formulae of continuous type. We also provide applications. Namely, for special semigroups, the results presented apply to asymptotic expansions for multivariate Feller operators. Also asymptotic expansions of univariate and multivariate Bernstein operators are obtained and studied. The methods used are deeply probabilistic. In the quantitative approach we employ a natural modulus of continuity. This treatment is based on [179].

### 14.1  Background

Let $\mathcal{X}$ be a Banach space with elements $f, g, \ldots$, having norms $\|f\|$, $\|g\|, \ldots$, and $\mathcal{E}(\mathcal{X})$ be the Banach algebra of endomorphisms of $\mathcal{X}$. For $T \in \mathcal{E}(\mathcal{X})$, $\|T\|$ also denotes the norm of $T$. Let $\mathbf{R}^m$ be the $m$-dimensional Euclidean space endowed with the usual definitions of arithmetic operations and metric, and

$$\mathbf{R}^m_+ = \{t = (t_1, \ldots, t_m) \in \mathbf{R}^m; t_i \geq 0, i = 1, \ldots, m\}$$

denotes the first closed $2^m$-ant in $\mathbf{R}^m$. $\mathbf{Z}_+$ is the set of all nonnegative integers and

$$\mathbf{Z}^m_+ = \{\mu = (\mu_1, \ldots, \mu_m); \mu_i \in \mathbf{Z}_+, i = 1, \ldots, m\},$$

while **N** is the set of all positive integers.

A family of bounded linear operators $\{T(t); t \in \mathbf{R}_+^m\}$ on $\mathcal{X}$ is called a $(C_0)$ m-*parameter operator semigroup* in $\mathcal{E}(\mathcal{X})$ when the following three conditions are met:

i) $T(t+s) = T(t)T(s)$, $t, s \in \mathbf{R}_+^m$; (14.1)

ii) $T(0) = I$ (identity operator); (14.2)

iii) $s - \lim\limits_{t \in \mathbf{R}_+^m, t \to 0} T(t)f = f$, $f \in \mathcal{X}$. (14.3)

It follows that $\{T(t); t \in \mathbf{R}_+^m\}$ is the direct product of m $(C_0)$ one-parameter operator semigroups in $\mathcal{E}(\mathcal{X})$:

$$T(t) = \prod_{i=1}^{m} T_i(t_i), \tag{14.4}$$

where $T_i(t_i) = T(t_i e_i)$ and $e_i = (0, \ldots, 1, \ldots, 0)$ with 1 in the $i$th place and 0 elsewhere is the $i$th canonical basis of $\mathbf{R}^m$.

Let $A_i$ be the infinitesimal generator of $\{T_i(t_i); 0 \le t_i < \infty\}$ with domain $D(A_i)$, $i = 1, \ldots, m$. Then, if $f \in D(A_i)$, so does $T(t)f$ for each $t \in \mathbf{R}_+^m$ and $A_i T(t) f = T(t) A_i f$.

To each $i = 1, \ldots, m$, there correspond two numbers $M_i \ge 1$ and $\omega_i \ge 0$ such that $\|T_i(t_i)\| \le M_i e^{\omega_i t_i}$, $0 \le t_i < \infty$. Hence it holds

$$\|T(t)\| \le M_1 \cdots M_m \exp(\omega(t_1 + \cdots + t_m)) = M e^{\omega \bar{t}}, \ t \in \mathbf{R}_+^m, \tag{14.5}$$

where $M = M_1 \cdots M_m$, $\omega = \max\{\omega_i, 1 \le i \le m\}$ and $\bar{t} = t_1 + \cdots + t_m$.

In the following text we always mean that $\{T(t); t \in \mathbf{R}_+^m\}$ satisfies (14.5) unless otherwise noted.

For more on the above we refer to Butzer-Berens [54], Hille-Phillips [101], W. Köhnen [109], or Zhou-Anastassiou [180].

The following notations are used a lot in this chapter.

Let $t = (t_1, \ldots, t_m) \in \mathbf{R}^m$, $s = (s_1, \ldots, s_m) \in \mathbf{R}_+^m$, $\mu = (\mu_1, \ldots, \mu_m) \in \mathbf{Z}_+^m$ and $A_1, \ldots, A_m$ are as above. We put

$$\bar{t} = t_1 + \cdots + t_m, \quad |t| = (|t_1|, \ldots, |t_m|),$$
$$A = (A_1, \ldots, A_m), \quad t \bullet A = t_1 A_1 + \cdots + t_m A_m,$$
$$t^\mu = t_1^{\mu_1} \cdots t_m^{\mu_m}, \quad A^\mu = A_1^{\mu_1} \cdots A_m^{\mu_m}$$

and

$$s < t \text{ means } s_i < t_i, \quad \text{all } i = 1, \ldots, m.$$

## 14. m-Parameter Operator Semigroups

Further, for $r \in \mathbf{N}$

$$D^r := \bigcap_{\bar{\mu}=r,\ \mu \in \mathbf{Z}_+^m} D(A^\mu),$$

where $D(A^\mu)$ is the domain of $A^\mu$.

Let $(\Omega, \mathcal{A}, P)$ be a probability space. For every real-valued random variable $X$ defined on $(\Omega, \mathcal{A}, P)$, $E(X)$ denotes its expectation. If $\xi = E(X)$ exists, then $\sigma^2 = \sigma^2(X) = E[(X - \xi)^2]$ is called the variance of $X$. Let further $\Psi_X(u) := E(u^X)$, $u \geq 0$ and $\Psi_X^*(u) := E(e^{uX})$, $u \in \mathbf{R}$, denote the probability-generating function and the moment-generating function of $X$ respectively.

Here we consider m-dimensional random vectors, also denoted by $X, Y, \ldots$, on $(\Omega, \mathcal{A}, P)$. For an m-dimensional random vector $X = (X_{01}, \ldots, X_{0m})$, we also use $E(X)$ to denote its expectation, i.e., $E(X) := (E(X_{01}), \ldots, E(X_{0m}))$.

We can extend the integration theory about extended Pettis-integral developed by Pfeifer in [143] to the multivariate case.

Let $\{T(t); t \in \mathbf{R}_+^m\}$ be as above and $X$ be an $\mathbf{R}_+^m$-valued random vector such that

$$\Psi_{\bar{X}}^*(\omega) < \infty, \quad \bar{X} = X_{01} + \cdots + X_{0m},$$

then for every $f \in \mathcal{X}$, define

$$E[T(X)f] := \int_\Omega T(x) f\, dP,$$

which exists in the Bochner sense in $\mathcal{X}$ by the strong continuity of $\{T(t); t \in \mathbf{R}_+^m\}$ and (14.5). Furthermore, the map $E[T(X)] : f \to E[T(X)f]$ on $\mathcal{X}$ defines a bounded linear operator $E[T(X)] \in \mathcal{E}(\mathcal{X})$ with $\|E[T(X)]\| \leq M\Psi_{\bar{X}}^*(\omega)$. $E[T(X)]$ is called the expectation of $T(X)$ and is meant as an extended Pettis integral, see [143].

If $X, Y$ are independent $\mathbf{R}_+^m$-valued random vectors such that $\Psi_{\bar{X}}^*(\omega) < \infty$, $\Psi_{\bar{Y}}^*(\omega) < \infty$, then $E[T(X)]$, $E[T(Y)]$ and $E[T(X+Y)]$ exist in $\mathcal{E}(\mathcal{X})$ and it holds

$$E[T(X) \circ T(Y)] = E[T(X+Y)] = E[T(X)] \circ E[T(Y)],$$

where "∘" denotes composition and many times is suppressed. This fact along with some known properties about linear operators (see [54], pp. 288–294) guarantee the legitimacy of the operations on integral in what follows.

For more on the above see [143], [144], and [145].

## 14.2 Basic Results

We use the following lemmas.

**LEMMA 14.1**
Let $X_i \stackrel{i.i.d.}{\sim} X$, $i = 1, 2, \ldots$, be a sequence of random variables with $E(X) = x$, $E[(X - x)^2] = \sigma^2$ and $E[|X|^r] < \infty$ for some fixed $r \in \mathbf{Z}_+$. Set $S_n = \sum_{i=1}^n X_i$. Then it holds

$$\left| E\left[ \left( \frac{S_n}{n} - x \right)^r \right] \right| \leq M(r, X) \frac{1}{n^{[\frac{r+1}{2}]}}, \qquad (14.6)$$

where $M(r, X) \geq 0$ is a polynomial in $E[|X - x|^l]$, $2 \leq l \leq r$, independent of $n$ and $[\cdot]$ in the exponent of (14.6) is the integral part.

Further, for even $r$, $r = 2k$, we obtain

$$E\left[ \left( \frac{S_n}{n} - x \right)^{2k} \right] = \frac{(2k)!}{2^k k!} \frac{\sigma^{2k}}{n^k} + \frac{\varepsilon(n, k, X)}{n^{k+1}}, \qquad (14.7)$$

where $\varepsilon(n, 1, X) = 0$, $|\varepsilon(n, k, X)| \leq N(k, X)$ for all $n \in \mathbf{N}$ and $N(k, X)$ is a polynomial in $E[|X - x|^l]$, $2 \leq l \leq 2k$, independent of $n$.

**PROOF** We proceed by induction. Clearly, (14.6) is true for $r = 0, 1$. Assume (14.6) is true for all nonnegative integers less than $r$, $r > 1$. Then

$$\left| E\left[ \left( \frac{S_n}{n} - x \right)^r \right] \right|$$

$$= \frac{1}{n^r} \left| E\left[ ((X_1 - x) + \cdots + (X_n - x)) \left( \sum_{i=1}^n (X_i - x) \right)^{r-1} \right] \right|$$

$$= \frac{1}{n^r} \left| n E\left[ (X_1 - x) \left( \sum_{i=1}^n (X_i - x) \right)^{r-1} \right] \right|$$

$$= \frac{1}{n^{r-1}} \left| E\left[ (X_1 - x) \left( (X_1 - x) + \sum_{i=2}^n (X_i - x) \right)^{r-1} \right] \right|$$

$$= \frac{1}{n^{r-1}} \left| E\left[ (X_1 - x) \left( \sum_{j=0}^{r-1} \binom{r-1}{j} (X_1 - x)^j \left( \sum_{i=2}^{n} (X_i - x) \right)^{r-1-j} \right) \right] \right|$$

$$= \frac{1}{n^{r-1}} \left| E\left[ \sum_{j=1}^{r-1} \binom{r-1}{j} (X_1 - x)^{j+1} \left( \sum_{i=2}^{n} (X_i - x) \right)^{r-1-j} \right] \right|$$

$$\leq \frac{1}{n^{r-1}} \sum_{j=1}^{r-1} \binom{r-1}{j} E[|X_1 - x|^{j+1}] \left| E\left[ \left( \sum_{i=2}^{n} (X_i - x) \right)^{r-1-j} \right] \right|$$

$$\leq \frac{1}{n^{r-1}} \sum_{j=1}^{r-1} \binom{r-1}{j} E[|X_1 - x|^{j+1}] M(r-1-j, X)(n-1)^{r-1-j-[\frac{r-j}{2}]}$$

$$\leq \frac{1}{n^{[\frac{r+1}{2}]}} \sum_{j=1}^{r-1} \binom{r-1}{j} E[|X_1 - x|^{j+1}] M(r-1-j, X)$$

$$= M(r, X)/n^{[\frac{r+1}{2}]}.$$

Notice that $[\frac{r+1}{2}] - [\frac{r-j}{2}] - j \leq 0$ for $j \geq 1$, and $M(r, X)$ is a polynomial in $E[|X - x|^l]$, $l = 2, \ldots, r$, under the induction hypothesis. Then (14.6) is true for all $r \in \mathbf{Z}_+$.

Next we prove (14.7) by induction on $k$. Clearly (14.7) is true for $k = 0, 1$. Assume that (14.7) is true for all nonnegative integers less than $k$, $k > 1$. As in the proof of (14.6), we can write

$$E\left[ \left( \frac{S_n}{n} - x \right)^{2k} \right]$$

$$= \frac{1}{n^{2k-1}} \sum_{j=1}^{2k-1} \binom{2k-1}{j} E[(X_1 - x)^{j+1}] E\left[ \left( \sum_{i=2}^{n} (X_i - x) \right)^{2k-1-j} \right]$$

$$= \frac{1}{n^{2k-1}} \binom{2k-1}{1} E[(X_1 - x)^2] E\left[ \left( \sum_{i=2}^{n} (X_i - x) \right)^{2k-2} \right]$$

$$+ \frac{1}{n^{2k-1}} \sum_{j=2}^{2k-1} \binom{2k-1}{j} E[(X_1-x)^{j+1}] E\left[\left(\sum_{i=2}^{n}(X_i-x)\right)^{2k-1-j}\right]$$

$$:= I_1 + I_2. \tag{14.8}$$

From (14.6), we have

$$|I_2| \leq \frac{1}{n^{2k-1}} \sum_{j=2}^{2k-1} \binom{2k-1}{j} |E[(X_1-x)^{j+1}]| (n-1)^{2k-1-j}$$

$$\cdot \frac{M(2k-1-j, X)}{(n-1)^{\left[\frac{2k-1-j+1}{2}\right]}}$$

$$\leq \frac{1}{n^{2k-1}} \sum_{j=2}^{2k-1} \binom{2k-1}{j} E[|X_1-x|^{j+1}] M(2k-1-j, X) n^{2k-1-j-\left[\frac{2k-j}{2}\right]}$$

$$\leq \frac{K(k,X)}{n^{k+1}} \text{ (notice that } k+1-j-\left[\frac{2k-j}{2}\right] \leq 0 \text{ when } j \geq 2), \tag{14.9}$$

where $K(k,X) = \sum_{j=2}^{2k-1} \binom{2k-1}{j} E[|X_1-x|^{j+1}] M(2k-1-j, X)$ is a polynomial in $E[|X-x|^j]$'s.

Moreover by the induction hypothesis we see that

$$I_1 = \frac{1}{n^{2k-1}} \binom{2k-1}{1} E[(X_1-x)^2] E\left[\left(\sum_{i=2}^{n}(X_i-x)\right)^{2k-2}\right]$$

$$= \frac{(2k-1)\sigma^2}{n^{2k-1}} \left[\frac{(2(k-1))!}{2^{k-1}(k-1)!} \sigma^{2k-2} \frac{1}{(n-1)^{k-1}}\right.$$

$$\left. + \frac{\varepsilon(n-1, k-1, X)}{(n-1)^k}\right] (n-1)^{2k-2}$$

$$= \frac{(2k)!}{2^k k!} \frac{\sigma^{2k}}{n^k} + \frac{(2k)!}{2^k k!} \frac{\sigma^{2k}}{n^k} \left(\frac{(n-1)^{k-1}}{n^{k-1}} - 1\right)$$

$$+ \frac{(2k-1)\sigma^2 \varepsilon(n-1, k-1, X)}{n^{2k-1}} (n-1)^{k-2}$$

## 14. m-Parameter Operator Semigroups

$$=: \frac{(2k)!\sigma^{2k}}{2^k k! n^k} + I_{12}. \tag{14.10}$$

However, we obtain

$$|I_{12}| \leq \frac{(2k)!}{2^k k!} \frac{\sigma^{2k}}{n^{2k-1}} \sum_{j=1}^{k-1} \binom{k-1}{j} n^{k-1-j} + \frac{(2k-1)\sigma^2 N(k-1, X)}{n^{k+1}}$$

$$\leq \frac{L(k, X)}{n^{k+1}}, \tag{14.11}$$

where

$$L(k, X) := \frac{(2k)!}{2^k k!} \sigma^{2k} \sum_{j=1}^{k-1} \binom{k-1}{j} + (2k-1)\sigma^2 N(k-1, X)$$

is a polynomial in $E[|X - x|^l]$, $2 \leq l \leq 2k - 2$.

So, by combining (14.8)–(14.11), we find

$$\left| E\left[\left(\frac{S_n}{n} - x\right)^{2k}\right] - \frac{(2k)!}{2^k k!} \frac{\sigma^{2k}}{n^k} \right| \leq |I_{12}| + |I_2| \leq \frac{K(k, X) + L(k, X)}{n^{k+1}}.$$

Thus

$$E\left[\left(\frac{S_n}{n} - x\right)^{2k}\right] = \frac{(2k)!}{2^k k!} \frac{\sigma^{2k}}{n^k} + \frac{\varepsilon(n, k, X)}{n^{k+1}},$$

where $|\varepsilon(n, k, X)| \leq N(k, X) := K(k, X) + L(k, X)$, $N(k, X)$ is a polynomial in $E[(X - x)^l]$, $2 \leq l \leq 2k$ and so (14.7) holds for all $k \in \mathbf{Z}_+$. ∎

The following Taylor's expansion for $(C_0)$ m-parameter operator semigroups is proved from the Taylor's formula with integral remainder for Banach space valued functions, by applying the linear functional method, and using the Hahn-Banach Theorem.

### LEMMA 14.2
$\{T(t); t \in \mathbf{R}_+^m\}$ is a $(C_0)$ m-parameter operator semigroup, $r \in \mathbf{N}$ is fixed. If $T(x_0)g \in D^r$ for some $x_0 \in \mathbf{R}_+^m$, then for all $s, t > x_0$ it holds

$$T(t)g - T(s)g = \sum_{j=1}^{r} \frac{1}{j!}[(t-s) \bullet A]^j T(s)g \tag{14.12}$$

$$+ \int_0^1 \frac{(1-v)^{r-1}}{(r-1)!}[(t-s) \bullet A]^r [T(s+v(t-s)) - T(s)]g\,dv.$$

To present a quantitative estimate for the asymptotic representation formulae for $(C_0)$ m-parameter operator semigroups, we use the following modulus of continuity $\omega(T, r, g; \delta)$. Its main property is in Lemma 14.3.

**DEFINITION 14.1** Let $\{T(t); t \in \mathbf{R}_+^m\}$ be as above. For $g \in D^r$ define

$$\omega(T, r, g; \delta) := \sup_{\bar{y} < \delta,\ \mu \in \mathbf{Z}_+^m,\ \bar{\mu}=r} \{\|(T(y) - I)A^\mu g\|\}.$$

Notice that when $\delta \to 0$, $\omega(T, r, g; \delta) \to 0$.

**LEMMA 14.3**
For positive real numbers $\lambda$, $\delta$, and $(C_0)$ m-parameter operator semigroup $\{T(t); t \in \mathbf{R}_+^m\}$ obeying (14.5) we get

$$\omega(T, r, g; \lambda\delta) \leq Me^{\omega\lambda\delta}(1+\lambda)\omega(T, r, g; \delta). \tag{14.13}$$

**PROOF** Assume $\bar{y} < \lambda\delta$, and $n - 1 \leq \lambda < n$ for some $n \in \mathbf{N}$, then

$$\|T(y)A^\mu g - A^\mu g\| \leq \sum_{j=1}^n \left\|\left(T\left(\frac{j}{n}y\right) - T\left(\frac{j-1}{n}y\right)\right)A^\mu g\right\|$$

$$\leq \sum_{j=1}^n Me^{\omega\frac{j}{n}\bar{y}} \left\|\left(T\left(\frac{1}{n}y\right) - I\right)A^\mu g\right\|$$

$$\leq Me^{\omega\lambda\delta} n\omega(T, r, g; \delta)$$

$$\leq Me^{\omega\lambda\delta}(1+\lambda)\omega(T, r, g; \delta).$$

By taking the supremum we obtain (14.13). ∎

We also need

**LEMMA 14.4**
Let $\{X_i\}_{i=1}^\infty$ be a sequence of $\mathbf{R}_+^m$-valued random vectors and $N$ be a $\mathbf{Z}_+$-valued random variable. Assume $X_i \overset{i.i.d.}{\sim} X$, with $E(X) = x$ and $\Psi_X^* :=$

$E(e^{\delta \bar{X}}) < \infty$ for some $\delta > 0$. Further, $N$ and $X_i$'s are also supposed to be independent. Take $Y = \frac{1}{n}\sum_{i=1}^{n} X_i$. Then for $u > 0$ we find

$$e^{u\bar{x}} \leq E[e^{u\bar{Y}}] = \Psi_{\bar{Y}}^*(u) \leq e^{u\bar{x}} \exp\left[\frac{2nu^2}{e^2(n\delta - u)^2}\Psi_{\bar{X}}^*(\delta)\right], \qquad (14.14)$$

when $u/n < \delta$, and

$$E\left[\left|\frac{1}{\tau}\sum_{i=1}^{N}(X_{il} - x_l)\right|^{\alpha}\right]$$

$$\leq 2\left(\frac{\alpha}{e\sqrt{\tau}}\right)^{\alpha} E\left\{\left(\exp\left[\frac{2\tau}{e^2(\sqrt{\tau}\delta - 1)^2}e^{\delta \bar{x}}\Psi_{\bar{X}}^*(\delta)\right]\right)^{\frac{N}{\tau}}\right\}, (14.15)$$

where $1 \leq l \leq m$ and $\tau, \alpha \in \mathbf{R}_+$.

**PROOF**  We see that

$$\Psi_{\bar{Y}}^*(u) = (E[e^{\frac{u}{n}\bar{X}}])^n \leq \left(1 + \frac{u}{n}E[\bar{X}] + E\left[\frac{u^2 \overline{X}^2}{2n^2}e^{\frac{u}{n}\bar{X}}\right]\right)^n$$

$$\leq \left(1 + \frac{u}{n}\bar{x} + \frac{u^2}{2n^2}\left(\frac{2}{\delta - u/n}\right)^2 e^{-2}E[e^{\delta \bar{X}}]\right)^n$$

$$\leq e^{u\bar{x}} \exp\left[\frac{2nu^2}{e^2(n\delta - u)^2}\Psi_{\bar{X}}^*(\delta)\right],$$

when $u/n < \delta$, and the right-hand side of inequality (14.14) follows. Above we used the inequalities

$$r^{\alpha}e^{\eta r} \leq \left(\frac{\alpha}{\delta - \eta}\right)^{\alpha} e^{-\alpha}e^{\delta r} \quad (\text{when } \eta < \delta, r > 0, \alpha > 0) \qquad (14.16)$$

and

$$(1 + r)^n \leq e^{nr}. \qquad (14.17)$$

The left-hand side of inequality (14.14) follows by Jensen's inequality.

Next, for any $l \in \{1, \ldots, m\}$, we observe

$$E\left[\left|\frac{1}{\tau}\sum_{i=1}^{N}(X_{il}-x_l)\right|^{\alpha}\right] \leq \left(\frac{\alpha}{\sqrt{\tau}}\right)^{\alpha} e^{-\alpha} E\left[e^{\frac{1}{\sqrt{\tau}}|\sum_{i=1}^{N}(X_{il}-x_l)|}\right] \quad \text{(by (14.16))}$$

$$\leq \left(\frac{\alpha}{e\sqrt{\tau}}\right)^{\alpha}\left\{E\left[e^{\frac{1}{\sqrt{\tau}}\sum_{i=1}^{N}(X_{il}-x_l)}\right] + E\left[e^{\frac{1}{\sqrt{\tau}}\sum_{i=1}^{N}(x_l-X_{il})}\right]\right\}$$

$$\leq \left(\frac{\alpha}{e\sqrt{\tau}}\right)^{\alpha}\left\{E\left[(E(e^{\frac{1}{\sqrt{\tau}}(X_{0l}-x_l)}))^N\right] + E\left[(E(e^{\frac{1}{\sqrt{\tau}}(x_l-X_{0l})}))^N\right]\right\}$$

$$\leq \left(\frac{\alpha}{e\sqrt{\tau}}\right)^{\alpha}\left\{E\left[\left(E\left(1+\frac{1}{\sqrt{\tau}}(X_{0l}-x_l)+\frac{1}{2\tau}(X_{0l}-x_l)^2 e^{\frac{1}{\sqrt{\tau}}|X_{0l}-x_l|}\right)\right)^N\right]\right.$$

$$\left. + E\left[\left(E\left(1+\frac{1}{\sqrt{\tau}}(x_l-X_{0l})+\frac{1}{2\tau}(x_l-X_{0l})^2 e^{\frac{1}{\sqrt{\tau}}|X_{0l}-x_l|}\right)\right)^N\right]\right\}$$

$$\leq 2\left(\frac{\alpha}{e\sqrt{\tau}}\right)^{\alpha} E\left\{(\exp[E((X_{0l}-x_l)^2 e^{\frac{1}{\sqrt{\tau}}|X_{0l}-x_l|})])^{N/2\tau}\right\} \quad \text{(by (14.17))}$$

$$\leq 2\left(\frac{\alpha}{e\sqrt{\tau}}\right)^{\alpha} E\left\{\left(\exp\left[\left(\frac{2}{\delta-1/\sqrt{\tau}}\right)^2 e^{-2} E(e^{\delta|X_{0l}-x_l|})\right]\right)^{N/2\tau}\right\}$$

(by (14.16), when $\frac{1}{\sqrt{\tau}} < \delta$)

$$\leq 2\left(\frac{\alpha}{e\sqrt{\tau}}\right)^{\alpha} E\left\{\left(\exp\left[\frac{2\tau}{e^2(\delta\sqrt{\tau}-1)^2}e^{\delta\bar{x}}\Psi_{\bar{X}}^*(\delta)\right]\right)^{N/\tau}\right\}.$$

Thus (14.15) follows. ∎

## 14.3 Quantitative Asymptotic Probabilistic Expansions

We present

## 14. m-Parameter Operator Semigroups

**THEOREM 14.1**
Let $X = (X_{01}, \ldots, X_{0m})$ be an $\mathbf{R}_+^m$-valued random vector with $E(X) = x = (x_1, \ldots, x_m)$ and there exists a $\delta > 0$ such that $\Psi_X^*(\delta) < \infty$. Let further that $X_i \overset{i.i.d.}{\sim} X$, $i = 1, 2, \ldots$, and denote $S_n = \sum_{i=1}^n X_i$. Assume $g \in D^{2k}$ for some fixed $k \in \mathbf{N}$. Then for sufficiently large $n$ we obtain

$$\left\| \{E[T(X/n)]\}^n g - T(x)g - \sum_{j=2}^{2k} \frac{1}{j!} E\left[\left(\left(\frac{1}{n} S_n - x\right) \bullet A\right)^j\right] T(x)g \right\|$$

$$\leq \frac{K}{n^k} \omega\left(T, 2k, g; \frac{1}{\sqrt{n}}\right), \tag{14.18}$$

where $K$ is a constant independent of $n$.

**PROOF** Because $\Psi_X^*(\delta) < \infty$, $E[\overline{X}^r] < \infty$ for all $r \in \mathbf{Z}_+$. From Lemma 14.2 (let $x_0 = 0$), we see that

$$T\left(\frac{S_n}{n}\right) g - T(x)g = \sum_{j=1}^{2k} \frac{1}{j!} \left[\left(\frac{S_n}{n} - x\right) \bullet A\right]^j T(x)g$$

$$+ \int_0^1 \frac{(1-v)^{2k-1}}{(2k-1)!} \left[T\left(x + v\left(\frac{S_n}{n} - x\right)\right) - T(x)\right]$$

$$\cdot \left[\left(\frac{S_n}{n} - x\right) \bullet A\right]^{2k} g \, dv.$$

Hence, by noticing that $E\left[\frac{S_n}{n} - x\right] = 0$ and using $S_{nr}$ as the $r$th component of $S_n$, we get the following estimate:

$$\Delta := \left\| E\left[T\left(\frac{S_n}{n}\right)\right] g - T(x)g - \sum_{j=2}^{2k} \frac{1}{j!} E\left[\left(\left(\frac{1}{n} S_n - x\right) \bullet A\right)^j\right] T(x)g \right\|$$

$$\leq E\left[\int_0^1 \frac{(1-v)^{2k-1}}{(2k-1)!} \sum_{\bar{\mu}=2k} \left(\left|\frac{S_n}{n} - x\right|\right)^{\mu}\right]$$

$$\cdot \left\| \left[T\left(x + v\left(\frac{S_n}{n} - x\right)\right) - T(x)\right] A^\mu g \right\| dv$$

(set next $y = (y_1, \ldots, y_m)$, where

$$y_r := \min\left\{x_r + v\left(\frac{S_{nr}}{n} - x_r\right), x_r\right\}, \quad \text{for } r = 1, \ldots, m)$$

$$\leq E\Bigg[\int_0^1 \frac{(1-v)^{2k-1}}{(2k-1)!} \sum_{\bar{\mu}=2k} \left(\left|\frac{S_n}{n} - x\right|\right)^{\mu}$$

$$\cdot \left(\|T(y)\| \left\|\left(T\left((x-y) + v\left(\frac{S_n}{n} - x\right)\right) - I\right) A^{\mu} g\right\|\right.$$

$$+ \|T(y)\| \|(T(x-y) - I)A^{\mu}g\|\Big)dv\Bigg]$$

$$\leq \frac{2}{(2k)!} E\left[\left(\left|\frac{S_n}{n} - x\right|\right)^{2k} M e^{\omega \bar{x}} e^{\omega |\frac{S_n}{n} - x|} \omega\left(T, 2k, g; \left|\frac{S_n}{n} - x\right|\right)\right]$$

$$\left(\text{notice that } \|T(y)\| \leq M e^{\omega(\bar{x} + |\frac{S_n}{n} - x|)}, \left|x - y - v\left(\frac{S_n}{n} - x\right)\right|\right.$$

$$\leq \left|\frac{S_n}{n} - x\right| \quad \text{and} \quad |x - y| \leq \left|\frac{S_n}{n} - x\right|\Bigg)$$

$$\leq \frac{2 M e^{\omega \bar{x}}}{(2k)!} \int_{|\frac{S_n}{n} - x| < \lambda} \left(\left|\frac{S_n}{n} - x\right|\right)^{2k}$$

$$\cdot e^{\omega |\frac{S_n}{n} - x|} \omega\left(T, 2k, g; \left|\frac{S_n}{n} - x\right|\right) dP$$

$$+ \frac{2 M e^{\omega \bar{x}}}{(2k)!} \int_{|\frac{S_n}{n} - x| \geq \lambda} \left(\left|\frac{S_n}{n} - x\right|\right)^{2k}$$

$$\cdot e^{\omega |\frac{S_n}{n} - x|} \omega\left(T, 2k, g; \left|\frac{S_n}{n} - x\right|\right) dP$$

$$:= \frac{2 M e^{\omega \bar{x}}}{(2k)!} I_1 + \frac{2 M e^{\omega \bar{x}}}{(2k)!} I_2, \quad (\lambda > 0).$$

## 14. m-Parameter Operator Semigroups

By Lemma 14.3, we see that

$$I_2 \leq \int_{|\overline{\frac{S_n}{n}-x}|\geq \lambda} \left(\left|\overline{\frac{S_n}{n}-x}\right|\right)^{2k} \left(1 + \frac{1}{\lambda}\left|\overline{\frac{S_n}{n}-x}\right|\right)$$

$$\cdot M e^{2\omega|\overline{\frac{S_n}{n}-x}|} \omega(T, 2k, g; \lambda)\, dP$$

$$\leq \frac{2M}{\lambda^2} E\left[\left(\left|\overline{\frac{S_n}{n}-x}\right|\right)^{2k+2} e^{2\omega|\overline{\frac{S_n}{n}-x}|}\right] \omega(T, 2k, g; \lambda)$$

$$\leq \frac{2M}{\lambda^2} \left(E\left[\left(\left|\overline{\frac{S_n}{n}-x}\right|\right)^{4k+4}\right]\right)^{\frac{1}{2}}$$

$$\cdot \left(E\left[e^{4\omega|\overline{\frac{S_n}{n}-x}|}\right]\right)^{\frac{1}{2}} \omega(T, 2k, g; \lambda), \qquad (14.19)$$

the last inequality is true by the Cauchy-Schwartz inequality.

Moreover, set $B(4k+4) := \{\mu = (\mu_1, \ldots, \mu_m) \in \mathbf{Z}_+^m;\ \overline{\mu} = 4k+4\}$, then by Lemma 14.1, we observe that

$$E\left[\left(\left|\overline{\frac{S_n}{n}-x}\right|\right)^{4k+4}\right]$$

$$= E\left[\sum_{\mu \in B(4k+4)} \binom{4k+4}{\mu} \left(\left|\overline{\frac{S_n}{n}-x}\right|\right)^{\mu}\right]$$

$$\leq \sum_{\mu \in B(4k+4)} \binom{4k+4}{\mu} \left(E\left[\left(\left|\frac{S_{n_1}}{n}-x_1\right|\right)^{m\mu_1}\right]\right)^{\frac{1}{m}}$$

$$\times \cdots \left(E\left[\left(\left|\frac{S_{nm}}{n}-x_m\right|\right)^{m\mu_m}\right]\right)^{\frac{1}{m}} \quad \text{(by Hölder's inequality)}$$

$$\leq \sum_{\mu \in B(4k+4)} \binom{4k+4}{\mu} [M(m\mu_1, X_{01}) \cdots M(m\mu_m, X_{0m})]^{\frac{1}{m}}$$

$$\times \left(\frac{1}{n}\right)^{\frac{1}{m}\left(\left[\frac{m\mu_1+1}{2}\right]+\cdots+\left[\frac{m\mu_m+1}{2}\right]\right)}$$

$$\leq K_1^2 \frac{1}{n^{2k+2}} \quad \left(\left[\frac{a+1}{2}\right] \geq \frac{a}{2}, \text{ for } a \in \mathbf{N}\right), \tag{14.20}$$

where the positive constant $K_1$ is independent of $n$.

In the following, apply (14.14) of Lemma 14.4 for $u = 4\omega$. Then we have for $n > \frac{4\omega}{\delta}$ that

$$E\left[e^{4\omega\left|\frac{S_n}{n}-x\right|}\right] \leq e^{8\omega\bar{x}} \exp\left[\frac{2n(4\omega)^2}{e^2(n\delta-4\omega)^2}\Psi_{\bar{X}}^*(\delta)\right] \leq K_2^2, \tag{14.21}$$

for some positive constant $K_2$ independent of $n$.

By combining (14.19)–(14.21) we have for sufficiently large $n$ that

$$I_2 \leq \frac{2M}{\lambda^2} \frac{K_1 K_2}{n^{k+1}} \omega(T, 2k, g; \lambda). \tag{14.22}$$

Similar to the proof of (14.20) and (14.21), we can prove that

$$\left(E\left[\left(\left|\frac{S_n}{n}-x\right|\right)^{4k}\right]\right)^{1/2} \leq \frac{K_3}{n^k} \quad \text{and} \quad E\left[e^{2\omega\left|\frac{S_n}{n}-x\right|}\right] \leq K_4^2,$$

for some positive constants $K_3$ and $K_4$ independent of $n$. So we get

$$I_1 \leq \left(\int_{\left|\frac{S_n}{n}-x\right|<\lambda} \left(\left|\frac{S_n}{n}-x\right|\right)^{2k} e^{\omega\left|\frac{S_n}{n}-x\right|} dP\right) \omega(T, 2k, g; \lambda)$$

$$\leq \left(E\left[\left(\left|\frac{S_n}{n}-n\right|\right)^{4k}\right]\right)^{1/2} \left(E\left[e^{2\omega\left|\frac{S_n}{n}-x\right|}\right]\right)^{1/2} \omega(T, 2k, g; \lambda)$$

$$\leq \frac{K_3 K_4}{n^k} \omega(T, 2k, g; \lambda). \tag{14.23}$$

Hence, by (14.22), (14.23), we have for sufficiently large $n$,

$$\Delta \leq \frac{2Me^{\omega\bar{x}}}{(2k)!} \left(K_3 K_4 + \frac{2M K_1 K_2}{\lambda^2 n}\right) \frac{1}{n^k} \omega(T, 2k, g; \lambda).$$

## 14. m-Parameter Operator Semigroups

Now pick $\lambda = 1/\sqrt{n}$, we get for sufficiently large $n$,

$$\Delta \leq \frac{K}{n^k} \omega\left(T, 2k, g; \frac{1}{\sqrt{n}}\right),$$

where $K := 2Me^{\omega \bar{x}}(K_3 K_4 + 2MK_1 K_2)/(2k)!$ is independent of $n$. Finally, by noticing that $E[T(S_n/n)] = \{E[T(X/n)]\}^n$, (14.18) follows. ∎

**COROLLARY 14.1**

In Theorem 14.1, take $m = 1$, and assume that $\{T(t); t \in \mathbf{R}_+\}$ is a $(C_0)$ contraction one-parameter operator semigroup, i.e., $\|T(t)\| \leq 1$ for all $t \in \mathbf{R}_+$. Assume $g \in D(A^{2k})$. Then, for sufficiently large $n$ we obtain

$$\left\| \{E[T(X/n)]\}^n g - T(x)g - \sum_{j=2}^{2k} \frac{1}{j!} E\left[\left(\frac{1}{n}S_n - x\right)^j\right] A^j T(x)g \right\|$$

$$\leq \left(\frac{(8k+5)\sigma^{2k}}{2^{k-1}k!} + \frac{L(k,X)}{n}\right) \frac{1}{n^k} \omega\left(T, 2k, g; \frac{\sigma}{\sqrt{n}}\right), \quad (14.24)$$

where $L(k, X)$ is a constant independent of $n$.

**PROOF** As in the proof of Theorem 14.1, now by using (14.7) in Lemma 14.1 we see that

$$\left\| E\left[T\left(\frac{S_n}{n}\right)\right] g - T(x)g - \sum_{j=2}^{2k} \frac{1}{j!} E\left[\left(\frac{S_n}{n} - x\right)^j\right] A^j T(x)g \right\|$$

$$\leq \left\{ \frac{2}{(2k)!} E\left[\left(\frac{S_n}{n} - x\right)^{2k}\right] + \frac{4}{(2k)!} E\left[\left(\frac{S_n}{n} - x\right)^{2k+2}\right] \frac{1}{\lambda^2} \right\} \omega(T, 2k, g; \lambda)$$

$$= \frac{2}{(2k)!} \left\{ \frac{(2k)!}{2^k k!} \frac{\sigma^{2k}}{n^k} + \frac{\varepsilon(n,k,X)}{n^{k+1}} + \frac{4(2k+2)!}{2^{k+1}(k+1)!} \frac{\sigma^{2k+2}}{n^{k+1}\lambda^2} \right.$$

$$\left. + \frac{\varepsilon(n,k+1,X)}{n^{k+2}\lambda^2} \right\} \omega(T, 2k, g; \lambda)$$

$$\leq \left(\frac{(8k+5)\sigma^{2k}}{2^{k-1}k!} + \frac{L(k,X)}{n}\right) \frac{1}{n^k} \omega\left(T, 2k, g; \frac{\sigma}{\sqrt{n}}\right),$$

(by choosing $\lambda = \frac{\sigma}{\sqrt{n}}$),

where $L(k, X) \geq 0$ is a constant independent of $n$. ∎

**REMARK 14.1** If we take $T(t)$ to be the left translation semigroup (see Butzer-Berens [54]), Corollary 14.1 improves a result obtained by R.A. Khan (see [107], Theorem 2) by reducing the conditions on $X_i$'s and giving quantitative results. Especially when $X$ is the two-point distribution: $P(X = 0) = 1 - x$, $P(X = 1) = x$, we have from Corollary 14.1 the asymptotic expansions of Bernstein operators. If we appropriately modify the proofs of Lemma 14.1 and Corollary 14.1 for this particular case, we can also find the so-called point-factor $\sigma^2 = x(1-x)$ on the right-hand side of (14.24), as stated in Wang [171] (also see Chen-Zhou [64] for the Bernstein polynomials case). ∎

The next Theorems 14.2 and 14.3 are the ramifications of Theorem 14.1 for the representation formulae in the forms of random sums of random vectors.

## THEOREM 14.2

Let $N$ be a $\mathbf{Z}_+$-valued random variable with $E(N) = \eta$, $\eta > 0$, and let $Y_l \stackrel{i.i.d.}{\sim} Y$, $l = 1, 2, \ldots$, where $Y = (Y_{01}, \ldots, Y_{0m})$ is an $\mathbf{R}_+^m$-valued random vector independent of $N$ with $E(Y) = \gamma = (\gamma_1, \ldots, \gamma_m)$. Suppose that there exists a $\delta > 0$ such that

$$\Psi_N(\Psi_Y^*(\delta)) < \infty.$$

Let further, $X_i \stackrel{i.i.d.}{\sim} X$, $i = 1, 2, \ldots$, where $X = \sum_{j=1}^N Y_j$, and assume $g \in D^{2k}$ for some fixed $k \in \mathbf{Z}_+$. Then for sufficiently large $n$ we obtain

$$\left\| \{\Psi_N(E[T(Y/n)])\}^n g - T(\eta\gamma)g \right.$$

$$\left. - \sum_{j=2}^{2k} \frac{1}{j!} E\left[\left(\left(\frac{1}{n}\sum_{i=1}^n X_i - \eta\gamma\right) \bullet A\right)^j\right] T(\eta\gamma)g \right\|$$

$$\leq \frac{K}{n^k} \omega\left(T, 2k, g; \frac{1}{\sqrt{n}}\right), \tag{14.25}$$

where $K$ is a constant independent of $n$.

**Note.** Throughout the chapter, the constant $K$ may have different values in each step in which it appears.

**PROOF** (of Theorem 14.2) Noticing now that

$$E\left[T\left(\frac{1}{n}X\right)\right] = \Psi_N\left(E\left(T\left(\frac{1}{n}Y\right)\right)\right), \quad E(X) = \eta\gamma, \quad \Psi_X^*(\delta) = \Psi_N(\Psi_Y^*(\delta)),$$

we obtain (14.25) from Theorem 14.1. ■

### THEOREM 14.3

Let $N = (N_1, \ldots, N_m)$ be a $\mathbf{Z}_+^m$-valued random vector with $E(N) = \eta = (\eta_1, \ldots, \eta_m)$. For each $r \in \{1, \ldots, m\}$, let $\{Y_{lr}\}_{l=1}^\infty$ be a sequence of i.i.d. $\mathbf{R}_+$-valued random variables distributed as $Y$, which is a fixed $\mathbf{R}_+$-valued random variable with $E(Y) = \gamma$. $N$ and $Y_{lr}$'s ($l = 1, 2, \ldots, r = 1, \ldots, m$) are supposed to be altogether independent. Also suppose that there exists a $\delta > 0$ such that

$$\Psi_{\bar{N}}(\Psi_Y^*(\delta)) < \infty.$$

Let further $X_i \stackrel{i.i.d.}{\sim} X$, $i = 1, 2, \ldots$, where

$$X := \left(\sum_{l_1}^{N_1} Y_{l_1 1}, \ldots, \sum_{l_m}^{N_m} Y_{l_m m}\right).$$

Assume $g \in D^{2k}$ for some fixed $k \in \mathbf{Z}_+$. Then for sufficiently large $n$ we find

$$\left\|\left\{E\left[T\left(\sum_{l_1=1}^{N_1} \frac{1}{n} Y_{l_1 1}, \ldots, \sum_{l_m=1}^{N_m} \frac{1}{n} Y_{l_m m}\right)\right]\right\}^n g - T(\gamma\eta)g\right.$$

$$\left. - \sum_{j=2}^{2k} \frac{1}{j!} E\left[\left(\left(\frac{1}{n}\sum_{i=1}^n X_i - \gamma\eta\right) \bullet A\right)^j\right] T(\gamma\eta)g\right\|$$

$$\leq \frac{K}{n^k} \omega\left(T, 2k, g; \frac{1}{\sqrt{n}}\right), \qquad (14.26)$$

where $K$ is a constant independent of $n$.

**PROOF**  We only need to see that

$$E(X) = \left(E\left[\sum_{l_1=1}^{N_1} Y_{l_1 1}\right], \ldots, E\left[\sum_{l_m=1}^{N_m} Y_{l_m m}\right]\right)$$

$$= (EN_1 EY, \ldots, EN_m EY) = \gamma \eta,$$

$$E\left[T\left(\frac{1}{n}X\right)\right] = E\left[T\left(\sum_{l_1=1}^{N_1} \frac{1}{n} Y_{l_1 1}, \ldots, \sum_{l_m=1}^{N_m} \frac{1}{n} Y_{l_m m}\right)\right],$$

and

$$\Psi_{\bar{X}}^*(\delta) = E\left[e^{\delta\left(\sum_{l_1=1}^{N_1} Y_{l_1 1} + \cdots + \sum_{l_m=1}^{N_m} Y_{l_m m}\right)}\right]$$

$$= \sum_{n_1=0}^{\infty} \cdots \sum_{n_m=0}^{\infty} P(N=(n_1,\ldots,n_m)) E\left[e^{\delta \sum_{l_1=1}^{n_1} Y_{l_1 1} + \cdots + \delta \sum_{l_m=1}^{n_m} Y_{l_m m}}\right]$$

$$= \sum_{n_1=0}^{\infty} \cdots \sum_{n_m=0}^{\infty} P(N=(n_1,\ldots,n_m)) E\left[e^{\delta \sum_{l_1=1}^{n_1} Y_{l_1 1}}\right] \cdots E\left[e^{\delta \sum_{l_m=1}^{n_m} Y_{l_m m}}\right]$$

$$= \sum_{n_1=0}^{\infty} \cdots \sum_{n_m=0}^{\infty} P(N=(n_1,\ldots,n_m))(E[e^{\delta Y}])^{n_1} \cdots (E[e^{\delta Y}])^{n_m}$$

$$= \sum_{n_1=0}^{\infty} \cdots \sum_{n_m=0}^{\infty} P(N=(n_1,\ldots,n_m))(E[e^{\delta Y}])^{n_1+\cdots+n_m}$$

$$= E[(E[e^{\delta Y}])^{N_1+\cdots+N_m}] = \Psi_{\bar{N}}(\Psi_Y^*(\delta)).$$

For the rest of the proof we just apply Theorem 14.1. ∎

## 14.4 Asymptotic Probabilistic Expansions for Continuous Type Formulae

In this section we study the asymptotic probabilistic expansions of the representation formulae of continuous type for $(C_0)$ m-parameter operator semigroups. The following result is without rates and is true under weaker assumptions on $g$.

**THEOREM 14.4**

*For each $\tau \in T \subseteq \mathbf{R}_+$, let $N_\tau$ be a $\mathbf{Z}_+$-valued random variable with $E(N_\tau) = \tau\eta$, where $\eta \in \mathbf{R}_+$ is fixed. Let $X_i \stackrel{i.i.d.}{\sim} X$, $i = 1, 2, \ldots$, be a sequence of $\mathbf{R}_+^m$-valued random vectors with $E(X) = \gamma = (\gamma_1, \ldots, \gamma_m)$, also independent of $N_\tau$. Assume $g \in \mathcal{X}$ satisfies $T(x_0)g \in D^{2k}$ for some fixed $k \in \mathbf{N}$ and $x_0 = (x_{01}, \ldots, x_{0m}) \in \mathbf{R}_+^m$ where $x_0 < x := \eta\gamma$. Suppose also that there exists a $\delta > 0$ such that $\Psi_{\bar{X}}^*(\delta) < \infty$, and further there are $p > 0$, $q > 0$ with $1/p + 1/q = 1$ such that*

$$\lim_{\tau\to\infty, \tau\in T} \sup \Psi_{N_\tau}\left(\Psi_{\bar{X}}^*\left(\frac{p\omega}{\tau}\right)\right) = d_1 < \infty, \tag{14.27}$$

$$\limsup_{\tau\to\infty, \tau\in T} \max_{j=k,k+1, r=1,q} \tau^j \left\{ E\left[\left(\frac{1}{\tau}N_\tau - \eta\right)^{2jr}\right]\right\}^{1/r} = d_2 < \infty \tag{14.28}$$

*and*

$$\limsup_{\tau\to\infty, \tau\in T} \Psi_{N_\tau}^*\left(\frac{2}{e^2(\sqrt{\tau\delta}-1)^2} e^{\delta\bar{\gamma}} \Psi_{\bar{X}}^*(\delta)\right) = d_3 < \infty. \tag{14.29}$$

*Then we obtain*

$$\|\Psi_{N_\tau}(E[T(X/\tau)])g - T(\eta\gamma)g$$

$$- \sum_{j=2}^{2k} \frac{1}{j!} E\left[\left(\left(\frac{1}{\tau}\sum_{i=1}^{N_\tau} X_i - \eta\gamma\right) \bullet A\right)^j\right] T(\eta\gamma)g \right\|$$

$$= o\left(\frac{1}{\tau^k}\right), \quad (as\ \tau \to \infty). \tag{14.30}$$

**PROOF** In the following we set for $r \in \mathbf{Z}_+$ that

$$B(r) := \{\mu = (\mu_1, \ldots, \mu_m) \in \mathbf{Z}_+^m; \bar{\mu} = r\}.$$

Because $T(x_0)g \in D^{2k}$, we have for $t = (t_1, \ldots, t_m) \in \mathbf{R}_+^m$ and $t \geq x_0$ (means $t_i \geq x_{0i}$, $i = 1, \ldots, m$) that $T(t)g \in D^{2k}$.

Given any $\varepsilon > 0$, by the strong continuity of $T(t)$, there exists an $\lambda$ with $0 < \lambda < 1$ such that when $|\overline{t - x}| < \lambda$ and $t \geq x_0$ it holds

$$\sup_{\mu \in B(2k), 0 \leq v \leq 1} \|A^\mu (T(x + v(t-x)) - T(x))g\| < \varepsilon. \tag{14.31}$$

We need to estimate $\|E(\Delta)\|$, where

$$\Delta := T\left(\frac{1}{\tau}\sum_{i=1}^{N_\tau} X_i\right) g - \sum_{j=0}^{2k} \frac{1}{j!} \left(\left(\frac{1}{\tau}\sum_{i=1}^{N_\tau} X_i - \eta\gamma\right) \bullet A\right)^j T(\eta\gamma)g.$$

Put

$$F(-) := \left\{\omega \in \Omega; \left|\overline{\frac{1}{\tau}\sum_{i=1}^{N_\tau} X_i - \eta\gamma}\right| < \lambda\right\}, \quad F(+) := \Omega - F(-).$$

Then

$$\|E(\Delta)\| \leq \left\|\int_{F(-)} \Delta\, dP\right\| + \left\|\int_{F(+)} \Delta\, dP\right\| := I_1 + I_2. \tag{14.32}$$

To continue, we see that, for $j = k, k+1$, and $r = 1$ or $q$,

$$E\left[\left(\left|\overline{\frac{1}{\tau}\sum_{i=1}^{N_\tau} X_i - \eta\gamma}\right|\right)^{2jr}\right]$$

$$\leq m^{2jr} \sum_{l=1}^{m} E\left[\left(\left|\frac{1}{\tau}\sum_{i=1}^{N_\tau} X_{il} - \eta\gamma_l\right|\right)^{2jr}\right]$$

$$\leq m^{2jr} \sum_{l=1}^{m} E\left[\left(\left|\frac{1}{\tau}\sum_{i=1}^{N_\tau} X_{il} - \frac{1}{\tau}N_\tau\gamma_l + \frac{1}{\tau}N_\tau\gamma_l - \eta\gamma_l\right|\right)^{2jr}\right]$$

$$\leq (2m)^{2jr} \sum_{l=1}^{m} \left\{ E\left[ \left| \frac{1}{\tau} \sum_{i=1}^{N_\tau} (X_{il} - \gamma_l) \right|^{2jr} \right] + \gamma_l^{2jr} E\left[ \left| \frac{1}{\tau} N_\tau - \eta \right|^{2jr} \right] \right\}$$

$$\leq (2m)^{2jr} m \left( 2 \left( \frac{2jr}{e\sqrt{\tau}} \right)^{2jr} E\left\{ \left( \exp\left[ \frac{2\tau}{e^2(\sqrt{\tau}\delta - 1)^2} e^{\delta\bar{\gamma}} \Psi_{\bar{X}}^*(\delta) \right] \right)^{\frac{N_\tau}{\tau}} \right\}$$

$$+ \bar{\gamma}^{2jr} \frac{d_2^r}{\tau^{jr}} \Bigg)$$

(by (14.15) of Lemma 14.4 and (14.28))

$$\leq (2m)^{2jr} m \left[ 2 \left( \frac{2jr}{e} \right)^{2jr} d_3 + \bar{\gamma}^{2jr} d_2^r \right] \frac{1}{\tau^{jr}} \quad \text{(by (14.29))}$$

$$=: P(j, r) \frac{1}{\tau^{jr}}, \qquad (14.33)$$

where $P(j, r)$ is a constant independent of $\tau$.

Now we start to estimate $\|E(\Delta)\|$. First, by Lemma 14.2, the opening argument of the proof and (14.33), we obtain

$$I_1 \leq \varepsilon E\left[ \left( \left| \frac{1}{\tau} \sum_{i=1}^{N_\tau} X_i - x \right| \right)^{2k} \right] \leq \frac{P(k, 1)}{\tau^k} \varepsilon. \qquad (14.34)$$

Next we notice that

$$I_2 \leq \int_{F(+)} \left\| T\left( \frac{1}{\tau} \sum_{i=1}^{N_\tau} X_i \right) g \right\| dP$$

$$+ \sum_{j=0}^{2k} \left[ \frac{1}{j!} \left( \int_{F(+)} \left( \left| \frac{1}{\tau} \sum_{i=1}^{N_\tau} X_i - x \right| \right)^j dP \right) \max_{\mu \in B(j)} \{\|A^\mu T(x)g\|\} \right]$$

$$=: I_{21} + I_{22}. \qquad (14.35)$$

We see that

$$I_{21} \leq \int_{F(+)} \frac{1}{\lambda^{2k+2}} M\|g\| e^{\frac{\omega}{\tau}(\sum_{i=1}^{N_\tau} \bar{X}_i)} \left(\left|\frac{1}{\tau}\sum_{i=1}^{N_\tau} X_i - x_i\right|\right)^{2k+2} dP$$

$$\leq \frac{M\|g\|}{\lambda^{2k+2}} (E[e^{\frac{p\omega}{\tau}(\sum_{i=1}^{N_\tau} \bar{X}_i)}])^{\frac{1}{p}} \left(E\left[\left(\left|\frac{1}{\tau}\sum_{i=1}^{N_\tau} X_i - x_i\right|\right)^{2qk+2q}\right]\right)^{\frac{1}{q}}$$

(from Hölder's inequality)

$$\leq \frac{M\|g\|}{\lambda^{2k+2}} \left(\Psi_{N_\tau}\left(\Psi^*_{\bar{X}}\left(\frac{p\omega}{\tau}\right)\right)\right)^{\frac{1}{p}} \left(\frac{P(k+1,q)}{\tau^{(k+1)q}}\right)^{\frac{1}{q}}$$

$$\leq \frac{M\|g\| d_1^{\frac{1}{p}} (P(k+1,q))^{\frac{1}{q}}}{\tau^k \lambda^{2k+2} \tau} \quad \text{(by (14.27))}.$$

So if

$$\tau > \tau_1 := \frac{M\|g\| d_1^{\frac{1}{p}} (P(k+1,q))^{\frac{1}{q}}}{\varepsilon \lambda^{2k+2}},$$

we get

$$I_{21} < \frac{\varepsilon}{\tau^k}. \tag{14.36}$$

Furthermore, we observe

$$I_{22} = \sum_{j=0}^{2k} \frac{1}{j!} \left(\int_{F(+)} \left(\left|\frac{1}{\tau}\sum_{i=1}^{N_\tau} X_i - x\right|\right)^j dP\right) \max_{\mu \in B(j)} \{\|A^\mu T(x)g\|\}$$

$$\leq \sum_{j=0}^{2k} \frac{1}{j!} \frac{1}{\lambda^{2k+2-j}} E\left[\left(\left|\frac{1}{\tau}\sum_{i=1}^{N_\tau} X_i - x\right|\right)^{2k+2}\right] \max_{\mu \in B(j)} \{\|A^\mu T(x)g\|\}$$

$$\leq \frac{1}{\lambda^{2k+2}} \frac{1}{\tau^{k+1}} P(k+1,1) \left(\sum_{j=0}^{2k} \frac{1}{j!} \max_{\mu \in B(j)} \{\|A^\mu T(x)g\|\}\right)$$

(by (14.33) and noticing that $\lambda < 1$).

So if
$$\tau > \tau_2 := \frac{P(k+1,1)}{\varepsilon \lambda^{2k+2}} \sum_{j=0}^{2k} \frac{1}{j!} \max_{\mu \in B(j)} \{\|A^\mu T(x)g\|\},$$

we obtain
$$I_{22} < \frac{\varepsilon}{\tau^k}. \tag{14.37}$$

Thus by (14.32), (14.34)–(14.37), for the given $\varepsilon > 0$, when $\tau > \tau_1 + \tau_2$, we obtain
$$\|E(\Delta)\| < \frac{P(k,1)}{\tau^k}\varepsilon + \frac{\varepsilon}{\tau^k} + \frac{\varepsilon}{\tau^k} = P\frac{\varepsilon}{\tau^k}, \tag{14.38}$$

where the constant $P$ is independent of $\varepsilon$ and $\tau$. Notice that

$$E\left[T\left(\frac{1}{\tau}\sum_{i=1}^{N_\tau} X_i\right)\right]g = \Psi_{N_\tau}(E[T(X/\tau)])g \text{ and } E\left[\frac{1}{\tau}\sum_{i=1}^{N_\tau} X_i\right] = \eta\gamma.$$

Now it is clear that (14.30) follows by (14.38) and the proof is finished. ∎

**REMARK 14.2** When taking $T = \mathbf{N}$, $N_\tau = n$ in Theorem 14.4, we obtain a result corresponding to the representation formulae using the arithmetic mean of i.i.d. random variables as in Theorem 14.1. ∎

## 14.5 Examples

In this section we apply the presented results to the study of multivariate operator approximation theory, and we give asymptotic expansions for multivariate Feller, Bernstein, and Szász-Mirakjan operators (see [165]).

Let $\mathcal{X} := BUC(\mathbf{R}^m) := \{f; f \text{ is a uniformly continuous bounded function from } \mathbf{R}^m \text{ to } \mathbf{R}\}$ and $\{T(t); t \in \mathbf{R}^m_+\}$ be an m-parameter left translation contraction semigroup defined by

$$T(t)f(x) := f(x+t) = f(x_1 + t_1, \ldots, x_m + t_m),$$

for $f \in \mathcal{X}$, $x \in \mathbf{R}^m_+$, $t \in \mathbf{R}^m_+$.

It is easy to prove that $\{T(t); t \in \mathbf{R}^m_+\}$ is strongly continuous and its infinitesimal generators are $A_i := \frac{\partial}{\partial x_i}$, $i = 1, \ldots, m$. For fixed $k \in \mathbf{Z}_+$ and

$g \in BUC(\mathbf{R}^m)$, if

$$P^\mu g := \frac{\partial^{\mu_1} \cdots \partial^{\mu_m}}{\partial x_1^{\mu_1} \cdots \partial x_m^{\mu_m}} g$$

exists and is in $BUC(\mathbf{R}^m)$ for all $\mu = (\mu_1, \ldots, \mu_m) \in \mathbf{Z}_+^m$ satisfying $\bar{\mu} = 2k$, then $g \in D^{2k}$ and the corresponding modulus of continuity collapses to

$$\omega(T, 2k, g; \delta) = \omega(g, 2k; \delta)$$

$$:= \sup_{\bar{y} < \delta, \mu \in \mathbf{Z}_+^m, \bar{\mu}=2k} \left\{ \sup_{x \in \mathbf{R}^m} |P^\mu g(y+x) - P^\mu g(x)| \right\}.$$

Furthermore, $\omega(g, 2k; \delta) \to 0$ as $\delta \to 0$.

Here comes the first example about multivariate Feller operators (see [165], [107]) as an application of Theorem 14.1.

**EXAMPLE 14.1**

Assume that the $\mathbf{R}_+^m$-valued random vector $X$ fulfills the conditions of Theorem 14.1, i.e., $\Psi^*_{\bar{X}}(\delta) < \infty$ for some $\delta > 0$, then for any fixed $k \in \mathbf{Z}$ and $g \in BUC(\mathbf{R}^m)$ satisfying $P^\mu g \in BUC(\mathbf{R}^m)$ for all $\bar{\mu} = 2k$ we get

$$\sup_{y=(y_1,\ldots,y_m) \in \mathbf{R}^m} \left| \int_{\mathbf{R}_+^m} g\left(y_1 + \frac{u_1}{n}, \ldots, y_m + \frac{u_m}{n}\right) dF_{n,x}(u_1, \ldots, u_m) \right.$$

$$- g(y_1 + x_1, \ldots, y_m + x_m) - \sum_{j=2}^{2k} \frac{1}{j!} \sum_{\mu=(\mu_1,\ldots,\mu_m) \in \mathbf{Z}_+^m, \bar{\mu}=j}$$

$$\times \left( \int_{\mathbf{R}_+^m} \left(\frac{u_1}{n} - x_1\right)^{\mu_1} \cdots \left(\frac{u_m}{n} - x_m\right)^{\mu_m} \right.$$

$$\left. \left. \cdot dF_{n,x}(u_1, \ldots, u_m) P^\mu g(x+y) \right) \right|$$

$$\leq \frac{K}{n^k} \omega\left(g, 2k; \frac{1}{n}\right) \quad (\to 0, \text{ when } n \to \infty), \tag{14.39}$$

where $F_{n,x}(u_1, \ldots, u_m)$ is the joint distribution function of $S_n = X_1 + \cdots +$

$X_n$. Especially when we set $y = 0$ we obtain

$$\lim_{n \to \infty} n^k \left( \int_{\mathbf{R}_+^m} g\left(\frac{u_1}{n}, \ldots, \frac{u_m}{n}\right) dF_{n,x}(u_1, \ldots, u_m) - g(x) \right.$$

$$- \sum_{j=2}^{2k} \frac{1}{j!} \sum_{\mu \in \mathbf{Z}_+^m, \bar{\mu}=j} \int_{\mathbf{R}_+^m} \left(\frac{u_1}{n} - x_1\right)^{\mu_1}$$

$$\left. \cdots \left(\frac{u_m}{n} - x_m\right)^{\mu_m} dF_{n,x}(u_1, \ldots, u_m) P^\mu g(x) \right)$$

$$= 0. \tag{14.40}$$

Next we study the m-variate Bernstein operators $B_n(f, x)$ defined over the simplex $\{x \in \mathbf{R}_+^m, \bar{x} \leq 1\}$ and given by

$$B_n(f, x) = \sum_{\mu \in \mathbf{Z}_+^m, \bar{\mu} \leq n} f\left(\frac{\mu_1}{n}, \ldots, \frac{\mu_m}{n}\right)$$

$$\cdot \frac{n!}{\mu_1! \cdots \mu_m!(n - \mu_1 - \cdots - \mu_m)!} x_1^{\mu_1} \cdots x_m^{\mu_m}$$

$$\times (1 - x_1 - \cdots - x_m)^{n - \mu_1 - \cdots - \mu_m}. \tag{14.41}$$

**EXAMPLE 14.2**

In Example 14.1, take $X$ to be the m-point distribution with $EX = x = (x_1, \ldots, x_m)$: $P(X = e_i) = x_i$ ($e_i = (0, \ldots 1, \ldots, 0)$), $i = 1, \ldots, m$, and

$$P(X = 0) = 1 - \bar{x}, \quad \text{where} \quad 0 < \bar{x} < 1, \quad (\bar{x} = x_1 + \cdots + x_m).$$

So that

$$\Psi_{\bar{X}}^*(\delta) = E(e^{\delta \bar{X}}) = P(\bar{X} = 0) + P(\bar{X} = 1)e^\delta = 1 - \bar{x} + \bar{x}e^\delta < \infty.$$

Therefore, the conditions for $X$ in Theorem 14.1 are satisfied. Moreover, we notice that for any $\mu = (\mu_1, \ldots, \mu_m) \in \mathbf{Z}_+^m$ it holds

$$P(X_1 + \cdots + X_n = \mu) = \frac{n!}{\mu_1! \cdots \mu_m!(n - \mu_1 - \cdots - \mu_m)!} x^{\mu_1} \cdots x^{\mu_m}$$

$$\times (1 - x_1 - \cdots - x_m)^{n - \mu_1 - \cdots - \mu_m}.$$

We observe that

$$\int_{\mathbf{R}_+^m} f\left(\frac{u_1}{n}, \ldots, \frac{u_m}{n}\right) dF_{n,x}(u_1, \ldots, u_m)$$

$$= \sum_{\mu \in \mathbf{Z}_+^m, \bar{\mu} \le n} f\left(\frac{\mu_1}{n}, \ldots, \frac{\mu_m}{n}\right) \frac{n!}{\mu_1! \cdots \mu_m!(n - \mu_1 - \cdots - \mu_m)!} x_1^{\mu_1} \cdots x_m^{\mu_m}$$

$$\times (1 - x_1 - \cdots - x_m)^{n - \mu_1 - \cdots - \mu_m}$$

$$= B_n(f, x).$$

By (14.40) of Example 14.1 we have for the same $g$ as there that

$$\lim_{n \to \infty} n^k \Bigg( B_n(g, x) - g(x) - \sum_{j=2}^{2k} \frac{1}{j!} \sum_{\mu \in \mathbf{Z}_+^m, \bar{\mu} = j} \qquad (14.42)$$

$$\times B_n((\cdot - x_1)^{\mu_1} \cdots (\cdot - x_m)^{\mu_m}, x) \frac{\partial^{\mu_1}}{\partial x_1^{\mu_1}} \cdots \frac{\partial^{\mu_m}}{\partial x_m^{\mu_m}} g(x) \Bigg) = 0.$$

Finally we discuss the m-variate Szász-Mirakjan operators introduced by Stancu in [165].

**EXAMPLE 14.3**

If $X$ has the m-variate Poisson distribution with $E[X] = x = (x_1, \ldots, x_m) \in \mathbf{R}_+^m$, where

$$P(X = \mu) = e^{-(x_1 + \cdots + x_m)} \frac{x_1^{\mu_1} \cdots x_m^{\mu_m}}{\mu_1! \cdots \mu_m!} \quad (\mu = (\mu_1, \ldots, \mu_m) \in \mathbf{Z}_+^m),$$

## 14. m-Parameter Operator Semigroups

then

$$\Psi_{\bar{X}}^*(\delta) = E[e^{\delta \bar{X}}] = \sum_{\mu \in \mathbf{Z}_+^m} e^{\delta \bar{\mu}} P(X = \mu) = \exp((e^\delta - 1)\bar{x}) < \infty.$$

Moreover we notice that (see [165], p. 337)

$$E\left[f\left(\frac{S_n}{n}\right)\right] = \sum_{\mu=(\mu_1,\ldots,\mu_m)\in\mathbf{Z}_+^m} e^{-n(x_1+\cdots+x_m)} \frac{(nx_1)^{\mu_1} \cdots (nx_m)^{\mu_m}}{\mu_1! \cdots \mu_m!}$$

$$\cdot f\left(\frac{\mu_1}{n}, \ldots, \frac{\mu_m}{n}\right)$$

$$=: P_n(f, x),$$

where $P_n(f,x)$ is the m-variate Szász-Mirakjan operators (see [165]). Hence (14.40) of Example 14.1, for the same $g$ as there, we get the next asymptotic expansion for the m-variate Szász-Mirakjan operators:

$$\lim_{n\to\infty} n^k \left( P_n(g, x) - g(x) - \sum_{j=2}^{2k} \frac{1}{j!} \right. \tag{14.43}$$

$$\left. \times \sum_{\mu \in \mathbf{Z}_+^m} P_n((\cdot - x_1)^{\mu_1} \cdots (\cdot - x_m)^{\mu_m}, x) \frac{\partial^{\mu_1}}{\partial x_1^{\mu_1}} \cdots \frac{\partial^{\mu_m}}{\partial x_m^{\mu_m}} g(x) \right) = 0.$$

# Part V

# On Stochastics

# Chapter 15

## Quantitative Probability Limit Theorems over Banach Spaces

In this chapter are presented quantitative results which involve the standard modulus of continuity in Banach spaces. These give the convergence in distribution for Banach space-valued martingale difference sequences and the weak convergence of the distributions of random polygonal lines to the Wiener-measure on $C([0,1])$. A general theorem is presented with applications to the central limit theorem and weak law of large numbers for Banach space-valued martingales. Also, a general theorem is given on the weak invariance principle with an application to a central limit theorem for real-valued martingales. This treatment is based on [16], [55] and [56]. In particular, the exhibited main inequalities here involve the standard modulus of continuity of certain Fréchet derivative of the acting function in the associated weak convergences.

**About** §15.1  Let $(X_i, \mathcal{F}_i)_{i \in \mathbf{Z}_+}$ be a martingale difference sequence of Banach space $B$-valued random variables, defined on the probability space $(\Omega, \mathcal{A}, P)$. Denote $S_n := \sum_{i=1}^n X_i$. Let $Z$ be a $B$-valued random variable which is $\varphi$-decomposable and $f : B \to \mathbf{R}$ be a $(r-1)$-times uniformly continuously Fréchet differentiable and bounded function, $r \in \mathbf{N} - \{1\}$. Here $E$ stands for the expectation. The random variables $X_i$, $i \in \mathbf{Z}_+$ do not have to be independent.

In Theorem 15.1, under the moment condition (15.21), we show an estimate for $|E(f(\varphi(n) \cdot S_n)) - E(f(Z))|$, see inequality (15.22), which involves $\omega_1(f^{(r-1)}, h)$—the modulus of continuity, $h$, depends on $X_i$, $Z$, and the normalizing function $\varphi$. Applications of this theorem are Theorem 15.2, the weak law of large numbers, and Theorem 15.3, the central limit theorem for martingale difference sequences on Banach spaces.

**About** §15.2  Here we study the weak convergence of the distributions of the random polygonal lines $S_n(\omega, t)$, see (15.46), to the Wiener-measure $W_R$ on $C = C([0,1])$. In here again, the random variables $X_i$ determining $S_n$

335

do not have to be independent. This is possibly due to a moment condition, that is, relation (15.65) of the main theorem, Theorem 15.4. There we estimate $|E(f(S_n(t))) - E(f(W(t)))|$, see inequality (15.67), where $f \in C_C^{r-1}$ and $W$ is the Wiener process. Inequality (15.67) involves $\omega_1(f^{(r-1)}, h)$, $r \in \mathbf{N} - \{1\}$, where $h$ depends on $X_i$, $W$, and some parameters.

When we treat a more concrete dependency structure among the random variables $X_i$, see Definition 15.3, we simplify the moment condition (15.65) and give Theorem 15.5, a central limit theorem for possibly dependent random variables. A direct application of this theorem is Theorem 15.6, which is the central limit theorem for a martingale difference sequence of real-valued random variables.

## 15.1 Limit Theorems for Martingales in Banach Spaces

### 15.1.1 Auxiliary Results

We use the following:

**LEMMA 15.1**
Let $(V_1, \|\cdot\|_1)$, $(V_2, \|\cdot\|_2)$ be real normed vector spaces and $Q$ be a subset of $V_1$ which is star-shaped relative to the fixed point $x_0 \in Q$. Consider $f : Q \to V_2$ such that

$$f(x_0) = 0 \quad \text{and} \quad \|s - t\|_1 \le h \Rightarrow \|f(s) - f(t)\|_2 \le w; w, h > 0. \quad (15.1)$$

Then it holds

$$\|f(t)\|_2 \le w \cdot \left\lceil \frac{\|t - x_0\|_1}{h} \right\rceil, \quad \forall t \in Q. \quad (15.2)$$

(Here $\lceil \cdot \rceil$ is the *ceiling* function, i.e., for $x \ge 0$, $\lceil x \rceil$ is defined to be the least integer greater than or equal to $x$.)

**PROOF** If $\|t - x_0\|_1 \le h$ then $\|f(t)\|_2 \le w$. For any other $t \in Q$ there is $n \in \mathbf{N}$: $(n-1)h < \|t - x_0\|_1 \le nh$. See that

$$t - x_0 = \sum_{k=0}^{n-1} \Delta_k,$$

where
$$\Delta_k := \frac{(n-k)t + kx_0}{n} - \frac{(n-(k+1))t + (k+1)x_0}{n} = \frac{t - x_0}{n}.$$

Now since $\|\Delta_k\|_1 \leq h$, we obtain

$$\|f(t)\|_2 \leq \sum_{k=0}^{n-1} \left\| f\left(\frac{(n-k)t + kx_0}{n}\right) - f\left(\frac{(n-(k+1))t + (k+1)x_0}{n}\right) \right\|_2 \leq n \cdot w.$$

∎

**LEMMA 15.2**
Let $\phi : \mathbf{R}_+ \to \mathbf{R}_+$ be a continuous convex function such that $\phi(0) = 0$. Then it holds
$$\phi(x) + \phi(y) \leq \phi(x+y), \quad \forall x, y \in \mathbf{R}_+. \tag{15.3}$$

I.e., $\phi$ is superadditive.

**PROOF** Obvious. ∎

For $t \in \mathbf{R}$ we set
$$\phi_0(t) := \left\lceil \frac{|t|}{h} \right\rceil, \quad h > 0,$$

and for $r \geq 2$ integer we define
$$\phi_{r-1}(t) := \int_0^{|t|} \left\lceil \frac{s}{h} \right\rceil \cdot \frac{(|t|-s)^{r-2}}{(r-2)!} \cdot ds. \tag{15.4}$$

$\phi_{r-1}$ is continuous and convex on $\mathbf{R}$ and strictly increasing on $\mathbf{R}^+$. Moreover,
$$\phi_{r-1}(t) \leq \left( \frac{|t|^r}{r!h} + \frac{|t|^{r-1}}{2(r-1)!} + \frac{h|t|^{r-2}}{8(r-2)!} \right), \quad \forall t \in \mathbf{R}. \tag{15.5}$$

For more on $\phi_{r-1}$ see Anastassiou [7].

**DEFINITION 15.1** Let $(V_1, \|\cdot\|_1)$, $(V_2, \|\cdot\|_2)$ be real normed vector spaces and $Q$ be a subset of $V_1$. For a continuous bounded function $f : Q \to V_2$ we define its (first) modulus of continuity

$$\omega_1(f, h) := \sup\{\|f(x) - f(y)\|_2 : \text{ all } x, y \in Q, \|x - y\|_1 \le h, h > 0\}. \quad (15.6)$$

Let $B$ be a real Banach space with a normalized basis $(e_k)_{k \in \mathbf{N}}$, and norm $\|\cdot\|_B$, $(X_i)_{i \in \mathbf{N}}$ be a sequence of $B$-valued integrable random variables (r.v.s) defined on a common probability space $(\underline{0}, \mathcal{A}, P)$, and let $(\mathcal{F}_i)_{i \in \mathbf{Z}_+}$ be an increasing sequence of sub-$\sigma$-algebras of $\mathcal{A}$ so that $X_i$ is $\mathcal{F}_i$-measurable $\forall i \in \mathbf{N}$. Then $(X_i, \mathcal{F}_i)_{i \in \mathbf{Z}_+}$, $X_0 = 0$ is called a *martingale difference sequence* (MDS) if

$$E(X_i \mid \mathcal{F}_{i-1}) = 0 \quad \text{a.s. } (i \in \mathbf{N}). \quad (15.7)$$

This implies $EX_i = 0$. Let $S_n := \sum_{i=1}^n X_i$; the above property is equivalent to $(S_n, \mathcal{F}_n)_{n \in \mathbf{Z}_+}$ being a martingale, that is

$$E(S_n \mid \mathcal{F}_{n-1}) = S_{n-1} \quad \text{a.s. } (n \in \mathbf{N}). \quad (15.8)$$

Because $B$ is a real Banach space with countable basis $(e_k)_{k \in \mathbf{N}}$, for each $x \in B$ there exists a unique sequence of real numbers $(d_k)_{k \in \mathbf{N}}$ such that

$$x = \sum_{k=1}^\infty d_k e_k. \quad (15.9)$$

This defines the sequence of coefficient functionals $(e_k^*)_{k \in \mathbf{N}}$ associated with the basis $(e_k)_{k \in \mathbf{N}}$, defined by $e_k^*(x) := d_k$, $k \in \mathbf{N}$, so that $e_k^* \in B^*$. These spaces $B$ are separable, e.g., $L_p[0,1]$, $l^p$, $1 \le p < \infty$, $C[0,1]$, $c_0$.

Let $B^j := B \times \cdots \times B$ be the $j$-fold product endowed with the max-norm $\|c\|_{B^j} := \max_{1 \le k \le j} \|x_k\|_B$, where $c := (x_1, \ldots, x_j) \in B^j$. Then the space $\mathcal{L}_j := \mathcal{L}_j(B^j, \mathbf{R})$ of all real-valued multilinear continuous functions $g : B^j \to \mathbf{R}$ is a Banach space with norm

$$\|g\|_{\mathcal{L}_j} := \sup_{\|c\|_{B^j} = 1} |g(c)| = \sup_{\substack{c \in B^j \\ x_k \ne 0}} \frac{|g(x)|}{\|x_1\|_B \cdots \|x_j\|_B}.$$

Let $f : B \to \mathbf{R}$ be a function with $\|f\|_\infty := \sup_{x \in B} |f(x)|$, whose Fréchet derivatives $f^{(j)} : B \to \mathcal{L}_j$ exist and are continuous for $1 \le j \le r$, $r \in \mathbf{N}$,
Then we get Taylor's formula

$$f(x+y) = f(y) + \sum_{j=1}^r \frac{f^{(j)}(y)[x]^j}{j!} + \frac{1}{(r-1)!} \cdot \int_0^1 (1-t)^{r-1}$$

## 15. Quantitative Probability Limit Theorems

$$\cdot \{f^{(r)}(y+tx)[x]^r - f^{(r)}(y)[x]^r\} \cdot dt, \tag{15.10}$$

where $x, y \in B$ and $[x]^j := (x, \ldots, x) \in B^j$. Moreover, by multilinearity and continuity of $f^{(j)}$ and (15.9), one has for a $j$-times continuously differentiable function $f$ that

$$f^{(j)}(y)(x,\ldots,x) = \sum_{v_1=1,\ldots,v_j=1}^{\infty} e_{v_1}^*(x) \cdots e_{v_j}^*(x) \cdot f^{(j)}(y)(e_{v_1},\ldots,e_{v_j}), \tag{15.11}$$

where $v_k \in \mathbf{N}$, $1 \le k \le j$, $y \in B$.

To shorten (15.11), we introduce the following notations for $v = (v_1, \ldots, v_j) \in \mathbf{N}^j$:

$$|v| := j, \quad x^v := \prod_{k=1}^{j} e_{v_k}^*(x), \quad f^{[v]}(\cdot) := f^{(j)}(\cdot)(e_{v_1},\ldots,e_{v_j}) : B \to \mathbf{R}. \tag{15.12}$$

Hence (15.11) is rewritten as

$$f^{(j)}(y)([x]^j) = \sum_{|v|=j} x^v \cdot f^{[v]}(y), \quad (x, y \in B). \tag{15.13}$$

We use the following families of functions ($r \in \mathbf{N}$ : $C_B^0 = C_B := \{f : B \to \mathbf{R};\ f$ uniformly continuous and bounded on $\mathbf{R}\}$, $C_B(\mathcal{L}_r) := \{g : B \to \mathcal{L}_r;\ g$ uniformly continuous and bounded on $\mathcal{L}_r\}$,

$$C_B^r := \{f \in C_B; f^{(j)} \in C_B(\mathcal{L}_j), 1 \le j \le r\}.$$

Smoothness of $f : B \to \mathbf{R}$ is estimated by ($r \in \mathbf{N}$)

$$\omega_1(f^{(r-1)}, h) := \sup_{\substack{x,y \in B \\ \|x-y\|_B \le h}} \|f^{(r-1)}(x) - f^{(r-1)}(y)\|_{\mathcal{L}_{r-1}}, \quad h > 0. \tag{15.14}$$

We suppose that

$$\omega_1(f^{(r-1)}, h) \le w, \quad w > 0. \tag{15.15}$$

Let $x_0 \in B$ be such that $f^{(r-1)}(x_0) = 0$. Then from Lemma 15.1 we obtain

$$\|f^{(r-1)}(x)\|_{\mathcal{L}_{r-1}} \le \omega_1(f^{(r-1)}, h) \cdot \left\lceil \frac{\|x - x_0\|_B}{h} \right\rceil, \quad \forall x \in B. \tag{15.16}$$

For an arbitrary probability space $(\underline{0}, \mathcal{A}, P)$, let us consider a $B$-valued r.v. $Z : \underline{0} \to B$; $B$ is endowed with the Borel $\sigma$-algebra $\mathcal{B}_B$. $Z$ has distribution $P_Z$ on $\mathcal{B}_B$ defined by $P_Z(B) := P(\{\omega \in \underline{0}: Z(\omega) \in B\})$ for any $B \in \mathcal{B}_B$, and expectation $E(Z) := \int_{\underline{0}} Z(\omega) P(d\omega)$, which is a Bochner integral.

For the main result in §15.1 we use the next lemma due to Butzer et al. [55].

**LEMMA 15.3**
Let $X, Y$ be two $B$-valued r.v.s with $E(\|X\|_B^j) < \infty$ for some $j \in \mathbf{N}$, and $X^v = \prod_{k=1}^{j} e_{v_k}^*(X)$. Then $f \in C_B^j$ implies that

$$E(f^{(j)}(Y)[X]^j) = \sum_{|v|=j} E(X^v \cdot f^{[v]}(Y)). \qquad (15.17)$$

**DEFINITION 15.2** Let $\varphi : \mathbf{N} \to \mathbf{R}^+$ such that $\varphi(n) \to 0$ as $n \to \infty$. The $B$-valued r.v. $Z$ is said to be $\varphi$-decomposable, if for each $n \in \mathbf{N}$ there exist $n$ independent r.v. $Z_i$, $1 \leq i \leq n$, not depending on $n$, such that

$$P_Z = P_{\varphi(n) \cdot \sum_{i=1}^n Z_i}. \qquad (15.18)$$

For a comparison of the concepts of $\varphi$-decomposability and infinite divisibility, see Butzer et al. [55]. One can easily observe that

$$E(X_i \mid \mathcal{F}_{i-1}) = \sum_{k=1}^{\infty} E(e_k^*(X_i) \mid \mathcal{F}_{i-1}) \cdot e_k \quad a.s.$$

Also $(X_i, \mathcal{F}_i)_{i \in \mathbf{Z}_+}$ is an MDS, iff

$$E(e_k^*(X_i) \mid \mathcal{F}_{i-1}) = 0 \quad (k, i \in \mathbf{N}) \quad a.s. \qquad (15.19)$$

iff (we write $(X_i)^v \equiv X_i^v$)

$$E(X_i^v \mid \mathcal{F}_{i-1}) = 0 \quad (|v| = 1, i \in \mathbf{N}) \quad a.s. \qquad (15.20)$$

### 15.1.2 Main Theorems

We present the main result of §15.1.

**THEOREM 15.1**
Let $(X_i, \mathcal{F}_i)_{i \in \mathbf{Z}_+}$ be an MDS, $Z$ be a $\varphi$-decomposable r.v. with $E(Z) = 0$, and $E(\|Z\|_B^r) < \infty$, $r \in \mathbf{N} - \{1\}$. Suppose that $E(\|X_i\|_B^r) < \infty$ $\forall i \in \mathbf{N}$;

also suppose that

$$E(X_i^v \mid \mathcal{F}_{i-1}) = E(Z_i^v) \quad a.s. \ (1 \le |v| \le r-1, i \in \mathbf{N}). \tag{15.21}$$

Let $f \in C_B^{r-1}$. Then it holds

$$|E(f(\varphi(n) \cdot S_n)) - E(f(Z))| \le \omega_1(f^{(r-1)}, h) \cdot h^{r-1} \cdot \left\{ \frac{\varphi^{r-1}(n)}{r!} \right.$$

$$\left. + \frac{(\varphi(n))^{(r-1)^2/r}}{2 \cdot (r-1)!} + \frac{(\varphi(n))^{((r^2-3r+2)/r)}}{8 \cdot (r-2)!} \right\}, \tag{15.22}$$

where

$$h := \left( \varphi(n) \cdot E\left( \left( \sum_{i=1}^n (\|X_i\|_B + \|Z_i\|_B) \right)^r \right) \right)^{1/r}. \tag{15.23}$$

We proceed as in the proof of Theorem 1 in Butzer et al. [55].

**PROOF** Notice that $f(\varphi(n) \cdot S_n)$ and $f(Z)$ are real integrable r.v.s for any $f \in C_B$. Let $Z_i$ be independent r.v.s chosen independently of $\mathcal{F}_i$ such that (15.18) is fulfilled. Setting

$$R_{n,i} := \sum_{k=1}^{i-1} X_k + \sum_{k=i+1}^{n} Z_k, \quad 1 \le i \le n, \ n \in \mathbf{N},$$

for $f \in C_B^{r-1}$, a double application of Taylor's formula (15.10) gives

$$f(\varphi(n) \cdot S_n) - f\left( \varphi(n) \cdot \sum_{i=1}^n Z_i \right)$$

$$= \sum_{i=1}^n \sum_{j=1}^{r-1} [f^{(j)}(\varphi(n) \cdot R_{n,i})[\varphi(n) \cdot X_i]^j$$

$$- f^{(j)}(\varphi(n) \cdot R_{n,i})[\varphi(n) \cdot Z_i]^j] + \mathcal{R}, \tag{15.24}$$

where

$$\mathcal{R} := I_1 - I_2, \tag{15.25}$$

with
$$I_1 := \sum_{i=1}^{n} I_{1i}, \quad I_2 := \sum_{i=1}^{n} I_{2i}. \tag{15.26}$$

Here
$$I_{1i} := \frac{1}{(r-2)!} \cdot \int_0^1 (1-t)^{r-2} \cdot [f^{(r-1)}(\varphi(n) \cdot R_{n,i}$$
$$+ t \cdot \varphi(n) \cdot X_i)[\varphi(n) \cdot X_i]^{r-1}$$
$$- f^{(r-1)}(\varphi(n) \cdot R_{n,i})[\varphi(n) \cdot X_i]^{r-1}] \cdot dt, \tag{15.27}$$

and
$$I_{2i} := \frac{1}{(r-2)!} \cdot \int_0^1 (1-t)^{r-2} \cdot [f^{(r-1)}(\varphi(n) \cdot R_{n,i}$$
$$+ t \cdot \varphi(n) \cdot Z_i)[\varphi(n) \cdot Z_i]^{r-1}$$
$$- f^{(r-1)}(\varphi(n) \cdot R_{n,i})[\varphi(n) \cdot Z_i]^{r-1}] \cdot dt.$$

We see the following
$$|[f^{(r-1)}(\varphi(n) \cdot R_{n,i} + t \cdot \varphi(n) \cdot X_i)[\varphi(n) \cdot X_i]^{r-1}$$
$$- f^{(r-1)}(\varphi(n) \cdot R_{n,i})][\varphi(n) \cdot X_i]^{r-1}|$$
$$\leq \|f^{(r-1)}(\varphi(n) \cdot R_{n,i} + t \cdot \varphi(n) \cdot X_i)$$
$$- f^{(r-1)}(\varphi(n) \cdot R_{n,i})\|_{\mathcal{L}_{r-1}} \cdot \|\varphi(n) \cdot X_i\|_B^{r-1}$$
$$\leq \omega_1(f^{(r-1)}, h) \cdot \left\lceil \frac{t \cdot \varphi(n) \cdot \|X_i\|_B}{h} \right\rceil \cdot (\varphi(n))^{r-1} \cdot \|X_i\|_B^{r-1}$$

by Lemma 15.1 and (15.16). That is,
$$|[f^{(r-1)}(\varphi(n) \cdot R_{n,i} + t \cdot \varphi(n) \cdot X_i)[\varphi(n) \cdot X_i]^{r-1}$$
$$- f^{(r-1)}(\varphi(n) \cdot R_{n,i})[\varphi(n) \cdot X_i]^{r-1}| \tag{15.28}$$

## 15. Quantitative Probability Limit Theorems

$$\leq (\varphi(n))^{r-1} \cdot \|X_i\|_B^{r-1} \cdot \omega_1(f^{(r-1)}, h) \cdot \left\lceil \frac{t \cdot \varphi(n) \cdot \|X_i\|_B}{h} \right\rceil, \quad h > 0.$$

Similarly, we get that

$$|[f^{(r-1)}(\varphi(n) \cdot R_{n,i} + t \cdot \varphi(n) \cdot Z_i)[\varphi(n) \cdot Z_i]^{r-1}$$

$$- f^{(r-1)}(\varphi(n) \cdot R_{n,i})[\varphi(n) \cdot Z_i]^{r-1}| \tag{15.29}$$

$$\leq (\varphi(n))^{r-1} \cdot \|Z_i\|_B^{r-1} \cdot \omega_1(f^{(r-1)}, h) \cdot \left\lceil \frac{t \cdot \varphi(n) \cdot \|Z_i\|_B}{h} \right\rceil, \quad h > 0.$$

Thus, from (15.28) we obtain that

$$|I_{1i}| \leq \frac{1}{(r-2)!} \cdot \int_0^1 (1-t)^{r-2} \cdot |[f^{(r-1)}(\varphi(n)$$

$$\cdot R_{n,i} + t \cdot \varphi(n) \cdot X_i)[\varphi(n) \cdot X_i]^{r-1}$$

$$- f^{(r-1)}(\varphi(n) \cdot R_{n,i})[\varphi(n) \cdot X_i]^{r-1}]| \cdot dt$$

$$\leq \omega_1(f^{(r-1)}, h) \cdot \left\{ (\varphi(n))^{r-1} \cdot \|X_i\|_B^{r-1} \right.$$

$$\left. \cdot \int_0^1 \frac{(1-t)^{r-2}}{(r-2)!} \cdot \left\lceil \frac{t \cdot \varphi(n) \cdot \|X_i\|_B}{h} \right\rceil \cdot dt \right\}$$

$$= \omega_1(f^{(r-1)}, h) \cdot \phi_{r-1}(\varphi(n) \cdot \|X_i\|_B);$$

the last is valid by change of variable on $\phi_{r-1}$, see (15.4).

We have established that

$$|I_{1i}| \leq \omega_1(f^{(r-1)}, h) \cdot \phi_{r-1}(\varphi(n) \cdot \|X_i\|_B), \quad i = 1, \ldots, n. \tag{15.30}$$

Similarly it is proved that

$$|I_{2i}| \leq \omega_1(f^{(r-1)}, h) \cdot \phi_{r-1}(\varphi(n) \cdot \|Z_i\|_B), \quad i = 1, \ldots, n. \tag{15.31}$$

Hence

$$|\mathcal{R}| \le |I_1| + |I_2| \le \sum_{i=1}^{n} |I_{1i}| + \sum_{i=1}^{n} |I_{2i}|$$

$$\le \sum_{i=1}^{n} \omega_1(f^{(r-1)}, h) \cdot \phi_{r-1}(\varphi(n) \cdot \|X_i\|_B)$$

$$+ \sum_{i=1}^{n} \omega_1(f^{(r-1)}, h) \cdot \phi_{r-1}(\varphi(n) \cdot \|Z_i\|_B)$$

$$= \omega_1(f^{(r-1)}, h) \cdot \left[ \sum_{i=1}^{n} (\phi_{r-1}(\varphi(n) \cdot \|X_i\|_B) + \phi_{r-1}(\varphi(n) \cdot \|Z_i\|_B)) \right]$$

$$\le \omega_1(f^{(r-1)}, h) \cdot \phi_{r-1}\left(\varphi(n) \cdot \sum_{i=1}^{n} (\|X_i\|_B + \|Z_i\|_B)\right),$$

the last being true by superadditivity of $\phi_{r-1}$, see Lemma 15.2, (15.4).

We have proved ($r \ge 2$)

$$|\mathcal{R}| \le \omega_1(f^{(r-1)}, h) \cdot \phi_{r-1}\left(\varphi(n) \cdot \sum_{i=1}^{n} (\|X_i\|_B + \|Z_i\|_B)\right). \quad (15.32)$$

Notice that by $\varphi$-decomposability of $Z$ we get that $E(\|Z_i\|_B^r) < \infty$. Also by $\varphi$-decomposability of $Z$, see (15.18), we have that

$$E(f(Z)) = E\left(f\left(\varphi(n) \cdot \sum_{i=1}^{n} Z_i\right)\right). \quad (15.33)$$

From Butzer et al. [55], and assumptions of this theorem, we see that

$$E[X_i^v \cdot f^{[v]}(\varphi(n) \cdot R_{n,i})] = E[Z_i^v \cdot f^{[v]}(\varphi(n) \cdot R_{n,i})]$$

for $1 \le i \le n$, $n \in \mathbf{N}$, $1 \le j \le r-1$, and $|v| = j$, and from Lemma 15.3 we find

$$E(f^{(j)}(\varphi(n) \cdot R_{n,i})[\varphi(n) \cdot X_i]^j) = E(f^{(j)}(\varphi(n) \cdot R_{n,i})[\varphi(n) \cdot Z_i]^j). \quad (15.34)$$

## 15. Quantitative Probability Limit Theorems

Integrating (15.24) against the probability measure $P$ and taking into account (15.33) and (15.34) we obtain that

$$|E(f(\varphi(n) \cdot S_n)) - E(f(Z))| = |E(\mathcal{R})| \leq E(|\mathcal{R}|). \tag{15.35}$$

From (15.32) we find $(r \geq 2)$

$$|E(f(\varphi(n) \cdot S_n)) - E(f(Z))| \leq \omega_1(f^{(r-1)}, h)$$

$$\cdot E\left(\phi_{r-1}\left(\varphi(n) \cdot \sum_{i=1}^{n}(\|X_i\|_B + \|Z_i\|_B)\right)\right), \quad h > 0. \tag{15.36}$$

From inequality (15.5) we get for the right-hand side of (15.36), $(r \geq 2)$ that

$$\text{R.H.S. (15.36)} \leq \omega_1(f^{(r-1)}, h) \cdot E\left\{\frac{\varphi^r(n) \cdot (\sum_{i=1}^{n}(\|X_i\|_B + \|Z_i\|_B))^r}{r! h}\right.$$

$$+ \frac{\varphi^{r-1}(n) \cdot (\sum_{i=1}^{n}(\|X_i\|_B + \|Z_i\|_B))^{r-1}}{2 \cdot (r-1)!}$$

$$\left. + \frac{h \cdot \varphi^{r-2}(n) \cdot (\sum_{i=1}^{n}(\|X_i\|_B + \|Z_i\|_B))^{r-2}}{8 \cdot (r-2)!}\right\}.$$

That is, from (15.36), Hölder's inequality, and linearity of $E$ we see that

$$|E(f(\varphi(n) \cdot S_n)) - E(f(Z))| \tag{15.37}$$

$$\leq \omega_1(f^{(r-1)}, h) \cdot \left\{\frac{\varphi^r(n)}{r! h} \cdot E\left(\left(\sum_{i=1}^{n}(\|X_i\|_B + \|Z_i\|_B)\right)^r\right)\right.$$

$$+ \frac{\varphi^{r-1}(n)}{2 \cdot (r-1)!} \cdot \left(E\left(\left(\sum_{i=1}^{n}(\|X_i\|_B + \|Z_i\|_B)\right)^r\right)\right)^{((r-1)/r)}$$

$$\left. + \frac{h \cdot \varphi^{r-2}(n)}{8 \cdot (r-2)!} \cdot \left(E\left(\left(\sum_{i=1}^{n}(\|X_i\|_B + \|Z_i\|_B)\right)^r\right)\right)^{((r-2)/r)}\right\}.$$

Picking $h$ as in (15.23) and noticing that $h < \infty$, from (15.37) we are able to establish (15.22). ∎

Theorem 15.1 includes the independent case, because a sequence of independent r.v.s $X_i$ such that $E(X_i) = 0$ forms an MDS.

An especially important case comes next.

**COROLLARY 15.1**
*Under the assumptions of Theorem 15.1 when $r = 2$ we get*

$$|E(f(\varphi(n) \cdot S_n)) - E(f(Z))| \le \omega_1(f', h) \cdot \frac{h}{2} \cdot \left\{\varphi(n) + \sqrt{\varphi(n)} + \frac{1}{4}\right\}, \quad (15.38)$$

*where*

$$h = \left(\varphi(n) \cdot E\left(\left(\sum_{i=1}^n (\|X_i\|_B + \|Z_i\|_B)\right)^2\right)\right)^{1/2}. \quad (15.39)$$

Next we give the weak law of large numbers for MDS on $B$-spaces with rates when $\|(\|x\|_B)'\|_{\mathcal{L}_1} \le A$, $x \in B - \{0\}$.

**THEOREM 15.2**
*Let $(X_i, \mathcal{F}_i)_{i \in \mathbf{Z}_+}$ be an MDS, and $E(\|X_i\|_B^2) < \infty$, $i \in \mathbf{N}$. Let $f \in C_B^1$. Then it holds*

$$|E(f(\varphi(n) \cdot S_n)) - f(0)| \le \omega_1(f', h) \cdot \frac{h}{2} \cdot \left\{\varphi(n) + \sqrt{\varphi(n)} + \frac{1}{4}\right\}, \quad (15.40)$$

*where*

$$h = \left(\varphi(n) \cdot E\left(\left(\sum_{i=1}^n \|X_i\|_B\right)^2\right)\right)^{1/2}. \quad (15.41)$$

**PROOF** We apply Corollary 15.1. Again we proceed as in the proof of Theorem 5 of Butzer et al. [55]. Here $Z$ is the degenerate Gaussian limiting r.v. and $Z_i$, $1 \le i \le n$ are independent r.v.s distributed as $Z$. Notice $EZ = 0$. When $r = 2$, equality (15.21) is always true. Also here $E(\|Z\|_B^s) = 0$ for all $s > 0$. Quantity $h$ comes from (15.39) and the Cauchy–Schwarz's inequality. ∎

## 15. Quantitative Probability Limit Theorems

If $X$ is a $B$-valued r.v. with $E(\|X\|_B^2) < \infty$ and $E(X) = 0$, the covariance functional of $X$ is given by $R_X(f^*, g^*) := E(f^*(X) \cdot g^*(X))$, $f^*, g^* \in B^*$. Let $X_R$ be the uniquely determined Gaussian r.v. with mean zero and covariance functional $R = R_{X_R}$. Because we are in a separable $B$-space, it is known that for $s \geq 0$ $E(\|X_R\|_B^s) < \infty$.

In the following we give the central limit theorem for martingales in Banach spaces with a countable basis.

### THEOREM 15.3

Let $(X_i, \mathcal{F}_i)_{i \in \mathbb{Z}_+}$ be an MDS, $R$ be the covariance functional of any mean zero Gaussian r.v., and $r \in \mathbb{N} - \{1\}$. Assume $E(\|X_i\|_B^r) < \infty$, $i \in \mathbb{N}$, and suppose that there exists a sequence of $a_i > 0$ such that

$$E(X_i^v \mid \mathcal{F}_{i-1}) = a_i^{|v|} \cdot E(X_R^v) \tag{15.42}$$

a.s. $(1 \leq |v| \leq r-1,\ i \in \mathbb{N})$. Let $f \in C_B^{r-1}$. Put

$$A_n := \left(\sum_{i=1}^n a_i^2\right)^{1/2}. \tag{15.43}$$

Then it holds

$$|E(f(A_n^{-1} \cdot S_n)) - E(f(X_R))| \leq \omega_1(f^{(r-1)}, h) \cdot h^{r-1}$$

$$\cdot \left\{\frac{A_n^{1-r}}{r!} + \frac{A_n^{-(r-1)^2/r}}{2 \cdot (r-1)!} + \frac{A_n^{-((r^2-3r+2)/r)}}{8 \cdot (r-2)!}\right\}, \tag{15.44}$$

where

$$h = \left(A_n^{-1} \cdot E\left(\left(\sum_{i=1}^n (\|X_i\|_B + a_i \cdot \|X_R\|_B)\right)^r\right)\right)^{1/r}. \tag{15.45}$$

**PROOF**  By Theorem 15.1 and as in Butzer et al. [55]; proof of their Theorem 3. Here we have $\varphi(n) = A_n^{-1}$, $Z = X_R$, $Z_i = a_i X_R$. ∎

## 15.2 On Weak Invariance Principle

### 15.2.1 Background

Here we study the *weak invariance principle*, that is, the convergence of the distributions of the random polygonal lines

$$S_n(\omega, t) := n^{-1/2} \cdot \left( \sum_{i=0}^{[nt]} X_i(\omega) + (nt - [nt]) \cdot X_{[nt]+1}(\omega) \right) \quad (15.46)$$

to the Wiener-measure $W_R$ on $C = C([0,1])$, the space of all real-valued, continuous functions on $[0,1]$ with the supremum-norm ($[a]$ denotes the largest integer $\leq a$). Here the random variables (r.v.s) $X_i$ need not be independent.

The expectation of a $C([0,1])$-valued random function (r.f.) $X : \underline{0} \to C$ on an arbitrary probability space $(\underline{0}, \mathcal{A}, P)$ is given by the Bochner integral

$$E(X) = \int_{\underline{0}} X(\omega) P(d\omega)$$

and, if $E(\|X\|_C^2) < \infty$, the covariance functional $R_X : C^* \times C^* \to \mathbf{R}$ by

$$R_X(f^*, g^*) := E(f^*(X) \cdot g^*(X)), \quad \forall f^*, g^* \in C^*.$$

Here $C^*$ denotes the topological dual of $C$. A stochastic process

$$\{W(t)\}_{0 \leq t \leq 1} = \{W(\omega, t)\}_{0 \leq t \leq 1}$$

is called a Wiener process or Brownian motion process (on $[0,1]$) given that (i) $W(\omega, 0) = 0$, $\forall \omega \in \underline{0}$, (ii) $W(\omega, \cdot)$ is a continuous function on $[0,1]$ $\forall \omega \in \underline{0}$, (iii) for $0 \leq t_1 < t_2 < \cdots < t_n = 1$ the differences $W(t_1)$, $W(t_2) - W(t_1), \ldots, W(t_n) - W(t_{n-1})$ are independent, Gaussian distributed r.v.s with expectation 0 and variance 1. $W(\omega, t)$ is a function of $\omega \in \underline{0}$ and $t \in [0,1]$, such that $W(\cdot, t)$ is $\mathcal{A}$-measurable function $\forall t \in [0,1]$ and $W(\omega, \cdot)$ belongs to $C$ $\forall \omega \in \underline{0}$. Thus the Wiener process is a r.f. taking values in $C$. Its distribution is called the Wiener-measure $W_R(t)$.

We use the next three lemmas. These are mentioned also in Butzer and Kirschfink [56].

## LEMMA 15.4

Let $X_0 = 0$, $X_1, X_2, \ldots$ be a sequence of identically distributed r.v.s such that $E(X_1) = 0$, $\text{Var}(X_1) = 1$. $S_n(\omega, t)$ is as in (15.46) $\forall n \in \mathbf{N}$, $\omega \in \underline{0}$, $t \in [0, 1]$. Consider the trajectories

$$X_{ni}(\omega, t) := \begin{cases} 0, & 0 < nt \leq i - 1 \\ n^{1/2} \cdot \left(t - \frac{i-1}{n}\right) \cdot X_i(\omega), & i - 1 < nt \leq i \\ n^{-1/2} \cdot X_i(\omega), & i < nt \leq n. \end{cases} \quad (15.47)$$

Then $(X_{ni})_{1 \leq i \leq n}$, $n \in \mathbf{N}$ is an array of r.f.s with values in $C$ such that

(i) $\quad S_n(\omega, t) = \sum_{i=1}^{n} X_{ni}(\omega, t),$ \hfill (15.48)

(ii) $\quad \|X_{ni}(\omega)\|_C = n^{-1/2} \cdot |X_i(\omega)|,$ \hfill (15.49)

where $\|\cdot\|_C$ is the supremum norm with respect to $t \in [0, 1]$.

**PROOF**   The proof of (15.48) follows from Giné [87]. ■

## LEMMA 15.5

Let $W(t)$ be a r.f. with Wiener-measure $W_R(t)$ as its distribution on $[0, 1]$, which has covariance functional $R$. If $(a_{ni})_{1 \leq i \leq n}$, $n \in \mathbf{N}$ is a triangular array of positive real numbers, and

$$A_n := \left(\sum_{i=1}^{n} a_{ni}^2\right)^{1/2}, \quad (15.50)$$

then there exist independent, Gaussian distributed r.f.s $W_{ni}$ given from $P_{W_{ni}} := P_{a_{ni} \cdot W}$ such that

$$P_{A_n^{-1} \cdot \sum_{i=1}^{n} W_{ni}(t)} = P_{W(t)} = W_R(t). \quad (15.51)$$

The proof is given in Butzer and Schulz [57].

The Feller-type condition on a triangular array of positive real numbers $(a_{ni})_{1 \leq i \leq n}$, $n \in \mathbf{N}$ is described by

$$\lim_{n \to \infty} \max_{1 \leq i \leq n} \frac{a_{ni}}{A_n} = 0, \quad (15.52)$$

where $A_n$ is given by (15.50).

Let $C^r$, $r \in \mathbf{N}$ the $r$-fold product space $C \times C \times \cdots \times C$ with norm

$$\|X\|_{C^r} := \max_{1 \le i \le r} \|X_i\|_C, \quad \text{where } X := (X_1, \ldots, X_r) \in C^r.$$

The space $L_r := L_r(C^r, \mathbf{R})$ of all real-valued, multilinear, continuous functions $g : C^r \to \mathbf{R}$ is a Banach space with norm

$$\|g\|_{L_r} := \sup_{\|X\|_{C^r}=1} |g(X)|.$$

We study the following classes of functions:

$$C_C := C_{C([0,1])} := \{g : C \to \mathbf{R}; g \text{ continuous and bounded on } \mathbf{R}\},$$

$$C_C(L_r) := \{f : C \to L_r; f \text{ continuous and bounded on } L_r\}$$

$$C_C^r := \{f \in C_C; \text{ Fréchet derivative } f^{(j)} \in C_C(L_j), 1 \le j \le r\}, \quad (15.53)$$

$$\overline{C}_C^r := \{f \in C_C^r; f^{(r)}$$

$$\text{is uniformly continuous and bounded on } L_r\}. \quad (15.54)$$

For $f \in C_C^r$, $\|f\|_\infty := \sup_{X \in C} |f(X)|$ may be infinite.

Let $(e_k)_{k \in \mathbf{N}}$ be a normalized countable basis for $C$. Then each $X \in C$ has a unique representation $X = \sum_{k=1}^\infty X^{(\nu)} \cdot e_k$, where $X^{(\nu)}$ are real components with respect to the given basis. Since $f^{(j)}(Y)$ are multilinear and continuous, we obtain

$$f^{(j)}(Y)[X]^j = \sum_{\nu_1=1,\ldots,\nu_j=1}^\infty X^{(\nu_1)} \cdot X^{(\nu_j)} \cdot f^{(j)}(Y)(e_{\nu_1}, \ldots, e_{\nu_j}), \quad (15.55)$$

where $\nu_k \in \mathbf{N}$, $1 \le k \le j$, $[X]^j = (X, \ldots, X) \in C^j$. To shorten (15.55) further, for every $j$-tuple $\nu = (\nu_1, \ldots, \nu_j) \in \mathbf{N}^j$ we put

$$|\nu| := j, X^\nu := \prod_{k=1}^i X^{(\nu_k)}, f^{[\nu]}(\cdot) := f^{[j]}(\cdot)(e_{\nu_1}, \ldots, e_{\nu_j}) : C \to \mathbf{R}. \quad (15.56)$$

Thus
$$f^{(j)}(Y)[X]^j = \sum_{|\nu|=j} X^\nu \cdot f^{[\nu]}(Y), \quad (X, Y \in C). \quad (15.57)$$

We use also Taylor's formula for $X, Y \in C([0,1])$, $f \in C_C^r$, $r \in \mathbf{N}$

$$f(X+Y) = f(Y) + \sum_{j=1}^{r} \frac{f^{(j)}(Y)}{j!}[X]^j + \frac{1}{(r-1)!} \cdot \int_0^1 (1-t)^{r-1}$$

$$\cdot \{f^{(r)}(Y + t \cdot X)[X]^r - f^{(r)}(Y)[X]^r\} dt. \qquad (15.58)$$

Smoothness of $C \to \mathbf{R}$ is measured by ($r \in \mathbf{N}$)

$$\omega_1(f^{(r-1)}, h) := \sup_{\substack{X,Y \in C \\ \|X-Y\|_C \leq h}} \|f^{(r-1)}(X) - f^{(r-1)}(Y)\|_{L_{r-1}}, \qquad (15.59)$$

where $h > 0$.

We suppose that

$$\omega_1(f^{(r-1)}, h) \leq w, \quad w > 0. \qquad (15.60)$$

Let $X_0 \in C$ be such that $f^{(r-1)}(X_0) = 0$. From Lemma 15.1, we get

$$\|f^{(r-1)}(X)\|_{L_{r-1}} \leq \omega_1(f^{(r-1)}, h) \cdot \left\lceil \frac{\|X - X_0\|_C}{h} \right\rceil, \quad \forall X \in C. \qquad (15.61)$$

**LEMMA 15.6**
Let $X, Y$ be r.f.s with values in $C$ such that $E(\|X\|_C^j) < \infty$ for $j \in \mathbf{N}$. Let $f \in C_C^j$. Then it holds

$$E(f^{(j)}(Y)[X]^j) = \sum_{|\nu|=j} E(X^\nu \cdot f^{[\nu]}(Y)). \qquad (15.62)$$

Here we mention that if $X$ is a normally distributed r.v. and $r > 0$, then $E(|X|^r) < \infty$. Also, if $W$ is a Brownian motion and

$$\|W\|_C := \|W(\omega)\|_C := \sup_{t \in [0,1]} |W(t, \omega)|, \quad r > 0,$$

then it is known that

$$l_r := E(\|W\|_C^r) < \infty. \qquad (15.63)$$

### 15.2.2 Central Results

The following result deals with the rate of approximation of the random polygonal lines $S_n$, see (15.46) to the Weiner-measure $W_R$. The next invariance principle is considered to be a central limit theorem (CLT) for $C$-valued r.f.s $X_{n,i}(\omega, t)$, $1 \leq i \leq n$, $n \in \mathbf{N}$ defined in Lemma 15.4, (15.47), so many people refer to the invariance principle as a functional CLT.

**THEOREM 15.4**
Let $(X_i)_{i \in \mathbf{Z}_+}$ be a sequence of identically distributed real r.v.s with mean 0 and variance 1, and let $(a_{n_i})_{1 \leq i \leq n} > 0$, $n \in \mathbf{N}$ and $A_n := (\sum_{i=1}^n a_{ni}^2)^{1/2}$. Suppose that
$$\zeta_r := E(|X_1|^r) < \infty \tag{15.64}$$
for $r \in \mathbf{N} - \{1\}$, and for the r.f.s $(X_{ni})_{1 \leq i \leq n}$, $n \in \mathbf{N}$ defined in (15.47) we assume
$$E(X_{ni}^\nu \mid \mathcal{A}_{ni}) = a_{ni}^j \cdot A_n^{-j} \cdot E(W^\nu) \quad a.s. \tag{15.65}$$
for all $|\nu| = j$, $1 \leq j \leq r - 1$, $1 \leq i \leq n$, and $n \in \mathbf{N}$. Here
$$\mathcal{A}_{ni} := \mathcal{A}(X_1, \ldots, X_{i-1}, W_{n,i+1}, \ldots, W_{nn}) \tag{15.66}$$
(the sub-$\sigma$-algebra generated by $\{X_1, \ldots, X_{i-1}, W_{n,i+1}, \ldots, W_{nn}\}$). Let $f \in C_C^{r-1}$. Then $P_{S_n}$ converges weakly to the Wiener-measure $W_R$, with rates given from
$$|E(f(S_n(t))) - E(f(W(t)))| \leq \omega_1(f^{(r-1)}, h) \tag{15.67}$$
$$\cdot \left\{ \frac{1}{r!} + \left( \frac{n^{((3-r)/2)} \cdot \zeta_r^{((r-1)/r)} + A_n^{1-r} \cdot (\sum_{i=1}^n a_{ni}^{r-1}) \cdot l_r^{((r-1)/r)}}{2 \cdot (r-1)!} \right) \right.$$
$$\left. + \frac{h}{8(r-2)!} \cdot \left\{ n^{((4-r)/2)} \cdot \zeta_r^{((r-2)/r)} + A_n^{2-r} \left( \sum_{i=1}^n a_{ni}^{r-2} \right) \cdot l_r^{((r-2)/r)} \right\} \right\},$$
where
$$h := h_n := n^{(2-r)/2} \cdot \zeta_r + A_n^{-r} \cdot \left( \sum_{i=1}^n a_{ni}^r \right) \cdot l_r < \infty. \tag{15.68}$$

**PROOF** We proceed as in the proof of Theorem 1 of Butzer and Kirschfink [56]. Notice that $f(S_n)$ and $f(W_R)$ are real integrable r.v.s

## 15. Quantitative Probability Limit Theorems

for each $f \in C_C$. Consider the r.f.

$$R_{ni} := \sum_{k=1}^{i-1} X_{nk} + A_n^{-1} \cdot \sum_{k=i+1}^{n} W_{nk}, \quad (1 \le i \le n; n \in \mathbf{N}). \tag{15.69}$$

By Lemma 15.5, we have by summing up and from double application of Taylor's formula (15.58) the following equality:

$$f(S_n) - f(W) = \sum_{i=1}^{n} \sum_{j=1}^{r-1} \frac{1}{j!} \cdot \{f^{(j)}(R_{ni})[X_{ni}]^j$$

$$- f^{(j)}(R_{ni})[A_n^{-1} \cdot W_{ni}]^j\} + \mathcal{R}^*, \tag{15.70}$$

where

$$\mathcal{R}^* := \sum_{i=1}^{n} I_i, \tag{15.71}$$

with

$$I_i = \frac{1}{(r-2)!} \cdot \int_0^1 (1-t)^{r-2} \cdot \{\{f^{(r-1)}(R_{ni} + t \cdot X_{ni})[X_{ni}]^{r-1}$$

$$- f^{(r-1)}(R_{ni})[X_{ni}]^{r-1}\}$$

$$- \{f^{(r-1)}(R_{ni} + t \cdot A_n^{-1} \cdot W_{ni})[A_n^{-1} \cdot W_{ni}]^{r-1}$$

$$- f^{(r-1)}(R_{ni})[A_n^{-1} \cdot W_{ni}]^{r-1}\}\} \cdot dt. \tag{15.72}$$

We see the following

$$|I_1| \le \frac{1}{(r-2)!} \cdot \int_0^1 (1-t)^{r-2} \cdot |\{f^{(r-1)}(R_{ni} + t \cdot X_{ni})$$

$$- f^{(r-1)}(R_{ni})\}[X_{ni}]^{r-1}| \cdot dt$$

$$+ \frac{1}{(r-2)!} \cdot \int_0^1 (1-t)^{r-2} \cdot |\{f^{(r-1)}(R_{ni} + t \cdot A_n^{-1} \cdot W_{ni})$$

$$- f^{(r-1)}(R_{ni})\}[A_n^{-1} \cdot W_{ni}]^{r-1}|dt\}$$

$$\leq \left\{ \frac{1}{(r-2)!} \cdot \int_0^1 (1-t)^{r-2} \cdot \|f^{(r-1)}(R_{ni} + t \cdot X_{ni}) \right.$$

$$- f^{(r-1)}(R_{ni})\|_{L_{r-1}} \cdot \|X_{ni}\|_C^{r-1} \cdot dt$$

$$+ \frac{1}{(r-2)!} \cdot \int_0^1 (1-t)^{r-2} \cdot \|f^{(r-1)}(R_{ni} + t \cdot A_n^{-1} \cdot W_{ni})$$

$$\left. - f^{(r-1)}(R_{ni})\|_{L_{r-1}} \cdot \|A_n^{-1} \cdot W_{ni}\|_C^{r-1} dt \right\}$$

$$\leq \frac{1}{(r-2)!} \cdot \int_0^1 (1-t)^{r-2} \cdot \omega_1(f^{(r-1)}, h) \cdot \left\lceil \frac{t \cdot \|X_{ni}\|_C}{h} \right\rceil$$

$$\cdot \|X_{ni}\|_C^{r-1} \cdot dt + \frac{1}{(r-2)!} \cdot \int_0^1 (1-t)^{r-2} \cdot \omega_1(f^{(r-1)}, h)$$

$$\cdot \left\lceil \frac{t \cdot A_n^{-1} \cdot \|W_{ni}\|_C}{h} \right\rceil$$

$$\left. \cdot (A_n^{-1})^{r-1} \cdot \|W_{ni}\|_C^{r-1} \cdot dt \right\}$$

(the last inequality is valid by Lemma 15.1)

$$= \omega_1(f^{(r-1)}, h) \cdot \{\phi_{r-1}(\|X_{ni}\|_C) + \phi_{r-1}(A_n^{-1} \cdot \|W_{ni}\|_C))\},$$

by change of variables and the definition of $\phi_{r-1}$, see (15.4).

We have established that

$$|I_i| \leq \omega_1(f^{(r-1)}, h) \cdot \{\phi_{r-1}(\|X_{ni}\|_C)$$

$$+ \phi_{r-1}(A_n^{-1} \cdot \|W_{ni}\|_C))\}, \quad i = 1, \ldots, n. \tag{15.73}$$

Hence

$$|\mathcal{R}^*| \leq \omega_1(f^{(r-1)}, h) \cdot \left\{ \sum_{i=1}^n (\phi_{r-1}(\|X_{ni}\|_C) + \phi_{r-1}(A_n^{-1} \cdot \|W_{ni}\|_C)) \right\}. \tag{15.74}$$

## 15. Quantitative Probability Limit Theorems

By Butzer and Kirschfink [56], and Lemma 15.6, we have that

$$E\{f^{(j)}(R_{ni})[X_{ni}]^j - f^{(j)}(R_{ni})[A_n^{-1} \cdot W_{ni}]^j\} = 0, \tag{15.75}$$

for $1 \leq i \leq n$, $n \in \mathbf{N}$ and $1 \leq j \leq r-1$. Integrating (15.70) against the probability measure $P$ and taking into account (15.75) we get that

$$|E(f(S_n)) - E(f(W))| = |E(\mathcal{R}^*)| \leq E(|\mathcal{R}^*|). \tag{15.76}$$

From (15.74) now we find ($r \geq 2$)

$$|E(f(S_n(t))) - E(f(W(t)))|$$

$$\leq \omega_1(f^{(r-1)}, h) \cdot \left\{ \sum_{i=1}^n (E(\phi_{r-1}(\|X_{ni}\|_C)) \right.$$

$$\left. + E(\phi_{r-1}(A_n^{-1} \cdot \|W_{ni}\|_C))) \right\}, \quad h > 0. \tag{15.77}$$

By inequality (15.5) we have for the right-hand side of (15.77) ($r \geq 2$) that

$$\text{RHS}(15.77) \leq \omega_1(f^{(r-1)}, h)$$

$$\cdot \left\{ \sum_{i=1}^n \left[ E \left( \frac{\|X_{ni}\|_C^r}{r!h} + \frac{\|X_{ni}\|_C^{r-1}}{2 \cdot (r-1)!} + \frac{h \cdot \|X_{ni}\|_C^{r-2}}{8 \cdot (r-2)!} \right) \right. \right.$$

$$+ E \left( \frac{A_n^{-r} \cdot \|W_{ni}\|_C^r}{r!h} + \frac{A_n^{1-r} \cdot \|W_{ni}\|_C^{r-1}}{2 \cdot (r-1)!} \right.$$

$$\left. \left. \left. + \frac{h \cdot A_n^{2-r} \cdot \|W_{ni}\|_C^{r-2}}{8 \cdot (r-2)!} \right) \right] \right\}$$

(from Hölder's inequality and linearity of $E$)

$$\leq \omega_1(f^{(r-1)}, h) \cdot \left\{ \frac{\sum_{i=1}^n (E(\|X_{ni}\|_C^r) + A_n^{-r} \cdot E(\|W_{ni}\|_C^r))}{r!h} \right.$$

$$\left. + \frac{\sum_{i=1}^n ((E(\|X_{ni}\|_C^r))^{((r-1)/r)} + A_n^{1-r} \cdot (E(\|W_{ni}\|_C^r))^{((r-1)/r)})}{2 \cdot (r-1)!} \right.$$

$$+ \frac{h}{8 \cdot (r-2)!} \cdot \left[ \sum_{i=1}^{n} ((E(\|X_{ni}\|_C^r))^{((r-2)/r)} \right.$$

$$\left. \left. + A_n^{2-r} \cdot (E(\|W_{ni}\|_C^r))^{((r-2)/r)} \right) \right] \right\} =: J.$$

By (15.49) and (15.64) we have that

$$E(\|X_{ni}\|_C^r) = E(\|X_{ni}(\omega)\|_C^r) = \frac{E(|X_i(\omega)|^r)}{n^{r/2}} = \frac{E(|X_i|^r)}{n^{r/2}} < \infty,$$

hence

$$E(\|X_{ni}\|_C^r) = \frac{E(|X_1|^r)}{n^{r/2}} = n^{-r/2} \cdot \zeta_r.$$

That is,

$$E(\|X_{ni}\|_C^r) = n^{-r/2} \cdot \zeta_r. \tag{15.78}$$

Also from Lemma 15.5 and (15.63) we get that

$$E(\|W_{ni}\|_C^r) = E(\|a_{ni} \cdot W\|_C^r) = a_{ni}^r \cdot E(\|W\|_C^r) = a_{ni}^r \cdot l_r < \infty;$$

i.e.,

$$E(\|W_{ni}\|_C^r) = a_{ni}^r \cdot l_r. \tag{15.79}$$

Thus from (15.77)–(15.79) we find that

$$|E(f(S_n(t))) - E(f(W(t)))| \le J = \omega_1(f^{(r-1)}, h)$$

$$\cdot \left\{ \frac{\sum_{i=1}^{n} (n^{-r/2} \cdot \zeta_r + A_n^{-r} \cdot a_{ni}^r \cdot l_r)}{r! h} \right.$$

$$+ \frac{\sum_{i=1}^{n} (n^{(1-r)/2} \cdot \zeta_r^{((r-1)/r)} + A_n^{1-r} \cdot a_{ni}^{r-1} \cdot l_r^{((r-1)/r)})}{2 \cdot (r-1)!}$$

$$\left. + \frac{h}{8(r-2)!} \cdot \left[ \sum_{i=1}^{n} (n^{((2-r)/2)} \cdot \zeta_r^{((r-2)/r)} + A_n^{2-r} \cdot a_{ni}^{r-2} \cdot l_r^{((r-2)/r)}) \right] \right\}.$$

Consequently,

$$|E(f(S_n(t))) - E(f(W(t)))| \le \omega_1(f^{(r-1)}, h)$$

## 15. Quantitative Probability Limit Theorems

$$\cdot \left\{ \frac{(n^{((2-r)/2)} \cdot \zeta_r + A_n^{-r} \cdot (\sum_{i=1}^{n} a_{ni}^r) \cdot l_r)}{r! h} \right.$$

$$+ \frac{(n^{((3-r)/2)} \cdot \zeta_r^{((r-1)/r)} + A_n^{1-r}(\sum_{i=1}^{n} a_{ni}^{r-1}) \cdot l_r^{((r-1)/r)})}{2 \cdot (r-1)!}$$

$$+ \frac{h}{8 \cdot (r-2)!} \cdot \left[ n^{((4-r)/2)} \cdot \zeta_r^{((r-2)/r)} + A_n^{2-r} \right.$$

$$\left. \left. \cdot \left( \sum_{i=1}^{n} a_{ni}^{r-2} \right) \cdot l_r^{((r-2)/r)} \right] \right\}. \tag{15.80}$$

Picking $h$ as in (15.68), inequality (15.67) is proved by inequality (15.80). It is clear that $h < \infty$.  ∎

**REMARK 15.1**  The right-hand side of (15.67) is of practical interest only if it converges to zero as $n \to \infty$. For this we need to suppose $r \geq 5$,

$$\lim_{n \to \infty} \left( A_n^{-r} \cdot \left( \sum_{i=1}^{n} a_{ni}^r \right) \right) = 0,$$

and

$$A_n^{1-r} \cdot \left( \sum_{i=1}^{n} a_{ni}^{r-1} \right), \quad A_n^{2-r} \cdot \left( \sum_{i=1}^{n} a_{ni}^{r-2} \right)$$

are at least bounded as $n \to \infty$.

For example, one could assume that

$$\left( \frac{\max_{1 \leq i \leq n} a_{ni}}{A_n} \right) \leq c \cdot \frac{1}{\sqrt{n}}, \quad n \in \mathbf{N}, \tag{15.81}$$

where $c$ is a constant, compare with (15.52). Then

$$\left( \frac{\max_{1 \leq i \leq n} a_{ni}}{A_n} \right)^{r-j} \leq c^{r-j} \cdot \frac{1}{n^{(r-j)/2}}, \quad j = 0, 1, 2; \ r \geq 5,$$

and

$$\frac{\sum_{i=1}^{n} a_{ni}^{r-j}}{A_n^{r-j}} \leq \frac{n \cdot (\max_{1 \leq i \leq n} a_{ni})^{r-j}}{A_n^{r-j}}$$

$$\leq n \cdot c^{r-j} \cdot \frac{1}{n^{(r-j)/2}} = c^{r-j} \cdot \frac{1}{n^{((r-j)/2)-1}}.$$

Hence

$$\frac{\sum_{i=1}^{n} a_{ni}^{r-j}}{A_n^{r-j}} \leq c^{r-j} \cdot \frac{1}{n^{(r-j-2)/2}} \to 0, \; j = 0, 1, 2; \; r \geq 5, \text{ as } n \to \infty. \quad (15.82)$$

In this case as $n \to \infty$, $h = h_n \to 0$ (see (15.68)) and $\omega_1(f^{(r-1)}, h) \to 0$, and the quantity within the braces on the right side of (15.67) converges to $1/r!$. Thus (RHS) (15.67) converges to zero as $n \to \infty$. Consequently, from (15.67) we get that $E(f(S_n(t))) \to E(f(W(t)))$, as $n \to \infty$. ■

In Theorem 15.4 we made no assumptions for the dependence or independence of the r.v.s $X_i$. This was made possible by assumption (15.65), which also makes applicable Theorem 15.4, in particular to a CLT to be presented in the following. There (15.65) becomes more specific. For this we use

**DEFINITION 15.3** (Butzer and Kirschfink [56]). *Let $(X_i)_{i \in \mathbf{R}}$ a sequence of real r.v.s.*

(i) *This sequence is said to be "dependent from below" if for each $1 \leq i \leq n$ and $n \in \mathbf{N}$*

$$E[X_i \mid \mathcal{A}(X_1, \ldots, X_{i-1}, X_{i+1}, \ldots, X_n)]$$

$$= E[X_i \mid \mathcal{A}(X_1, \ldots, X_{i-1})] \quad \text{a.s.} \quad (15.83)$$

(ii) *This sequence is "dependent from above" if*

$$E[X_i \mid \mathcal{A}(X_1, \ldots, X_{i-1}, X_{i+1}, \ldots, X_n)]$$

$$= E[X_i \mid \mathcal{A}(X_{i+1}, \ldots, X_n)] \quad \text{a.s.} \quad (15.84)$$

Let $X^*$ be a standard normally distributed r.v.

**THEOREM 15.5**
*Let $(X_i)_{i \in \mathbf{N}}$ be a sequence of identically distributed, possibly dependent, real r.v.s for which $E(X_1) = 0$, $\mathrm{Var}(X_1) = 1$, and $\zeta_r := E(|X_1|^r) < \infty$, for $r \in \mathbf{N} - \{1\}$.*

## 15. Quantitative Probability Limit Theorems

(i) Suppose that $(X_i)_{i \in \mathbf{N}}$ is dependent from below and that

$$E(X_i^j \mid \mathcal{E}_{i-1}) = E(X^{*j}) \quad \text{a.s.} \quad (1 \leq j \leq r-1, i \in \mathbf{N}), \tag{15.85}$$

where

$$\mathcal{E}_{i-1} := \mathcal{A}(X_1, \ldots, X_{i-1}).$$

Let $f : \mathbf{R} \to \mathbf{R}$ for which $f^{(j)}$ is bounded and continuous on $\mathbf{R}$ for all $0 \leq j \leq r-1$. Then it holds

$$\left| E\left(f\left(\frac{\sum_{i=1}^n X_i}{\sqrt{n}}\right)\right) - E(f(X^*)) \right|$$

$$\leq \omega_1(f^{(r-1)}, h) \cdot \left\{ \frac{1}{r!} + \frac{n^{((3-r)/2)} \cdot (\zeta_r^{((r-1)/r)} + l_r^{((r-1)/r)})}{2 \cdot (r-1)!} \right.$$

$$\left. + \frac{h}{8 \cdot (r-2)!} \cdot n^{((4-r)/2)} \cdot (\zeta_r^{((r-2)/r)} + l_r^{((r-2)/r)}) \right\}. \tag{15.86}$$

Here

$$l_r = E(|X^*|^r) < \infty$$

and

$$h := h_n = n^{((2-r)/2)} \cdot (\zeta_r + l_r).$$

(ii) Suppose that the sequence $(X_i)_{i \in \mathbf{N}}$ is dependent from above and that

$$E(X_i^j \mid F_{i+1}^*) = E(X^{*j}) \quad \text{a.s.} \quad (1 \leq j \leq r-1, i \in \mathbf{N}), \tag{15.87}$$

where

$$F_{i+1}^* := \mathcal{A}(X_{i+1}, \ldots, X_n)$$

and $f$ as in part (i). Then (15.86) is true again. The RHS (15.86) converges to zero as $n \to \infty$, when $r \geq 5$.

**PROOF** From Theorem 15.4 and Theorem 3 of Butzer and Kirschfink [56], it is the case $t = 1$. Here the r.f.s $X_{ni}$, $1 \leq i \leq n$, $n \in \mathbf{N}$ become real r.v.s $X_{ni}(1, \omega) = n^{-1/2} \cdot X_i(\omega)$. The r.f. has the Wiener-measure as its distribution reduces to the standard normal distribution $P_{X^*}$. Also, here $a_{ni} := 1$, $1 \leq i \leq n$; $n \in \mathbf{N}$ and $A_n = n^{1/2}$. Notice that

$$A_n^{-r} \cdot \left(\sum_{i=1}^n a_{ni}^r\right) = n^{((2-r)/2)}, \quad A_n^{1-r} \cdot \left(\sum_{i=1}^n a_{ni}^{r-1}\right) = n^{((3-r)/2)},$$

and
$$A_n^{2-r} \cdot \left(\sum_{i=1}^n a_{ni}^{r-2}\right) = n^{((4-r)/2)}.$$

Take into account also the proof of Theorem 4 of Butzer and Kirschfink [56]. ∎

If $(X_i)_{i \in \mathbf{N}}$ is a sequence of real r.v.s on $(\underline{0}, \mathcal{A}, P)$, and $(\mathcal{F}_i)_{i \in \mathbf{Z}_+}$ is a monotone increasing sequence of sub-$\sigma$-algebras of $\mathcal{A}$ such that $\mathcal{F}_i := \mathcal{A}(X_1, \ldots, X_i)$, $i \in \mathbf{N}$ and $\mathcal{F}_0 := \{\phi, \underline{0}\}$, $X_0 = 0$, then $(X_i, \mathcal{F}_i)_{i \in \mathbf{Z}_+}$ is called a martingale difference sequence (MDS) iff

$$E(X_i \mid \mathcal{F}_{i-1}) = 0, \quad \text{a.s. } (i \in \mathbf{N}). \tag{15.88}$$

Butzer and Kirschfink [56], in Lemma 8, establish that for the case of MDS, $F_i$ is dependent from below.

Finally comes the CLT for MDS with rates.

**THEOREM 15.6**
Let $(X_i \mid \mathcal{F}_i)_{i \in \mathbf{Z}_+}$ be an MDS such that $X_i$ are identically distributed, $\operatorname{Var}(X_1) = 1$, $\zeta_r := E(|X_1|^r) < \infty$, for $r \in \mathbf{N} - \{1\}$. Suppose that

$$E(X_i^j \mid \mathcal{F}_{i-1}) = E(X^{*j}) \quad \text{a.s. } (1 \le j \le r-1, i \in \mathbf{N}). \tag{15.89}$$

Let $f$ be as in Theorem 15.5. Then inequality (15.86) is valid.

**PROOF** From Theorem 15.5(i). Notice that $E(X_i) = E(X_i \mid \mathcal{F}_{i-1}) = 0$ a.s. all $i \in \mathbf{N}$, by Lemma 4 of Butzer and Kirschfink [56]. ∎

# Chapter 16

## Quantitative Study of Bias Convergence for Generalized L-Statistics

Employing methods from Approximation Theory, we determine the rates of bias convergence for sequences of generalized $L$-statistics based on i.i.d. samples under mild smoothness conditions on the weight function and simple moment conditions on the score function. Except for standard methods of weighting, we present and analyze $L$-statistics with possibly random coefficients defined with the help of positive linear functionals acting on the weight function. This treatment is based on [39].

## 16.1 Background

Let $X_n$, $n \in \mathbf{N}$, be a sequence of independent identically distributed (i.i.d.) random variables with a common distribution function $F$. For a sample of size $n : X_1, \ldots, X_n$, the sequence of order statistics is denoted by $X_{1:n}, \ldots, X_{n:n}$. The last are obtained by reordering $X_1, \ldots, X_n$ according to increasing magnitude. A sequence $L_n$, $n \in \mathbf{N}$, of (generalized) $L$-statistics is defined by

$$L_n = \sum_{i=1}^{n} c_{i,n} g(X_{i:n}), \quad n \in \mathbf{N}, \tag{16.1}$$

where the real number coefficients $c_{i,n}$, $1 \leq i \leq n < \infty$, and a measurable score function $g$ are given. $L$-statistics have numerous applications in the statistical inference (see, e.g., Balakrishnan and Cohen [48], and David

[72]). The expectation of (16.1) is given by

$$E_F L_n = \sum_{i=1}^{n} nc_{i,n} \int_0^1 g(F^{-1}(y))p_{n-1,i-1}(y)\,dy, \qquad (16.2)$$

where

$$p_{n,i}(x) = \binom{n}{i} x^i (1-x)^{n-i}, \quad i = 0, 1, \ldots, n \qquad (16.3)$$

form the Bernstein basis of the linear space of polynomials on $[0, 1]$ of degree $n$ at most, for $n \in \mathbf{N} \cup \{0\}$. When asymptotic properties of $L$-statistics are studied, it is usually assumed that the coefficients are defined by means of some weight function $f : [0, 1] \mapsto \mathbf{R}$ in two ways resulting in the following definitions of $L$-estimate sequences

$$L_n(f, g) = \sum_{i=1}^{n} l_{i,n} g(X_{i:n})$$

$$= n^{-1} \sum_{i=1}^{n} f\left(\frac{i-1}{n-1}\right) g(X_{i:n}), \quad n \in \mathbf{N} \setminus \{1\}, \qquad (16.4)$$

(cf. e.g., Chernoff et al. [65], Shorack [160], [161], Mason [127]) and

$$K_n(f, g) = \sum_{i=1}^{n} k_{i,n} g(X_{i:n}) = \sum_{i=1}^{n} \left[ \int_{(i-1)/n}^{i/n} f(t)\,dt \right] g(X_{i:n}), \quad n \in \mathbf{N}, \qquad (16.5)$$

(see, e.g., Boos [51], van Zwet [181], Mason and Shorack [129]). Under some regularity conditions on the weight, score and distribution functions $f$, $g$ and $F$, respectively, both (16.4) and (16.5) converge to

$$\mu(f, g, F) = E_F[g(X_1)f(F(X_1))] = \int_0^1 g(F^{-1}(y))f(y)\,dy \qquad (16.6)$$

in various modes of convergence.

The aim of this chapter is to present rates of convergence for sequences of expected $L$-statistics (16.2) defined by (16.4), (16.5) and other formulae to (16.6) under mild conditions on $f$, $g$ and $F$. The problem was investigated also by Stigler [166], Mason [127] and Xiang [177]. Here we impose no smoothness conditions on the score and distribution functions; we assume only finiteness of moments $E_F|g(X_1)|^q$ for some $q \geq 1$. Another natural requirement that will be assumed throughout this chapter and makes possible defining (16.6) is that

$$E_F|g(X_1)f(F(X_1))| = \int_0^1 |g(F^{-1}(y))f(y)|\,dy < \infty.$$

Here the only regularity conditions concern the weight function $f$. In many approximation problems, measuring the smoothness by differentiability is too crude, and the measures of smoothness enable us to do that more subtly for larger classes of possibly nondifferentiable functions.

For $f \in C[0,1]$, we define the first (order) modulus of continuity and the second (order) modulus of continuity (smoothness) as

$$\omega_1(f,h) = \sup\{|f(u) - f(v)| : u, v \in [0,1], |u-v| \le h\},$$

$$\omega_2(f,h) = \sup\{|f(u) - 2f(\frac{u+v}{2}) + f(v)| : u, v \in [0,1],$$

$$|u-v| \le 2h\}, \tag{16.7}$$

$h > 0$, respectively. In general, for $f \in L_p[a,b]$, $1 \le p \le \infty$, we define the $r$-th forward difference

$$\Delta_h^r(f,x) = \sum_{k=0}^r \binom{r}{k}(-1)^{r-k}f(x+kh),$$

and $A_{rh} = [a, b-rh]$. The $r$-th modulus of smoothness for $f \in L_p[a,b]$, $1 \le p < \infty$, and $f \in C[a,b]$, $p = \infty$, is defined by

$$\omega_r(f,t)_p = \sup_{0 < h \le t} \|\Delta_h^r(f,\cdot)\|_{p, A_{rh}}, \quad t \ge 0, \tag{16.8}$$

(see DeVore and Lorentz ([74], pp. 40–46) and Schumaker ([154], pp. 53–55)). The latter index of the norm describes the domain of each element of $L_p$-space, and it will be dropped if it coincides with the unit interval. One can check that $\omega_r(f,t)_p$ is a finite, continuous and increasing function in $t$, with $\omega_r(f,0)_p = 0$ and $\omega_r(f,t)_p \to 0$ as $t \to 0$.

Loosely speaking, $K$-functionals enable us to describe parametrically the accuracy of approximating a function of a space by elements of a subspace. Below we define the Ditzian–Totik version of $K$-functional. Define the second-order symmetric difference as

$$\tilde{\Delta}_s^2(f,x) = \begin{cases} f(x+s) - 2f(x) + f(x-s), & \text{if } [x-s, x+s] \subset [0,1], \\ 0, & \text{otherwise.} \end{cases}$$

Consider $\tilde{\Delta}^2_{h\varphi(\cdot)}(f,\cdot)$, for $h > 0$ and $\varphi(x) = \sqrt{x(1-x)}$, $x \in [0,1]$, and

$$W_\infty^2(\varphi) = \{g \in C[0,1] : g' \text{ is absolutely continuous on } [0,1]$$

$$\text{and } ||\varphi^2 g''||_\infty < \infty\}.$$

We define the Ditzian–Totik $K$-functional as

$$K(f,t) = K(f,t; C, W_\infty^2(\varphi)) = \inf_{g \in W_\infty^2(\varphi)} \{||f-g||_\infty + t||\varphi^2 g''||_\infty\} \quad (16.9)$$

and the Ditzian–Totik modulus of smoothness as

$$\omega_2^\varphi(f,t) = \sup_{0 \leq h \leq t} ||\tilde{\Delta}_{h\varphi(\cdot)}(f,\cdot)||_\infty \quad (16.10)$$

(see DeVore and Lorentz ([74], p. 322)). The last is a representative of weighted moduli of smoothness, with less emphasis laid on the smoothness at the borders of the domain.

In Section 16.2 we present rates of convergence for the expectations of (16.4) and (16.5) to (16.6) in terms of moduli of smoothness and $K$-functionals, making use of their mutual relations to the Bernstein and Kantorovich operators, respectively. In Section 16.3, we also give nonstandard methods of constructing $L$-statistics with possibly randomized coefficients determined by a given weight function. Using a general notion of positive linear functionals, we establish that the nonstandard $L$-statistics converge to (16.6) and evaluate the rates of their bias decrease. The rates will be specified more precisely for $L$-statistics related to Bernstein–Durrmeyer, Mache and Stancu operators. The rates given here are optimal for the wide classes we study and best constants of approximation are also given for certain cases. In Section 16.4 we refer to saturation theorems that indicate classes of weight functions generating $L$-statistics with faster vanishing bias.

## 16.2 Quantitative Results for $L$-Statistics with Standard Weights

Let
$$(B_n f)(x) = \sum_{k=0}^{n} \binom{n}{k} x^k (1-x)^{n-k} f\left(\frac{k}{n}\right) \tag{16.11}$$

the $n$th Bernstein operator for $f \in C[0,1]$. In Theorems 16.1–16.3 we use the approximation properties of this operator for estimating rates of bias convergence of (16.4).

### THEOREM 16.1
For an i.i.d. sequence $X_i$, $i \in \mathbf{N}$, with a common distribution function $F$ and $\mathrm{E}|g(X_1)| < \infty$, and $L$-statistics defined by (16.4), we get

$$\mathrm{E}_F L_n(f,g) = \int_0^1 g(F^{-1}(y))(B_{n-1}f)(y)\,dy, \tag{16.12}$$

If $f \in C^1[0,1]$, then for $n \in \mathbf{N} \setminus \{1\}$, it holds

$$|\mathrm{E}_F L_n(f,g) - \mu(f,g,F)| \le \frac{25}{32}(n-1)^{-1/2} \omega_1(f', \frac{1}{4}(n-1)^{-1/2}) \mathrm{E}_F |g(X_1)|.$$

**PROOF**  Combining (16.2), (16.4) and (16.11), we find

$$\mathrm{E}_F L_n(f,g) = \mathrm{E}_F n^{-1} \sum_{i=1}^{n} f\left(\frac{i-1}{n-1}\right) g(X_{i:n})$$

$$= \int_0^1 g(F^{-1}(y)) \sum_{i=1}^{n} f\left(\frac{i-1}{n-1}\right) p_{n-1,i-1}(y)\,dy$$

$$= \int_0^1 g(F^{-1}(y))(B_{n-1}f)(y)\,dy,$$

which proves (16.12). Therefore, using Anastassiou ([12], Corollary 7.3.4, p. 230), we get

$$|\mathrm{E}_F L_n(f,g) - \mu(f,g,F)| = \left| \int_0^1 g(F^{-1}(y))[(B_{n-1}f)(y) - f(y)]\,dy \right|$$

$$\leq ||B_{n-1}f - f||_\infty \int_0^1 |g(F^{-1}(y))|\,dy$$

$$\leq \frac{25}{32}(n-1)^{-1/2}\omega_1(f', \frac{1}{4}(n-1)^{-1/2})\mathrm{E}_F|g(X_1)|,$$

$n \in \mathbf{N} \setminus \{1\}$, which finishes the proof. ∎

**THEOREM 16.2**
*Under the assumptions and notations of Theorem 16.1, with $f \in C[0,1]$, we have for all $n \in \mathbf{N} \setminus \{1\}$*

$$|\mathrm{E}_F L_n(f,g) - \mu(f,g,F)| \leq ||B_{n-1}f - f||_\infty \mathrm{E}_F|g(X_1)|$$

$$\leq \left\{\begin{array}{l} \frac{4306+837\sqrt{6}}{5832}\omega_1(f, \frac{1}{\sqrt{n-1}}) \\ \frac{35}{32}\omega_2(f, \frac{1}{\sqrt{n-1}}) \end{array}\right\} \mathrm{E}_F|g(X_1)|. \qquad (16.13)$$

*Furthermore, the first estimate in (16.13) is optimal.*

**PROOF** From Sikkema [162] for the best constant $\frac{4306+837\sqrt{6}}{5832} < 1.09$ and see Paltanea [138] for the constant $\frac{35}{32} < 1.094$, which is not optimal. ∎

**REMARK 16.1** We have the following approximations for $f \in C[0,1]$ by the associated Bernstein operators, expressed in terms of the Ditzian–Totik $K$-functional and modulus of smoothness (see (16.9) and (16.10))

$$||B_n f - f||_\infty \leq 2K(f, n^{-1}),$$

$$||B_n f - f||_\infty \leq C\omega_2^\varphi(f, n^{-1/2}),$$

respectively, where $n \in \mathbf{N}$, and $C > 0$ is an absolute constant (see DeVore and Lorentz ([74], pp. 323–325)). ∎

A direct application of Remark 16.1 gives

**THEOREM 16.3**
*All notations and assumptions here are as in Theorem 16.2. For all $n \in$*

$\mathbf{N} \setminus \{1\}$, we obtain

$$|\mathrm{E}_F L_n(f,g) - \mu(f,g,F)| \leq \left\{ \begin{array}{l} 2K(f,(n-1)^{-1}), \\ C\omega_2^\varphi(f,(n-1)^{-1/2}) \end{array} \right\} \mathrm{E}_F|g(X_1)|.$$

**REMARK 16.2** By applying (16.3), we mention the Kantorovich operators

$$(K_n f)(x) = (n+1) \sum_{k=0}^{n} \left( \int_{k/(n+1)}^{(k+1)/(n+1)} f(t)\,dt \right) p_{n,k}(x), \quad n \in \mathbf{N} \cup \{0\},$$
(16.14)

for either $f \in L_p[0,1]$ or $f \in C[0,1]$ in cases $1 \leq p < \infty$ and $p = \infty$, respectively. We also mention the Gonska–Zhou version of $K$-functional

$$K^*(f,t)_p = \inf\{\|f-g\|_p + t^2\|(\varphi^2 g')'\|_p : g \in C^2[0,1]\}, \qquad (16.15)$$

where $\varphi(x) = \sqrt{x(1-x)}$, $x \in [0,1]$. ■

We use the following three results:

**THEOREM 16.4**
(Gonska and X.-L. Zhou [94]). There exists $C > 0$ such that

$$C^{-1} K^*(f, n^{-1/2})_p \leq \|K_n f - f\|_p \leq C K^*(f, n^{-1/2})_p, \quad 1 \leq p \leq \infty.$$

**THEOREM 16.5**
(Gonska and X.-L. Zhou [94]). It holds

$$K^*(f,t)_p \sim \omega_2^\varphi(f,t)_p + \omega_1(f,t^2)_p, \quad 1 \leq p \leq \infty.$$

**THEOREM 16.6**
(Gonska and D.-X. Zhou [93]). Let $f \in C[0,1]$ and $n \in \mathbf{N}$. Then we have

$$\|K_n f - f\|_\infty \leq \frac{M}{n} \left[ \int_{n^{-1/2}}^{1/2} \omega_2^\varphi(f,t)_\infty t^{-3}\,dt + E_0(f)_\infty \right],$$

where $M > 0$ is independent of $f$ and $n$, and

$$E_0(f)_\infty = \inf_{c \in \mathbf{R}} \|f - c\|_\infty. \qquad (16.16)$$

We are now ready to analyze $L$-statistics (16.5) with coefficients defined by integrals.

**THEOREM 16.7**
Let $X_i$, $i \in \mathbf{N}$, be an i.i.d. sequence of random variables. We follow the assumptions and notation of Remark 16.2 and suppose that $\mathrm{E}|g(X_1)|^q < \infty$ for $q = p/(p-1)$ and $p > 1$. Then it holds

$$\mathrm{E}_F K_n(f,g) = \int_0^1 g(F^{-1}(y))(K_{n-1}f)(y)\,dy$$

and

$$|\mathrm{E}_F K_n(f,g) - \mu(f,g,F)| \leq \|K_{n-1}f - f\|_p (\mathrm{E}_F|g(X_1)|^q)^{1/q}, \quad n \in \mathbf{N} \setminus \{1\}.$$

**PROOF** By (16.2), (16.6) and (16.14) we have

$$\mathrm{E}_F K_n(f,g) = \mathrm{E}_F \sum_{i=1}^n k_{i,n} g(X_{i:n})$$

$$= \int_0^1 g(F^{-1}(y)) n \left[\sum_{i=1}^n \int_{(i-1)/n}^{i/n} f(t)\,dt\right] p_{n-1,i-1}(y)\,dy$$

$$= \int_0^1 g(F^{-1}(y))(K_{n-1}f)(y)\,dy.$$

Thus,

$$\mathrm{E}_F K_n(f,g) - \mu(f,g,F) = \int_0^1 g(F^{-1}(y))[(K_{n-1}f)(y) - f(y)]\,dy,$$

and, by Hölder's inequality, we obtain

$$|\mathrm{E}_F K_n(f,g) - \mu(f,g,F)| \leq \left(\int_0^1 |(K_{n-1}f)(y) - f(y)|\,dy\right)^p$$

$$\cdot \left(\int_0^1 |g(F^{-1}(y))|\,dy\right)^q$$

$$= \|K_{n-1}f - f\|_p (\mathrm{E}_F|g(X_1)|^q)^{1/q}.$$

From Theorems 16.4–16.7, we give

**THEOREM 16.8**
Under the assumptions and notations of Theorem 16.7, we obtain

$$|E_F K_n(f,g) - \mu(f,g,F)|$$

$$\leq \begin{cases} CK^*(f,(n-1)^{-1/2})_p \\ C^*[\omega_2^\varphi(f,(n-1)^{-1/2})_p + \omega_1(f,(n-1)^{-1})_p] \end{cases} (E_F|g(X_1)|^q)^{1/q},$$

for $1 < p < \infty$ and universal constants $C, C^* > 0$. Furthermore, if $p = \infty$, then

$$|E_F K_n(f,g) - \mu(f,g,F)| \leq ||K_{n-1}f - f||_\infty E_F|g(X_1)|$$

$$\leq \begin{cases} CK^*(f,(n-1)^{-1/2})_\infty \\ C^*[\omega_2^\varphi(f,(n-1)^{-1/2})_\infty + \omega_1(f,(n-1)^{-1})_\infty] \\ \frac{M}{n-1}\left[\int_{(n-1)^{-1/2}}^{1/2} \omega_2^\varphi(f,t)_\infty t^{-3}\,dt + E_0(f)_\infty\right] \end{cases} E_F|g(X_1)|,$$

where $M > 0$ and $E_0(f)_\infty$ is defined by (16.16).

## 16.3 Quantitative Results for *L*-Statistics with Nonstandard Weights

**REMARK 16.3** For $f \in L_1[0,1]$ we define the Bernstein–Durrmeyer operators by

$$(D_n f)(x) = (n+1) \sum_{k=0}^n \left(\int_0^1 f(t) p_{n,k}(t)\,dt\right) p_{n,k}(x) \qquad (16.17)$$

for $x \in [0,1]$ and $n \in \mathbf{N} \cup \{0\}$. Then there exists universal constant $C > 0$ such that

$$C^{-1} K^*(f, n^{-1/2})_p \leq ||D_n f - f||_p \leq C K^*(f, n^{-1/2})_p \qquad (16.18)$$

for any $1 \leq p \leq \infty$ (see Gonska and D.-X. Zhou [93]). ■

We introduce a sequence of $L$-statistics by

$$M_n(f,g) = \sum_{i=1}^n m_{i,n} g(X_{i:n}) = \sum_{i=1}^n \left( \int_0^1 p_{n-1,i-1}(t) f(t) \, dt \right) g(X_{i:n}). \tag{16.19}$$

This is a modification of (16.5) that results by replacing step weight function $\mathbf{1}_{((i-1)/n, i/n]}$ in the integral representation of coefficients $k_{i,n}$, $1 \leq i \leq n < \infty$, by smooth ones $p_{i-1,n-1}$. Under the above assumptions and notation, we present

**THEOREM 16.9**
*For an i.i.d. sequence of random variables $X_i$, $i \in \mathbf{N}$, with a common distribution function $F$ and a weight function $f \in L_p[0,1]$, $1 < p \leq \infty$, we suppose that $\mathbf{E}_F |g(X_1)|^q < \infty$ for $q = p/(p-1)$ and $p < \infty$, and $\mathbf{E}_F |g(X_1)| < \infty$ for $p = \infty$. Then we have*

$$\mathbf{E}_F M_n(f,g) = \int_0^1 g(F^{-1}(y))(D_{n-1}f)(y) \, dy, \quad n \in \mathbf{N}. \tag{16.20}$$

*If $1 < p < \infty$, then it holds*

$$|\mathbf{E}_F M_n(f,g) - \mu(f,g,F)| \leq \|D_{n-1}f - f\|_p (\mathbf{E}_F |g(X_1)|^q)^{1/q}$$

$$\leq C K^*(f, (n-1)^{-1/2})_p (\mathbf{E}_F |g(X_1)|^q)^{1/q}. \tag{16.21}$$

*If $p = \infty$, then it holds*

$$|\mathbf{E}_F M_n(f,g) - \mu(f,g,F)| \tag{16.22}$$

$$\leq \left\{ \begin{array}{l} C K^*(f, (n-1)^{-1/2})_\infty \\ \frac{M}{n-1} \left[ \int_{(n-1)^{-1/2}}^{1/2} \omega_2^\varphi(f,t)_\infty t^{-3} \, dt + E_0(f)_\infty \right] \end{array} \right\} \mathbf{E}_F |g(X_1)|.$$

**PROOF** Formula (16.20) can be immediately concluded by (16.2), (16.17) and (16.19). Hence

$$\mathbf{E}_F M_n(f,g) - \mu(f,g,F) = \int_0^1 g(F^{-1}(y))[(D_{n-1}f)(y) - f(y)] \, dy.$$

By applying the Hölder's inequality, and then using (16.18), we get (16.21). In a similar way, we can find the first relation in (16.22). The last inequality is an implication of Theorem 3 in Gonska and D.-X. Zhou [93] that gives us

$$||D_n f - f||_\infty \le \frac{M}{n}\left[\int_{n^{-1/2}}^{1/2} \omega_2^\varphi(f,t)_\infty t^{-3}\,dt + E_0(f)_\infty\right].$$

∎

**REMARK 16.4** One may consider $f \in L_p[0,1]$, $1 \le p \le \infty$, such that either

$$\omega_1(f,h)_p = \mathcal{O}(h^\alpha), \quad 0 < \alpha \le 1,$$

or

$$\omega_2^\varphi(f,h)_p = \mathcal{O}(h^\alpha), \quad 0 < \alpha \le 2,$$

(i.e., $K(f,h)_p = \mathcal{O}(h^\beta)$, $0 < \beta \le 1$, cf. (16.9) and (16.10). It could also be

$$K^*(f,h)_p = \mathcal{O}(h^\alpha), \quad 0 < \alpha \le 2, \quad 1 < p \le \infty$$

(see (16.15)). All the above are various forms of Lipschitz-type conditions for $f$, and can simplify previous results when applicable. From DeVore and Lorentz ([74], p. 327) we have

$$||D_n f - f||_\infty \le 3\omega_1(f,(3/n)^{1/2}), \quad n \in \mathbf{N}, \quad f \in C[0,1],$$

where $\omega_1$ is the first (ordinary) modulus of continuity (see (16.7)). Consequently, following the assumptions and notations of Theorem 16.9, we find

$$|\mathrm{E}_F M_n(f,g) - \mu(f,g,F)| \le ||D_{n-1}f - f||_\infty \mathrm{E}_F|g(X_1)|$$

$$\le 3\omega_1(f,(3/n)^{1/2})\mathrm{E}_F|g(X_1)|. \quad (16.23)$$

∎

We see that the coefficients of $L$-statistic in (16.19) can be expressed as

$$m_{i,n} = \mathrm{E}f(U_{i:n}), \quad 1 \le i \le n < \infty,$$

(cf. (16.2)), where $U_{i:n}$, $1 \le i \le n$, are the order statistics from a standard uniform i.i.d. sample of size $n$, which can be easily generated. Therefore, replacing the $L$-statistics by their randomized modifications

$$\tilde{M}_n(f,g) = \sum_{i=1}^{n} f(U_{i:n})g(X_{i:n}), \quad n \in \mathbf{N}, \qquad (16.24)$$

we maintain all the conclusions of Theorem 16.9 and formula (16.23). Also, (16.5) can be substituted by randomized counterparts

$$\tilde{K}_n(f,g) = \sum_{i=1}^{n} f(V_{i,n})g(X_{i:n}), \quad n \in \mathbf{N}, \qquad (16.25)$$

with the same expectations, if $V_{i,n}$, $1 \leq i \leq n < \infty$, are uniformly distributed on $[(i-1)/n, i/n]$. In fact, it suffices to set $V_{i,n} = (i-1+U)/n$ for a single random variable uniformly distributed on $[0,1]$. Formulae (16.24) and (16.25) reveal numerous possibilities for nonstandard choice of randomized coefficients with desired expectations. Below we give even more general constructions, based on positive linear operators that generalize the notion of the expectation operator.

**REMARK 16.5** Here we refer to Gavrea and Mache [86]. Let $T_{n,k} : C[0,1] \mapsto \mathbf{R}$, $n \in \mathbf{N}$, $k = 0, 1, \ldots, n$, be positive linear functionals such that $T_{n,k}1 = 1$. Then

$$(A_n f)(x) = \sum_{k=0}^{n} T_{n,k} f \; p_{n,k}(x)$$

is a positive linear operator acting on $f \in C[0,1]$. Put

$$\Delta_n(x) = \sum_{k=0}^{n} T_{n,k}(\cdot - k/n)^2 \; p_{n,k}(x).$$

Notice that $\Delta_n(x) \geq 0$. Denote

$$\tilde{\Delta}_h^2 f(x) = f(x+h) - 2f(x) + f(x-h), \quad x \in [h, 1-h], \; 0 < h < 1,$$

and define

$$\omega_2^*(f,t) = \sup_{0 < h \leq t, \, x \in [h, 1-h]} |\tilde{\Delta}_h^2 f(x)|.$$

We use

## THEOREM 16.10

(Gavrea and Mache [86]). *Suppose that* $\Delta_n(x) \leq C/n^{2\beta}$ *for some* $C > 0$, $1 < \beta < 2$ *and all* $x \in [0,1]$ *(i.e.,* $\Delta_n(x) = \mathcal{O}(n^{-2\beta})$*). Then it holds*

$$|(A_n f)(x) - f(x)| \leq C\{\Delta_n^{1/2}(x) + [\Delta_n(x) + n^{-1}x(1-x)]^{\beta/2}\}$$

*for all* $x \in [0,1]$ *iff* $\omega_2^*(f,t) = \mathcal{O}(t^\beta)$.

## COROLLARY 16.1

*If* $\Delta_n(x) \leq C/n^{2\beta}$ *for some* $C > 0$, $1 < \beta < 2$ *and all* $x \in [0,1]$, *and* $\omega_2^*(f,t) = \mathcal{O}(t^\beta)$ *for* $f \in C[0,1]$, *then we get*

$$\|A_n f - f\|_\infty \leq C\left[\frac{C^{1/2}}{n^\beta} + \left(\frac{C}{n^{2\beta}} + \frac{1}{4n}\right)^{\beta/2}\right], \quad n \in \mathbf{N}.$$

Because

$$E_F g(X_{i:n}) = n \int_0^1 g(F^{-1}(y)) p_{n-1,i-1}(y) \, dy,$$

$$E_F \sum_{i=1}^n (T_{n-1,i-1} f) g(X_{i:n}) = n \int_0^1 g(F^{-1}(y))(A_{n-1} f)(y) \, dy, \ f \in C[0,1],$$

we can generally define $L$-statistics $T_n(f,g)$, $n \in \mathbf{N} \setminus \{1\}$, with coefficients $t_{i,n} = n^{-1} T_{n-1,i-1} f$, $i = 1, \ldots, n$, that fulfill

$$E_F T_n(f,g) = \int_0^1 g(F^{-1}(y))(A_{n-1} f)(y) \, dy,$$

and

$$E_F T_n(f,g) - \mu(f,g,F) = \int_0^1 g(F^{-1}(y))[(A_{n-1} f)(y) - f(y)] \, dy. \quad (16.26)$$

Thus we obtain

## THEOREM 16.11

*Let* $X_i$, $i \in \mathbf{N}$, *be i.i.d. random variables of common distribution function* $F$, *and* $f \in C[0,1]$. *We suppose that* $E_F |g(X_1)| < \infty$, $\Delta_n(x) \leq Cn^{-2\beta}$ *for* some $C > 0$, $1 < \beta < 2$, and all $x \in [0,1]$, and $\omega_2^*(f,t) = \mathcal{O}(t^\beta)$, using the notions and notations of Remark 16.5. Then it holds

$$|E_F T_n(f,g) - \mu(f,g,F)| \le \|A_{n-1}f - f\|_\infty E_F|g(X_1)|$$

$$\le C\left[\frac{C^{1/2}}{(n-1)^\beta} + \left(\frac{C}{(n-1)^{2\beta}}\right.\right.$$

$$\left.\left. + \frac{1}{4(n-1)}\right)^{\beta/2}\right] E_F|g(X_1)|.$$

**PROOF** From (16.26) we have

$$|E_F T_n(f,g) - \mu(f,g,F)| \le \int_0^1 |g(F^{-1}(y))|\,|(A_{n-1}f)(y) - f(y)|\,dy$$

$$\le \|A_{n-1}f - f\|_\infty \left(\int_0^1 |g(F^{-1}(y))|\,dy\right)$$

$$= \|A_{n-1}f - f\|_\infty E_F|g(X_1)|.$$

By applying Corollary 16.1, we end the proof. ∎

**REMARK 16.6** Here we follow Mache [120]. We recall the beta function

$$B(p,q) = \int_0^1 x^{p-1}(1-x)^{q-1}\,dx, \quad p,q > 0.$$

Let $a, b > -1$, $\alpha \ge 0$ and $c = c_n = [n^\alpha]$ ($[\cdot]$ denotes the integral part) for $n \in \mathbf{N}$. Define the positive linear functionals $T_{\alpha,k,n} : C[0,1] \mapsto \mathbf{R}$, $k = 0, \ldots, n$, as follows

$$T_{\alpha,k,n}f = \int_0^1 f(t) t^{ck+a}(1-t)^{c(n-k)+b}\,dt / B(ck+a+1, c(n-k)+b+1). \tag{16.27}$$

We next define the positive linear operators

$$(M_n^\alpha f)(x) = \sum_{k=0}^n T_{\alpha,k,n}f\,p_{n,k}(x), \quad n \in \mathbf{N},\ \alpha \ge 0. \tag{16.28}$$

When $a = b = 0$, we get the so-called Durrmeyer operators with Legendre weights. If, moreover, $\alpha = 0$, we get the standard Bernstein–Durrmeyer operators. In (16.27) and (16.28), we ignore in notation the dependence of defined notions on $a$ and $b$, because these do not affect rates of approximation given below. ∎

**THEOREM 16.12**
(Mache [120]). *Case 1 (Durrmeyer operators with Jacobi weights): For $\alpha = 0$, we obtain*

$$\|M_n^\alpha f - f\|_\infty \leq Cn^{-1} \left[ \int_{n^{-1/2}}^{1/2} \omega_2^\varphi(f,t)_\infty t^{-3}\, dt + \|f\|_\infty \right].$$

*Case 2: For $0 < \alpha < 1$, we find*

$$\|M_n^\alpha f - f\|_\infty \leq C \left\{ n^{-1-\alpha} \left[ \int_{n^{-1/2}}^{1/2} \omega_2^\varphi(f,t)_\infty t^{-3}\, dt + \|f\|_\infty \right] + \omega_2^\varphi(f, n^{-1/2})_\infty \right\}.$$

*Case 3: For $\alpha \geq 1$, we get*

$$\|M_n^\alpha f - f\|_\infty \leq C \left[ n^{-1-\alpha} \|f\|_\infty + \omega_2^\varphi(f, n^{-1/2})_\infty \right].$$

*Case 4 (Bernstein operators): For $\alpha \to \infty$, it holds*

$$\|M_n^\alpha f - f\|_\infty \leq C \omega_2^\varphi(f, n^{-1/2})_\infty,$$

*where $C > 0$ is independent of $n, \alpha$ and $f$.*

We also need

**THEOREM 16.13**
(Mache [120]). *Let $\alpha \geq 1$ and $0 < \beta < 1$. Then we have $\|M_n^\alpha f - f\|_\infty = \mathcal{O}(n^{-\beta})$ iff $\omega_2^\varphi(f,t)_\infty = \mathcal{O}(t^{2\beta})$.*

Formula (16.28) describes a large parametric class of generalized $L$-statistics

$$M_n^\alpha(f,g) = n^{-1} \sum_{i=1}^n T_{\alpha,i-1,n-1} f \, g(X_{i:n}), \quad n \in \mathbf{N} \setminus \{1\},$$

with coefficients defined by specific positive linear functionals (16.27). The $L$-statistics can be determined randomly by means of a probabilistic model of generalized uniform order statistics introduced by Kamps [104]. Putting $\alpha = 0$, we get the so-called fractional order statistics with nonintegral sample sizes, studied in Stigler [167], and Rohatgi and Saleh [151]. For some choices of parameters, they have practical interpretations as sequential order statistics and certain records (see Kamps [104]).

See that

$$E_F M_n^\alpha(f,g) = \int_0^1 g(F^{-1}(y))(M_{n-1}^\alpha f)(y) \, dy,$$

and of course

$$E_F M_n^\alpha(f,g) - \mu(f,g,F) = \int_0^1 g(F^{-1}(y))[(M_{n-1}^\alpha f)(y) - f(y)] \, dy. \quad (16.29)$$

Next we give

**THEOREM 16.14**
*Let $X_i$, $i \in \mathbf{N}$, be i.i.d. random variables of common distribution function $F$ and $f \in C[0,1]$. We suppose that $E_F |g(X_1)| < \infty$, and use the notions and notations of Remark 16.6. Then it holds*

$$|E_F M_n^\alpha(f,g) - \mu(f,g,F)| \le \|M_{n-1}^\alpha f - f\|_\infty \int_0^1 |g(F^{-1}(y))| \, dy$$

$$\le C \begin{cases} (n-1)^{-1} \left[ \int_{(n-1)^{-1/2}}^{1/2} \omega_2^\varphi(f,t)_\infty t^{-3} \, dt + \|f\|_\infty \right], & (\alpha = 0) \\ \{(n-1)^{-1-\alpha} \left[ \int_{(n-1)^{-1/2}}^{1/2} \omega_2^\varphi(f,t)_\infty t^{-3} \, dt \right. \\ \left. + \|f\|_\infty \right] + \omega_2^\varphi(f,(n-1)^{-1/2})_\infty \}, & (0 < \alpha < 1) \\ [(n-1)^{-1-\alpha} \|f\|_\infty + \omega_2^\varphi(f,(n-1)^{-1/2})_\infty], & (\alpha \ge 1) \\ \omega_2^\varphi(f,(n-1)^{-1/2})_\infty, & (\alpha \to \infty) \end{cases} \times$$

$$\times E_F|g(X_1)|. \quad (16.30)$$

## 16. Generalized L-Statistics

**PROOF** From (16.29) we get that

$$|\mathrm{E}_F M_n^\alpha(f,g) - \mu(f,g,F)| \leq \int_0^1 |g(F^{-1}(y))|\,|(M_{n-1}^\alpha f)(y) - f(y)|\,dy$$

$$\leq \|M_{n-1}^\alpha f - f\|_\infty \int_0^1 |g(F^{-1}(y))|\,dy.$$

Then we use Theorem 16.12. ∎

### THEOREM 16.15
Under the assumptions of Theorem 16.14, with $\omega_2^\varphi(f,t)_\infty = \mathcal{O}(t^{2\beta})$ for some $0 < \beta < 1$, it holds

$$|\mathrm{E}_F M_n^\alpha(f,g) - \mu(f,g,F)| = \mathcal{O}((n-1)^{-\beta}).$$

**PROOF** Use (16.30) and Theorem 16.13. ∎

**REMARK 16.7** Here we refer to Gonska and Meier [92]. For $f \in C[0,1]$, $m \in \mathbf{N}$, and $0 \leq \beta \leq \gamma$, we define the Stancu-type positive linear operators

$$(L_{m0}^{<0\beta\gamma>} f)(x) = \sum_{k=0}^m f\left(\frac{k+\beta}{m+\gamma}\right) p_{m,k}(x), \quad x \in [0,1].$$

∎

We apply the following theorem

### THEOREM 16.16
(Gonska and Meier [92]). For $f \in C[0,1]$, $h > 0$, $0 \leq \beta \leq \gamma$, $m \in \mathbf{N}$ and $x \in [0,1]$ it holds

$$|(L_{m0}^{<0\beta\gamma>} f)(x) - f(x)|$$

$$\leq \left[3 + \max\{h^{-2},1\}\frac{(\gamma^2-m)x^2 + mx + \beta^2}{(m+\gamma)^2}\right]\omega_2(f,h)$$

$$+ \frac{2|\beta - \gamma x|}{m + \gamma} \max\{h^{-1}, 1\}\omega_1(f, h). \qquad (16.31)$$

Maximizing the right-hand side of (16.31), we find that

**COROLLARY 16.2**
For sufficiently large $m \in \mathbf{N}$,

$$\|L_{m0}^{<0\beta\gamma>} f - f\|_\infty \leq \left[3 + \frac{m^3 + 4m^2\beta(\beta - \gamma)}{4(m - \gamma^2)(m + \gamma)^2}\right] \omega_2(f, m^{-1/2})$$

$$+ \frac{2(\beta + \gamma)m^{1/2}}{m + \gamma} \omega_1(f, m^{-1/2}).$$

At the end, we consider some generalizations of (16.4). Many people studied modifications of the $L$-statistics that consist in replacing arguments of the weight function in coefficients $l_{i,n} = n^{-1}f((i-1)/(n-1))$. The most common choices were $i/n$ and $i/(n+1)$. These two and many other cases can be examined simultaneously if we define

$$L_n^{\beta\gamma}(f, g) = n^{-1} \sum_{i=1}^{n} f\left(\frac{i - 1 + \beta}{n - 1 + \gamma}\right) g(X_{i:n}),$$

and apply the statements of Remark 16.7. See that

$$\mathrm{E}_F L_n^{\beta\gamma}(f, g) = \int_0^1 g(F^{-1}(y))(L_{n-1,0}^{<0\beta\gamma>} f)(y)\, dy$$

and

$$\mathrm{E}_F L_n^{\beta\gamma}(f, g) - \mu(f, g, F) = \int_0^1 g(F^{-1}(y))[(L_{n-1,0}^{<0\beta\gamma>} f)(y) - f(y)]dy. \qquad (16.32)$$

Consequently,

**THEOREM 16.17**
Assume that $X_i$, $i \in \mathbf{N}$, are i.i.d. random variables with a common distribution function $F$. Let $f \in C[0,1]$ and $\mathrm{E}_F|g(X_1)| < \infty$. Then for sufficiently large $n$, we obtain

$$|\mathrm{E}_F L_n^{\beta\gamma}(f, g) - \mu(f, g, F)| \leq \|L_{n-1,0}^{<0\beta\gamma>} f - f\|_\infty \int_0^1 |g(F^{-1}(y))|\, dy$$

## 16. Generalized L-Statistics

$$\leq \left\{ \left[ 3 + \frac{(n-1)^3 + 4(n-1)^2\beta(\beta-\gamma)}{4(n-1-\gamma^2)(n-1+\gamma)^2} \right] \right.$$

$$\times \omega_2(f, (n-1)^{-1/2}) + \frac{2(\beta+\gamma)(n-1)^{1/2}}{n-1+\gamma}$$

$$\left. \times \omega_1(f, (n-1)^{-1/2}) \right\} E_F |g(X_1)|.$$

**PROOF** Application of (16.32) and Corollary 16.2. ∎

## 16.4 More Conclusions

We conclude this chapter by discussing rates of convergence of the expectations of $L$-statistics (16.4), (16.5), and (16.19) to (16.6) for various classes of weight functions. These rates coincide with those of the Bernstein, Kantorovich and Bernstein–Durrmeyer operators, respectively, to the unit in $L_p$-norms, $1 \leq p \leq \infty$. In general, $n^{-1/2}$ is the best rate for these operators [see Knoop and Zhou [108], DeVore and Lorentz ([74], formula (7.3)), and Gonska and X.-L. Zhou [94], and Gonska and D.-X. Zhou [93], respectively].

**REMARK 16.8** For sufficiently large $n$, (16.4) has a better constant of approximation than the ones presented in Theorem 16.1. In fact, the right-hand side of (16.13) can be replaced by $\omega_2(f, (n-1)^{-1/2})E|X_1|$, which is also optimal (see Paltanea [139]). ∎

**REMARK 16.9** Because $B_n$ reproduces linear functions, $B_n f - f = 0$ for linear $f$. This means that (16.4) provides unbiased estimates for linear weight functions. The sample mean and Gini mean difference are typical examples here. ∎

**REMARK 16.10** If $f \in C^2[0, 1]$, then

$$\|B_n f - f\|_\infty = \mathcal{O}(n^{-1}),$$

(see Gonska and Meier [92]), and also

$$\|K_n f - f\|_\infty = \mathcal{O}(n^{-1}),$$

$$\|D_n f - f\|_\infty = \mathcal{O}(n^{-1}),$$

(see Cao and Gonska [61] and Gonska and Kovacheva [91]). ∎

**REMARK 16.11** Since defining the Bernstein operators for discontinuous functions does not make sense, we have $L_p$-norm estimates, $p < \infty$, for the Kantorovich and Bernstein–Durrmeyer ones, only. Setting $\mathcal{L}_n$ for either $K_n$ or $D_n$, the following statements hold true. ∎

**THEOREM 16.18**
(see Totik [170], Ditzian and Ivanov [77]). Let $1 \leq p < \infty$. Then for $0 < \alpha < 2$, we get

$$\|\mathcal{L}_n f - f\|_p = \mathcal{O}(n^{-\alpha/2})$$

*iff*

$$\omega_2^\varphi(f, t)_p = \mathcal{O}(t^\alpha).$$

Here the Ditzian–Totik modulus of smoothness

$$\omega_2^\varphi(f, t)_p = \sup_{0 \leq h \leq t} \|\tilde{\Delta}_{h\varphi(\cdot)}^k(f, \cdot)\|_p, \quad f \in L_p[0, 1],$$

is defined analogously to (16.10). For the saturation case, we obtain

**THEOREM 16.19**
(see Maier [121], [122], Riemenschneider [150], Totik [169], Heilmann [97]). It holds

$$\|\mathcal{L}_n f - f\|_p = \mathcal{O}(n^{-1}), \quad 1 \leq p < \infty,$$

*iff either*

$$\omega_2^\varphi(f, t)_p = \mathcal{O}(t^2)$$

for $1 < p < \infty$ or

$$f(x) = K + \int_y^x \frac{h(t)}{t(1-t)}\,dt, \quad a.e. \text{ on } [0, 1]$$

for $p = 1$ with $y \in (0, 1)$, $h \in BV[0, 1]$ and $h(0) = h(1) = 0$.

# Part VI

# On Functional Analysis

# Chapter 17

# Quantitative Korovkin-Type Results for Vector Valued Functions

We consider here the space of continuous functions from a compact and convex subset of a normed vector space into an abstract Banach lattice. Also we consider here lattice homomorphisms from the above space into itself or into the associated space of vector valued bounded functions. The uniform convergence of such operators to the unit operator with rates is mainly studied in this chapter. The presented quantitative results are inequalities which engage the modulus of continuity of the involved continuous function or of its high-order Fréchet derivative. This treatment relies on [19].

## 17.1 Background

The study of the convergence of positive linear operators became more intensive and attractive when P. Korovkin (1953) proved his famous theorem (see [110], p. 14).

**Korovkin's First Theorem** *Let $[a,b]$ be a compact interval in $\mathbf{R}$ and $(L_n)_{n \in \mathbf{N}}$ be a sequence of positive linear operators $L_n$ mapping $C([a,b])$ into itself. Assume that $(L_n f)$ converges uniformly to $f$ for the three test functions $f = 1, x, x^2$. Then $(L_n f)$ converges uniformly to $f$ on $[a,b]$ for all functions of $f \in C([a,b])$.*

So a lot of authors since then have worked on the theoretical aspects of the above convergence. But R. A. Mamedov (1959) (see [124]) was the first to set Korovkin's theorem in a quantitative scheme.

**Mamedov's Theorem** *Let $\{L_n\}_{n \in \mathbf{N}}$ be a sequence of positive linear op-*

erators in the space $C([a,b])$, for which $L_n 1 = 1$, $L_n(t,x) = x + \alpha_n(x)$, $L_n(t^2, x) = x^2 + \beta_x(x)$. Then it holds

$$\|L_n(f,x) - f(x)\|_\infty \leq 3\omega_1(f, \sqrt{d_n}),$$

where $\omega_1$ is the first modulus of continuity and $d_n := \|\beta_n(x) - 2x\alpha_n(x)\|_\infty$.

An improvement of the last result was the following.

**Shisha and Mond's Theorem** (1968, see [159]). *Let $[a,b] \subset \mathbf{R}$ be a compact interval. Let $\{L_n\}_{n \in \mathbf{N}}$ be a sequence of positive linear operators acting on $C([a,b])$. For $n = 1, 2, \ldots$, suppose $L_n(1)$ is bounded. Let $f \in C([a,b])$. Then for $n = 1, 2, \ldots$, it holds*

$$\|L_n f - f\|_\infty \leq \|f\|_\infty \cdot \|L_n 1 - 1\|_\infty + \|L_n(1) + 1\|_\infty \cdot \omega_1(f, \mu_n),$$

*where*

$$\mu_n := \|(L_n((t-x)^2))(x)\|_\infty^{1/2}.$$

Shisha–Mond inequality generated and inspired a lot of research done by many authors worldwide on the rate of *convergence of a sequence of positive linear operators to the unit operator*, always producing similar inequalities however in many different directions, e.g., see the important work of H. Gonska of 1983 in [88], etc.

The author (see [12]) in his 1993 research monograph, produces in many directions best upper bounds for $|(L_n f)(x_0) - f(x_0)|$, $x_0 \in Q \subseteq \mathbf{R}^n$, $n \geq 1$, compact and convex, which lead for the first time to sharp/attained inequalities of Shisha–Mond type. The method of proving is probabilistic from the theory of moments. His pointwise approach is closely related to the study of the weak convergence with rates of a sequence of finite positive measures to the unit measure at a specific point.

All of the above have inspired and motivated the work in this chapter. Here is our setting: Let $X$ be a normed vector space, $Y$ be a Banach lattice; $M \subset X$ is a compact and convex subset. Consider the space of continuous functions from $M$ into $Y$, denoted by $C(M,Y)$; also consider the space of bounded functions $B(M,Y)$. Here we study the rate of the uniform convergence of lattice homomorphisms $T : C(M,Y) \to C(M,Y)$ or $T : C(M,Y) \to B(M,Y)$ to the unit operator $I$.

In our main results here we assume that $X$ is a Banach space. Our presented inequalities are of Shisha–Mond type, i.e., of Korovkin type.

In there we find upper bounds to $\||Tf - f|\|_\infty$, $f \in C(M,Y)$, and $\||TP - P|\|_\infty$, $P \in C^n(M,Y)$, $n \in \mathbf{N}$, (space of $n$-times continuously Fréchet differentiable functions), where $\|| \cdot \||_\infty$ is the supremum norm in $C(M,Y)$

or $B(M,Y)$. These inequalities involve the modulus of continuity of $f$ or $P^{(n)}$. At the end we give several applications.

We note that because $C(M,Y)$ is a Banach lattice, the lattice homomorphism $T$ as described above is also a positive operator and therefore a continuous linear operator.

## 17.2 Auxiliary Results

Let $(X, \|\cdot\|)$, $(Y, \|\cdot\|)$ real normed vector spaces, and let $M \subset X$ be a set. Suppose that $(Y, \|\cdot\|, <)$ is a Banach lattice (see [4], p. 197). Denote by $C(M,Y)$ the vector space of continuous functions from $M$ into $Y$.

**LEMMA 17.1**
*$C(M,Y)$ is a vector lattice.*

**PROOF** Let $f, g \in C(M, Y)$ and $x_n, x \in M$, such that $x_n \to x$, then $f(x_n) \to f(x)$, and $g(x_n) \to g(x)$, as $n \to +\infty$. I.e., $\|f(x_n) - f(x)\| < \varepsilon$, $\|g(x_n) - g(x)\| < \varepsilon$ for $\varepsilon > 0$ arbitrarily small, iff $|f(x_n) - f(x)| < \varepsilon \cdot i$, $|g(x_n) - g(x)| < \varepsilon \cdot i$, $i \in Y^+$, positive cone of $Y$, such that $\|i\| = 1$. Here $|f| := f \vee (-f)$, where $\vee$, $\wedge$ stand for the supremum and infimum, respectively. Denote by $\circ$ either of $\vee$, $\wedge$,

We introduce the order $f < g$ iff $f(x) < g(x)$, all $x \in M$. It holds that

$$(f \vee g)(x) = f(x) \vee g(x),$$

$$(f \wedge g)(x) = f(x) \wedge g(x), \qquad (17.1)$$

$$|f|(x) = |f(x)|, \quad \text{all } x \in M.$$

Clearly here $|f(x)| < |g(x)|$ iff

$$\|f(x)\| < \|g(x)\|, \quad \text{all } x \in M.$$

We see that (cf. Theorem 24.1, [4], p. 194)

$$|f(x_n) \circ g(x_n) - f(x) \circ g(x)|$$

$$\leq |f(x_n) \circ g(x_n) - f(x) \circ g(x_n)| + |f(x) \circ g(x_n) - f(x) \circ g(x)|$$

$$\leq |f(x_n) - f(x)| + |g(x_n) - g(x)| \leq 2\varepsilon \cdot i.$$

Thus
$$|f(x_n) \circ g(x_n) - f(x) \circ g(x)| \leq 2\varepsilon \cdot i,$$
iff
$$\|f(x_n) \circ g(x_n) - f(x) \circ g(x)\| \leq 2\varepsilon.$$
That is,
$$\|(f \circ g)(x_n) - (f \circ g)(x)\| \leq 2\varepsilon,$$
for any $\varepsilon > 0$ small.

Hence $(f \circ g)(x_n) \to (f \circ g)(x)$, i.e., $f \vee g$, $f \wedge g$ are continuous functions. Therefore $C(M, Y)$ is a vector lattice. ∎

From now on we suppose that $M$ is compact. Define for $f \in C(M, Y)$
$$\||f\||_\infty := \sup\{\|f(x)\| : x \in M\}. \tag{17.2}$$

We can easily see that $\||\cdot\||_\infty$ defines a norm on $C(M, Y)$. For $f, g \in C(M, Y)$ we have that, $|f| \leq |g|$ iff $|f|(x) \leq |g|(x)$ iff $|f(x)| \leq |g(x)|$ iff $\|f(x)\| \leq \|g(x)\|$, all $x \in M$. The last implies $\||f\||_\infty \leq \||g\||_\infty$, i.e., if
$$|f| \leq |g| \implies \||f\||_\infty \leq \||g\||_\infty.$$

Thus $\||\cdot\||_\infty$ is a lattice norm, and $C(M, Y)$ is a normed vector lattice, where $M$ is a compact subset of $X$.

**PROPOSITION 17.1**
$C(M, Y)$ is a Banach lattice.

**PROOF**  Let $\{f_n\}_{n \in \mathbb{N}}$ be a Cauchy sequence in $C(M, Y)$. Then given $\varepsilon > 0$ there exists $n_0 \in \mathbb{N}$ such that $\||f_n - f_m\||_\infty < \varepsilon$, all $n, m > n_0$. Therefore, for any $x \in M$ we have
$$\|f_n(x) - f_m(x)\| \leq \||f_n - f_m\||_\infty < \varepsilon,$$
which implies that $\{f_n(x)\}_{n \in \mathbb{N}}$ is a Cauchy sequence in the Banach lattice $Y$. Hence, by completeness of $Y$, $\{f_n(x)\}$ converges in $Y$ for every $x \in M$; let $f(x) := \lim_{n \to +\infty} f_n(x)$, all $x \in M$. Because $\|f_n(x) - f_m(x)\| < \varepsilon$, all $n, m > n_0$, for a fixed $n > n_0$ we obtain that
$$\|f_n(x) - f_m(x)\| < \varepsilon, \quad \text{all } m > n_0.$$

17. Quantitative Korovkin-Type Results

By continuity of $\|\cdot\|$ and taking the limit in the last inequality as $m \to +\infty$ we obtain $\|f_n(x) - f(x)\| < \varepsilon$, true for all $n > n_0$ and all $x \in M$. That is,

$$\||f_n - f|\|_\infty < \varepsilon, \quad \text{all } n > n_0,$$

i.e., $\lim_{n \to +\infty} f_n = f$ in $\|| \cdot \|\|_\infty$.

Let $x_N$, $x \in M$ be such that $x_N \to x$, then $f_n(x_N) \to f_n(x)$, by $f_n \in C(M, Y)$. Next observe that

$$\|f(x_N) - f(x)\| \leq \|f(x_N) - f_n(x_N)\| + \|f_n(x_N) - f_n(x)\|$$

$$+ \|f_n(x) - f(x)\| \leq \varepsilon + \varepsilon + \varepsilon,$$

by $f_n \to f$ and $f_n$-continuity. I.e., $\|f(x_N) - f(x)\| \leq 3\varepsilon$, $\varepsilon > 0$ small. Thus $f(x_N) \to f(x)$, as $N \to +\infty$, i.e., $f \in C(M, Y)$. That is, $C(M, Y)$ is complete, proving that it is a Banach lattice. ∎

**DEFINITION 17.1** Let $T : C(M, Y) \to C(M, Y)$ a linear operator. $T$ is called a lattice homomorphism iff it fulfills one of the following equivalent statements:

(i) $T(f \vee g) = T(f) \vee T(f)$, all $f, g \in C(M, Y)$,

(ii) $T(f \wedge g) = T(f) \wedge T(g)$, all $f, g \in C(M, Y)$,

(iii) $T(f) \wedge T(g) = 0$ holds whenever $f \wedge g = 0$,

(iv) $|T(f)| = T(|f|)$, all $f \in C(M, Y)$, see [4], p. 202.

Clearly a lattice homomorphism is a positive one, i.e., whenever $f \geq g$ we get that $T(f) \geq T(g)$, $f, g \in C(M, Y)$. Because $C(M, Y)$ is a Banach lattice, then a positive operator $T$ from $C(M, Y)$ into itself is a continuous one, see [4], p. 200.

In this chapter, we are dealing mainly with lattice homomorphisms $T : C(M, Y) \to C(M, Y)$. We use the following auxiliary results.

**PROPOSITION 17.2**
Let $f \in C(M, Y)$ and $T : C(M, Y) \to C(M, Y)$ a continuous linear operator. Then $(T(f(x_0)))(x_0)$ is a continuous function of $x_0 \in M$.

**PROOF** Let $x_n, x_0 \in M$ be such that $x_n \to x_0$, then $\rho_n(x) := f(x_n) \to f(x_0) =: \rho(x)$, all $x \in M$, i.e., $\rho_n \to \rho$ uniformly. By continuity of $T$ we have that $T(\rho_n) \to T(\rho)$ uniformly, i.e., $T(f(x_n)) \to T(f(x_0))$ uniformly. (Here $T(f(x_n)), T(f(x_0)) \in C(M, Y)$.) That is, $T(f(x_n)) - T(f(x_0)) \xrightarrow{u} 0$, i.e., given $\varepsilon > 0$ there exists $n_0 \in \mathbf{N}$ such that $|||T(f(x_n)) - T(f(x_0))|||_\infty < \varepsilon$, all $n > n_0$.

Thus given $\varepsilon > 0$ there exists $n_0 \in \mathbf{N}$ such that

$$\|(T(f(x_n)))(x_n) - T(f(x_0)))(x_n)\|$$

$$\leq |||T(f(x_n)) - T(f(x_0))|||_\infty < \varepsilon,$$

all $n > n_0$. Notice that (as $x_n \to x_0$)

$$\|(T(f(x_n)))(x_n) - (T(f(x_0)))(x_0)\| \leq \|(T(f(x_n)))(x_n) - (T(f(x_0)))(x_n)\|$$

$$+ \|T(f(x_0)))(x_n) - (T(f(x_0)))(x_0)\| < 2\varepsilon.$$

The last proves the claim of the proposition. ■

**REMARK 17.1** From Proposition 17.2 we find that $(T(f - f(x_0)))(x_0)$ is a continuous function of $x_0 \in M$ with values in $Y$. Also let $i \in Y^+$ be such that $\|i\| = 1$, then $T(i) \in C(M, Y)$. ■

### PROPOSITION 17.3

Let $T : C(M, Y) \to C(M, Y)$ a continuous linear operator and $r > 0$. Then $(T(\|x - x_0\|^r \cdot i))(x_0)$ is a continuous function of $x_0 \in M$ with values in $Y$, where $i \in Y^+$ is such that $\|i\| = 1$.

**PROOF** Let $x_n, x_0 \in M$ be such that $x_n \to x_0$, as $n \to +\infty$. See that

$$\|(T(\|x - x_n\|^r \cdot i))(x_n) - (T(\|x - x_0\|^r \cdot i))(x_0)\|$$

$$\leq \|(T(\|x - x_n\|^r \cdot i))(x_n) - (T(\|x - x_0\|^r \cdot i))(x_n)\|$$

$$+ \|(T(\|x - x_0\|^r \cdot i))(x_n) - (T(\|x - x_0\|^r \cdot i))(x_0)\| =: (*).$$

Notice that $\|x - x_0\|^r \cdot i$ is a continuous function of $x$ and so is $(T(\|x - x_0\|^r \cdot i))(x)$. That is,

$$(T(\|x - x_0\|^r \cdot i))(x_n) \to (T(\|x - x_0\|^r \cdot i))(x_0),$$

i.e.,
$$\|(T(\|x-x_0\|^r \cdot i))(x_n) - (T(\|x-x_0\|^r \cdot i))(x_0)\| < \varepsilon_1,$$

where $\varepsilon_1 > 0$ is small.

Hence
$$(*) < \||T(\|x-x_n\|^r \cdot i) - T(\|x-x_0\|^r \cdot i)|\|_\infty + \varepsilon_1 =: (**).$$

Observe that
$$\sup_{x \in M} \big|\|x-x_n\| - \|x-x_0\|\big| \leq \|x_n - x_0\| \to 0,$$

thus
$$\big\|\|x-x_n\| - \|x-x_0\|\big\|_{\infty,x} \to 0, \quad (\|\cdot\|_{\infty,x} \text{ supremum in } x),$$

i.e.,
$$\|x-x_n\| \xrightarrow{u} \|x-x_0\|, \quad \text{uniformly}.$$

Here $\|x-y\| \leq \Delta$ — the diameter of $M$, all $x, y \in M$, i.e., $\|x-x_n\|$, $\|x-x_0\| \leq \Delta$.

It was proved in [38], pp. 20–21, that $\|x-x_n\|^r \xrightarrow{u} \|x-x_0\|^r$, uniformly, i.e., for $\varepsilon > 0$ there exists $n_0 \in \mathbf{N}$ such that
$$\big\|\|x-x_n\|^r - \|x-x_0\|^r\big\|_{\infty,r} < \varepsilon,$$

for all $n > n_0$.

See that
$$\big\|\|x-x_n\|^r \cdot i - \|x-x_0\|^r \cdot i\big\|_\infty = \sup_{x \in M} \|\|x-x_n\|^r \cdot i - \|x-x_0\|^r \cdot i\|$$

$$= \sup_{x \in M} \big|\|x-x_n\|^r - \|x-x_0\|^r\big| \cdot \|i\|$$

$$= \big\|\|x-x_n\|^r - \|x-x_0\|^r\big\|_{\infty,x} < \varepsilon.$$

Therefore,
$$\big\|\|x-x_n\|^r \cdot i - \|x-x_0\|^r \cdot i\big\|_\infty < \varepsilon, \quad \text{all } n > n_0,$$

i.e.,
$$\|id - x_n\|^r \cdot i \xrightarrow{u} \|id - x_0\|^r \cdot i, \quad \text{uniformly}$$

($id$ is the identity map). By the continuity of operator $T$ we obtain that

$$T(\|x - x_n\|^r \cdot i) \xrightarrow{u} T(\|x - x_0\|^r \cdot i), \quad \text{uniformly.}$$

I.e.,
$$\||T(\|x - x_n\|^r \cdot i) - T(\|x - x_0\|^r \cdot i)|\|_\infty < \varepsilon_1,$$

where $\varepsilon_1 > 0$ is small as above. Consequently

$$(**) < 2\varepsilon_1.$$

We have shown that

$$(T(\|x - x_n\|^r) \cdot i))(x_n) \to (T(\|x - x_0\|^r \cdot i))(x_0),$$

as $x_n \to x_0$ and $n \to +\infty$. That is we have established the claim of the proposition. ∎

From now on we suppose that $M$ is a compact and convex subset of $X$.

**DEFINITION 17.2** Let $f \in C(M, Y)$, its (first) modulus of continuity is defined by

$$\omega_1(f, h) := \sup\{\|f(x) - f(x)\| :$$

$$\text{all } x, y \in M \text{ such that } \|x - y\| \le h\}, \quad (17.3)$$

$h > 0$, Here $Y$ can be just a normed vector space.

**LEMMA 17.2**
(From Lemma 7.11, [12], p. 208). Let $f \in C(M, Y)$, $h > 0$ and fixed $x_0 \in M$. Then

$$\|f(x) - f(x_0)\| \le \omega_1(f, h) \cdot \left\lceil \frac{\|x - x_0\|}{h} \right\rceil$$

$$\le \omega_1(f, h) \cdot \left(1 + \frac{\|x - x_0\|}{h}\right), \quad \text{all } x \in M. \quad (17.4)$$

## 17. Quantitative Korovkin-Type Results

*Here $\lceil \cdot \rceil$ stands for the ceiling of the number. Here $Y$ can be just a normed vector space. Clearly one can get that*

$$\|f(x) - f(x_0)\| \le \left\| \omega_1(f, h) \cdot \left(1 + \frac{\|x - x_0\|}{h}\right) \cdot i \right\|, \qquad (17.5)$$

*where $i \in Y^+$ such that $\|i\| = 1$. Hence, for $f \in C(M, Y)$ and $h > 0$ we find that*

$$\||f - f(x_0)|\|_\infty \le \left\|\left\| \omega_1(f, h) \cdot \left(1 + \frac{\|id - x_0\|}{h}\right) \cdot i \right\|\right\|_\infty, \qquad (17.6)$$

$x_0 \in M$ be fixed.

From now on in this section we assume that $X$ is a Banach space and that $P$ maps $X$ into the Banach lattice $Y$. Here $M \subset X$ still is a compact and convex subset. Moreover we assume that $P$ is $n$-times continuously Fréchet differentiable on $M$, i.e., $P|_M \in C^n(M, Y)$. Clearly $P|_M \in C(M, Y)$, which is a Banach lattice. That is, here $(P|_M)^{(k)}$ is a continuous map from $M$ into the space of all $k$-linear bounded operators from $X$ into $Y$, all $k = 1, \ldots, n$. It is obvious that all $(P|_M)^{(k)}$, $k = 0, 1, \ldots, n$ are norm bounded by continuity, and thus they are integrable. Here for any $x_0, x_1 \in M$ we put $x(\theta) := \theta x_1 + (1 - \theta) x_0 \in M$, $0 \le \theta \le 1$, and we identify $P(\theta) \equiv P(x(\theta))$. That is, $P^{(k)}(\theta) = P^{(k)}(x(\theta))(x_1 - x_0)^k$, where $(x_1 - x_0)^k := (x_1 - x_0, \ldots, x_1 - x_0)$ is a $k$-tuple, $k = 1, \ldots, n$. We are following [149], pp. 87–127, Chapter 3.

In particular from [149], p. 124, Theorem 20.2 (Taylor's Theorem) for any $x_0, x_1 \in M$ we find that

$$P(x_1) = P(x_0) + \sum_{k=1}^n \frac{1}{k!} \cdot P^{(k)}(x_0)(x_1 - x_0)^k + \mathcal{R}_n(x_0, x_1), \qquad (17.7)$$

where

$$\mathcal{R}_n(x_0, x_1) := \int_0^1 (P^{(n)}(\theta x_1 + (1-\theta)x_0)(x_1 - x_0)^n$$

$$- P^{(n)}(x_0)(x_1 - x_0)^n) \cdot \frac{(1-\theta)^{n-1}}{(n-1)!} \cdot d\theta, \qquad (17.8)$$

is a vector valued abstract Riemann integral. Set

$$\Delta(x_1, x_0) := P(x_1) - P(x_0) - \sum_{k=1}^{n} \frac{1}{k!} P^{(k)}(x_0)(x_1 - x_0)^k. \qquad (17.9)$$

Notice that
$$\Delta(x_1, x_0) = \mathcal{R}_n(x_0, x_1). \qquad (17.10)$$

It is obvious that
$$\Delta(\bullet, x_0) = \mathcal{R}_n(x_0, \bullet) \in C(M, Y).$$

Consider also
$$\phi_n(t) := \int_0^{|t|} \left\lceil \frac{s}{h} \right\rceil \cdot \frac{(|t| - s)^{n-1}}{(n-1)!} \cdot ds, \quad n \in \mathbf{N}, \ t \in \mathbf{R}, \qquad (17.11)$$

which is a continuous function in $t$. From [12], p. 210, we observe that

$$\phi_n(\|x_1 - x_0\|)$$
$$\leq \left( \frac{\|x_1 - x_0\|^{n+1}}{(n+1)! \cdot h} + \frac{\|x_1 - x_0\|^n}{2 \cdot n!} + \frac{h \cdot \|x_1 - x_0\|^{n-1}}{8 \cdot (n-1)!} \right). \qquad (17.12)$$

Clearly, $i \cdot \phi_n(\|x - x_0\|) \in C(M, Y)$, where $i \in Y^+$ such that $\|i\| = 1$.
We see the following

$$\|\mathcal{R}_n(x_0, x_1)\| = \left\| \int_0^1 (P^{(n)}(\theta x_1 + (1-\theta)x_0) \right.$$

$$\left. - P^{(n)}(x_0))(x_1 - x_0)^n \cdot \frac{(1-\theta)^{n-1}}{(n-1)!} \cdot d\theta \right\|$$

$$\leq \int_0^1 \|(P^{(n)}(\theta x_1 + (1-\theta)x_0)$$

$$- P^{(n)}(x_0))(x_1 - x_0)^n\| \cdot \frac{(1-\theta)^{n-1}}{(n-1)!} \cdot d\theta$$

$$\leq \int_0^1 \|(P^{(n)}(\theta x_1 + (1-\theta)x_0)$$

## 17. Quantitative Korovkin-Type Results

$$- P^{(n)}(x_0) \| \cdot \|x_1 - x_0\|^n \cdot \frac{(1-\theta)^{n-1}}{(n-1)!} \cdot d\theta \leq (*).$$

Remember that $P^{(n)}(x)$, $x \in M$, is an $n$-linear bounded operator from $X$ into $Y$.

We define $(h > 0)$

$$\omega_1(P^{(n)}, h) := \sup\{\|P^{(n)}(x) - P^{(n)}(y)\| :$$

$$\text{all } x, y \in M \text{ such that } \|x - y\| \leq h\}. \tag{17.13}$$

By Lemma 17.2 we get that

$$(*) \leq \int_0^1 \omega_1(P^{(n)}, h) \cdot \left\lceil \frac{\theta \cdot \|x_1 - x_0\|}{h} \right\rceil \cdot \|x_1 - x_0\|^n \cdot \frac{(1-\theta)^{n-1}}{(n-1)!} \cdot d\theta$$

$$= \omega_1(P^{(n)}, h) \cdot \|x_1 - x_0\|^n \cdot \int_0^1 \left\lceil \frac{\theta \cdot \|x_1 - x_0\|}{h} \right\rceil \cdot \frac{(1-\theta)^{n-1}}{(n-1)!} \cdot d\theta$$

$$= \omega_1(P^{(n)}, h) \cdot \phi_n(\|x_1 - x_0\|).$$

I.e., we have shown that

$$\|\mathcal{R}_n(x_0, x_1)\| \leq \omega_1(P^{(n)}, h) \cdot \phi_n(\|x_1 - x_0\|). \tag{17.14}$$

Now by (17.12) we see that

$$\|\mathcal{R}_n(x_0, x_1)\| \leq \omega_1(P^{(n)}, h) \cdot \left\{ \frac{\|x_1 - x_0\|^{n+1}}{(n+1)! \cdot h} + \frac{\|x_1 - x_0\|^n}{2 \cdot n!} \right.$$

$$\left. + \frac{h \cdot \|x_1 - x_0\|^{n-1}}{8 \cdot (n-1)!} \right\} < +\infty, \tag{17.15}$$

for all $x_0, x_1 \in M$. Here $0 < \omega_1(P^{(n)}, h) < +\infty$, by $M$ being compact and $P^{(n)}$ being continuous. Notice also that R.H.S. (17.15) is continuous in $x_1 \in M$. Hence

$$\|\Delta(x, x_0)\| \leq \left\| \omega_1(P^{(n)}, h) \cdot \left\{ \frac{\|x - x_0\|^{n+1}}{(n+1)! \cdot h} + \frac{\|x - x_0\|^n}{2 \cdot n!} \right. \right.$$

$$+ \frac{h \cdot \|x - x_0\|^{n-1}}{8 \cdot (n-1)!} \biggr\} \cdot i \biggr\|, \tag{17.16}$$

all $x \in M$, $i \in Y^+$ such that $\|i\| = 1$. Clearly the function within the long $\|\cdot\|$ in the R.H.S. (17.16) belongs to $C(M,Y)$. Thus

$$\|\|\Delta(\cdot, x_0)\|\|_\infty \leq \|\|\omega_1(P^{(n)}, h) \cdot \biggl\{ \frac{\|id - x_0\|^{n+1}}{(n+1)! \cdot h}$$

$$+ \frac{\|id - x_0\|^n}{2 \cdot n!} + \frac{h \cdot \|id - x_0\|^{n-1}}{8 \cdot (n-1)!} \biggr\} \cdot i\|\|_\infty < +\infty. \tag{17.17}$$

Obviously $C^n(M,Y) \subset C(M,Y)$.

As a reminder, we are going to be dealing with lattice homomorphisms $T$ from $C(M,Y)$ into itself. We use the following

### PROPOSITION 17.4

Let $T : C(M,Y) \to C(M,Y)$ be a continuous linear operator, $P : X \to Y$ and $P|_M \in C^n(M,Y)$ (in the Fréchet sense). Then $(T(P^{(k)}(x_0)(x - x_0)^k))(x_0)$ is a continuous function for any $x_0 \in M$, where $(x - x_0)^k := (x - x_0, \ldots, x - x_0)$, k-tuple; $1 \leq k \leq n$, $x \in M$.

**PROOF** Let $x_n, x_0 \in M$ such that $x_n \to x_0$ as $n \to +\infty$. We prove that

$$(T(P^{(n)}(x_n)(x - x_n)^k))(x_n) \to (T(P^{(k)}(x_0)(x - x_0)^k))(x_0).$$

For that we see

$$\|(T(P^{(k)}(x_n)(x - x_n)^k))(x_n) - (T(P^{(k)}(x_0)(x - x_0)^k))(x_0)\|$$

$$\leq \|(T(P^{(k)}(x_n)(x - x_n)^k))(x_n) - (T(P^{(k)}(x_0)(x - x_0)^k))(x_n)\|$$

$$+ \|(T(P^{(k)}(x_0)(x - x_0)^k))(x_n) - (T(P^{(k)}(x_0)(x - x_0)^k))(x_0)\|$$

$$=: (*).$$

Here $j(x) := (x - x_0)^k$ is continuous in $x \in X$ and the continuous linear

## 17. Quantitative Korovkin-Type Results

operator $P^{(k)}(x_0)$ maps $X^k$ into $Y$, i.e., $P^{(k)}(x_0)(x-x_0)^k \in Y$. Notice that

$$(P^{(k)}(x_0) \circ j)(x) = P^{(k)}(x_0)(x-x_0)^k \text{ is continuous in } x \in X.$$

I.e., $P^{(k)}(x_0)(x-x_0)^k := P^{(k)}(x_0)(x-x_0)^k\big|_M \in C(M,Y)$, therefore

$$T(P^{(k)}(x_0)(x-x_0)^k) \in C(M,Y).$$

Hence for arbitrarily small $\varepsilon_1 > 0$ we have that

$$\|(T(P^{(k)}(x_0)(x-x_0)^k))(x_n) - (T(P^{(k)}(x_0)(x-x_0)^k))(x_0)\| < \varepsilon_1.$$

Thus

$$(*) < \||(T(P^{(k)}(x_n)(x-x_n)^k)) - (T(P^{(k)}(x_0)(x-x_0)^k))\||_\infty + \varepsilon_1. \quad (17.18)$$

We see that

$$\||P^{(k)}(x_n)(x-x_n)^k - P^{(k)}(x_0)(x-x_0)^k\||_\infty$$

$$\leq \||P^{(k)}(x_n)(x-x_n)^k - P^{(k)}(x_n)(x-x_0)^k\||_\infty$$

$$+ \||P^{(k)}(x_n)(x-x_0)^k - P^{(k)}(x_0)(x-x_0)^k\||_\infty =: A + B.$$

By assumption $P^{(k)}$ maps $M$ into the space of $k$-linear bounded operators from $X^k$ into $Y$ and $P^{(k)}$ is assumed to be continuous. I.e., $P^{(k)}(x_n) \to P^{(k)}(x_0)$ as $x_n \to x_0$, that is, $\|P^{(k)}(x_n) - P^{(k)}(x_0)\| \to 0$. Also it holds $\|x - x_0\| < \text{diameter}(M) =: d(M) < +\infty$. Consequently we get that

$$\|P^{(k)}(x_n)(x-x_0)^k - P^{(k)}(x_0)(x-x_0)^k\|$$

$$= \|(P^{(k)}(x_n) - P^{(k)}(x_0))(x-x_0)^k\|$$

$$\leq \|P^{(k)}(x_n) - P^{(k)}(x_0)\| \cdot \|x - x_0\|^k$$

$$\leq \|P^{(k)}(x_n) - P^{(k)}(x_0)\| \cdot (d(M))^k \to 0.$$

That is, for arbitrary small $\varepsilon > 0$ we find that

$$B := \||P^{(k)}(x_n)(x-x_0)^k - P^{(k)}(x_0)(x-x_0)^k\||_\infty < \varepsilon,$$

for all $n > n_0 \in \mathbf{N}$. By assumption we see that $\|P^{(k)}(x)\| < \gamma < +\infty$, for all $x \in M$. Therefore,

$$\|P^{(k)}(x_n)(x - x_n)^k - P^{(k)}(x_n)(x - x_0)^k\|$$

$$= \|P^{(k)}(x_n)((x - x_n)^k - (x - x_0)^k)\| = \|P^{(k)}(x_n)(x_0 - x_n)^k\|$$

$$\leq \|P^{(k)}(x_n)\| \cdot \|x_n - x_0\|^k < \gamma \cdot \|x_n - x_0\|^k \to 0.$$

Finally, we can get that

$$A := \||P^{(k)}(x_n)(x - x_n)^k - P^{(k)}(x_n)(x - x_0)^k\|_\infty \leq \varepsilon,$$

for all $n > n_1 \in \mathbf{N}$. Furthermore

$$\||P^{(k)}(x_n)(x - x_n)^k - P^{(k)}(x_0)(x - x_0)^k\|_\infty \leq 2\varepsilon,$$

for all $n > \max(n_0, n_1)$, i.e.,

$$P^{(k)}(x_n)(x - x_n)^k \xrightarrow{u} P^{(k)}(x_0)(x - x_0)^k,$$

uniformly. Because $T$ is continuous, we obtain that

$$T(P^{(k)}(x_n)(x - x_n)^k) \xrightarrow{u} T(P^{(k)}(x_0)(x - x_0)^k),$$

uniformly. So for sufficiently large $n$ we find that

$$\||T(P^{(k)}(x_n)(x - x_n)^k) - T(P^{(k)}(x_0)(x - x_0)^k)\|_\infty \leq \varepsilon_1.$$

Consequently from (17.18) we get $(*) \leq 2\varepsilon_1$. Hence

$$\|(T(P^{(k)}(x_n)(x - x_n)^k))(x_n) - (T(P^{(k)}(x_0)(x - x_0)^k))(x_0)\| \to 0,$$

as $x_n \to x_0$. We have proved the claim of the proposition. ■

## 17.3 Quantitative General Theorems

Next comes the first general result

## THEOREM 17.1

Let $M$ be a compact and convex subset of $(X, \|\cdot\|)$ and $(Y, \|\cdot\|, <)$ is a Banach lattice. Let $T$ be a lattice homomorphism from $C(M, Y)$ into itself, and $f \in C(M, Y)$. Then it holds

$$\||Tf - f\||_\infty \leq \||((T(f(x_0))) - f)(x_0)\||_{\infty, x_0}$$

$$+ \omega_1(f, \||(T(\|x - x_0\| \cdot i))(x_0)\||_{\infty, x_0}) \cdot (1 + \||T(i)\||_\infty), \quad (17.19)$$

where $\|\|\cdot\|\|_\infty$, $\|\|\cdot\|\|_{\infty, x_0}$ are the supremum norms taken over $M$ and over all $x_0 \in M$, respectively, and $i \in Y^+$ is such that $\|i\| = 1$.

**REMARK 17.2** Notice that R.H.S. (17.19) is finite. This comes from the definition of $T$, $M$ being compact, $f \in C(M, Y)$, Proposition 17.2 and Proposition 17.3. ∎

**PROOF** (of Theorem 17.1) We see that

$$(Tf)(x_0) - f(x_0)$$

$$= (Tf)(x_0) - (T(f(x_0))(x_0) + (T(f(x_0)))(x_0) - f(x_0)$$

$$= [(T(f - f(x_0)))(x_0)] + [(T(f(x_0)))(x_0) - f(x_0)].$$

Hence

$$\|(Tf)(x_0) - f(x_0)\|$$

$$\leq \|(T(f - f(x_0)))(x_0)\| + \|(T(f(x_0)))(x_0) - f(x_0)\|.$$

Thus

$$\||Tf - f\||_\infty \leq \||(T(f - f(x_0)))(x_0)\||_{\infty, x_0}$$

$$+ \||(T(f(x_0)))(x_0) - f(x_0)\||_{\infty, x_0}. \quad (17.20)$$

From Remark 17.1 we have that $(T(f - f(x_0)))(x_0)$ is a continuous function for any $x_0 \in M$ with values in $Y$, therefore its supremum norm is finite.

We notice also that ($h > 0$)

$$\left(T\left(\left(1 + \frac{\|x - x_0\|}{h}\right) \cdot i\right)\right)(x_0)$$

$$= (T(i))(x_0) + \frac{1}{h} \cdot (T(\|x - x_0\| \cdot i))(x_0))$$

is a continuous function for any $x_0 \in M$. By Lemma 17.2 we obtain that

$$\|f(x) - f(x_0)\| \le \omega_1(f, h) \cdot \left(1 + \frac{\|x - x_0\|}{h}\right), \quad \text{all } x, x_0 \in M.$$

That is,

$$\|f(x) - f(x_0)\| \le \left\|\omega_1(f, h) \cdot \left(1 + \frac{\|x - x_0\|}{h}\right) \cdot i\right\|,$$

iff

$$|f(x) - f(x_0)| \le \left|\omega_1(f, h) \cdot \left(1 + \frac{\|x - x_0\|}{h}\right) \cdot i\right|,$$

iff

$$|f - f(x_0)|(x) \le \left|\omega_1(f, h) \cdot \left(1 + \frac{\|id - x_0\|}{h}\right) \cdot i\right|(x), \quad \text{all } x \in M.$$

Hence by the positivity of $T$ and being a lattice homomorphism we get that

$$T|f - f(x_0)| \le T\left|\omega_1(f, h) \cdot \left(1 + \frac{\|id - x_0\|}{h}\right) \cdot i\right|$$

and

$$|T(f - f(x_0))| \le \left|T\left(\omega_1(f, h) \cdot \left(1 + \frac{\|id - x_0\|}{h}\right)i\right)\right|,$$

iff

$$|T(f - f(x_0))|(x)$$

$$\le \left|T\left(\omega_1(f, h) \cdot \left(1 + \frac{\|id - x_0\|}{h}\right) \cdot i\right)\right|(x), \quad \text{all } x \in M.$$

## 17. Quantitative Korovkin-Type Results

Moreover we have

$$|(T(f - f(x_0)))(x)|$$

$$\leq \left|\left(T\left(\omega_1(f,h)\cdot\left(1 + \frac{\|id - x_0\|}{h}\right)\cdot i\right)\right)(x)\right|, \text{ all } x \in M.$$

Since Lemma 17.2 is true for any $x_0 \in M$, we obtain that

$$|(T(f - f(x_0)))(x_0)|$$

$$\leq \left|\left(T\left(\omega_1(f,h)\cdot\left(1 + \frac{\|id - x_0\|}{h}\right)\cdot i\right)\right)(x_0)\right|,$$

true for any $x_0 \in M$. And because $Y$ is a normed vector lattice, the last inequality implies

$$\|(T(f - f(x_0)))(x_0)\|$$

$$\leq \omega_1(f,h)\cdot\left\|\left(T\left(\left(1 + \frac{\|id - x_0\|}{h}\right)\cdot i\right)\right)(x_0)\right\|,$$

which consequently implies

$$\||(T(f - f(x_0)))(x_0)\||_{\infty, x_0}$$

$$\leq \omega_1(f,h)\cdot\left\|\left|\left(T\left(\left(1 + \frac{\|id - x_0\|}{h}\right)\cdot i\right)\right)(x_0)\right|\right\|_{\infty, x_0}$$

$$= \omega_1(f,h)\cdot\left\|\left|(T(i))(x_0) + \frac{1}{h}\cdot(T(\|id - x_0\|\cdot i))(x_0)\right|\right\|_{\infty, x_0}$$

$$\leq \omega_1(f,h)\cdot\left(\||T(i)\||_{\infty} + \frac{1}{h}\cdot\||T(\|id - x_0\|\cdot i))(x_0)\||_{\infty, x_0}\right).$$

Choosing
$$h := \||(T(\|id - x_0\|\cdot i))(x_0)\||_{\infty, x_0}, \tag{17.21}$$

we obtain that

$$\||(T(f - f(x_0)))(x_0)\||_{\infty, x_0}$$

$$\leq \omega_1(f, |||(T(\|id - x_0\| \cdot i))(x_0)|||_{\infty,x_0}) \cdot (1 + |||T(i)|||_{\infty}). \quad (17.22)$$

It is clear now that inequalities (17.20) and (17.22) imply inequality (17.19). The theorem is now established. ∎

In the following we present the second general result.

**THEOREM 17.2**

*Let $X$ be a Banach space, $Y$ be a Banach lattice, $M$ be a compact and convex subset of $X$, $h > 0$, and $T$ is a lattice homomorphism from $C(M,Y)$ into itself. Consider a function $P$ from $X$ into $Y$ such that $P\big|_M \in C^n(M,Y)$, $n \in \mathbf{N}$. Then it holds*

$$|||TP - P|||_{\infty}$$

$$\leq |||(T(P(x_0)))(x_0) - P(x_0)|||_{\infty,x_0}$$

$$+ \sum_{k=1}^{n} \frac{1}{k!} \cdot |||(T(P^{(k)}(x_0)(x - x_0)^k))(x_0)|||_{\infty,x_0}$$

$$+ \omega_1(P^{(n)}, h) \cdot \left[ \frac{1}{(n+1)! \cdot h} \cdot |||(T(\|x - x_0\|^{n+1} i))(x_0)|||_{\infty,x_0} \right.$$

$$+ \frac{1}{2 \cdot n!} \cdot |||(T(\|x - x_0\|^n \cdot i))(x_0)|||_{\infty,x_0}$$

$$\left. + \frac{h}{8 \cdot (n-1)!} \cdot |||(T(\|x - x_0\|^{n-1} \cdot i))(x_0)|||_{\infty,x_0} \right]. \quad (17.23)$$

*Here $i \in Y^+$ is such that $\|i\| = 1$, and $|||\cdot|||_{\infty}$, $|||\cdot|||_{\infty,x_0}$ are the supremum norms taken over $M$ and over all $x_0 \in M$, respectively.*

**REMARK 17.3** Observe that R.H.S. (17.23) is finite. This is implied from the definition of $T$, $M$ being compact, $P\big|_M \in C^n(M,Y)$, Proposition 17.2, Proposition 17.3 and Proposition 17.4. ∎

**COROLLARY 17.1**
(Same setting and assumptions as in Theorem 17.2) *Pick*

$$h := h^* := \frac{1}{(n+1)!} \cdot \max\{|||(T(\|x-x_0\|^{n+1} \cdot i))(x_0)|||_{\infty,x_0},$$

$$|||(T(\|x-x_0\|^{n} \cdot i))(x_0)|||_{\infty,x_0},$$

$$|||(T(\|x-x_0\|^{n-1} \cdot i))(x_0)|||_{\infty,x_0}\}. \tag{17.24}$$

*Then it holds*

$$|||TP - P|||_\infty$$

$$\leq |||(T(P(x_0)))(x_0) - P(x_0)|||_{\infty,x_0}$$

$$+ \sum_{k=1}^{n} \frac{1}{k!} \cdot |||(T(P^{(k)}(x_0)(x-x_0)^k))(x_0)|||_{\infty,x_0}$$

$$+ \omega_1(P^{(n)}, h^*) \cdot \left[1 + \left(\frac{n+1}{2}\right) \cdot h^* + \frac{n \cdot (n+1)}{8} \cdot h^{*2}\right]. \tag{17.25}$$

**PROOF**  Notice that

$$\psi := \frac{1}{(n+1)!} \cdot |||(T(\|x-x_0\|^{n+1} \cdot i))(x_0)|||_{\infty,x_0} \leq h^*,$$

i.e., $\frac{\psi}{h^*} \leq 1$. Also we observe that

$$\frac{1}{2 \cdot n!} \cdot |||(T(\|x-x_0\|^{n} \cdot i))(x_0)|||_{\infty,x_0} \leq \frac{(n+1)}{2} \cdot h^*,$$

and

$$\frac{h^*}{8 \cdot (n-1)!} \cdot |||T(\|x-x_0\|^{n-1} \cdot i))(x_0)|||_{\infty,x_0} \leq \frac{(h^*)^2}{8} \cdot n \cdot (n+1).$$

Thus,

$$\text{Remainder}(17.23) \leq \omega_1(P^{(n)}, h^*)$$

$$\cdot \left[1 + \left(\frac{n+1}{2}\right) \cdot h^* + \frac{n \cdot (n+1)}{8} \cdot (h^*)^2 \right].$$

∎

**PROOF** (of Theorem 17.2) From (17.7) we have on $M$ that

$$P(\bullet) = P(x_0) + \sum_{k=1}^{n} \frac{1}{k!} \cdot P^{(k)}(x_0)(\bullet - x_0)^k + \mathcal{R}_n(x_0, \bullet), \qquad (17.26)$$

where $x_0 \in M$. Hence

$$(T(P))(x_0) - (T(P(x_0)))(x_0) - \sum_{k=1}^{n} \frac{1}{k!} \cdot (T(P^{(k)}(x_0)(id - x_0)^k))(x_0)$$

$$= (T(\mathcal{R}_n(x_0, \cdot)))(x_0) \in C(M, Y), \qquad (17.27)$$

by Propositions 17.2 and 17.4. Also we obtain

$$(TP)(x_0) - P(x_0)$$

$$= (TP)(x_0) - (T(P(x_0)))(x_0) + (T(P(x_0)))(x_0) - P(x_0)$$

$$= [(T(P - P(x_0)))(x_0)] + [(T(P(x_0)))(x_0) - P(x_0)].$$

Therefore,

$$\|(TP)(x_0) - P(x_0)\| \le \|(T(P - P(x_0)))(x_0)\|$$

$$+ \|(T(P(x_0)))(x_0) - P(x_0)\|, \qquad (17.28)$$

and so we obtain

$$\||TP - P\||_\infty \le \||(T(P - P(x_0)))(x_0)|\|_{\infty, x_0}$$

$$+ \||(T(P(x_0)))(x_0) - P(x_0)|\|_{\infty, x_0}. \qquad (17.29)$$

Notice that all functions involved in (17.26) belong to $C(M, Y)$, which is a normed vector lattice.

Consequently,

$$|(T(P - P(x_0)))(x)|$$

$$= \left|\left(T\left(\sum_{k=1}^{n} \frac{1}{k!} \cdot P^{(k)}(x_0)(id - x_0)^k + \mathcal{R}_n(x_0, x)\right)\right)(x)\right|, \quad (17.30)$$

valid for all $x \in M$, for arbitrary $x_0 \in M$. I.e., we find that

$$|(T(P - P(x_0)))(x_0)|$$

$$= \left|\left(T\left(\sum_{k=1}^{n} \frac{1}{k!} \cdot P^{(k)}(x_0)(id - x_0)^k + \mathcal{R}_n(x_0, x)\right)\right)(x_0)\right| \quad (17.31)$$

is true for any $x_0 \in M$. Because $Y$ is a normed vector lattice the last implies that

$$\|(T(P - P(x_0)))(x_0)\|$$

$$= \left\|\left(T\left(\sum_{k=1}^{n} \frac{1}{k!} \cdot P^{(k)}(x_0)(id - z_0)^k + \mathcal{R}_n(x_0, x)\right)\right)(x_0)\right\|. \quad (17.32)$$

Also it holds

$$|||(T(P - P(x_0)))(x_0)|||_{\infty, x_0}$$

$$= \left|\left|\left|\left(T\left(\sum_{k=1}^{n} \frac{1}{k!} \cdot P^{(k)}(x_0)(id - x_0)^k + \mathcal{R}_n(x_0, x)\right)\right)(x_0)\right|\right|\right|_{\infty, x_0}$$

$$\leq \sum_{k=1}^{n} \frac{1}{k!} \cdot |||(T(P^{(k)}(x_0)(id - x_0)^k))(x_0)|||_{\infty, x_0}$$

$$+ |||(T(\mathcal{R}_n(x_0, x)))(x_0)|||_{\infty, x_0}. \quad (17.33)$$

From (17.15) we get that

$$|||\mathcal{R}_n(x_0, x)|||_\infty \leq |||i \cdot \omega_1(P^{(n)}, h) \cdot \left\{\frac{\|x - x_0\|^{n+1}}{(n+1)! \cdot h} + \frac{\|x - x_0\|^n}{2 \cdot n!}\right.$$

$$+ \frac{h \cdot \|x - x_0\|^{n-1}}{8 \cdot (n-1)!} \bigg\} \|\|_\infty < +\infty, \qquad (17.34)$$

valid for arbitrary $x_0 \in M$. Since $C(M,Y)$ is a Banach lattice by (17.15), again we obtain

$$|\mathcal{R}_n(x_0, x)| \leq \bigg| \bigg\{ \frac{\|x - x_0\|^{n+1}}{(n+1)! \cdot h} + \frac{\|x - x_0\|^n}{2 \cdot n!}$$

$$+ \frac{h \cdot \|x - x_0\|^{n-1}}{8 \cdot (n-1)!} \bigg\} \cdot \omega_1(P^{(n)}, h) \cdot i \bigg| =: |\varphi|. \qquad (17.35)$$

Hence, by positivity of $T$ we see that

$$T|\mathcal{R}_n(x_0, x)| \leq T|\varphi|, \qquad (17.36)$$

i.e.,

$$(T|\mathcal{R}_n(x_0, x)|)(x) \leq (T|\varphi|)(x), \qquad (17.37)$$

for all $x \in M$, for any $x_0 \in M$.

Because $T$ is a lattice homomorphism we find

$$|T(\mathcal{R}_n(x_0, x))|(x) \leq |T(\varphi)|(x), \qquad (17.38)$$

true for all $x \in M$, for any $x_0 \in M$. That is,

$$|T(\mathcal{R}_n(x_0, x))|(x_0) \leq |T(\varphi)|(x_0), \qquad (17.39)$$

for any $x_0 \in M$. Furthermore,

$$|(T(\mathcal{R}_n(x_0, x)))(x_0)| \leq |(T(\varphi))(x_0)| \qquad (17.40)$$

for any $x_0 \in M$. Since $C(M,Y)$ is a normed vector lattice, and both sides of inequality (17.40) in $|\bullet|$'s belong to $C(M,Y)$ (for the last statement see Propositions 17.2, 17.3, 17.4 and the definition of $T$) we get that

$$\||(T(\mathcal{R}_n(x_0, x)))(x_0)\||_{\infty, x_0} \leq \||(T(\varphi))(x_0)\||_{\infty, x_0}. \qquad (17.41)$$

I.e.,

$$\||(T(\mathcal{R}_n(x_0, x)))(x_0)\||_{\infty, x_0}$$

$$\leq |||\left(T\left(\left\{\frac{\|x-x_0\|^{n+1}}{(n+1)!\cdot h}+\frac{\|x-x_0\|^n}{2\cdot n!}+\frac{h\cdot\|x-x_0\|^{n-1}}{8\cdot(n-1)!}\right\}\right.\right.$$
$$\left.\left.\cdot\omega_1(P^{(n)},h)\cdot i\right)\right)(x_0)|||_{\infty,x_0}. \tag{17.42}$$

That is,

$$|||(T(\mathcal{R}_n(x_0,x)))(x_0)|||_{\infty,x_0}$$

$$\leq \omega_1(P^{(n)},h)\cdot\left[\frac{1}{(n+1)!\cdot h}\cdot|||(T((\|x-x_0\|^{n+1})\cdot i))(x_0)|||_{\infty,x_0}\right.$$

$$+\frac{1}{2\cdot n!}\cdot|||(T((\|x-x_0\|^n)\cdot i))(x_0)|||_{\infty,x_0}$$

$$\left.+\frac{h}{8\cdot(n-1)!}\cdot|||(T((\|x-x_0\|^{n-1})\cdot i))(x_0)|||_{\infty,x_0}\right]$$

$$=:\lambda<+\infty. \tag{17.43}$$

The last quantity $\lambda$ is finite by Proposition 17.3.

Next, from (17.33) and (17.43), we obtain that

$$|||(T(P-P(x_0)))(x_0)|||_{\infty,x_0}$$

$$\leq \sum_{k=1}^{n}\frac{1}{k!}\cdot|||(T(P^{(k)}(x_0)(x-x_0)^k))(x_0)|||_{\infty,x_0}+\lambda. \tag{17.44}$$

Finally, inequality (17.23) comes from inequalities (17.29) and (17.44). The proof of Theorem 17.2 now has been finished. ∎

## 17.4 Further Results

Here the whole setting is the same as that introduced in Section 17.2. We however consider the vector space

$$B(M,Y):=\{F:M\to Y\mid \exists\theta_F>0:\|F(x)\|\leq\theta_F,\forall x\in M\}, \tag{17.45}$$

which is the space of normed bounded functions on $M$. Clearly here the associated norm $|||\cdot|||_\infty$ is again a lattice norm. The completeness of $B(M,Y)$ is established in a similar manner as in Proposition 17.1. That is $B(M;Y)$ is a Banach lattice.

For the last we still need to establish that

### LEMMA 17.3
$B(M,Y)$ is a vector lattice.

**PROOF** Let $f, g \in B(M,Y)$, it is enough to show that $f \vee g$, $f \wedge g \in B(M,Y)$. There exists $M^* > 0$ such that $\|f(x)\| < M^*$, $\|g(x)\| < M^*$, for all $x \in M$. That is

$$\pm f(x) < |f(x)| < M^* \cdot i,$$

$$\pm g(x) < |g(x)| < M^* \cdot i,$$

where $i \in Y^+$ such that $\|i\| = 1$. Here $Y^+$ is the positive cone of $Y$.

i) Hence
$$f(x) \vee g(x) < M^* \cdot i$$

and
$$-(f(x) \vee g(x)) = (-f(x)) \wedge (-g(x)) < M^* \cdot i.$$

Thus
$$(f \vee g)(x) < M^* \cdot i$$

and
$$-(f \vee g)(x) < M^* \cdot i$$

imply
$$|(f \vee g)(x)| < M^* \cdot i.$$

And since $Y$ is a Banach lattice we have that

$$\|(f \vee g)(x)\| < M^*, \quad \text{all } x \in M.$$

That is,
$$|||f \vee g|||_\infty < M^*,$$

i.e., $f \vee g \in B(M,Y)$.

ii) Clearly
$$f(x) \wedge g(x) < M^* \cdot i$$

and
$$-(f(x) \wedge g(x)) = (-f(x)) \vee (-g(x)) < M^* \cdot i.$$

Thus
$$(f \wedge g)(x) < M^* \cdot i$$

and
$$-(f \wedge g)(x) < M^* \cdot i$$

imply that
$$|(f \wedge g)(x)| < M^* \cdot i.$$

That is,
$$\|(f \wedge g)(x)\| < M^*, \quad \text{all } x \in M,$$

and so $\||f \wedge g|\|_\infty < M^*$. I.e., $f \wedge g \in B(M,Y)$. ∎

Here we consider *lattice homomorphisms* $T$ from $C(M,Y)$ into $B(M,Y)$ and produce similar results as in Section 17.3. Obviously such a $T$ is a *positive operator*, and because $C(M,Y)$ is a Banach lattice we have that $T$ is *continuous*.

To proceed we use the following auxiliary results.

### LEMMA 17.4
Let $f \in C(M,Y)$ and the lattice homomorphism $T : C(M,Y) \to B(M,Y)$. Then
$$(T(f(x_0)))(x_0) \in B(M,Y),$$
as a function of $x_0 \in M$.

**PROOF** Because $f \in C(M,Y)$ and $M$ is a compact subset of $X$ we have for any $x_0 \in M$ that
$$\|f(x_0)\| \le \theta_f = \|\theta_f \cdot i\| < +\infty,$$

where $i \in Y^+$ such that $\|i\| = 1$.

Equivalently we get
$$|f(x_0)| \le |\theta_f \cdot i|.$$

Set $\varphi(x) := f(x_0)$, $\rho(x) := \theta_f \cdot i$, all $x \in M$. I.e., $|\varphi(x)| \le |\rho(x)|$, that is,
$$|\varphi|(x) \le |\rho|(x), \quad \text{all } x \in M,$$

iff
$$|\varphi| \le |\rho|.$$

Since $T$ is a positive operator we have that

$$T|\varphi| \leq T|\rho|.$$

And because $T$ is a lattice homomorphism, we get

$$|T(\varphi)| \leq |T(\rho)|,$$

and

$$|T(\varphi)|(x) \leq |T(\rho)|(x), \quad \text{all } x \in M.$$

In particular, it holds true that

$$|(T(\varphi))(x_0)| \leq |T(\rho))(x_0)|.$$

Since $Y$ is a Banach lattice we find that

$$\|(T(\varphi))(x_0)\| \leq \|(T(\rho))(x_0)\|.$$

That is,

$$\|(T(f(x_0)))(x_0)\| \leq \|(T(\theta_f \cdot i))(x_0)\|$$

$$= \theta_f \cdot \|(T(i))(x_0)\|$$

$$\leq \theta_f \cdot \||T(i)|\|_\infty =: \Omega_{f,T} < +\infty.$$

That is,

$$\|(T(f(x_0)))(x_0)\| \leq \Omega_{f,T} < +\infty,$$

for any arbitrary $x_0 \in M$. We have shown that $(T(f(\cdot)))(\cdot) \in B(M,Y)$.
∎

Clearly $(T(f - f(x_0)))(x_0) \in B(M,Y)$.

### LEMMA 17.5

Let $r > 0$, $i \in Y^+$ such that $\|i\| = 1$, $T : C(M,Y) \to B(M,Y)$ a lattice homomorphism and $x_0 \in M$ be arbitrary. Then $(T(\|x - x_0\|^r \cdot i))(x_0) \in B(M,Y)$ as a function of $x_0$.

**PROOF** For any $x, x_0 \in M$, by the compactness of $M$ we see that

$$\|x - x_0\|^r \leq \ell, \quad \text{for some } \ell > 0.$$

Here $\|x - x_0\|^r \cdot i \in C(M, Y)$, thus

$$\| \|x - x_0\|^r \cdot i \|_{\infty, x} \leq \ell,$$

and

$$\left| \|x - x_0\|^r \cdot i \right| \leq |\ell \cdot i|.$$

Therefore

$$T\left| \|x - x_0\|^r \cdot i \right| \leq T|\ell \cdot i|$$

and

$$|(T(\|x - x_0\|^r \cdot i))| \leq |(T(\ell \cdot i))|.$$

Moreover,

$$|(T(\|x - x_0\|^r \cdot i))|(x) \leq |T(\ell i)|(x), \quad \text{all } x \in M.$$

In particular

$$|(T(\|x - x_0\|^r \cdot i))|(x_0) \leq |T(\ell i)|(x_0),$$

for any arbitrary $x_0 \in M$. So that

$$|(T(\|x - x_0\|^r \cdot i))(x_0)| \leq |(T(\ell i))(x_0)|,$$

for any $x_0 \in M$. The last implies that

$$\|(T(\|x - x_0\|^r \cdot i))(x_0)\| \leq \|(T(\ell i))(x_0)\|$$

$$= \ell \cdot \|(T(i))(x_0)\| \leq \ell \cdot \|\|T(i)\|\|_\infty < \theta_{\ell, T} < +\infty,$$

for some constant $\theta_{\ell, T} > 0$. We have shown that

$$\|\|(T(\|x - x_0\|^r \cdot i))(x_0)\|\|_{\infty, x_0} < +\infty,$$

i.e.,

$$(T(\|x - x_0\|^r \cdot i))(x_0) \in B(M, Y)$$

as a function of $x_0 \in M$. ∎

The last result we use here is

**LEMMA 17.6**
Let $X$ be a Banach space, $Y$ a Banach lattice, $P : X \to Y$ such that $P|_M \in C^n(M, Y)$, $n \in \mathbf{N}$, where $M$ is a convex and compact subset of $X$. Let $T$ be a lattice homomorphism from $C(M, Y)$ into $B(M, Y)$. Here $k = 1, \ldots, n$. Then it holds

$$(T(P^{(k)}(x_0)(x - x_0)^k))(x_0) \in B(M, Y)$$

as a function of $x_0 \in M$.

**PROOF** We have that

$$P^{(k)}(x_0)(x - x_0)^k \in Y, \quad \text{for all } x \in X,$$

and $P^{(k)}(x_0)(x - x_0)^k$ is continuous in $x \in X$.

Because $P^{(k)}(x_0)$ is a bounded $k$-linear operator and $P|_M \in C^n(M, Y)$ we have that

$$\|P^{(k)}(x_0)(x - x_0)^k\| \leq \|P^{(k)}(x_0)\| \cdot \|x - x_0\|^k < +\infty,$$

i.e., there exists $\mathcal{D}_P > 0$ such that

$$\|P^{(k)}(x_0)(x - x_0)^k\| \leq \mathcal{D}_P < +\infty,$$

for all $x \in M$ and any $x_0 \in M$. That is,

$$\|P^{(k)}(x_0)(x - x_0)^k\| \leq \|\mathcal{D}_P \cdot i\|,$$

where $i \in Y^+$ such that $\|i\| = 1$. Equivalently

$$|P^{(k)}(x_0)(x - x_0)^k| \leq |\mathcal{D}_p \cdot i|.$$

Because $T$ is a lattice homomorphism we have that

$$T|P^{(k)}(x_0)(x - x_0)^k| \leq T|\mathcal{D}_P \cdot i|,$$

i.e.,

$$(T|P^{(k)}(x_0)(x - x_0)^k|)(x) \leq (T|\mathcal{D}_P \cdot i|)(x), \quad \text{all } x \in M.$$

Thus
$$|T(P^{(k)}(x_0)(x-x_0)^k)|(x) \le |T(\mathcal{D}_P \cdot i)|(x),$$
and
$$|(T(P^{(k)}(x_0)(x-x_0)^k))(x)| \le |(T(\mathcal{D}_P \cdot i))(x)|,$$
all $x \in M$, and for any $x_0 \in M$.

In particular it holds
$$|(T(P^{(k)}(x_0)(x-x_0)^k))(x_0)| \le |(T(\mathcal{D}_P \cdot i))(x_0)|,$$
for any arbitrary $x_0 \in M$. Hence,
$$\|(T(P^{(k)}(x_0)(x-x_0)^k))(x_0)\| \le \|(T(\mathcal{D}_P \cdot i))(x_0)\|,$$
for any $x_0 \in M$. Consequently,
$$\| \|(T(P^{(k)}(x_0)(x-x_0)^k))(x_0)\| \|_{\infty, x_0}$$
$$\le \zeta_{P,T} := \mathcal{D}_P \cdot \| \|T(i)\| \|_\infty < +\infty.$$

We have proved that $(T(P^{(k)}(x_0)(x-x_0)^k))(x_0) \in B(M,Y)$ as a function of $x_0 \in M$, all $k = 1, \ldots, n$. ∎

Next one can establish again in exactly the same manner inequalities (17.19), (17.23), and (17.25) of Theorems 17.1, 17.2, and of Corollary 17.1, respectively, within the same settings—*except that now $T$ is a lattice homomorphism from $C(M,Y)$ into $B(M,Y)$*.

The valid inequalities (17.19), (17.23), (17.25), under the extended $T$ as above, have again finite right-hand sides. The last is proved by the use of Lemmas 17.4, 17.5 and 17.6.

## 17.5 Applications

Next we show that the set of lattice homomorphisms where this theory can be applied is not a void one.

1) Let $\tau : Y \to Y$ be a lattice homomorphism and $f \in C(M,Y)$, then $F : C(M,Y) \to C(M,Y)$ defined by $F(f) := \tau \circ f$ is a lattice homomorphism.

2) Let $\varphi : \mathbf{N} \to \mathbf{R}_+ - \{0\}$ be such that $\varphi(n) \to 1$ as $n \to +\infty$ (e.g., $\varphi(n) = 1 + \frac{1}{n}$). Let $f \in C(M,Y)$, then $T_n : C(M,Y) \to C(M,Y)$ defined

by $(T_n f)(x) := \varphi(n) \cdot f(x)$, all $x \in M$ is a lattice homomorphism. Moreover, $T_n \to I$-unit operator, as $n \to +\infty$.

3) Let $\gamma_n : M \to \mathbf{R}_+ - \{0\}$ be such that $\|\gamma_n\|_\infty \le \alpha_n$, $\alpha_n > 0$, $\forall n \in \mathbf{N}$ and $\lim_{n \to +\infty} \gamma_n = 1$, uniformly (e.g., $\gamma_n(x) := e^{-\|x\|/n}$). Let $f \in C(M, Y)$, define $(T_n f)(x) := \gamma_n(x) \cdot f(x)$, all $x \in M$. Then $T_n$ determines a lattice homomorphism from $C(M, Y)$ into $B(M, Y)$ so that $T_n \to I$.

4) Let $T_n$, $n \in \mathbf{N}$, be a positive linear operator from $C(M, Y)$ into itself such that $T_n \to 0$. Suppose that $T_n$ is an *orthomorphism* (see [3], p. 109). Then (by Exercise 2, p. 124 of [3]) $E_n := I + T_n$ *is a lattice homomorphism*, where $I$ is the unit operator. This theory (§17.3) when applied to $E_n$ produces the convergence of $T_n \to 0$ with rates.

5) Let $\alpha_n > 0$ be such that $\alpha_n \to 1$ as $n \to +\infty$. Let $f \in C(X, Y)$ and define $(T_n f)(x) := f(\alpha_n x)$, all $x \in X$. Then $T_n$ is a lattice homomorphism from $C(X, Y)$ into itself so that $T_n f \to f$ pointwise, all $f \in C(X, Y)$.

6) Let $0 < \alpha_n \to 0$ as $n \to +\infty$ and $j \in X$ such that $\|j\| = 1$. Let $f \in C(X, Y)$ and define

$$(T_n f)(x) := f(x + \alpha_n \cdot j) \quad (\to f(x)),$$

all $x \in X$. Then $T_n$ is a lattice homomorphism from $C(X, Y)$ into itself so that $T_n f \to f$ pointwise, all $f \in C(X, Y)$.

# Chapter 18

## Quantitative $L_p$-Results for Positive Linear Operators

The author has proved several quantitative type results for determining the rate of convergence of sequences of positive linear operators to the unit. These engage the modulus of continuity of the associated function or its derivative of certain order, and they are *pointwise Korovkin type inequalities*, most of them sharp. Using these inequalities, we present a big variety of general $L_p$ ($1 \leq p \leq +\infty$) analogs, covering most of the expected cases of the convergence of positive linear operators with rates to the unit. In the same inequality we manage to combine different $L_p$-norms. This treatment relies on [14].

Here we work over one-dimensional, multi-dimensional, and abstract compact convex domains. In this study we include also positive stochastic operators. The functions we consider are either differentiable or just continuous. Let $M$ be a compact convex subset of $(V, \|\cdot\|)$ a real normed vector space. Let $L$ be a positive linear operator from $C(M)$ into itself and $f \in C(M)$. In this chapter we estimate $\|Lf - f\|_p$ ($1 \leq p \leq +\infty$) in many different directions. The $\|\cdot\|_p$-norm is taken over a general measure space of finite positive measure. One of the techniques used in the proofs here is transferring via the Riesz representation theorem, Anastassiou-weak convergence related results on finite positive measures to positive linear operators.

The exposed inequalities involve the standard modulus of continuity of $f$ or its certain order derivative (when $M = [a,b] \subset \mathbf{R}$), and mix different $L_p$-norms at a time. This work has been greatly motivated by the very important related article of Popov [146]. Other important and interesting articles on the topic are of Quak [147], [148]. In their papers they use the averaged modulus of smoothness, the so-called $\tau$-modulus of orders one and two, and they work on real domains with bounded Lebesgue measurable functions.

## 18.1 Background

We will use the following lemmas and remarks.

**LEMMA 18.1**
*Let $M$ be a compact subset of $(V, \|\cdot\|)$ a real normed vector space. Let $L$ be a positive linear operator from $C(M)$ into itself. Here $x, x_0 \in M$. Consider $r > 0$. Then $(L(\|x - x_0\|^r))(x_0)$ is a continuous function in $x_0$.*

**PROOF** Let $x_n, x_0 \in M$ be such that $x_n \to x_0$, as $n \to +\infty$. See that

$$|(L(\|x - x_n\|^r))(x_n) - (L(\|x - x_0\|^r))(x_0)| \le |(L(\|x - x_0\|^r))(x_n)$$

$$- (L(\|x - x_0\|^r))(x_n)| + |(L(\|x - x_0\|^r))(x_n)$$

$$- (L(\|x - x_0\|^r))(x_0)| := (*).$$

Notice that $\|x - x_0\|^r$ is a continuous function of $x$, so is $(L(\|x - x_0\|^r))(x)$. That is $(L(\|x - x_0\|^r)(x_n) \to (L(\|x - x_0\|^r))(x_0)$, as $x_n \to x_0$, i.e.,

$$|(L(\|x - x_0\|^r))(x_n) - (L(\|x - x_0\|^r))(x_0)| < \varepsilon_1,$$

where $\varepsilon_1 > 0$ is small. Thus

$$(*) < \|L(\|x - x_n\|^r) - L(\|x - x_0\|^r)\|_\infty + \varepsilon_1 =: (**).$$

Let $x_n, x_0, x \in M$ such that $x_n \to x_0$, as $n \to +\infty$. Note that

$$\sup_{x \in M} |\|x - x_n\| - \|x - x_0\|| \le \|x_n - x_0\| \to 0,$$

hence

$$\|\|x - x_n\| - \|x - x_0\|\|_\infty \to 0,$$

that is,

$$\|x - x_n\| \xrightarrow{u} \|x - x_0\|, \quad \text{uniformly.}$$

Here $\|x - y\| \le \Delta$ – the diameter of $M$, $\forall x, y \in M$, i.e., $\|x - x_n\|, \|x - x_0\| \le \Delta$. Call

$$f_n(x) := \|x - x_n\|, \quad f(x) := \|x - x_0\|,$$

then
$$0 \leq f_n(x), f(x) \leq \Delta, \quad \forall x \in M.$$

Consider $g(t) := t^r$, $t \in [0, \Delta]$, $r > 0$. Notice that $g$ is a uniformly continuous function on $[0, \Delta]$, i.e., given $\varepsilon_2 > 0$ $\exists \delta > 0$: whenever $|u_1 - u_2| < \delta$, $u_1, u_2 \in [0, \Delta]$ we have that $|g(u_1) - g(u_2)| < \varepsilon_2$. For the above $\delta > 0$ $\exists \mathcal{N}_0$: $\forall n \geq \mathcal{N}_0$ we get $|f_n(x) - f(x)| < \delta$, $\forall x \in M$. Hence for $\varepsilon_2 > 0$ $\exists \mathcal{N}_0$: $\forall n \geq \mathcal{N}_0$ we obtain
$$|g(f_n(x)) - g(f(x))| < \varepsilon_2.$$

That is, $\|x - x_n\|^r \xrightarrow{u} \|x - x_0\|^r$, uniformly. Here $L$ is a positive linear operator from $C(M)$ into itself, therefore $L$ is a continuous operator and $L(\|x - x_n\|^r) \xrightarrow{u} L(\|x - x_0\|^r)$, uniformly. I.e., for sufficiently large $n$ we get that
$$\|L(\|x - x_n\|^r) - L(\|x - x_0\|^r)\|_\infty < \varepsilon_1,$$

$\varepsilon_1 > 0$ as above small.

Finally, we find
$$(**) < 2 \cdot \varepsilon_1.$$

I.e.,
$$|(L(\|x - x_n\|^r))(x_n) - (L(\|x - x_0\|^r))(x_0)| < 2\varepsilon_1$$

$\varepsilon_1 > 0$ arbitrarily small, as $x_n \to x_0$. Hence
$$(L(\|x - x_n\|^r))(x_n) \to (L(\|x - x_0\|^r))(x_0),$$

as $x_n \to x_0$. ■

**REMARK 18.1** (i) Let $r \geq 1$, obviously from Lemma 18.1, we have that $((L(\|x - x_0\|^r))(x_0))^{1/r}$ is a continuous function in $x_0 \in M$. Therefore it is Borel measurable on $M \subseteq (V, \|\cdot\|)$, and hence Lebesgue measurable on $M \subseteq \mathbf{R}^k$, $k \geq 1$.

(ii) Let $L$ be a positive linear operator from $C([a,b])$ into itself. Let $x, t \in [a,b]$, $n \in \mathbf{N}$. Then from Lemma 18.1, we have that $(L(|t - x|^n))(x)$ is a continuous function in $x$ and hence Lebesgue measurable on $[a,b] \subseteq \mathbf{R}$. ■

**LEMMA 18.2**
*Let $L$ be a positive linear operator from $C([a,b])$ into itself, and $x, t \in [a,b]$, $n \in \mathbf{N}$. Then $(L((t-x)^n))(x)$ is a continuous function in $x$.*

**PROOF** By

$$(t-x)^n = \sum_{k=0}^{n} \binom{n}{k} \cdot (-1)^k \cdot t^{n-k} \cdot x^k,$$

we have that

$$(L((t-x)^n))(x) = \sum_{k=0}^{n} (-1)^k \cdot \binom{n}{k} \cdot x^k \cdot (L(t^{n-k}))(x)$$

is continuous in $x \in [a, b]$. ∎

**REMARK 18.2** Let $f \in C([0,1])$ and

$$B_n(f; x) := \sum_{k=0}^{n} f\left(\frac{k}{n}\right) \cdot \binom{n}{k} \cdot x^k \cdot (1-x)^{n-k}$$

the $n$th Bernstein polynomial for $f$, $n \in \mathbf{N}$, which is a positive linear operator from $C([0,1])$ into itself. It is known that

$$(B_n(1))(x_0) = 1 > 0, \quad \forall x_0 \in [0,1],$$
$$(B_n(x))(x_0) = x_0,$$

and

$$(B_n(x^2))(x_0) = \frac{x_0}{n} + \left(1 - \frac{1}{n}\right) \cdot x_0^2, \quad n \in \mathbf{N}.$$

Notice that

$$(B_n((x-x_0)^2))(x_0) = \frac{x_0 \cdot (1-x_0)}{n} \begin{cases} = 0, & \text{when } x_0 = 0, 1 \\ > 0, & x_0 \in (0,1). \end{cases}$$

Thus, for $L$ a positive linear operator from $C(M)$ into itself, where $M$ is a compact subset of $(V, \|\cdot\|)$, it is reasonable to suppose that

(i) $(L(1))(x_0) > 0$, $\forall x_0 \in M$,

(ii) $(L(\|x - x_0\|^r))(x_0) > 0$, $\forall x_0 \in M$, $r > 0$, or

(ii)' $(L(\|x - x_0\|^r))(x_0) > 0$, $\mu$-a.e., i.e.,

$$\mu(x_0 \in M : (L(\|x - x_0\|^r))(x_0) = 0) = 0,$$

where $\mu$ is a finite positive measure on $M$ with $\mu(M) > 0$.

From (ii)', we have that $(1 \le p \le +\infty)$

$$0 < \|(L(\|x - x_0\|^r))(x_0)\|_{L_p(\mu)} < +\infty.$$

∎

By the Riesz Representation Theorem (see Aliprantis and Burkinshaw [4], pp. 246–249), we have

### LEMMA 18.3

*Let $M$ be a compact subset of a real normed vector space $(V, \|\cdot\|)$. Let $C(M)$ be the space of continuous real valued functions on $M$ and $F$ be a positive linear functional on $C(M)$. Then there exists a unique regular Borel positive finite measure $\mu$ on $M$ such that*

$$F(f) = \int_M f\, d\mu, \quad \forall f \in C(M).$$

**REMARK 18.3** Let $L \not\equiv 0 : C(M) \hookrightarrow C(M)$ be a positive linear operator. Then $(L(\cdot))(x_0)$ is a positive linear functional on $C(M)$, where $x_0 \in M$ is fixed. From Lemma 18.3 we get that

$$(L(f))(x_0) = \int_M f\, d\mu_{x_0}, \quad \forall f \in C(M),$$

where $\mu_{x_0}$ is a unique regular Borel positive finite measure on $M$. ∎

### LEMMA 18.4

*Let $(M, \mathcal{B}, \mu)$ be a measure space, where $M$ is a compact subset of $(V, \|\cdot\|)$, $\mathcal{B}$ is the Borel $\sigma$-algebra on $M$ and $\mu$ is a positive finite measure on $M$. Consider $C(M)\ (\subset L_p(M, \mathcal{B}, \mu),$ any $p > 1)$. Let $f, g \in C(M)$. Then it holds*

$$\|f \cdot g\|_p \le \|f\|_{p \cdot q} \cdot \|g\|_{p^2}, \tag{18.1}$$

*where $p, q > 1$ such that $\frac{1}{p} + \frac{1}{q} = 1$.*

**PROOF**  Easy by Hölder's inequality. ∎

**REMARK 18.4** Inequality (18.1) is valid in general, when $f \in L_{p \cdot q}(M, \mathcal{B}, \mu)$ and $g \in L_{p^2}(M, \mathcal{B}, \mu)$ then $f \cdot g \in L_p(M, \mathcal{B}, \mu)$. ∎

**REMARK 18.5** From Anastassiou [5], pp. 251–252, we obtain the following useful function $\phi_n$ and its properties: Let $h > 0$ be fixed. Define

$$\phi_n(x) := \int_0^{|x|} \left\lceil \frac{t}{h} \right\rceil \cdot \frac{(|x| - t)^{n-1}}{(n-1)!} \cdot dt, \quad (x \in \mathbf{R}), \ n \in \mathbf{N}, \tag{18.2}$$

where $\lceil \cdot \rceil$ is the ceiling of the number.

Define $\phi_0(t) := \lceil \frac{t}{h} \rceil$, $t \geq 0$ then

$$\phi_n(x) = \int_0^x \phi_{n-1}(t) \, dt, \quad (x \in \mathbf{R}^+, n \in \mathbf{N}). \tag{18.3}$$

The function $\phi_n$ is even, continuous and convex on $\mathbf{R}$, and it is strictly increasing on $\mathbf{R}^+$, $n \in \mathbf{N}$; $\phi_n(0) = 0$. One can easily show that $\phi_n(u^{1/n})$ is a convex function in $u \geq 0$, $n \in \mathbf{N}$. ∎

We have the following inequality valid

$$\phi_n(x) \leq \left( \frac{|x|^{n+1}}{(n+1)! \cdot h} + \frac{|x|^n}{2 \cdot n!} + \frac{h \cdot |x|^{n-1}}{8 \cdot (n-1)!} \right), \quad \forall x \in \mathbf{R}, n \in \mathbf{N}. \tag{18.4}$$

Hence for $x \neq 0$ we obtain

$$\frac{\phi_n(x)}{|x|^n} \leq \left( \frac{|x|}{(n+1)! \cdot h} + \frac{1}{2 \cdot n!} + \frac{h}{8 \cdot |x| \cdot (n-1)!} \right). \tag{18.5}$$

We also use the following result from Anastassiou [5], p. 259.

**COROLLARY A**
Consider the positive linear operator

$$L : C([a,b]) \longrightarrow C([a,b]).$$

For a fixed $x \in [a,b]$ define

$$c(x) := \max(x - a, b - x) \tag{18.6}$$

## 18. Quantitative $L_p$-Results for Positive Linear Operators

(i.e., $\left(\frac{b-a}{2}\right) \leq c(x) \leq b - a$). Let $f \in C^n([a,b])$, $n \in \mathbf{N}$, such that the first modulus of continuity $\omega_1(f^{(n)}, h) \leq w$, where $w$, $h$ are fixed positive numbers, $0 < h < b - a$. Then we get the upper bound

$$|L(f,x) - f(x)| \leq |f(x)| \cdot |L(1,x) - 1| + \sum_{k=1}^{n} \frac{|f^{(k)}(x)|}{k!}$$

$$\cdot |L((t-x)^k, x)| + w \cdot \frac{\phi_n(c(x))}{(c(x))^n} \cdot L(|t-x|^n, x), \text{ all } x \in [a,b]. \quad (18.7)$$

Inequality (18.7) is sharp.

## 18.2 Univariate Results

These are inequalities with respect to basic norms for differentiable functions.

Next comes the first main result.

**THEOREM 18.1**
*Consider the positive linear operator*

$$L : C([a,b]) \longrightarrow C([a,b]).$$

*Define* ($n \in \mathbf{N}$)
$$D_n := \|(L(|t - \cdot|^n))(\cdot)\|_\infty^{1/n}, \quad (18.8)$$

*where* $\|\cdot\|_\infty$ *is the supremum norm. Let* $f \in C^n([a,b])$. *Then we get that*

$$\|Lf - f\|_\infty \leq \|f\|_\infty \cdot \|L1 - 1\|_\infty + \sum_{k=1}^{n} \frac{\|f^{(k)}\|_\infty}{k!} \cdot \|(L(t - \cdot)^k)(\cdot)\|_\infty$$

$$+ \omega_1(f^{(n)}, D_n) \cdot D_n^{n-1} \left( \frac{(b-a)}{(n+1)!} + \frac{D_n}{2 \cdot n!} \right.$$

$$\left. + \frac{D_n^2}{8 \cdot (b-a) \cdot (n-1)!} \right). \quad (18.9)$$

*Notice that by Remark 18.1(ii), $D_n < +\infty$.*

**PROOF** From $\phi_n(u^{1/n})$ being convex in $u \geq 0$ and $\phi_n(0) = 0$, we have that $\frac{\phi_n(u^{1/n})}{u}$ is increasing in $u > 0$. Therefore by $c^n(x) \leq (b-a)^n$ we obtain

$$\frac{\phi_n(c(x))}{c^n(x)} \leq \frac{\phi_n(b-a)}{(b-a)^n}.$$

Consequently,

$$\text{R.H.S.}(18.7) \leq \|f\|_\infty \cdot \|L1 - 1\|_\infty + \sum_{k=1}^n \frac{\|f^{(k)}\|_\infty}{k!} \cdot \|(L(t-\cdot)^k)(\cdot)\|_\infty$$

$$+ w \cdot \frac{\phi_n(b-a)}{(b-a)^n} \cdot \|L(|t-\cdot|^n)(\cdot)\|_\infty.$$

Thus

$$\|Lf - f\|_\infty \leq \|f\|_\infty \cdot \|L1 - 1\|_\infty + \sum_{k=1}^n \frac{\|f^{(k)}\|_\infty}{k!} \cdot \|(L(t-\cdot)^k)(\cdot)\|_\infty$$

$$+ w \cdot \frac{\phi_n(b-a)}{(b-a)^n} \cdot \|L(|t-\cdot|^n)(\cdot)\|_\infty. \tag{18.10}$$

In particular from (18.5) we have

$$\frac{\phi_n(b-a)}{(b-a)^n} \leq \left(\frac{(b-a)}{(n+1)! \cdot h} + \frac{1}{2 \cdot n!} + \frac{h}{8 \cdot (b-a) \cdot (n-1)!}\right).$$

Pick $h = D_n := \|(L(|t-\cdot|^n))(\cdot)\|_\infty^{1/n}$, i.e., $h^n = D_n^n = \|(L(|t-\cdot|^n))(\cdot)\|_\infty$. We obtain that

$$\text{Remainder}(18.10) = w \cdot \frac{\phi_n(b-a)}{(b-a)^n} \cdot D_n^n$$

$$\leq w \cdot \left(\frac{(b-a) \cdot D_n^n}{(n+1)! \cdot h} + \frac{D_n^n}{2 \cdot n!} + \frac{h \cdot D_n^n}{8 \cdot (b-a) \cdot (n-1)!}\right)$$

$$= w \cdot \left(\frac{(b-a) \cdot D_n^n}{(n+1)! \cdot D_n} + \frac{D_n^n}{2 \cdot n!} + \frac{D_n \cdot D_n^n}{8 \cdot (b-a) \cdot (n-1)!}\right)$$

$$= w \cdot D_n^{n-1} \cdot \left(\frac{(b-a)}{(n+1)!} + \frac{D_n}{2 \cdot n!} + \frac{D_n^2}{8 \cdot (b-a) \cdot (n-1)!}\right).$$

Also choose $w = \omega_1(f^{(n)}, D_n)$. ∎

**COROLLARY 18.1**
Let $L$ be a positive linear operator from $C([a,b])$ into itself. Here

$$D_1 := \|(L(|t - \cdot|))(\cdot)\|_\infty < +\infty. \tag{18.11}$$

Let $f \in C^1([a,b])$. Then it holds

$$\|Lf - f\|_\infty \leq \|f\|_\infty \cdot \|L1 - 1\|_\infty + \|f'\|_\infty \cdot \|(L(t - \cdot))(\cdot)\|_\infty$$

$$+ \frac{1}{2} \cdot \omega_1(f', D_1) \cdot \left((b-a) + D_1 + \frac{D_1^2}{4 \cdot (b-a)}\right). \tag{18.12}$$

**COROLLARY 18.2**
Let $L$ be a positive linear operator from $C([a,b])$ into itself. Here

$$D_2 := \|(L((t - \cdot)^2))(\cdot)\|_\infty^{1/2} < +\infty. \tag{18.13}$$

Let $f \in C^2([a,b])$. Then it holds

$$\|Lf - f\|_\infty \leq \|f\|_\infty \cdot \|L1 - 1\|_\infty + \|f'\|_\infty \cdot \|(L(t - \cdot))(\cdot)\|_\infty$$

$$+ \frac{\|f''\|_\infty}{2} \cdot \|(L(t - \cdot)^2)(\cdot)\|_\infty$$

$$+ \frac{1}{2} \cdot \omega_1(f'', D_2) \cdot D_2 \cdot \left(\frac{(b-a)}{3} + \frac{D_2}{2} + \frac{D_2^2}{4 \cdot (b-a)}\right). \tag{18.14}$$

Next follows an application of Corollary 18.2.

**EXAMPLE 18.1**
Let $B_n$ be the Bernstein polynomials (see Remark 18.2) acting on $C([0,1])$. Let $f \in C^2([0,1])$. Then we easily get

$$\|B_n f - f\|_\infty \leq \frac{\|f''\|_\infty}{8 \cdot n} + \frac{1}{4\sqrt{n}} \cdot \omega_1\left(f'', \frac{1}{2\sqrt{n}}\right) \cdot \left(\frac{1}{3} + \frac{1}{4\sqrt{n}} + \frac{1}{16n}\right). \tag{18.15}$$

An improved result for $f \in C^1([a,b])$ with respect to $\|\cdot\|_\infty$ comes next.

## THEOREM 18.2
Let $L \not\equiv 0$ be a positive linear operator from $C([a,b])$ into itself. Set

$$\rho := \|(L(t-x_0)^2)(x_0)\|_\infty^{1/2}, \tag{18.16}$$

and consider $r > 0$. Let $f \in C^1([a,b])$. Then it holds

$$\|Lf - f\|_\infty - \|f\|_\infty \cdot \|L1 - 1\|_\infty - \|f'\|_\infty \tag{18.17}$$

$$\cdot \|(L((t-x_0))(x_0)\|_\infty$$

$$\leq \begin{cases} \frac{1}{8 \cdot r} \cdot (2 + \sqrt{\|L(1)\|_\infty} \cdot r)^2 \cdot \omega_1(f', r \cdot \rho) \cdot \rho, & \text{if } r \leq \frac{2}{\sqrt{\|L(1)\|_\infty}}; \\ \sqrt{\|L(1)\|_\infty} \cdot \omega_1(f', r \cdot \rho) \cdot \rho, & \text{if } r > \frac{2}{\sqrt{\|L(1)\|_\infty}}. \end{cases}$$

Notice that by Lemma 18.2 we have $\rho < +\infty$.

**PROOF** Let $L \not\equiv 0$ be a positive linear operator from $C([a,b])$ into itself. Let $x_0 \in [a,b]$. By Remark 18.3 we see that

$$(L(f))(x_0) = \int_{[a,b]} f(t) \cdot \mu_{x_0}(dt), \quad \text{all } f \in C([a,b]),$$

where $\mu_{x_0}$ is a finite nonnegative measure on $[a,b]$. Let $r > 0$ and put

$$m_{x_0} := (L(1))(x_0) = \mu_{x_0}([a,b]),$$

$$d_2(x_0) := \left( \int_{[a,b]} (t-x_0)^2 \cdot \mu_{x_0}(dt) \right)^{1/2}.$$

Then, working similarly as in the proof of Theorem 2.19, pp. 263–265, Anastassiou [5], we find

$$\left| \int_{[a,b]} f \, d\mu_{x_0} - f(x_0) \right| \leq |m_{x_0} - 1| \cdot |f(x_0)| + |f'(x_0)| \cdot \left| \int_{[a,b]} (t-x_0) \cdot \mu_{x_0}(dt) \right|$$

$$+ \frac{1}{8 \cdot r} \cdot (2 + \sqrt{m_{x_0}} \cdot r)^2 \cdot \omega_1(f', r \cdot d_2(x_0)) \cdot d_2(x_0) =: (*).$$

Note
$$d_2^2(x_0) = \int (t-x_0)^2 \cdot \mu_{x_0}(dt) = (L((x-x_0)^2)))(x_0) \leq \|(L(x-x_0)^2)(x_0)\|_\infty.$$

I.e.,
$$d_2(x_0) \leq \|(L(x-x_0)^2)(x_0)\|_\infty^{1/2} := \rho.$$

Thus
$$(*) \leq \|L1 - 1\|_\infty \cdot \|f\|_\infty + \|f'\|_\infty \cdot \|(L(t-x_0))(x_0)\|_\infty$$
$$+ \frac{1}{8 \cdot r} \cdot \left(2 + \sqrt{\|L(1)\|_\infty} \cdot r\right)^2 \cdot \omega_1(f', r \cdot \rho) \cdot \rho.$$

We have shown that
$$\|Lf - f\|_\infty \leq \|f\|_\infty \cdot \|L1 - 1\|_\infty + \|f'\|_\infty \cdot \|(L(t-x_0))(x_0)\|_\infty$$
$$+ \frac{1}{8 \cdot r} \cdot \left(2 + \sqrt{\|L(1)\|_\infty} \cdot r\right)^2 \cdot \omega_1(f', r \cdot \rho) \cdot \rho, \quad r > 0. \quad (18.18)$$

Set $M := \|L(1)\|_\infty > 0$, by $L \not\equiv 0$, and
$$g(r) := \frac{1}{8 \cdot r} \cdot (2 + \sqrt{M} \cdot r)^2 > 0, \quad \text{all } r > 0.$$

Notice that $g(r)$ is a convex function in $r$. Moreover,
$$\min_{r>0} g(r) = g\left(\frac{2}{\sqrt{M}}\right) = \sqrt{M}.$$

Thus the remainder of (18.18), when $r = 2/\sqrt{M}$, equals
$$\left(\sqrt{\|L(1)\|_\infty} \cdot \omega_1\left(f', \frac{2}{\sqrt{\|L(1)\|_\infty}} \cdot \rho\right) \cdot \rho\right)$$
$$\leq \left(\sqrt{\|L(1)\|_\infty} \cdot \omega_1(f', r \cdot \rho) \cdot \rho\right), \quad \text{for all } r \geq \frac{2}{\sqrt{\|L(1)\|_\infty}}.$$

Using Corollary A we get some $\|\cdot\|_{L_1} - \|\cdot\|_{L_2}$ results with respect to Lebesgue measure that follow.

**THEOREM 18.3**
Consider the positive linear operator
$$L : C([a,b]) \longrightarrow C([a,b]), \quad a \ne b.$$

Call $(n \in \mathbf{N})$
$$h_n := \|L(|t-x|^n)(x)\|_2^{1/n}. \tag{18.19}$$

Let $f \in C^n([a,b])$. Then it holds

$$\|Lf - f\|_1 \le \|f\|_2 \cdot \|L1 - 1\|_2 + \sum_{k=1}^{n} \frac{1}{k!} \cdot \|f^{(k)}\|_2 \cdot \|L((t-\cdot)^k)(\cdot)\|_2$$

$$+ \frac{1}{2} \cdot \omega_1(f^{(n)}, h_n) \cdot h_n^{n-1} \cdot \left\{ \left( \frac{b-a}{(n+1)!} \right) \right.$$

$$\cdot \sqrt{\frac{7}{3} \cdot (b-a)} + \frac{\sqrt{b-a}}{n!} \cdot h_n$$

$$\left. + \frac{h_n^2}{4 \cdot (n-1)!} \cdot \sqrt{\frac{2}{b-a}} \right\}. \tag{18.20}$$

From Lemmas 18.1 and 18.2, $h_n < +\infty$ and R.H.S.(18.20) $< +\infty$.

**PROOF**  Note that from (18.5) we have that

$$\text{Remainder}(18.7) = w \cdot \frac{\phi_n(c(x))}{(c(x))^n} \cdot L(|t-x|^n, x)$$

$$\le w \cdot \left( \frac{c(x)}{(n+1)! \cdot h} + \frac{1}{2 \cdot n!} + \frac{h}{8 \cdot c(x) \cdot (n-1)!} \right)$$

$$\cdot L(|t-x|^n, x), \quad \forall x \in [a,b],$$

where $c(x) := \max(x-a, b-x)$. Therefore, from (18.7) we get that

$$|L(f,x) - f(x)| \leq |f(x)| \cdot |L(1,x) - 1| + \sum_{k=1}^{n} \frac{|f^{(k)}(x)|}{k!} \cdot |L((t-x)^k, x)|$$

$$+ w \cdot \left( \frac{c(x)}{(n+1)! \cdot h} + \frac{1}{2 \cdot n!} + \frac{h}{8 \cdot c(x) \cdot (n-1)!} \right) \cdot L(|t-x|^n, x).$$

Because all functions involved in the last inequality are $L_1$-, $L_2$-Lebesgue integrable functions (see Lemmas 18.1 and 18.2, and assumptions of the theorem), we obtain

$$\|Lf - f\|_1 \leq \| |f(x)| \cdot |L(1,x) - 1| \|_1 + \sum_{k=1}^{n} \left\| \frac{|f^{(k)}(x)|}{k!} \cdot |L((t-x)^k, x)| \right\|_1$$

$$+ w \cdot \left\| \left( \frac{c(x)}{(n+1)! \cdot h} + \frac{1}{2 \cdot n!} + \frac{h}{8 \cdot c(x) \cdot (n-1)!} \right) \cdot L(|t-x|^n, x) \right\|_1.$$

Then, by using Cauchy–Schwarz's inequality repeatedly, we have

$$\|Lf - f\|_1 \leq \|f\|_2 \cdot \|L1 - 1\|_2 + \sum_{k=1}^{n} \frac{1}{k!} \|f^{(k)}\|_2 \|L((t-\cdot)^k)(\cdot)\|_2$$

$$+ w \cdot \left\{ \frac{\|c(x)\|_2}{(n+1)! \cdot h} + \frac{1}{2 \cdot n!} \cdot \sqrt{b-a} + \frac{h}{8 \cdot (n-1)!} \cdot \left\| \frac{1}{c(x)} \right\|_2 \right\}$$

$$\cdot \|L(|t - \cdot|^n)(\cdot)\|_2. \tag{18.21}$$

We see that

$$\|c(x)\|_2 = \sqrt{\int_a^b c^2(x) dx} = \sqrt{\int_a^{\frac{a+b}{2}} (b-x)^2 \cdot dx + \int_{\frac{a+b}{2}}^b (x-a)^2 \cdot dx}$$

$$= \left( \frac{b-a}{2} \right) \cdot \sqrt{\frac{7}{3} \cdot (b-a)}.$$

I.e.,

$$\|c(x)\|_2 = \left( \frac{b-a}{2} \right) \cdot \sqrt{\frac{7}{3} \cdot (b-a)}. \tag{18.22}$$

Also we obtain

$$\left\|\frac{1}{c(x)}\right\|_2 = \sqrt{\int_a^b \frac{1}{c^2(x)}dx} = \sqrt{\int_a^b (c(x))^{-2}dx}$$

$$= \sqrt{\int_a^{\frac{a+b}{2}} (b-x)^{-2}dx + \int_{\frac{a+b}{2}}^b (x-a)^{-2}dx} = \sqrt{\frac{2}{b-a}}.$$

I.e.,

$$\left\|\frac{1}{c(x)}\right\|_2 = \sqrt{\frac{2}{b-a}}. \tag{18.23}$$

Putting (18.22), (18.23) into (18.21), we get

$$\|Lf - f\|_1 \leq \|f\|_2 \cdot \|L1 - 1\|_2 + \sum_{k=1}^n \frac{1}{k!} \cdot \|f^{(k)}\|_2 \cdot \|L((t - \cdot)^k)(\cdot)\|_2$$

$$+ w \cdot \left\{ \frac{(\frac{b-a}{2}) \cdot \sqrt{\frac{7}{3} \cdot (b-a)}}{(n+1)! \cdot h} + \frac{1}{2 \cdot n!} \cdot \sqrt{b-a} \right.$$

$$\left. + \frac{h}{8 \cdot (n-1)!} \cdot \sqrt{\frac{2}{b-a}} \right\} \cdot \|L(|t - \cdot|^n)(\cdot)\|_2.$$

Picking

$$h = h_n := \|L(|t - x|^n)(x)\|_2^{1/n}, \quad n \in \mathbf{N},$$

i.e., $h^n = \|L(|t - x|^n)(x)\|_2$, we establish (18.20). ∎

### COROLLARY 18.3
*Consider the positive linear operator*

$$L : C([a,b]) \longrightarrow C([a,b]).$$

Call

$$h_1 := \|L(|t - x|)(x)\|_2. \tag{18.24}$$

Let $f \in C^1([a,b])$. Then it holds

$$\|Lf - f\|_1 \leq \|f\|_2 \cdot \|L1 - 1\|_2 + \|f'\|_2$$

$$\cdot \|L((t-\cdot))(\cdot)\|_2 + \frac{\omega_1(f', h_1)}{2}$$

$$\cdot \left\{ \left(\frac{b-a}{2}\right) \cdot \sqrt{\frac{7}{3}(b-a)} + (\sqrt{b-a}) \cdot h_1 \right.$$

$$\left. + \frac{h_1^2}{4} \cdot \sqrt{\frac{2}{b-a}} \right\}. \qquad (18.25)$$

**COROLLARY 18.4**
Consider the positive linear operator

$$L : C([a,b]) \longrightarrow C([a,b]).$$

Put

$$h_2 := \|L((t-x)^2)(x)\|_2^{1/2}. \qquad (18.26)$$

Let $f \in C^2([a,b])$. Then it holds

$$\|Lf - f\|_1 \le \|f\|_2 \cdot \|L1 - 1\|_2 + \|f'\|_2$$

$$\cdot \|L((t-\cdot))(\cdot)\|_2 + \frac{\|f''\|_2}{2} \cdot h_2^2$$

$$+ \frac{1}{4} \cdot \omega_1(f'', h_2) \cdot h_2 \cdot \left\{ \left(\frac{b-a}{3}\right) \cdot \sqrt{\frac{7}{3} \cdot (b-a)} \right.$$

$$\left. + (\sqrt{b-a}) \cdot h_2 + \frac{h_2^2}{2} \cdot \sqrt{\frac{2}{b-a}} \right\}. \qquad (18.27)$$

**REMARK 18.6** (on Corollary 18.4). Using Lemma 18.3 and Cauchy–Schwarz's inequality we have

$$((L(t-x))(x))^2 \le (L((t-x)^2)(x)) \cdot ((L(1))(x)).$$

Then

$$\int_a^b ((L(t-x))(x))^2 \cdot dx \le \int_a^b ((L(t-x)^2)(x) \cdot (L(1))(x)) \cdot dx$$

$$\le \|(L(t-\cdot)^2)(\cdot)\|_2 \cdot \|L(1)\|_2,$$

the last inequality comes by Cauchy–Schwarz's inequality. We have proved that
$$\|(L(t-\cdot))(\cdot)\|_2 \le h_2 \cdot \sqrt{\|L(1)\|_2}. \tag{18.28}$$

∎

Next we give some general $L_p$ results.

### THEOREM 18.4
Let $([a,b], \mathcal{B}, \mu)$, $a \ne b$, be a measure space, where $\mathcal{B}$ is the Borel $\sigma$-algebra on $[a,b]$ and $\mu$ is a positive finite measure on $[a,b]$. (Note that $C([a,b]) \subset L_p([a,b], \mathcal{B}, \mu)$, any $p > 1$.) Here $\|\cdot\|_p$ stands for the related $L_p$ norm with respect to $\mu$. Let $p, q > 1$ such that $\frac{1}{p} + \frac{1}{q} = 1$. Consider the positive linear operator
$$L : C([a,b]) \longrightarrow C([a,b]).$$

Set $(n \in \mathbb{N})$
$$h_n := \|(L(|t-\cdot|^n))(\cdot)\|_p^{1/n}. \tag{18.29}$$

Let $f \in C^n([a,b])$. Then it holds
$$\|Lf - f\|_p \le \|f\|_{p\cdot q} \cdot \|L1 - 1\|_{p^2}$$

$$+ \sum_{k=1}^{n} \frac{1}{k!} \cdot \|f^{(k)}\|_{p\cdot q} \cdot \|L(t-\cdot)^k(\cdot)\|_{p^2} + \omega_1(f^{(n)}, h_n)$$

$$\cdot h_n^{n-1} \cdot \left[ \frac{(b-a)}{(n+1)!} + \frac{h_n}{2 \cdot n!} + \frac{h_n^2}{4 \cdot (n-1)! \cdot (b-a)} \right]. \tag{18.30}$$

Notice that from Lemma 18.1, and 18.2, $h_n < +\infty$ and R.H.S.(18.30) $< +\infty$.

**PROOF**   In the proof of Theorem 18.3 we got that
$$|L(f,x) - f(x)| \le |f(x)| \cdot |L(1,x) - 1|$$

$$+ \sum_{k=1}^{n} \frac{|f^{(k)}(x)|}{k!} \cdot |L((t-x)^k, x)|$$

$$+ w \cdot \left( \frac{c(x)}{(n+1)! \cdot h} + \frac{1}{2 \cdot n!} + \frac{h}{8 \cdot c(x) \cdot (n-1)!} \right)$$

## 18. Quantitative $L_p$-Results for Positive Linear Operators

$$\cdot L(|t-x|^n, x). \tag{18.31}$$

Note that $0 < c(x) \le b - a$ and

$$\frac{1}{c(x)} \le \frac{2}{b-a}, \quad \forall x \in [a, b].$$

Therefore, from (18.31) we get

$$|L(f,x) - f(x)| \le |f(x)| \cdot |L(1,x) - 1|$$

$$+ \sum_{k=1}^{n} \frac{|f^{(k)}(x)|}{k!} \cdot |L((t-x)^k, x)|$$

$$+ w \cdot \left( \frac{(b-a)}{(n+1)! \cdot h} + \frac{1}{2 \cdot n!} + \frac{h}{4 \cdot (b-a) \cdot (n-1)!} \right)$$

$$\cdot L(|t-x|^n, x). \tag{18.32}$$

Thus

$$\|Lf - f\|_p \le \| |f(x)| \cdot |L(1,x) - 1| \|_p$$

$$+ \sum_{k=1}^{n} \frac{1}{k!} \cdot \| |f^{(k)}(x)| \cdot |L((t-x)^k, x)| \|_p$$

$$+ w \cdot \left[ \frac{(b-a)}{(n+1)! \cdot h} + \frac{1}{2 \cdot n!} + \frac{h}{4 \cdot (n-1)! \cdot (b-a)} \right]$$

$$\cdot \|(L(|t-\cdot|^n))(\cdot)\|_p. \tag{18.33}$$

In (18.33) we apply Lemma 18.4, namely inequality (18.1), and we obtain

$$\|Lf - f\|_p \le \|f\|_{p \cdot q} \cdot \|L1 - 1\|_{p^2}$$

$$+ \sum_{k=1}^{n} \frac{1}{k!} \cdot \|f^{(k)}\|_{p \cdot q} \cdot \|L(t-\cdot)^k(\cdot)\|_{p^2}$$

$$+ w \cdot \left[\frac{(b-a)}{(n+1)! \cdot h} + \frac{1}{2 \cdot n!} + \frac{h}{4 \cdot (n-1)! \cdot (b-a)}\right]$$

$$\cdot \|(L(|t - \cdot|^n))(\cdot)\|_p. \tag{18.34}$$

Choose
$$h = h_n := \|(L(|t - \cdot|^n))(\cdot)\|_p^{1/n}, \quad n \in \mathbf{N},$$

i.e., $h^n = h_n^n = \|(L(|t - \cdot|^n))(\cdot)\|_p$. From (18.34) we now have that

$$\|Lf - f\|_p \leq \|f\|_{p \cdot q} \cdot \|L1 - 1\|_{p^2} + \sum_{k=1}^{n} \frac{1}{k!} \cdot \|f^{(k)}\|_{p \cdot q}$$

$$\cdot \|L(t - \cdot)^k(\cdot)\|_{p^2} + \omega_1(f^{(n)}, h_n)$$

$$\cdot \left[\frac{(b-a)}{(n+1)!} \cdot h_n^{n-1} + \frac{1}{2 \cdot n!} \cdot h_n^n + \frac{h_n^{n+1}}{4 \cdot (n-1)! \cdot (b-a)}\right].$$

∎

**REMARK 18.7** (on Theorem 18.4). (i) Case of $n = 1$. Then it holds

$$\|Lf - f\|_p \leq \|f\|_{p \cdot q} \cdot \|L1 - 1\|_{p^2} + \|f'\|_{p \cdot q} \cdot \|L(t - \cdot)(\cdot)\|_{p^2}$$

$$+ \frac{\omega_1(f', h_1)}{2} \cdot \left[(b-a) + h_1 + \frac{h_1^2}{2 \cdot (b-a)}\right]. \tag{18.35}$$

(ii) Case of $n = 2$. Then it holds

$$\|Lf - f\|_p \leq \|f\|_{p \cdot q} \cdot \|L1 - 1\|_{p^2} + \|f'\|_{p \cdot q}$$

$$\cdot \|L(t - \cdot)(\cdot)\|_{p^2} + \frac{\|f''\|_{p \cdot q}}{2} \cdot \|L(t - \cdot)^2(\cdot)\|_{p^2}$$

$$+ \frac{\omega_1(f'', h_2)}{2} \cdot h_2 \cdot \left[\frac{(b-a)}{3} + \frac{h_2}{2} + \frac{h_2^2}{2 \cdot (b-a)}\right]. \tag{18.36}$$

∎

## 18. Quantitative $L_p$-Results for Positive Linear Operators

**THEOREM 18.5**
Let $([a,b], \mathcal{B}, \mu)$, $a \neq b$, be a measure space, where $\mathcal{B}$ is the Borel $\sigma$-algebra on $[a,b]$ and $\mu$ is a positive finite measure on $[a,b]$. (Note that $C([a,b]) \subset L_p([a,b], \mathcal{B}, \mu)$, any $1 \leq p < +\infty$.) Here $\|\cdot\|_p$ stands for the related $L_p$ norm with respect to $\mu$ and $\|\cdot\|_\infty$ stands for the supremum norm. Consider the positive linear operator

$$L : C([a,b]) \longrightarrow C([a,b]).$$

Set $(n \in \mathbf{N})$
$$h_n := \|L(|t-x|^n, x)\|_p^{1/n}. \tag{18.37}$$

Let $f \in C^n([a,b])$. Then it holds

$$\|Lf - f\|_p \leq \|f\|_\infty \cdot \|L1 - 1\|_p$$

$$+ \sum_{k=1}^{n} \frac{\|f^{(k)}\|_\infty}{k!} \cdot \|L((t-x)^k, x)\|_p + \omega_1(f^{(n)}, h_n)$$

$$\cdot h_n^{n-1} \cdot \left[ \frac{(b-a)}{(n+1)!} + \frac{h_n}{2 \cdot n!} + \frac{h_n^2}{4 \cdot (n-1)! \cdot (b-a)} \right]. \tag{18.38}$$

Note that from Lemmas 18.1 and 18.2, $h_n < +\infty$ and R.H.S (18.38) $< +\infty$.

**PROOF**  From (18.32), in the proof of Theorem 18.4, we get

$$|L(f,x) - f(x)| \leq \|f\|_\infty \cdot |L(1,x) - 1| + \sum_{k=1}^{n} \frac{\|f^{(k)}\|_\infty}{k!} \cdot |L((t-x)^k, x)|$$

$$+ w \cdot \left[ \frac{(b-a)}{(n+1)! \cdot h} + \frac{1}{2 \cdot n!} + \frac{h}{4 \cdot (n-1)! \cdot (b-a)} \right]$$

$$\cdot (L(|t-x|^n, x)), \quad \forall x \in [a,b], \ n \in \mathbf{N}. \tag{18.39}$$

Therefore $(1 \leq p < +\infty)$,

$$\|Lf - f\|_p \leq \|f\|_\infty \cdot \|L1 - 1\|_p + \sum_{k=1}^{n} \frac{\|f^{(k)}\|_\infty}{k!} \cdot \|L((t-x)^k, x)\|_p$$

$$+ w \cdot \left[ \frac{(b-a)}{(n+1)! \cdot h} + \frac{1}{2 \cdot n!} + \frac{h}{4 \cdot (n-1)! \cdot (b-a)} \right]$$

$$\cdot \|L(|t-x|^n, x)\|_p. \tag{18.40}$$

Pick
$$h = h_n := \|L(|t-x|^n, x)\|_p^{1/n},$$

i.e.,
$$h^n = h_n^n = \|L(|t-x|^n, x)\|_p.$$

Then we find that

$$\|Lf - f\|_p \le \|f\|_\infty \cdot \|L1 - 1\|_p + \sum_{k=1}^n \frac{\|f^{(k)}\|_\infty}{k!} \cdot \|L((t-x)^k, x)\|_p$$

$$+ \omega_1(f^{(n)}, h_n) \cdot \left[ \frac{(b-a) \cdot h_n^{n-1}}{(n+1)!} + \frac{h_n^n}{2 \cdot n!} + \frac{h_n^{n+1}}{4 \cdot (n-1)! \cdot (b-a)} \right].$$

∎

**DEFINITION 18.1** *Let $f, g \in C([a,b])$ such that $|f(x)| \le |g(x)|$, $\forall x \in [a,b]$. A norm $\|\cdot\|$ on $C([a,b])$ is called monotone iff $\|f\| \le \|g\|$. We denote a monotone norm by $\|\cdot\|_m$, e.g., $L_p$ norms in general ($1 \le p \le +\infty$), Orlicz norms, etc.*

In a similar way as in the proof of Theorem 18.5 we have

**COROLLARY 18.5**
*Consider the positive linear operator*

$$L : C([a,b]) \longrightarrow C([a,b]), \quad a \ne b.$$

*Let $\|\cdot\|_m$ be a monotone norm on $C([a,b])$. Call ($n \in \mathbf{N}$)*

$$h_n := \|L(|t-x|^n, x)\|_m^{1/n}. \tag{18.41}$$

*Let $f \in C^n([a,b])$. Then it holds*

$$\|Lf - f\|_m \le \|f\|_\infty \cdot \|L1 - 1\|_m + \sum_{k=1}^n \frac{\|f^{(k)}\|_\infty}{k!} \cdot \|L((t-x)^k, x)\|_m$$

$$+ \omega_1(f^{(n)}, h_n) \cdot h_n^{n-1}$$

$$\cdot \left[ \frac{(b-a)}{(n+1)!} + \frac{h_n}{2 \cdot n!} + \frac{h_n^2}{4 \cdot (n-1)! \cdot (b-a)} \right]. \tag{18.42}$$

**REMARK 18.8** (on Corollary 18.5). Special case of $n = 1$. We obtain $(f \in C^1([a, b])))$

$$\|Lf - f\|_m \le \|f\|_\infty \cdot \|L1 - 1\|_m + \|f'\|_\infty \cdot \|L((t-x), x)\|_m$$

$$+ \frac{\omega_1(f', h_1)}{2} \cdot \left[(b-a) + h_1 + \frac{h_1^2}{2 \cdot (b-a)}\right], \tag{18.43}$$

where

$$h_1 := \|L(|t-x|, x)\|_m. \tag{18.44}$$

∎

## 18.3  Abstract Theory

Next we give results just for continuous functions.
We use

**DEFINITION 18.2**  Let $M$ be a non-empty compact subset of the real normed vector space $(V, \|\cdot\|)$. Let $f : M \to \mathbf{R}$ be a Borel measurable function and $h > 0$. We call

$$\omega_1(f, h) := \sup_{\substack{x, y \in M \\ \|x-y\| \le h}} |f(x) - f(y)|$$

the (first) modulus of continuity of $f$.

We also need the following result from Anastassiou [6], p. 44.

**COROLLARY B**
  Let $\mu$ be a positive finite measure of mass $m \ne 0$ on the Borel $\sigma$-algebra of the nonempty convex and compact subset $M$ of the real normed vector

space $(V, \|\cdot\|)$. Consider $x_0 \in M$. Put

$$\left(\int_M \|x - x_0\|^r \mu(dx)\right)^{1/r} := D_r(x_0). \qquad (18.45)$$

where $r \geq 1$. Suppose that $D_r(x_0) > 0$. Let $f : M \to \mathbf{R}$ be a Borel measurable function with $\omega_1(f, h)$ its first modulus of continuity, where $h > 0$ is given. Then it holds

$$\left|\int_M f\, d\mu - f(x_0)\right| \leq |m - 1| \cdot |f(x_0)| + \omega_1(f, h)$$

$$\cdot \left(m + \frac{D_r(x_0)}{h} \cdot m^{(1-\frac{1}{r})}\right). \qquad (18.46)$$

**Note.** If $D_r(x_0) = 0$, then $\mu(M - \{x_0\}) = 0$ and $\mu = m \cdot \delta_{x_0}$. Then (18.46) holds trivially. It is also trivial when $m = 0$. When $r = 1$, then one can prove (independently) in an easier way inequality (18.46).

**REMARK 18.9** In Corollary B, we need to consider only regular Borel measures $\mu$ on $M$; these among others are positive measures such that $\mu(M) < +\infty$. Let $L$ be a positive linear operator from $C(M)$ into itself, where $M$ is a nonempty convex and compact subset of $(V, \|\cdot\|)$. Consider $x_0 \in M$. By Remark 18.3 we see that

$$(Lf)(x_0) = \int_M f\, d\mu_{x_0}, \quad \forall f \in C(M), \qquad (18.47)$$

where $\mu_{x_0}$ is a unique regular Borel positive finite measure on $M$. Notice that
$$\infty + > m_{x_0} := \mu_{x_0}(M) = (L(1))(x_0) \geq 0.$$
We suppose
$$(L(1))(x_0) > 0, \quad \forall x_0 \in M. \qquad (18.48)$$
Also observe that

$$(L(\|x - x_0\|^r))(x_0) = \int_M \|x - x_0\|^r \cdot \mu_{x_0}(dx) := D_r^r(x_0), \quad r \geq 1.$$

We also suppose that

$$(L(\|x - x_0\|^r))(x_0) > 0, \quad \forall x_0 \in M,\ r \geq 1. \qquad (18.49)$$

## 18. Quantitative $L_p$-Results for Positive Linear Operators

For a justification of the assumptions (18.48), (18.49), see Remark 18.2. By (18.47), (18.48), (18.49), and Corollary B, we obtain for $f \in C(M)$ that

$$|(L(f))(x_0) - f(x_0)| \le |(L(1))(x_0) - 1| \cdot |f(x_0)| + \omega_1(f, h) \quad (18.50)$$

$$\cdot \left[ (L(1))(x_0) + \frac{((L(\|x - x_0\|^r))(x_0))^{1/r}}{h} \cdot ((L(1))(x_0))^{(1 - \frac{1}{r})} \right],$$

$h > 0$. When $r = 1$ we have that

$$|(Lf)(x_0) - f(x_0)| \le |(L(1))(x_0) - 1| \cdot |f(x_0)| + \omega_1(f, h)$$

$$\cdot \left[ (L(1))(x_0) + \frac{L(\|x - x_0\|)(x_0)}{h} \right]. \quad (18.51)$$

Thus

$$|(Lf)(x_0) - f(x_0)| \le |(L(1))(x_0) - 1| \cdot \|f\|_\infty + \omega_1(f, h)$$

$$\cdot \left[ (L(1))(x_0) + \frac{L(\|x - x_0\|)(x_0)}{h} \right]. \quad (18.52)$$

For our convenience we set

$$\varphi_{x_0}(x) := \left[ \frac{(L(\|x - x_0\|^r))(x)}{(L(1))(x)} \right]^{1/r}, \quad \forall x \in M, x_0 \in M \text{ fixed}, r \ge 1. \quad (18.53)$$

Notice that $\varphi_{x_0} \in C(M)$, $\forall x_0 \in M$.

We also call

$$F(x_0) := \varphi_{x_0}(x_0), \quad \forall x_0 \in M. \quad (18.54)$$

I.e.,

$$F(x_0) = \left( \frac{(L(\|x - x_0\|^r)(x_0)}{(L(1))(x_0)} \right)^{1/r}, \quad \forall x_0 \in M. \quad (18.55)$$

From Lemma 18.1 and assumption (18.48), we have that $F \in C(M)$. By assumptions (18.48) and (18.49), we have that $F > 0$, thus $\|F\|_\infty > 0$.

Next follows the $\| \cdot \|_\infty$-result for continuous functions.

## THEOREM 18.6
Let $M$ be a nonempty convex and compact subset of the real normed vector space $(V, \|\cdot\|)$. Let $L$ be a positive linear operator from $C(M)$ into itself. Suppose that $(L(1))(x_0)$, $(L(\|x-x_0\|^r))(x_0) > 0$, $\forall x_0 \in M$, $r \geq 1$. Put

$$F(x_0) := \left( \frac{(L(\|x-x_0\|^r))(x_0)}{(L(1))(x_0)} \right)^{1/r}, \quad \forall x_0 \in M.$$

Consider $f \in C(M)$. Then it holds

$$\|Lf - f\|_\infty \leq \|f\|_\infty \cdot \|L1 - 1\|_\infty + \omega_1(f, r \cdot \|F\|_\infty) \cdot \|L(1)\|_\infty \cdot \left(1 + \frac{1}{r}\right). \tag{18.56}$$

Note that $0 < \|F\|_\infty < +\infty$.

**PROOF** Directly from (18.50) by taking the supremum norms everywhere and by picking $h := r \cdot \|F\|_\infty$. ∎

It follows the case $r = 1$.

## COROLLARY 18.6
Using the same terms and assumptions of Theorem 18.6, we find

$$\|Lf - f\|_\infty \leq \|f\|_\infty \cdot \|L1 - 1\|_\infty$$

$$+ \omega_1(f; \|L(\|x-x_0\|)(x_0)\|_\infty)$$

$$\cdot (1 + \|L1\|_\infty), \quad \forall f \in C(M). \tag{18.57}$$

(Notice that $0 < \|L(\|x-x_0\|)(x_0)\|_\infty < +\infty$.)

**PROOF** Directly from (18.51) by taking the supremum norms everywhere and by choosing $h := \|L(\|x-x_0\|)(x_0)\|_\infty$. ∎

Next we get some general $L_p$ results for continuous functions $(1 \leq p < +\infty)$.

## PROPOSITION 18.1
Let $M$ be a nonempty compact convex subset of the real normed vector space $(V, \|\cdot\|)$. Let $(M, \mathcal{B}, \mu)$ be a measure space, where $\mathcal{B}$ is the Borel $\sigma$-

# 18. Quantitative $L_p$-Results for Positive Linear Operators

algebra on $M$ and $\mu$ is a positive finite measure on $M$. (Note that $C(M) \subset L_p(M, \mathcal{B}, \mu)$, any $1 \leq p < +\infty$.) Here $\|\cdot\|_p$ stands for the related $L_p$ norm with respect to $\mu$ and $\|\cdot\|_\infty$ stands for the supremum norm. Consider the positive linear operator

$$L : C(M) \longrightarrow C(M).$$

Suppose that $(L(\|x - x_0\|))(x_0) > 0$, $\forall x_0 \in M$. Set

$$h^* := \|L(\|x - \cdot\|)(\cdot)\|_p. \tag{18.58}$$

Let $f \in C(M)$. Then it holds

$$\|Lf - f\|_p \leq \|f\|_\infty \cdot \|L1 - 1\|_p + \omega_1(f, h^*) \cdot (1 + \|L1\|_p). \tag{18.59}$$

Note that by Lemma 18.1 we obtain $0 < h^* < +\infty$.

**PROOF** By monotonicity, subadditivity and linearity properties of $\|\cdot\|_p$ applied on (18.52), we have that

$$\|Lf - f\|_p \leq \|L1 - 1\|_p \cdot \|f\|_\infty + \omega_1(f, h)$$

$$\cdot \left[\|L1\|_p + \frac{1}{h} \cdot \|L(\|x - \cdot\|)(\cdot)\|_p\right]. \tag{18.60}$$

Then pick $h = h^*$ as in (18.58). ∎

## THEOREM 18.7
Let $M$ be a nonempty compact convex subset of the real normed vector space $(V, \|\cdot\|)$. Let $(M, \mathcal{B}, \mu)$ be a measure space, where $\mathcal{B}$ is the Borel $\sigma$-algebra on $M$ and $\mu$ is a positive finite measure on $M$. (Note that $C(M) \subset L_p(M, \mathcal{B}, \mu)$, any $1 \leq p < +\infty$.) Consider the positive linear operator

$$L : C(M) \longrightarrow C(M).$$

Suppose
$$\begin{aligned}(L(1))(x_0) &> 0, & \forall x_0 \in M, \\ (L(\|x - x_0\|^r))(x_0) &> 0, & \forall x_0 \in M,\ r > 1.\end{aligned} \tag{18.61}$$

Set

$$F(x_0) := \left(\frac{(L(\|x - x_0\|^r))(x_0)}{(L(1))(x_0)}\right)^{1/r}, \quad \forall x_0 \in M. \tag{18.62}$$

Let $f \in C(M)$. Then it holds

$$\|Lf - f\|_p \leq \|f\|_\infty \cdot \|L1 - 1\|_p + \omega_1(f; r \cdot \|F\|_p) \cdot \|L(1)\|_\infty \cdot \left[(\mu(M))^{1/p} + \frac{1}{r}\right]. \quad (18.63)$$

Notice that $0 < \|F\|_p < +\infty$.

**PROOF** Let $f \in C(M)$. Then from inequality (18.50) we observe that

$$|(Lf)(x_0) - f(x_0)| \leq |(L1)(x_0) - 1| \cdot |f(x_0)| + \omega_1(f, h) \cdot (L1)(x_0)$$

$$\cdot \left[1 + \frac{1}{h} \cdot F(x_0)\right], \quad \forall x_0 \in M,\ h > 0,\ r > 1. \quad (18.64)$$

Thus

$$|(Lf)(x_0) - f(x_0)| \leq |(L1)(x_0) - 1| \cdot \|f\|_\infty + \omega_1(f, h) \cdot \|L(1)\|_\infty$$

$$\cdot \left[1 + \frac{1}{h} \cdot F(x_0)\right], \quad \forall x_0 \in M,\ h > 0,\ r > 1. \quad (18.65)$$

Hence

$$\|Lf - f\|_p \leq \|f\|_\infty \cdot \|L1 - 1\|_p + \omega_1(f, h) \cdot \|L(1)\|_\infty$$

$$\cdot \left[(\mu(M))^{1/p} + \frac{1}{h} \cdot \|F\|_p\right], \quad h > 0,\ r > 1. \quad (18.66)$$

Finally choose in (18.66)
$$h := r \cdot \|F\|_p.$$

∎

## 18.4 Multidimensional Theory

Here we use Theorem 5 from Anastassiou [7], p. 97.

## THEOREM A

Take $Q \subset \mathbf{R}^k$ of the form $Q := \{x \in \mathbf{R}^k : \|x\| \leq 1\}$, where $\|\cdot\|$ the $\ell_1$-norm in $\mathbf{R}^k$, $k \geq 1$, and let $\mathbf{x}_0 := (x_{01}, \ldots, x_{0k}) \in Q$ be fixed. Let the positive measure $\mu$ satisfy $\mu(Q) = 1$ and

$$\int_Q \|\mathbf{x} - \mathbf{x}_0\| \cdot \mu(d\mathbf{x}) := d^*.$$

Also, let $f \in C^n(Q)$ and assume that each of its nth partial derivatives $f_\alpha$ has a modulus of continuity $\omega_1(f_\alpha, h) \leq w$, where $h$ and $w$ are fixed positive numbers. Then it holds

$$\left| \int_Q f \, d\mu - f(\mathbf{x}_0) \right| \leq \left| \sum_{j=1}^n \frac{1}{j!} \cdot \int_Q g_\mathbf{x}^{(j)}(0) \cdot \mu(d\mathbf{x}) \right|$$

$$+ w \cdot d^* \cdot \frac{\phi_n(1 + \|\mathbf{x}_0\|)}{(1 + \|\mathbf{x}_0\|)}, \qquad (18.67)$$

where $g_\mathbf{x}(t) := f(\mathbf{x}_0 + t \cdot (\mathbf{x} - \mathbf{x}_0))$, $t \geq 0$.

**REMARK A** From $n = 1$, (18.67) yields the inequality ($\mathbf{x}_0 \in Q$):

$$\left| \int_Q f \, d\mu - f(\mathbf{x}_0) \right| \leq \left| \sum_{i=1}^k \frac{\partial f}{\partial x_i}(\mathbf{x}_0) \cdot \int_Q (x_i - x_{0i}) \cdot \mu(d\mathbf{x}) \right|$$

$$+ w \cdot d^* \cdot \frac{\phi_1(1 + \|\mathbf{x}_0\|)}{(1 + \|\mathbf{x}_0\|)}. \qquad (18.68)$$

Next follow the related results in $\mathbf{R}^k$, $k \geq 1$.

## THEOREM 18.8

Take $Q \subset \mathbf{R}^k$ of the form $Q := \{\mathbf{x} \in \mathbf{R}^k : \|\mathbf{x}\| \leq 1\}$, where $\|\cdot\|$ is the $\ell_1$-norm in $\mathbf{R}^k$, $k \geq 1$. Let $(Q, \mathcal{B}, \mu)$ be a measure space, where $\mathcal{B}$ is the Borel $\sigma$-algebra on $Q$ and $\mu$ is a positive finite measure on $Q$. (Note that $C(Q) \subset L_p(Q, \mathcal{B}, \mu)$, $1 \leq p < +\infty$). Consider the positive linear operator

$$L : C(Q) \longrightarrow C(Q),$$

such that

$$(L(1))(\mathbf{x}_0) = 1, \quad \forall \mathbf{x}_0 \in Q. \qquad (18.69)$$

*Suppose that*

$$h_1 := \|(L(\|\mathbf{x} - \mathbf{x}_0\|))(\mathbf{x}_0)\|_\infty > 0, \tag{18.70}$$

$$h_2 := \|(L(\|\mathbf{x} - \mathbf{x}_0\|))(\mathbf{x}_0)\|_p > 0, \quad 1 \leq p < +\infty. \tag{18.71}$$

*(Notice that $h_1, h_2 < +\infty$.)*

*Let $f \in C^n(Q)$, $n \geq 2$, and assume that each of its nth partial derivatives $f_\alpha$ has a modulus of continuity $\omega_1(f_\alpha, h_i) \leq w_i$, where $w_i$, $i = 1, 2$ are given positive numbers.*

*Set*

$$g_{(\mathbf{x},\mathbf{x}_0)}(t) := f(\mathbf{x}_0 + t \cdot (\mathbf{x} - \mathbf{x}_0)), \quad t \geq 0, \quad \mathbf{x}_0, \mathbf{x} \in Q. \tag{18.72}$$

*Then we have the following estimates:*
*(i)*

$$\|Lf - f\|_\infty \leq \sum_{j=1}^n \frac{1}{j!} \cdot \|(L(g^{(j)}_{(\mathbf{x},\mathbf{x}_0)}(0)))(\mathbf{x}_0)\|_\infty$$

$$+ w_1 \cdot \left( \frac{2^n}{(n+1)!} + \frac{h_1 \cdot 2^{n-2}}{n!} + \frac{h_1^2 \cdot 2^{n-5}}{(n-1)!} \right), \tag{18.73}$$

*(ii)*

$$\|Lf - f\|_p \leq \sum_{j=1}^n \frac{1}{j!} \cdot \|(L(g^{(j)}_{(\mathbf{x},\mathbf{x}_0)}(0)))(\mathbf{x}_0)\|_p$$

$$+ w_2 \cdot \left( \frac{2^n}{(n+1)!} + \frac{h_2 \cdot 2^{n-2}}{n!} + \frac{h_2^2 \cdot 2^{n-5}}{(n-1)!} \right). \tag{18.74}$$

*Note that R.H.S.'s (18.73) and (18.74) $< +\infty$.*

**PROOF** Here $n \geq 2$, $n \in \mathbf{N}$ and the fixed point $\mathbf{x}_0 \in Q := \{x \in \mathbf{R}^k : \|\mathbf{x}\| \leq 1\}$, where $\|\cdot\|$ is the $\ell_1$ norm in $\mathbf{R}^k$. Note that $1 + \|\mathbf{x}_0\| \leq 2$. Thus, from (18.4), we obtain

$$\frac{\phi_n(1+\|\mathbf{x}_0\|)}{(1+\|\mathbf{x}_0\|)} \leq \left( \frac{(1+\|\mathbf{x}_0\|)^n}{(n+1)! \cdot h} + \frac{(1+\|\mathbf{x}_0\|)^{n-1}}{2 \cdot n!} + \frac{h \cdot (1+\|\mathbf{x}_0\|)^{n-2}}{8 \cdot (n-1)!} \right)$$

$$\leq \left( \frac{2^n}{(n+1)! \cdot h} + \frac{2^{n-1}}{2 \cdot n!} + \frac{h \cdot 2^{n-2}}{8 \cdot (n-1)!} \right).$$

That is,

$$\frac{\phi_n(1+\|\mathbf{x}_0\|)}{(1+\|\mathbf{x}_0\|)} \leq \left( \frac{2^n}{(n+1)! \cdot h} + \frac{2^{n-2}}{n!} + \frac{h \cdot 2^{n-5}}{(n-1)!} \right), \quad h > 0. \quad (18.75)$$

Let $L$ be a positive linear operator from $C(Q)$ into itself, such that

$$(L(1))(\mathbf{x}_0) = 1, \quad \forall \mathbf{x}_0 \in Q.$$

Let $f \in C^n(Q)$ such that $\omega_1(f_\alpha, h_i) \leq w_i$. By Remark 18.3 we see that

$$(L(f))(\mathbf{x}_0) = \int_Q f(t) \cdot \mu_{\mathbf{x}_0}(dt), \quad \forall f \in C(Q),$$

where $\mu_{\mathbf{x}_0}(Q) = (L(1))(\mathbf{x}_0) = 1$. We put

$$(L(\|\mathbf{x} - \mathbf{x}_0\|))(\mathbf{x}_0) := d^*(\mathbf{x}_0)$$

and

$$g_{(\mathbf{x},\mathbf{x}_0)}(t) := f(\mathbf{x}_0 + t \cdot (\mathbf{x} - \mathbf{x}_0)), \quad t \geq 0, \ \mathbf{x} \in Q.$$

Note that $g_{(\mathbf{x},\mathbf{x}_0)}^{(j)}(0)$ is a continuous function of $\mathbf{x}$ over $Q$. Furthermore, $(L(g_{(\mathbf{x},\mathbf{x}_0)}^{(j)}(0)))(\mathbf{x}_0)$ is a continuous function in $\mathbf{x}_0$ over $Q$. From Theorem A and (18.75) we get that

$$|(Lf)(\mathbf{x}_0) - f(\mathbf{x}_0)| \leq \sum_{j=1}^n \frac{1}{j!} \cdot \left| \int_Q g_{(\mathbf{x},\mathbf{x}_0)}^{(j)}(0) \cdot \mu_{\mathbf{x}_0}(dx) \right|$$

$$+ w_i \cdot d^*(\mathbf{x}_0) \cdot \left( \frac{2^n}{(n+1)! \cdot h_i} + \frac{2^{n-2}}{n!} + \frac{h_i \cdot 2^{n-5}}{(n-1)!} \right).$$

That is,

$$|(Lf)(\mathbf{x}_0) - f(\mathbf{x}_0)| \leq \sum_{j=1}^n \frac{1}{j!} \cdot \left| (L(g_{(\mathbf{x},\mathbf{x}_0)}^{(j)}(0)))(\mathbf{x}_0) \right| + w_i \cdot (L(\|\mathbf{x} - \mathbf{x}_0\|))(\mathbf{x}_0)$$

$$\cdot \left( \frac{2^n}{(n+1)! \cdot h_i} + \frac{2^{n-2}}{n!} + \frac{h_i \cdot 2^{n-5}}{(n-1)!} \right), \quad h_i > 0, \ i = 1, 2. \quad (18.76)$$

### PROPOSITION 18.2

Take $Q := \{\mathbf{x} \in \mathbf{R}^k : \|\mathbf{x}\| \leq 1\}$, where $\|\cdot\|$ the $\ell_1$-norm in $\mathbf{R}^k$. Let $(Q, \mathcal{B}, \mu)$ be a measure space, where $\mathcal{B}$ is the Borel $\sigma$-algebra on $Q$ and $\mu$ is a positive finite measure on $Q$. Let the positive linear operator $L$ from $C(Q)$ into itself, such that

$$(L(1))(\mathbf{x}_0) = 1, \quad \forall \mathbf{x}_0 \in Q.$$

Suppose that

$$h_1 := \|(L(\|\mathbf{x} - \mathbf{x}_0\|))(\mathbf{x}_0)\|_\infty > 0, \tag{18.77}$$

$$h_2 := \|(L(\|\mathbf{x} - \mathbf{x}_0\|))(\mathbf{x}_0)\|_p > 0, \quad 1 \leq p < +\infty. \tag{18.78}$$

Let $f \in C^1(Q)$, and assume that each of its first partial derivatives $f_j$, $j = 1, \ldots, k$ has a modulus of continuity $\omega_1(f_j, h_i) \leq w_i$, where $w_i$, $i = 1, 2$ are given positive numbers. Then it holds

(i)

$$\|Lf - f\|_\infty \leq \sum_{j=1}^{k} \left\|\frac{\partial f}{\partial x_j}\right\|_\infty \cdot \|(L(x_j - x_{0j}))(\mathbf{x}_0)\|_\infty$$

$$+ w_1 \cdot \left(1 + \frac{h_1}{2} + \frac{h_1^2}{8}\right), \tag{18.79}$$

and

(ii)

$$\|Lf - f\|_p \leq \sum_{j=1}^{k} \left\|\frac{\partial f}{\partial x_j}\right\|_\infty \cdot \|(L(x_j - x_{0j}))(\mathbf{x}_0)\|_p$$

$$+ w_2 \cdot \left(1 + \frac{h_2}{2} + \frac{h_2^2}{8}\right). \tag{18.80}$$

Notice that R.H.S.'s (18.79) and (18.80) $< +\infty$.

## 18. Quantitative $L_p$-Results for Positive Linear Operators

**PROOF** Let $x > 0$, then by (18.4) we have

$$\frac{\phi_1(x)}{x} \leq \left(\frac{x}{2 \cdot h} + \frac{1}{2} + \frac{h}{8 \cdot x}\right).$$

Because $1 + \|\mathbf{x}_0\| \geq 1$, then $\frac{1}{1+\|\mathbf{x}_0\|} \leq 1$. Thus

$$\frac{\phi_1(1 + \|\mathbf{x}_0\|)}{(1 + \|\mathbf{x}_0\|)} \leq \left(\frac{(1 + \|\mathbf{x}_0\|)}{2 \cdot h} + \frac{1}{2} + \frac{h}{8 \cdot (1 + \|\mathbf{x}_0\|)}\right) \leq \left(\frac{1}{h} + \frac{1}{2} + \frac{h}{8}\right).$$

That is,

$$\frac{\phi_1(1 + \|\mathbf{x}_0\|)}{(1 + \|\mathbf{x}_0\|)} \leq \left(\frac{1}{h} + \frac{1}{2} + \frac{h}{8}\right), \quad h > 0. \tag{18.81}$$

Let $L$ be a positive linear operator from $C(Q)$ into itself, such that $(L(1))(\mathbf{x}_0) = 1$, $\forall \mathbf{x}_0 \in Q$. Let $f \in C^1(Q)$ such that $\omega_1(f_j, h_i) \leq w_i$, $j = 1, \ldots, k$; ($f_j$ the first partial derivative of $f$), $i = 1, 2$.

From Remark 18.3 we see that

$$(Lf)(\mathbf{x}_0) = \int_Q f(t) \cdot \mu_{\mathbf{x}_0}(dt), \quad \forall f \in C(Q),$$

where $\mu_{\mathbf{x}_0}(Q) = (L(1))(\mathbf{x}_0) = 1$. We set

$$(L(\|\mathbf{x} - \mathbf{x}_0\|))(\mathbf{x}_0) := d^*(\mathbf{x}_0)$$

and

$$pr_j(x) := x_j, \quad j = 1, \ldots, k$$

the projection map, which is continuous.

From (18.68) and (18.81) we find that

$$|(Lf)(\mathbf{x}_0) - f(\mathbf{x}_0)| \leq \sum_{j=1}^{k} \left\|\frac{\partial f}{\partial x_j}\right\|_\infty \cdot |L(pr_j(\mathbf{x} - \mathbf{x}_0))(\mathbf{x}_0)|$$

$$+ w \cdot ((L(\|\mathbf{x} - \mathbf{x}_0\|))(\mathbf{x}_0))$$

$$\cdot \frac{\phi_1(1 + \|\mathbf{x}_0\|)}{(1 + \|\mathbf{x}_0\|)} \leq \sum_{j=1}^{k} \left\|\frac{\partial f}{\partial x_j}\right\|_\infty \cdot |L(pr_j(\mathbf{x} - \mathbf{x}_0))(\mathbf{x}_0)|$$

$$+ w \cdot ((L(\|\mathbf{x} - \mathbf{x}_0\|))(\mathbf{x}_0)) \cdot \left(\frac{1}{h} + \frac{1}{2} + \frac{h}{8}\right).$$

That is,

$$|(Lf)(\mathbf{x}_0) - f(\mathbf{x}_0)| \le \sum_{j=1}^{k} \left\|\frac{\partial f}{\partial x_j}\right\|_\infty \cdot |L(pr_j(\mathbf{x} - \mathbf{x}_0))(\mathbf{x}_0)|$$

$$+ w_i \cdot \left(\frac{1}{h_i} + \frac{1}{2} + \frac{h_i}{8}\right) \cdot ((L(\|\mathbf{x} - \mathbf{x}_0\|))(\mathbf{x}_0)), \ i = 1, 2. \quad (18.82)$$

Notice that $(L(pr_j(\mathbf{x}-\mathbf{x}_0)))(\mathbf{x}_0)$, $(L(\|\mathbf{x}-\mathbf{x}_0\|))(\mathbf{x}_0)$ are continuous functions in $\mathbf{x}_0$. Then it holds

$$\|Lf - f\|_\infty \le \sum_{j=1}^{k} \left\|\frac{\partial f}{\partial x_j}\right\|_\infty \cdot \|(L(x_j - x_{0j}))(\mathbf{x}_0)\|_\infty$$

$$+ w_1 \cdot \left(\frac{1}{h_1} + \frac{1}{2} + \frac{h_1}{8}\right) \cdot \|L(\|\mathbf{x} - \mathbf{x}_0\|)(\mathbf{x}_0)\|_\infty, (18.83)$$

and also $(1 \le p < +\infty)$

$$\|Lf - f\|_p \le \sum_{j=1}^{k} \left\|\frac{\partial f}{\partial x_j}\right\|_\infty \cdot \|(L(x_j - x_{0j}))(\mathbf{x}_0)\|_p$$

$$+ w_2 \cdot \left(\frac{1}{h_2} + \frac{1}{2} + \frac{h_2}{8}\right) \cdot \|(L(\|\mathbf{x} - \mathbf{x}_0\|))(\mathbf{x}_0)\|_p. (18.84)$$

∎

## 18.5 Stochastic Theory

Let $(\underline{0}, \mathcal{A}, P)$ denote a probability space and $L^1(\underline{0}, \mathcal{A}, P)$ be the space of all real-valued random variables $Y = Y(\omega)$ with

$$\int_{\underline{0}} |Y(\omega)| \cdot P(d\omega) < +\infty.$$

## 18. Quantitative $L_p$-Results for Positive Linear Operators

Let $X = X(t, \omega)$ denote a stochastic process with index set $K$, a compact convex subset of $(V, \|\cdot\|)$ a real normed vector space, and real state space $(\mathbf{R}, \mathcal{B})$, $\mathcal{B}$ is the Borel $\sigma$-algebra on $\mathbf{R}$.

Denote by $C_{\underline{0}}(K)$ the space of $L^1$-continuous stochastic processes in $t$ and

$$B_{\underline{0}}(K) := \left\{ X : \sup_{t \in K} \int_{\underline{0}} |X(t,\omega)| \cdot P(d\omega) < +\infty \right\};$$

clearly $C_{\underline{0}}(K) \subset B_{\underline{0}}(K)$, $C(K) \subset C_{\underline{0}}(K)$.

With $E$ we denote the expectation operator

$$(EX)(t) := \int_{\underline{0}} X(t,\omega) \cdot P(d\omega).$$

Consider the linear operator

$$T : C_{\underline{0}}(K) \longrightarrow B_{\underline{0}}(K).$$

If $X \in C_{\underline{0}}(K)$ is nonnegative and $TX$, too, then $T$ is called a positive operator. If $ET = TE$, then $T$ is called an $E$-commutative operator.

If $T$ is $E$-commutative then $T(C(K)) \subset B(K)$ — the bounded real valued functions on $K$. Also, if $X(t,\omega) \in C_{\underline{0}}(K)$ then $(EX)(t) \in C(K)$. When $K = [a, b] \subset \mathbf{R}$, we set $C_{\underline{0}}^n([a,b]) := \{X : \text{there exists } X^{(k)}(t,\omega) \in C_{\underline{0}}([a,b])$ and it is continuous in $t$ for each $\omega \in \underline{0}$, $k = 0, 1, \ldots, n\}$.

From Anastassiou [9], p. 368, Theorem 1, we have

**THEOREM B**

*Consider the positive $E$-commutative linear operator*

$$T : C_{\underline{0}}([a,b]) \longrightarrow C_{\underline{0}}([a,b]).$$

*Let*

$$c(t_0) := \max(t_0 - a, b - t_0), \quad t_0 \in [a, b],$$

*and*

$$d_n(t_0) := ((T(|t - t_0|^n))(t_0))^{1/n}, \quad n \in \mathbf{N}.$$

*Let $X \in C_{\underline{0}}^n([a,b])$ such that $\omega_1(EX^{(n)}, h) \leq w$, where $w, h$ are given positive numbers, $0 < h \leq b - a$. Then it holds*

$$|(E(TX))(t_0) - (EX)(t_0)| \leq |(EX)(t_0)| \cdot |(T(1))(t_0) - 1|$$

$$+ \sum_{k=1}^{n} \frac{|(EX^{(k)})(t_0)|}{k!} \cdot |(T(t-t_0)^k)(t_0)|$$

$$+ w \cdot \frac{\phi_n(c(t_0))}{(c(t_0))^n} \cdot (d_n(t_0))^n, \quad \forall t_0 \in [a,b]. \tag{18.85}$$

**REMARK 18.10** (on Theorem B). From (18.85) we get that

$$|(E(TX))(t_0) - (EX)(t_0)| \le \|EX\|_\infty \cdot |(T(1))(t_0) - 1| + \sum_{k=1}^{n} \frac{\|EX^{(k)}\|_\infty}{k!}$$

$$\cdot |(T(t-t_0)^k)(t_0)| + w \cdot \frac{\phi_n(c(t_0))}{(c(t_0))^n} \cdot d_n^n(t_0), \quad \forall t_0 \in [a,b]. \tag{18.86}$$

Set

$$R_n(t_0) := w \cdot \frac{\phi_n(c(t_0))}{(c(t_0))^n} \cdot d_n^n(t_0). \tag{18.87}$$

Let $f \in C([a,b]) \subset C_{\underline{0}}([a,b])$, and assume $T$ as $E$-commutative. Then $Ef = f$ and $T(Ef) = T(f)$, i.e., $E(Tf) = Tf$, here $Tf \in C_{\underline{0}}([a,b])$. Thus $Tf \in C([a,b])$. We have shown that if

$$T: C_{\underline{0}}([a,b]) \longrightarrow C_{\underline{0}}([a,b]),$$

then

$$T: C([a,b]) \longrightarrow C([a,b]).$$

Next let $1 \le p \le +\infty$.

Let $([a,b], \mathcal{B}, \mu)$ be a measure space, where $\mathcal{B}$ is the Borel $\sigma$-algebra on $[a,b]$, $\mu$ is a positive finite measure on $[a,b]$. Here $\|\cdot\|_p$ ($1 \le p < +\infty$) stands for the $L_p([a,b], \mathcal{B}, \mu)$-norm, while $\|\cdot\|_\infty$ stands for the supremum norm.

From (18.86) and (18.87) we get

$$\|E(TX) - EX\|_p \le \|EX\|_\infty \cdot \|T1 - 1\|_p + \sum_{k=1}^{n} \frac{\|EX^{(k)}\|_\infty}{k!}$$

$$\cdot \|(T(t-t_0)^k)(t_0)\|_p + \|R_n(t_0)\|_p. \tag{18.88}$$

The remainder $\|R_n(t_0)\|_p$ ($1 \le p \le +\infty$) now can be treated exactly the same way as in Section 18.2 and obtain similar $\|\cdot\|_p$-results.

In general, let $T$ be an $E$-commutative operator from $C_{\underline{0}}(K)$ into itself, then $T$ maps $C(K)$ into itself. Also notice that

$$|E(TX)(t_0) - (EX)(t_0)| = |(T(EX))(t_0) - (EX)(t_0)|,$$

and if $X \in C_{\underline{0}}^n([a,b])$, then $X^{(n)}(t,\omega) \in C_{\underline{0}}([a,b])$, which is continuous in $t$ for every $\omega \in \underline{0}$, hence for each $\omega \in \underline{0}$ we have $X(t,\omega) \in C^n([a,b])$.

In that case, one can establish various estimates for $\|(E(TX))(t_0) - (EX)(t_0)\|_p$, $1 \leq p \leq +\infty$, using results from Anastassiou [9], Theorem 3, p. 374, Theorem 6, p. 378, Corollary 9', p. 381, in a similar way as in Sections 18.2–18.4, etc. ∎

# Part VII

# On Approximation Theory

# Chapter 19

# On Monotone Approximation Theory

## 19.1 High Order Quantitative Monotone Approximation with Linear Differential Operators

Let $f \in C^r([-1,1])$, $r \geq 0$ and let $L$ be a linear differential operator such that $L(f) \geq 0$ throughout $[-1,1]$. We can find a sequence of polynomials $Q_n(x)$ of degree $\leq n$ such that $L(Q_n) \geq 0$ over $[-1,1]$. Moreover, $Q_n^{(k)}$ approximates with rates uniformly $f^{(k)}$, $k = 0, 1, \ldots, r$. The degree of these approximations is given by inequalities using a higher order modulus of continuity for $f^{(r)}$. This section relies on [13].

One of the major subjects of approximation theory is that of monotone approximation, started by Shisha [158]. The initial problem was that, given a positive integer $\nu$, approximate with rates, that is with certain speed, a given function whose $\nu$th derivative is $\geq 0$ by polynomials having the same property.

This problem was generalized by Anastassiou and Shisha [40] by replacing the $\nu$th derivative with a linear differential operator of order $\nu$. The rate of the related convergence was given via the first modulus of continuity.

In this section we deal with the same generalized problem by giving a higher order of monotone approximation, that is a much more precise monotone approximation, with the use of the modulus of smoothness $w_s$, $s \geq 1$.

Without loss of generality and for our convenience in this section we use $I = [-1, 1]$ as a reference interval.

To prove the main result, Theorem 19.2, we use

**THEOREM 19.1**
(Gonska and Hinnemann [90]). *Let $r \geq 0$ and $s \geq 1$. Then there exists*

a sequence $Q_n = Q_n^{(r,s)}$ of linear polynomial operators mapping $C^r(I)$ into $P_n$, such that for all $f \in C^r(I)$, all $|x| \leq 1$ and all $n \geq \max(4(r+1), r+s)$ we have

$$|f^{(k)}(x) - (Q_n f)^{(k)}(x)| \leq M_{r,s}(\Delta_n(x))^{r-k} \cdot \omega_s(f^{(r)}, \Delta_n(x)), \quad 0 \leq k \leq r, \tag{19.1}$$

where $\Delta_n(x) = \frac{\sqrt{1-x^2}}{n} + \frac{1}{n^2}$, and $M_{r,s}$ is a constant independent of $f$, $x$, and $n$.

The following result is an immediate consequence of the previous theorem.

## COROLLARY 19.1

Let $r \geq 0$ and $s \geq 1$. Then there exists a sequence $Q_n = Q_n^{(r,s)}$ of linear polynomial operators mapping $C^r(I)$ into $P_n$, such that for all $f \in C^r(I)$ and all $n \geq \max(4(r+1), r+s)$ we have

$$\|f^{(k)} - (Q_n f)^{(k)}\|_\infty \leq \frac{C_{r,s}}{n^{r-k}} \omega_s\left(f^{(r)}, \frac{1}{n}\right), \quad k = 0, 1, \ldots, r, \tag{19.2}$$

where $C_{r,s}$ is a constant independent of $f$ and $n$.

Corollary 19.1 is used in the proof of the next main result.

## THEOREM 19.2

Let $h$, $\nu$, $r$ be integers, $0 \leq h \leq \nu \leq r$ and let $f \in C^r(I)$, with $f^{(r)}$ having modulus of smoothness $\omega_s(f^{(r)}, \delta)$ there, $s \geq 1$. Let $\alpha_j(x)$, $j = h, h+1, \ldots, \nu$ be real functions, defined and bounded on $I$ and suppose $\alpha_h(x)$ is either $\geq \alpha > 0$ or $\leq \beta < 0$ throughout $I$.

Take the operator

$$L = \sum_{j=h}^{\nu} \alpha_j(x) \cdot \left[\frac{d^j}{dx^j}\right]$$

and assume, throughout $I$,

$$L(f) \geq 0. \tag{19.3}$$

Then, for every integer $n \geq \max(4(r+1), r+s)$, there exists a real polynomial $Q_n(x)$ of degree $\leq n$ such that

$$L(Q_n) \geq 0 \quad \text{throughout } I$$

and
$$\|f^{(k)} - Q_n^{(k)}\|_\infty \leq \frac{C}{n^{r-\nu}} \omega_s\left(f^{(r)}, \frac{1}{n}\right), \quad 0 \leq k \leq h. \tag{19.4}$$

Moreover, we get
$$\|f^{(k)} - Q_n^{(k)}\|_\infty \leq \frac{C}{n^{r-k}} \omega_s\left(f^{(r)}, \frac{1}{n}\right), \quad h+1 \leq k \leq r. \tag{19.5}$$

where $C$ is a constant independent of $f$ and $n$.

**PROOF** Let $n$ be an integer $\geq \max(4(r+1), r+s)$. Put
$$l_j \equiv \sup_{x \in I} |\alpha_h^{-1}(x) \cdot \alpha_j(x)|, \quad h \leq j \leq \nu$$

and
$$\rho_n \equiv C_{r,s} \omega_s\left(f^{(r)}, \frac{1}{n}\right) \cdot \sum_{j=h}^{\nu} \frac{l_j}{n^{r-j}},$$

where the constant $C_{r,s}$ appears in inequality (19.2).

(i) Assume, throughout $I$, $\alpha_h(x) \geq \alpha > 0$. By Corollary 19.1, let $Q_n(x)$ be the real polynomial of degree $\leq n$ such that
$$\max_{x \in I} \left| \left(f(x) + \rho_n \frac{x^h}{h!}\right)^{(k)} - Q_n^{(k)}(x) \right| \leq \frac{C_{r,s}}{n^{r-k}} \omega_s\left(f^{(r)}, \frac{1}{n}\right), \quad 0 \leq k \leq r. \tag{19.6}$$

When $0 \leq k \leq h$, (19.6) becomes
$$\left\| f^{(k)}(x) + \rho_n \cdot \frac{x^{h-k}}{(h-k)!} - Q_n^{(k)}(x) \right\|_\infty \leq \frac{C_{r,s}}{n^{r-k}} \omega_s\left(f^{(r)}, \frac{1}{n}\right).$$

Hence
$$\|f^{(k)} - Q_n^{(k)}\|_\infty \leq \frac{\rho_n}{(h-k)!} + \frac{C_{r,s}}{n^{r-k}} \omega_s\left(f^{(r)}, \frac{1}{n}\right)$$

$$= \frac{C_{r,s}}{(h-k)!} \omega_s\left(f^{(r)}, \frac{1}{n}\right) \cdot \sum_{j=h}^{\nu} \frac{l_j}{n^{r-j}} + \frac{C_{r,s}}{n^{r-k}} \omega_s\left(f^{(r)}, \frac{1}{n}\right)$$

$$= C_{r,s} \omega_s\left(f^{(r)}, \frac{1}{n}\right) \cdot \left[\frac{1}{n^{r-k}} + \frac{1}{(h-k)!} \sum_{j=h}^{\nu} \frac{l_j}{n^{r-j}}\right]$$

$$\leq \frac{C_{r,s}}{\eta^{r-\nu}}\left(1 + \sum_{j=h}^{\nu} l_j\right) \cdot \omega_s\left(f^{(r)}, \frac{1}{n}\right).$$

Thus for $0 \leq k \leq h$, we obtain

$$\|f^{(k)} - Q_n^{(k)}\|_\infty \leq \frac{C}{n^{r-\nu}}\omega_s\left(f^{(r)}, \frac{1}{n}\right), \qquad (19.7)$$

where $C \equiv C_{r,s}\left(1 + \sum_{j=h}^{\nu} l_j\right)$ is a constant independent of $f$ and $n$ and depends only on $r$, $s$, $L$.

Also if $x \in I$, then

$$\alpha_h^{-1}(x)L(Q_n(x)) = \alpha_h^{-1}(x)L(f(x)) + \rho_n + \sum_{j=h}^{\nu} \alpha_h^{-1}(x) \cdot \alpha_j(x)$$

$$\cdot [Q_n(x) - f(x) - \rho_n(h!)^{-1}x^h]^{(j)}$$

$$\geq \rho_n - \sum_{j=h}^{\nu} l_j \frac{C_{r,s}}{n^{r-j}}\omega_s\left(f^{(r)}, \frac{1}{n}\right) = 0.$$

Consequently, $L(Q_n(x)) \geq 0$.

(ii) Assume, throughout $I$, $\alpha_h(x) \leq \beta < 0$. By Corollary 19.1, let $Q_n(x)$ be the real polynomial of degree $\leq n$ such that

$$\max_{x \in I}\left|\left(f(x) - \rho_n \cdot \frac{x^h}{h!}\right)^{(k)} - Q_n^{(k)}(x)\right|$$

$$\leq \frac{C_{r,s}}{n^{r-k}}\omega_s\left(f^{(r)}, \frac{1}{n}\right), \qquad 0 \leq k \leq r. \qquad (19.8)$$

Again (19.7) is valid. Also if $x \in I$, then

$$\alpha_h^{-1}(x)L(Q_n(x)) = \alpha_h^{-1}(x)L(f(x)) - \rho_n + \sum_{j=h}^{\nu} \alpha_h^{-1}(x) \cdot \alpha_j(x)$$

$$\cdot [Q_n(x) - f(x) + \rho_n(h!)^{-1}x^h]^{(j)}$$

$$\le -\rho_n + \sum_{j=h}^{\nu} l_j \frac{C_{r,s}}{n^{r-j}} \omega_s\left(f^{(r)}, \frac{1}{n}\right) = 0.$$

Hence, again $L(Q_n(x)) \geq 0$. Notice that from (19.6) and (19.8), for $h+1 \leq k \leq r$ we find

$$\|f^{(k)} - Q_n^{(k)}\|_\infty \le \frac{C_{r,s}}{n^{r-k}} \omega_s\left(f^{(r)}, \frac{1}{n}\right)$$

and $C_{r,s} \leq C$. The proof of the theorem is now finished. ∎

**REMARK 19.1** (to Theorem 19.2) Suppose that $\alpha_j(x)$, $j = h, h+1, \ldots, \nu$ are continuous functions on $I$ and (19.3) is replaced by $L(f) > 0$. Relax the condition that $\alpha_h(x)$ is either $\geq \alpha > 0$ or $\leq \beta < 0$ throughout $I$. Let $Q_n$ be the polynomial of degree $\leq n$ corresponding to $f$ from Corollary 19.1, then $Q_n^{(k)}$ converges to $f^{(k)}$ at a higher rate given by inequality (19.2), $0 \leq k \leq h$. Moreover, because $L(Q_n)$ converges uniformly to $L(f)$ on $I$, $L(Q_n) > 0$ throughout $I$ for all sufficiently large $n$. ∎

## 19.2 Quantitative Monotone Approximation with Pseudo-Polynomials

In this section we deal with the next general two-dimensional problem: Let $f$ be a two variable continuously differentiable real valued function of a given order, let $L$ be a linear differential operator involving mixed partial derivatives, and suppose that $L(f) \geq 0$. Then, for sufficiently large $n, m \in \mathbf{N}$, we can find a sequence of pseudo-polynomials $Q_{n,m}$ in two variables with the property $L(Q_{n,m}) \geq 0$ such that $f$ is approximated with rates simultaneously by $Q_{n,m}$ in the uniform norm. This approximation is given via inequalities involving the mixed modulus of smoothness $\omega_{s,q}$, $s, q \in \mathbf{N}$. This section is based on [10].

Let $I = [-1,1]$, $n, m \in \mathbf{Z}_+$, $P_\theta$ denote the space of algebraic polynomials of degree $\leq \theta$. Consider the tensor product spaces $P_n \otimes C(I)$, $C(I) \otimes P_m$ and their sum $P_n \otimes C(I) + C(I) \otimes P_m$, that is

$$P_n \otimes C(I) + C(I) \otimes P_m$$

$$= \left\{ \sum_{i=0}^{n} x^i A_i(y) + \sum_{j=0}^{m} B_j(x) y^j; A_i, B_j \in C(I), x, y \in I \right\}.$$

This is the space of pseudo-polynomials of degree $\leq (n,m)$, first introduced by A. Marchaud in 1924–1927 (see [125], [126]). Here $f^{(k,l)}$ denotes $\partial^{k+l} f / \partial x^k \partial y^l$, the $(k,l)$-partial derivative of $f$.

In this section we consider the space $C^{r,p}(I^2) = \{f : I^2 \to \mathbf{R}; \; f^{(k,l)} \text{ is continuous for } 0 \leq k \leq r, \; 0 \leq l \leq p\}$. Let $f \in C(I^2)$; for $\delta_1, \delta_2 \geq 0$, define the mixed modulus of smoothness of order $(s,q)$, $s,q \in \mathbf{N}$ (see [154], pp. 516–517) by

$$\omega_{s,q}(f; \delta_1, \delta_2) \equiv \sup\{|_x\Delta^s_{h_1} \circ {_y}\Delta^q_{h_2} f(x,y)| : (x,y),$$

$$(x + sh_1, y + qh_2) \in I^2, |h_i| \leq \delta_i, i = 1,2\}.$$

Here

$$_x\Delta^s_{h_1} \circ {_y}\Delta^q_{h_2} f(x,y) \equiv \sum_{\sigma=0}^{s} \sum_{\mu=0}^{q} (-1)^{s+q-\sigma-\mu}$$

$$\cdot \binom{s}{\sigma} \binom{q}{\mu} f(x + \sigma h_1, y + \mu h_2)$$

is a mixed difference of order $(s,q)$.

**THEOREM 19.3**
(H. H. Gonska [89]). Let $r, p \in \mathbf{Z}_+$, $s, q \in \mathbf{N}$, and $f \in C^{r,p}(I^2)$. Let $n, m \in \mathbf{N}$ with $n \geq \max\{4(r+1), r+s\}$ and $m \geq \max\{4(p+1), p+q\}$. Then there exists a linear operator $Q_{n,m}$ from $C^{r,p}(I^2)$ into the space of pseudopolynomials $(P_n \otimes C(I) + C(I) \otimes P_m)$ such that

$$|(f - Q_{n,m}(f))^{(k,l)}(x,y)| \leq M_{r,s} \cdot M_{p,q} (\Delta_n(x))^{r-k} \cdot (\Delta_m(y))^{p-l}$$

$$\cdot \omega_{s,q}(f^{(r,p)}; \Delta_n(x), \Delta_m(y)), \qquad (19.9)$$

for all $(0,0) \leq (k,l) \leq (r,p)$, $x, y \in I$, where

$$\Delta_\theta(z) = \frac{\sqrt{1-z^2}}{\theta} + \frac{1}{\theta^2} \quad \theta = n, m; \; z = x, y \in I.$$

The constants $M_{r,s}$, $M_{p,q}$, are independent of $f$, $(x,y)$ and $(n,m)$; they depend only on $(r,s)$, $(p,q)$, respectively.

## 19. On Monotone Approximation Theory

The following result is an easy consequence of the last theorem (see [154], p. 517).

**COROLLARY 19.2**
Let $r, p \in \mathbf{Z}_+$, $s, q \in \mathbf{N}$, and $f \in C^{r,p}(I^2)$. Let $n, m \in \mathbf{N}$ with $n \geq \max\{4(r+1), r+s\}$ and $m \geq \max\{4(p+1), p+q\}$. Then there exists a pseudopolynomial $Q_{n,m} \equiv Q_{n,m}(f) \in (P_n \otimes C(I) + C(I) \otimes P_m)$ such that

$$\|f^{(k,l)} - Q_{n,m}^{(k,l)}\|_\infty \leq \frac{\dot{C}}{n^{r-k}m^{p-l}} \cdot \omega_{s,q}\left(f^{(r,p)}; \frac{1}{n}, \frac{1}{m}\right), \tag{19.10}$$

for all $(0,0) \leq (k,l) \leq (r,p)$. Here the constant $\dot{C}$ depends only on $r$, $p$, $s$, $q$.

Corollary 19.2 is used in the proof of the main result that follows.

**THEOREM 19.4**
Let $h_1$, $h_2$, $\nu_1$, $\nu_2$, $r$, $p$ be integers, $0 \leq h_1 \leq \nu_1 \leq r$, $0 \leq h_2 \leq \nu_2 \leq p$ and let $f \in C^{r,p}(I^2)$, with $f^{(r,p)}$ having a mixed modulus of smoothness $\omega_{s,q}(f^{(r,p)}; x, y)$ there, $s, q \in \mathbf{N}$. Let $\alpha_{ij}(x,y)$, $i = h_1, h_1 + 1, \ldots \nu_1$; $j = h_2, h_2 + 1, \ldots, \nu_2$ be real valued functions, defined and bounded in $I^2$ and suppose $\alpha_{h_1 h_2}$ is either $\geq \alpha > 0$ or $\leq \beta < 0$ throughout $I^2$. Take the operator

$$L = \sum_{i=h_1}^{\nu_1} \sum_{j=h_2}^{\nu_2} \alpha_{ij}(x,y) \frac{\partial^{i+j}}{\partial x^i \partial y^j}$$

and assume, throughout $I^2$, that

$$L(f) \geq 0. \tag{19.11}$$

Then, for any integers $n, m$ with $n \geq \max\{4(r+1), r+s\}$, $m \geq \max\{4(p+1), p+q\}$, there exists a pseudopolynomial

$$Q_{n,m} \in (P_n \otimes C(I) + C(I) \otimes P_m)$$

such that $L(Q_{n,m}) \geq 0$ throughout $I^2$ and

$$\|f^{(k,l)} - Q_{n,m}^{(k,l)}\|_\infty \leq \frac{C}{n^{r-\nu_1} m^{p-\nu_2}} \cdot \omega_{s,q}\left(f^{(r,p)}; \frac{1}{n}, \frac{1}{m}\right) \tag{19.12}$$

for all $(0,0) \le (k,l) \le (h_1, h_2)$. Moreover we get

$$\|f^{(k,l)} - Q_{n,m}^{(k,l)}\|_\infty \le \frac{C}{n^{r-k} m^{p-l}} \cdot \omega_{s,q}\left(f^{(r,p)}; \frac{1}{n}, \frac{1}{m}\right) \quad (19.13)$$

for all $(h_1+1, h_2+1) \le (k,l) \le (r,p)$. Also (19.13) is valid whenever $0 \le k \le h_1$, $h_2+1 \le l \le p$, or $h_1+1 \le k \le r$, $0 \le l \le h_2$. Here $C$ is a constant independent of $f$ and $n, m$. It depends only on $r, p, s, q, L$.

**PROOF** Let $n, m$ be integers such that $n \ge \max\{4(r+1), r+s\}$, $m \ge \max\{4(p+1), p+q\}$. Put

$$l_{ij} \equiv \sup_{(x,y) \in I^2} |\alpha_{h_1 h_2}^{-1}(x,y) \cdot \alpha_{ij}(x,y)| < \infty$$

for all $h_1 \le i \le \nu_1$, $h_2 \le j \le \nu_2$, and also put

$$\rho_{n,m} \equiv \dot{C} \cdot \omega_{s,q}\left(f^{(r,p)}; \frac{1}{n}, \frac{1}{m}\right) \cdot \sum_{i=h_1}^{\nu_1} \sum_{j=h_2}^{\nu_2} \frac{l_{ij}}{n^{r-i} m^{p-j}},$$

where the constant $\dot{C}$ appears in inequality (19.10).

(i) Assume throughout $I^2$, $\alpha_{h_1 h_2}(x) \ge \alpha > 0$. By Corollary 19.2, let $Q_{n,m}(x,y)$ be the pseudopolynomial in $(P_n \otimes C(I) + C(I) \otimes P_m)$ such that

$$\left\|\left(f(x,y) + \rho_{n,m} \frac{x^{h_1}}{h_1!} \cdot \frac{y^{h_2}}{h_2!}\right)^{(k,l)} - Q_{n,m}^{(k,l)}(x,y)\right\|_\infty$$

$$\le \frac{\dot{C}}{n^{r-k} m^{p-l}} \cdot \omega_{s,q}\left(f^{(r,p)}; \frac{1}{n}, \frac{1}{m}\right), \quad (19.14)$$

for all $(0,0) \le (k,l) \le (r,p)$ (see [154], p. 517). When $(0,0) \le (k,l) \le (h_1, h_2)$, (19.14) becomes

$$\left\|f^{(k,l)}(x,y) + \rho_{n,m} \frac{x^{h_1-k}}{(h_1-k)!} \cdot \frac{y^{h_2-l}}{(h_2-l)!} - Q_{n,m}^{(k,l)}(x,y)\right\|_\infty$$

$$\le \frac{\dot{C}}{n^{r-k} m^{p-l}} \cdot \omega_{s,q}\left(f^{(r,p)}; \frac{1}{n}, \frac{1}{m}\right).$$

Therefore

$$\|f^{(k,l)} - Q_{n,m}^{(k,l)}\|_\infty \leq \frac{\rho_{n,m}}{(h_1 - k)!(h_2 - l)!}$$

$$+ \frac{\dot{C}}{n^{r-k} m^{p-l}} \cdot \omega_{s,q}\left(f^{(r,p)}; \frac{1}{n}, \frac{1}{m}\right)$$

$$= \frac{\dot{C} \cdot \omega_{s,q}(f^{(r,p)}; 1/n, 1/m)}{(h_1 - k)!(h_2 - l)!} \cdot \sum_{i=h_1}^{\nu_1} \sum_{j=h_2}^{\nu_2} \frac{l_{ij}}{n^{r-i} m^{p-j}}$$

$$+ \frac{\dot{C}}{n^{r-k} m^{p-l}} \cdot \omega_{s,q}\left(f^{(r,p)}; \frac{1}{n}, \frac{1}{m}\right) = \dot{C} \cdot \omega_{s,q}\left(f^{(r,p)}; \frac{1}{n}, \frac{1}{m}\right)$$

$$\cdot \left[\frac{1}{n^{r-k} m^{p-l}} + \frac{1}{(h_1 - k)!(h_2 - l)!} \cdot \sum_{i=h_1}^{\nu_1} \sum_{j=h_2}^{\nu_2} \frac{l_{ij}}{n^{r-i} m^{p-j}}\right]$$

$$\leq \frac{\dot{C}}{n^{r-\nu_1} m^{p-\nu_2}} \cdot \left(1 + \sum_{i=h_1}^{\nu_1} \sum_{j=h_2}^{\nu_2} l_{ij}\right) \cdot \omega_{s,q}\left(f^{(r,p)}; \frac{1}{n}, \frac{1}{m}\right).$$

Thus for $(0, 0) \leq (k, l) \leq (h_1, h_2)$ we obtain

$$\|f^{(k,l)} - Q_{n,m}^{(k,l)}\|_\infty \leq \frac{C}{n^{r-\nu_1} m^{p-\nu_2}} \cdot \omega_{s,q}\left(f^{(r,p)}; \frac{1}{n}, \frac{1}{m}\right), \tag{19.12}$$

where

$$C \equiv \dot{C} \cdot \left(1 + \sum_{i=h_1}^{\nu_1} \sum_{j=h_2}^{\nu_2} l_{ij}\right)$$

is a constant independent of $f$ and $n, m$ and dependent only on $r, p, s, q$, $L$. Furthermore, if $(x, y) \in I^2$, then it holds that

$$\alpha_{h_1 h_2}^{-1}(x, y) \cdot L(Q_{n,m}(x, y)) = \alpha_{h_1 h_2}^{-1}(x, y) \cdot L(f(x, y))$$

$$+ \rho_{n,m} + \sum_{i=h_1}^{\nu_1} \sum_{j=h_2}^{\nu_2} \alpha_{h_1 h_2}^{-1}(x, y) \cdot \alpha_{ij}(x, y)$$

$$\cdot [Q_{n,m}(x,y) - f(x,y) - \rho_{n,m} \cdot (h_1!)^{-1} \cdot (h_2!)^{-1} \cdot x^{h_1} \cdot y^{h_2}]^{(i,j)}$$

$$\geq \rho_{n,m} - \sum_{i=h_1}^{\nu_1} \sum_{j=h_2}^{\nu_2} l_{ij} \cdot \frac{\dot{C}}{n^{r-i}m^{p-j}} \cdot \omega_{s,q}\left(f^{(r,p)}; \frac{1}{n}, \frac{1}{m}\right) = 0.$$

Consequently,
$$L(Q_{n,m}(x,y)) \geq 0.$$

(ii) Assume throughout $I^2$, $\alpha_{h_1 h_2}(x,y) \leq \beta < 0$. By Corollary 19.2, let $Q_{n,m}(x,y)$ be the pseudopolynomial in $(P_n \otimes C(I) + C(I) \otimes P_m)$ such that

$$\left\| \left( f(x,y) - \rho_{n,m} \cdot \frac{x^{h_1} y^{h_2}}{h_1! h_2!} \right)^{(k,l)} - Q_{n,m}^{(k,l)}(x,y) \right\|_\infty$$

$$\leq \frac{\dot{C}}{n^{r-k}m^{p-l}} \cdot \omega_{s,q}\left(f^{(r,p)}; \frac{1}{n}, \frac{1}{m}\right) \tag{19.15}$$

for all $(0,0) \leq (k,l) \leq (r,p)$ (see [154], p. 517). Again (19.12) is valid. Also if $(x,y) \in I^2$, then it holds

$$\alpha_{h_1 h_2}^{-1}(x,y) \cdot L(Q_{n,m}(x,y)) = \alpha_{h_1 h_2}^{-1}(x,y) \cdot L(f(x,y))$$

$$- \rho_{n,m} + \sum_{i=h_1}^{\nu_1} \sum_{j=h_2}^{\nu_2} \alpha_{h_1 h_2}^{-1}(x,y) \cdot \alpha_{ij}(x,y)$$

$$\cdot [Q_{n,m}(x,y) - f(x,y) + \rho_{n,m} \cdot (h_1!)^{-1} \cdot (h_2!)^{-1} \times x^{h_1} \times y^{h_2}]^{(i,j)}$$

$$\leq -\rho_{n,m} + \sum_{i=h_1}^{\nu_1} \sum_{j=h_2}^{\nu_2} l_{ij} \cdot \frac{\dot{C}}{n^{r-i}m^{p-j}} \cdot \omega_{s,q}\left(f^{(r,p)}; \frac{1}{n}, \frac{1}{m}\right) - 0.$$

Consequently, again
$$L(Q_{n,m}(x,y)) \geq 0.$$

Moreover, we can observe from (19.14) and (19.15) when $(h_1 + 1, h_2 + 1) \leq (k,l) \leq (r,p)$ that it holds

$$\|f^{(k,l)} - Q_{n,m}^{(k,l)}\|_\infty \leq \frac{\dot{C}}{n^{r-k}m^{p-l}} \cdot \omega_{s,q}\left(f^{(r,p)}; \frac{1}{n}, \frac{1}{m}\right) \tag{19.16}$$

and $\dot{C} \leq C$, hence establishing (19.13). Inequality (19.16) remains valid also in the cases of $0 \leq k \leq h_1$, $h_2+1 \leq l \leq p$ or $h_1+1 \leq k \leq r$, $0 \leq l \leq h_2$. The proof of the theorem is now finished. ∎

## 19.3  Quantitative Bivariate Monotone Approximation

This treatment follows [11].

In this section we deal with the following general two-dimensional problem: Let $f$ be a two variable continuously differentiable real-valued function of given order and let $L$ be a linear differential operator involving mixed partial derivatives and assume that $L(f) \geq 0$. Then find a sequence of bivariate polynomials $Q_{m,n}(x, y)$ with the property $L(Q_{m,n}) \geq 0$ so that $f$ is approximated simultaneously by $Q_{m,n}$ in the uniform norm. This approximation is given with rates through inequalities involving the bivariate first modulus of continuity of Stancu (see [164]). We prove this result with the use of a bivariate simultaneous approximation theorem by I. Badea and C. Badea (see [46]).

We need to mention

**DEFINITION 19.1** (Stancu [164]). *Let $f \in C([0,1]^2)$, $[0,1]^2 = [0,1] \times [0,1]$, where $(x_1, y_1)$, $(x_2, y_2) \in [0,1]^2$ and $\delta_1, \delta_2 \geq 0$. The first modulus of continuity of $f$ is defined as follows:*

$$\omega(f, \delta_1, \delta_2) = \sup_{\substack{|x_1-x_2|\leq \delta_1 \\ |y_1-y_2|\leq \delta_2}} |f(x_1, y_1) - f(x_2, y_2)|.$$

**DEFINITION 19.2**  *Let $f$ be a real-valued function defined on $[0,1]^2$ and let $m$, $n$ be two positive integers. Let $B_{m,n}$ be the Bernstein (polynomial) operator of order $(m, n)$ given by*

$$B_{m,n}(f; x, y) = \sum_{i=0}^{m} \sum_{j=0}^{n} f\left(\frac{i}{m}, \frac{j}{n}\right) \cdot \binom{m}{i} \cdot \binom{n}{j}$$

$$\cdot x^i \cdot (1-x)^{m-i} \cdot y^j \cdot (1-y)^{n-j}.$$

For integers $r, s \geq 0$, we denote by $f^{(r,s)}$ the differential operator of order $(r, s)$, given by

$$f^{(r,s)}(x, y) = \partial^{r+s} f(x, y) / \partial x^r \partial y^s.$$

We need the following simultaneous approximation result:

**THEOREM 19.5**
(I. Badea, C. Badea [46]). *It holds that*

$$\|f^{(k,l)} - (B_{m,n}f)^{(k,l)}\|_\infty \leq t(k,l) \cdot \omega\left(f^{(k,l)}; \frac{1}{\sqrt{m-k}}, \frac{1}{\sqrt{n-l}}\right)$$

$$+ \max\left\{\frac{k(k-1)}{m}, \frac{l(l-1)}{n}\right\} \cdot \|f^{(k,l)}\|_\infty, \tag{19.17}$$

where $m > k \geq 0$, $n > l \geq 0$ are integers, $f$ is a real-valued function on $[0,1]^2$ such that $f^{(k,l)}$ is continuous, and $t$ is a positive real-valued function on $\mathbf{N}^2$, $\mathbf{N} = \{0,1,2,\ldots\}$. Here $\|\cdot\|_\infty$ is the supremum norm.

The main result follows here.

**THEOREM 19.6**
Let $h_1, h_2, \nu_1, \nu_2, r, p$ be integers, $0 \leq h_1 \leq \nu_1 \leq r$, $0 \leq h_2 \leq \nu_2 \leq p$ and let $f \in C^{r,p}([0,1]^2)$. Let $\alpha_{ij}(x,y)$, $i = h_1, h_1+1, \ldots, \nu_1$; $j = h_2, h_2+1, \ldots, \nu_2$ be real-valued functions, defined and bounded in $[0,1]^2$ and suppose $\alpha_{h_1 h_2}$ is either $\geq \alpha > 0$ or $\leq \beta < 0$ throughout $[0,1]^2$. Take the operator

$$L = \sum_{i=h_1}^{\nu_1} \sum_{j=h_2}^{\nu_2} \alpha_{ij}(x,y) \partial^{i+j}/\partial x^i \partial y^j$$

and assume that throughout $[0,1]^2$,

$$L(f) \geq 0. \tag{19.18}$$

Then for integers $m, n$ with $m > r$, $n > p$, there exists a polynomial $Q_{m,n}(x,y)$ of degree $(m,n)$ such that $L(Q_{m,n}(x,y)) \geq 0$ throughout $[0,1]^2$ and

$$\|f^{(k,l)} - Q_{m,n}^{(k,l)}\|_\infty \leq \frac{P_{m,n}(L,f)}{(h_1-k)!(h_2-l)!} + M_{m,n}^{k,l}(f), \tag{19.19}$$

all $(0,0) \leq (k,l) \leq (h_1, h_2)$. Furthermore, we obtain

$$\|f^{(k,l)} - Q_{m,n}^{(k,l)}\|_\infty \leq M_{m,n}^{k,l}(f), \tag{19.20}$$

## 19. On Monotone Approximation Theory

for all $(h_1 + 1, h_2 + 1) \leq (k, l) \leq (r, p)$. Also (19.20) is valid whenever $0 \leq k \leq h_1$, $h_2 + 1 \leq l \leq p$ or $h_1 + 1 \leq k \leq r$, $0 \leq l \leq h_2$. Here

$$M_{m,n}^{k,l} \equiv M_{m,n}^{k,l}(f) \equiv t(k,l) \cdot \omega\left(f^{(k,l)}; \frac{1}{\sqrt{m-k}}, \frac{1}{\sqrt{n-l}}\right)$$

$$+ \max\left\{\frac{k(k-1)}{m}, \frac{l(l-1)}{n}\right\} \cdot \|f^{(k,l)}\|_\infty$$

and

$$P_{m,n} \equiv P_{m,n}(L, f) \equiv \sum_{i=h_1}^{\nu_1} \sum_{j=h_2}^{\nu_2} l_{ij} \cdot M_{m,n}^{i,j},$$

where $t$ is a positive real-valued function on $\mathbf{N}^2$ and

$$l_{ij} \equiv \sup_{(x,y) \in [0,1]^2} |\alpha_{h_1 h_2}^{-1}(x, y) \cdot \alpha_{ij}(x, y)| < \infty.$$

**PROOF** Let $m, n$ be integers such that $m > r$, $n > p$.

Case (i). Suppose that throughout $[0,1]^2$, $\alpha_{h_1 h_2} \geq \alpha > 0$. By Theorem 19.5 we get

$$\left\|\left(f + P_{m,n} \cdot \frac{x^{h_1}}{h_1!} \cdot \frac{y^{h_2}}{h_2!}\right)^{(k,l)} - (Q_{m,n}(x,y))^{(k,l)}\right\|_\infty \leq M_{m,n}^{k,l}, \quad (19.21)$$

all $0 \leq u \leq r$, $0 \leq l \leq p$, where

$$Q_{m,n}(x, y) \equiv B_{m,n}(f; x, y) + P_{m,n} \cdot \frac{x^{h_1}}{h_1!} \cdot \frac{y^{h_2}}{h_2!}.$$

When $(0,0) \leq (k, l) \leq (h_1, h_2)$, (19.21) becomes

$$\left\|f^{(k,l)}(x,y) + P_{m,n} \frac{x^{h_1-k}}{(h_1-k)!} \cdot \frac{y^{h_2-l}}{(h_2-l)!} - Q_{m,n}^{(k,l)}(x,y)\right\|_\infty \leq M_{m,n}^{k,l}.$$

Then, by the triangle inequality property of $\|\cdot\|_\infty$ and $(x,y) \in [0,1]^2$ we have the validity of inequality (19.19). Moreover, if $(x,y) \in [0,1]^2$, then

$$\alpha_{h_1 h_2}^{-1}(x, y) \cdot L(Q_{m,n}(x, y)) = \alpha_{h_1 h_2}^{-1}(x, y) \cdot L(f(x, y))$$

$$+ P_{m,n} + \sum_{i=h_1}^{\nu_1} \sum_{j=h_2}^{\nu_2} \alpha_{h_1 h_2}^{-1}(x,y) \cdot \alpha_{ij}(x,y)$$

$$\cdot \left[ Q_{m,n}(x,y) - f(x,y) - P_{m,n} \frac{x^{h_1}}{h_1!} \cdot \frac{y^{h_2}}{h_2!} \right]^{(i,j)}$$

$$\geq P_{m,n} - \sum_{i=h_1}^{\nu_1} \sum_{j=h_2}^{\nu_2} l_{ij} \cdot M_{m,n}^{i,j} = 0,$$

the last is true by inequality (19.21). Thus $L(Q_{m,n}(x,y)) \geq 0$.

Case (ii). Suppose that throughout $[0,1]^2$, $\alpha_{h_1 h_2} \leq \beta < 0$. By Theorem 19.5 we find

$$\left\| \left( f - P_{m,n} \frac{x^{h_1}}{h_1!} \frac{y^{h_2}}{h_2!} \right)^{(k,l)} - (Q_{m,n}(x,y))^{(k,l)} \right\|_\infty \leq M_{m,n}^{k,l}, \qquad (19.22)$$

all $0 \leq u \leq r$, $0 \leq l \leq p$, where

$$Q_{m,n}(x,y) \equiv B_{m,n}(f;x,y) - P_{m,n} \frac{x^{h_1}}{h_1!} \cdot \frac{y^{h_2}}{h_2!}.$$

When $(0,0) \leq (k,l) \leq (h_1, h_2)$, (19.22) becomes

$$\left\| f^{(k,l)}(x,y) - P_{m,n} \frac{x^{h_1-k}}{(h_1-k)!} \cdot \frac{y^{h_2-l}}{(h_2-l)!} - Q_{m,n}^{(k,l)}(x,y) \right\|_\infty \leq M_{m,n}^{k,l}.$$

Therefore, by the triangle inequality property of $\|\cdot\|_\infty$ and $(x,y) \in [0,1]^2$ we get the validity of inequality (19.19). Moreover, if $(x,y) \in [0,1]^2$, then

$$\alpha_{h_1 h_2}^{-1}(x,y) \cdot L(Q_{m,n}(x,y)) = \alpha_{h_1 h_2}^{-1}(x,y) \cdot L(f(x,y))$$

$$- P_{m,n} + \sum_{i=h_1}^{\nu_1} \sum_{j=h_2}^{\nu_2} \alpha_{h_1 h_2}^{-1}(x,y) \cdot \alpha_{ij}(x,y)$$

$$\cdot \left[ Q_{m,n}(x,y) - f(x,y) + P_{m,n} \frac{x^{h_1}}{h_1!} \cdot \frac{y^{h_2}}{h_2!} \right]^{(i,j)}$$

$$\le -P_{m,n} + \sum_{i=h_1}^{\nu_1} \sum_{j=h_2}^{\nu_2} l_{ij} \cdot M_{m,n}^{i,j} = 0,$$

the last is true by inequality (19.22). Thus, again holds $L(Q_{m,n}(x,y)) \ge 0$.

In the cases of either $(h_1+1, h_2+1) \le (k,l) \le (r,p)$ or $(0 \le k \le h_1, h_2+1 \le l \le p)$ or $(h_1+1 \le k \le r, 0 \le l \le h_2)$, we see that

$$\left(f \pm P_{m,n} \frac{x^{h_1}}{h_1!} \cdot \frac{y^{h_2}}{h_2!}\right)^{(k,l)} = f^{(k,l)}.$$

Hence inequalities (19.21) and (19.22) imply inequality (19.20). The theorem is now proved. ∎

## 19.4 Quantitative Spline Monotone Approximation with Linear Differential Operators

Let $f \in C^s[a,b]$ and $L$ be a linear differential operator such that $L(f) \ge 0$. Then there exists a sequence $Q_n$, $n \ge 1$, of polynomial splines with equally spaced knots, such that $Q_n^{(\gamma)}$ approximates $f^{(\gamma)}$, $0 \le \gamma \le s$, simultaneously in the uniform norm. This approximation is given through inequalities with rates, involving a modulus of smoothness of $f^{(s)}$, so that $L(Q_n) \ge 0$. The examined cases are the continuous, periodic and discrete. Here we follow [8].

### 19.4.1 Quantitative Monotone Approximation by Polynomial Splines

Let $[a,b] \subset \mathbf{R}$ and for $n \ge 1$ consider the partition $\Delta_n$ with points $x_{in} = a + i\left(\frac{b-a}{n}\right)$, $i = 0, 1, \ldots, n$. Hence $\overline{\Delta}_n \equiv \max_{1 \le i \le n}(x_{in} - x_{i-1,n}) = \frac{b-a}{n}$.

Let $S_m(\Delta_n)$ be the space of polynomial splines of order $m > 0$ with simple knots at the points $x_{in}$, $i = 1, \ldots, n-1$. Then there exists a linear operator $Q_n : Q_n \equiv Q_n(f)$, mapping $B[a,b]$: the space of bounded real valued functions $f$ on $[a,b]$, into $S_m(\Delta_n)$ (see [154], p. 224, Theorem 6.18).

From the same reference, p. 227, Corollary 6.21, we get

**COROLLARY 19.3**
Let $1 \leq \sigma \leq m$, $n \geq 1$. Then for all $f \in C^{\sigma-1}[a,b]$; $r = 0, \ldots, \sigma-1$,

$$\|f^{(r)} - Q_n^{(r)}\|_\infty \leq C_1 \left(\frac{b-a}{n}\right)^{\sigma-r-1} \omega_{m-\sigma+1}\left(f^{(\sigma-1)}, \frac{b-a}{n}\right),$$

where $C_1$ depends only on $m$.

By denoting $C_2 = C_1 \max_{0 \leq r \leq \sigma-1}(b-a)^{\sigma-r-1}$ we obtain

**LEMMA 19.1**
Let $1 \leq \sigma \leq m$, $n \geq 1$. Then for all $f \in C^{\sigma-1}[a,b]$; $r = 0, \ldots, \sigma-1$,

$$\|f^{(r)} - Q_n^{(r)}\|_\infty \leq \frac{C_2}{n^{\sigma-r-1}} \omega_{m-\sigma+1}\left(f^{(\sigma-1)}, \frac{b-a}{n}\right), \qquad (19.23)$$

where $C_2$ depends only on $m$, $\sigma$ and $b - a$. Here $\omega_{m-\sigma+1}$ is the usual modulus of smoothness of order $m - \sigma + 1$.

The first main result follows.

**THEOREM 19.7**
Let $h$, $k$, $\sigma$, $m$ be integers, $0 \leq h \leq k \leq \sigma-1$, $\sigma \leq m$ and let $f \in C^{\sigma-1}[a,b]$. Let $\alpha_j(x) \in B[a,b]$, $j = h, h+1, \ldots, k$ and suppose that either $\alpha_n(x) \geq \alpha > 0$ or $\alpha_n(x) \leq \beta < 0$ for all $x \in [a,b]$. Take the linear differential operator

$$L = \sum_{j=h}^{k} \alpha_j(x) \left[\frac{d^j}{dx^j}\right]$$

and assume, throughout $[a,b]$,

$$L(f) \geq 0. \qquad (19.24)$$

Then, for every integer $n \geq 1$, there exists a polynomial spline function $Q_n(x)$ of order $m$ with simple knots at $\{a + i(\frac{b-a}{n}), i = 1, \ldots, n-1\}$ such that $L(Q_n) \geq 0$ throughout $[a,b]$ and

$$\|f^{(r)} - Q_n^{(r)}\|_\infty \leq \frac{C}{n^{\sigma-k-1}} \omega_{m-\sigma+1}\left(f^{(\sigma-1)}, \frac{b-a}{n}\right), \quad 0 \leq r \leq h. \quad (19.25)$$

Moreover, we find

$$\|f^{(r)} - Q_n^{(r)}\|_\infty \le \frac{C}{n^{\sigma-r-1}}\omega_{m-\sigma+1}\left(f^{(\sigma-1)}, \frac{b-a}{n}\right), \quad h+1 \le r \le \sigma-1, \tag{19.26}$$

where $C$ is a constant independent of $f$ and $n$. It depends only on $m$, $\sigma$, $L$, $a$, $b$.

**PROOF** Put

$$e_j \equiv \sup_{a \le x \le b} |\alpha_n^{-1}(x)\alpha_j(x)|, \quad h \le j \le k$$

and

$$\rho_n = C_2 \omega_{m-\sigma+1}\left(f^{(\sigma-1)}, \frac{b-a}{n}\right) \sum_{j=h}^{k} \frac{l_j}{n^{\sigma-j-1}},$$

where $C_2$ is the constant appearing in inequality (19.23).

(i) Assume, throughout $[a, b]$, $\alpha_h(x) \ge \alpha > 0$. By Lemma 19.1, let $Q_n(x)$ be the polynomial spline function of order $m$ with simple knots at $\Delta_n$ such that

$$\|(f(x) + \rho_n(h!)^{-1}x^h)^{(r)} - Q_n^{(r)}(x)\|_\infty$$

$$\le \frac{C_2}{n^{\sigma-r-1}}\omega_{m-\sigma+1}\left(f^{(\sigma-1)}, \frac{b-a}{n}\right), \quad 0 \le r \le \sigma-1. \tag{19.27}$$

Set $\gamma \equiv \max(|a|, |b|)$. When $0 \le r \le h$, (19.27) becomes

$$\left\|f^{(r)}(x) + \rho_n \frac{x^{h-r}}{(h-r)!} - Q_n^{(r)}(x)\right\|_\infty \le \frac{C_2}{n^{\sigma-r-1}}\omega_{m-\sigma+1}\left(f^{(\sigma-1)}, \frac{b-a}{n}\right).$$

Hence

$$\|f^{(r)} - Q_n^{(r)}\|_\infty \le \rho_n \frac{\gamma^{h-r}}{(h-r)!} + \frac{C_2}{n^{\sigma-r-1}}\omega_{m-\sigma+1}\left(f^{(\sigma-1)}, \frac{b-a}{n}\right)$$

$$= C_2 \cdot \omega_{m-\sigma+1}\left(f^{(\sigma-1)}, \frac{b-a}{n}\right) \cdot \left(\sum_{j=h}^{k} \frac{l_j}{n^{\sigma-j-1}}\right) \cdot \frac{\gamma^{h-r}}{(h-r)!}$$

$$+ \frac{C_2}{n^{\sigma-r-1}} \cdot \omega_{m-\sigma+1}\left(f^{(\sigma-1)}, \frac{b-a}{n}\right)$$

$$= C_2 \omega_{m-\sigma+1}\left(f^{(\sigma-1)}, \frac{b-a}{n}\right) \cdot \left[\frac{1}{n^{\sigma-r-1}} + \frac{\gamma^{h-r}}{(h-r)!} \cdot \sum_{j=h}^{k} \frac{l_j}{n^{\sigma-j-1}}\right]$$

$$\leq C_2 \cdot \left[1 + \nu^{h-r}/(h-r)! \sum_{j=h}^{k} l_j\right] \cdot \frac{1}{n^{\sigma-k-1}} \cdot \omega_{m-\sigma+1}\left(f^{(\sigma-1)}, \frac{b-a}{n}\right).$$

Set $C \equiv \max_{0 \leq r \leq h} C_2\left[1 + \frac{\gamma^{h-r}}{(h-r)!} \sum_{j=h}^{k} l_j\right]$. Thus

$$\|f^{(r)} - Q_n^{(r)}\|_\infty \leq \frac{C}{n^{\sigma-k-1}} \omega_{m-\sigma+1}\left(f^{(\sigma-1)}, \frac{b-a}{n}\right), \quad 0 \leq r \leq h. \quad (19.28)$$

Also if $a \leq x \leq b$, then

$$\alpha_h^{-1}(x) L(Q_n(x)) = \alpha_h^{-1}(x) L(f(x)) + \rho_n$$

$$+ \sum_{j=h}^{k} \alpha_h^{-1}(x) \cdot \alpha_j(x) \cdot [Q_n(x) - f(x) - \rho_n(h!)^{-1} x^h]^{(j)}$$

$$\geq \rho_n - \sum_{j=h}^{k} l_j \cdot \frac{C_2}{n^{\sigma-j-1}} \cdot \omega_{m-\sigma+1}\left(f^{(\sigma-1)}, \frac{b-a}{n}\right) = 0.$$

Then $L(Q_n(x)) \geq 0$, $a \leq x \leq b$.

(ii) Assume, throughout $[a, b]$, $\alpha_h(x) \leq \beta < 0$. By Lemma 19.1, let $Q_n(x)$ be the polynomial spline function of order $m$ with simple knots at $\Delta_n$ such that

$$\|(f(x) - \rho_n(h!)^{-1} x^h)^{(r)} - Q_n^{(r)}(x)\|_\infty$$

$$\leq \frac{C_2}{n^{\sigma-r-1}} \omega_{m-\sigma+1}\left(f^{(\sigma-1)}, \frac{b-a}{n}\right), \quad 0 \leq r \leq \sigma - 1. \quad (19.29)$$

Again (19.28) is valid. Also if $a \leq x \leq b$, then,

$$\alpha_h^{-1}(x) L(Q_n(x)) = \alpha_h^{-1}(x) L(f(x)) - \rho_n$$

$$+ \sum_{j=h}^{k} \alpha_h^{-1}(x) \cdot \alpha_j(x) \cdot [Q_n(x) - f(x) + \rho_n(h!)^{-1} x^h]^{(j)}$$

$$\leq -\rho_n + \sum_{j=h}^{k} l_j \cdot \frac{C_2}{n^{\sigma-j-1}} \cdot \omega_{m-\sigma+1}\left(f^{(\sigma-1)}, \frac{b-a}{n}\right) = 0.$$

Thus again $LQ_n(x) \geq 0$, $a \leq x \leq b$.
From (19.27) and (19.29), for $h+1 \leq r \leq \sigma - 1$ we obtain

$$\|f^{(r)} - Q_n^{(r)}\|_\infty \leq \frac{C_2}{n^{\sigma-r-1}} \omega_{m-\sigma+1}\left(f^{(\sigma-1)}, \frac{b-a}{n}\right)$$

and $C_2 \leq C$.
The proof of the theorem is now finished. ∎

### 19.4.2 Quantitative Monotone Approximation by Periodic Polynomial Splines

Let $[a,b] \subset \mathbf{R}$ and for $n \geq 1$ consider the partition $\Delta_n$ with points $x_{in} = a + i\left(\frac{b-a}{n}\right)$, $i = 0, 1, \ldots, n$. Hence $\overline{\Delta}_n \equiv \max_{1 \leq i \leq n}(x_{in} - x_{i-1,n}) = \frac{b-a}{n}$. Let $\dot{S}_m(\Delta_n)$ be the space of periodic polynomial splines of order $m > 0$ with simple knots at the points $x_{in}$, $i = 1, \ldots, n-1$. Then there exists a linear operator $\dot{Q}_n : \dot{Q}_n \equiv \dot{Q}_n(f)$, mapping the periodic functions $f$ on $[a,b]$ into $\dot{S}_m(\Delta_n)$, see [154], p. 306, Theorem 8.11. From the same reference, p. 307, Theorem 8.12, we get

### COROLLARY 19.4
Let $1 \leq \sigma \leq m$, $n \geq 1$. Then for all $f \in \dot{C}^{\sigma-1}[a,b]$, the space of $(\sigma - 1)$-times continuously differentiable periodic functions on $[a,b]$; $r = 0, \ldots, \sigma - 1$,

$$\|f^{(r)} - \dot{Q}_n^{(r)}\|_\infty \leq C_1 \left(\frac{b-a}{n}\right)^{\sigma-r-1} \dot{\omega}_1\left(f^{(\sigma-1)}, \frac{b-a}{n}\right),$$

where $C_1$ depends only on $m$. Here $\dot{\omega}_1$ is the periodic first modulus of continuity (see [154], p. 75, §2.11). More precisely,

$$\dot{\omega}_1(f,t) := \sup_{\substack{0 < h \leq t \\ x \in [a,b]}} |f(x+h) - f(x)|, \quad 0 < t \leq b-a.$$

By letting $C_2 = C_1 \max_{0 \leq r \leq \sigma-1}(b-a)^{\sigma-r-1}$ we obtain

**LEMMA 19.2**
Let $1 \leq \sigma \leq m$, $n \geq 1$. Then for all $f \in \dot{C}^{\sigma-1}[a,b]$; $r = 0, \ldots, \sigma-1$,

$$\|f^{(r)} - \dot{Q}_n^{(r)}\|_\infty \leq \frac{C_2}{n^{\sigma-r-1}} \dot{\omega}_1\left(f^{(\sigma-1)}, \frac{b-a}{n}\right), \qquad (19.30)$$

where $C_2$ depends only on $m$, $\sigma$ and $b-a$.

The next main result follows.

**THEOREM 19.8**
Let $1 \leq \sigma \leq m$ and $f \in \dot{C}^{\sigma-1}[a,b]$. Let $\alpha_j(x) \in B[a,b]$, $j = 0, 1, \ldots, k$, where $k \leq \sigma - 1$ and suppose that either $\alpha_0(x) \geq \alpha > 0$ or $\alpha_0(x) \leq \beta < 0$ for all $x \in [a,b]$.
Take the linear differentiable operator

$$L = \sum_{j=0}^k \alpha_j(x) \left[\frac{d^j}{dx^j}\right]$$

and assume, throughout $[a,b]$,

$$L(f) \geq 0. \qquad (19.31)$$

Then, for every integer $n \geq 1$, there exists a periodic polynomial spline function $\dot{Q}_n(x)$ of order $m$ with simple knots at $\{a+i(\frac{b-a}{n}), i = 1, \ldots, n-1\}$ such that $L(\dot{Q}_n) \geq 0$ throughout $[a,b]$ and

$$\|f - \dot{Q}_n^{(r)}\|_\infty \leq \frac{C}{n^{\sigma-k-1}} \dot{\omega}_1\left(f^{(\sigma-1)}, \frac{b-a}{n}\right). \qquad (19.32)$$

Moreover, we obtain

$$\|f^{(r)} - \dot{Q}_n^{(r)}\|_\infty \leq \frac{C}{n^{\sigma-r-1}} \dot{\omega}_1\left(f^{(\sigma-1)}, \frac{b-a}{n}\right), \quad 1 \leq r \leq \sigma - 1, \quad (19.33)$$

where $C$ is a constant independent of $f$ and $n$. It depends only on $m$, $\sigma$, $L$, $a$, $b$.

**PROOF** Similar to the proof of Theorem 19.7 by the use of Lemma 19.2. ∎

### 19.4.3 Quantitative Monotone Approximation by Discrete Polynomial Splines

Let $[a,b] \subset \mathbf{R}$ and $h > 0$ so that $b = a + Nh$ for some integer $N \geq 1$. Consider the discrete interval $[a,b]_h = \{a, a+h, a+2h, \ldots, b\}$ and let $\Delta = \{a = x_0 < x_1 < \cdots < x_{k+1} = b\} \subset [a,b]_h$ be a partition of points with $\overline{\Delta} \equiv \max_{1 \leq i \leq k+1}(x_i - x_{i-1})$.

Denote by $S_m(\Delta; h)$ the space of discrete polynomial splines of order $m > 0$ with simple knots at the points $x_1, \ldots, x_k$. For the real valued function $f$ on $[a,b]_h$ the difference operator $D_h^j$ is defined as follows:

$$D_h^j f(x) = \frac{\Delta_h^j f(x)}{h^j} = \frac{\sum_{v=0}^{j} \binom{j}{v}(-1)^{j-v} f(x+vh)}{h^j}, \quad D_h^0 f(x) = f(x).$$

Also define

$$\|f\|_{\ell_\infty[a,b]_h} = \max |f(x)|, \quad x \in [a,b]_h,$$

and

$$\omega_1(f;t)_{\ell_\infty[a,b]_h} = \sup |f(x) - f(y)|, \quad x, y \in [a,b]_h : |x-y| \leq t,$$

which is the discrete first modulus of continuity with similar properties as the usual one defined over $[a,b]$. For the above see [154], pp. 343, 344, 357.

There exists a linear operator $Q : Q \equiv Q(f)$, mapping $B[a,b]_h$: the space of bounded real valued functions $f$ on $[a,b]_h$, into $S_m(\Delta;h)$, see [154], p. 358, Theorem 8.70.

From the same reference, p. 360, Theorem 8.72, we get

**LEMMA 19.3**
Let $f \in B[a,b]_h$, and assume $1 \leq \sigma \leq m$. Then for $r = 0, 1, \ldots, \sigma - 1$

$$\|D_h^r(f - Q)\|_{\ell_\infty[a,b]_h} \leq C_1(\overline{\Delta})^{\sigma-r-1} \cdot \omega_1(D_n^{\sigma-1}f; \overline{\Delta})_{\ell_\infty[a,b]_h}, \quad (19.34)$$

where $C_1$ is a constant depending only on $m$.

The last main result of this section comes next.

**THEOREM 19.9**
Let $\theta, k, \sigma, m$ be integers, $0 \leq \theta \leq k \leq \sigma - 1$, $\sigma \leq m$ and $[a,b]_h$, $\Delta$, $\overline{\Delta}$ as above with $\overline{\Delta} \leq 1$ and $f \in B[a,b]_h$. Let $\alpha_j(x) \in B[a,b]_h$, $j = \theta, \theta + 1, \ldots, k$ and suppose that either $\alpha_\theta(x) \geq \alpha > 0$ or $\alpha_\theta(x) \leq \beta < 0$ throughout $[a,b]_h$.

*Take the linear divided difference operator*

$$L = \sum_{j=\theta}^{k} \alpha_j(x) D_h^j$$

*and assume, throughout* $[a,b]_h$,

$$L(f) \geq 0. \tag{19.35}$$

*Then there exists a discrete polynomial spline function $Q(x)$ of order $m$ with simple knots at $\{x_1 \cdots x_k\} \subset [a,b]_h$ such that $L(Q) \geq 0$ throughout $[a,b]_h$, and*

$$\|D_h^r(f-Q)\|_{\ell_\infty[a,b]_h} \leq C(\overline{\Delta})^{\sigma-k-1} \omega_1(D_h^{\sigma-1}f; \overline{\Delta})_{\ell_\infty[a,b]_h}, \quad 0 \leq r \leq \theta. \tag{19.36}$$

*Moreover, we obtain*

$$\|D_h^r(f-Q)\|_{\ell_\infty[a,b]_h} \leq C(\overline{\Delta})^{\sigma-r-1} \omega_1(D_h^{\sigma-1}f; \overline{\Delta})_{\ell_\infty[a,b]_h}, \quad \theta+1 \leq r \leq \sigma-1, \tag{19.37}$$

*where $C$ is a constant independent of $f$ and $\Delta$, it depends only on $m$, $L$, $a$, $b$.*

**PROOF** Similar to the proof of Theorem 19.7 by the use of Lemma 19.3. ∎

# Chapter 20

# Comparisons for Local Moduli of Continuity

Comparison results are exhibited here for the first and second local moduli of continuity of a continuously differentiable function and its higher order derivatives. The multivariate case is also discussed. Here we follow [27].

## 20.1 Univariate Results

We use the following notion.

**DEFINITION 20.1** *Let $f \in C([a,b])$ and $x_0 \in [a,b]$. Then the first local modulus of continuity of $f$ at $x_0$ is defined as follows:*

$$\omega_1(f, x_0, h) := \sup_{\substack{x \in [a,b] \\ |x-x_0| \leq h}} |f(x) - f(x_0)|, \qquad (20.1)$$

*where $0 \leq h \leq b - a$.*

We use the next auxiliary result.

**LEMMA 20.1**
*Let $[a,b] \subset \mathbf{R}$, $x_0 \in (a,b)$. Consider $f \in C^n([a,b])$, $n \geq 1$, and denote $\phi(x) := f^{(n)}(x) - f^{(n)}(x_0)$. Then $(a \leq x \leq b)$ it holds*

$$f(x) - f(x_0) = \sum_{k=1}^{n} \frac{f^{(k)}(x_0)}{k!} \cdot (x - x_0)^k + \mathcal{R}_n(x_0, x), \qquad (20.2)$$

where
$$\mathcal{R}_n(x_0, x) := \int_{x_0}^{x} \phi(t) \cdot \frac{(x-t)^{n-1}}{(n-1)!} \cdot dt. \tag{20.3}$$

**PROOF**  By Taylor's Theorem.  ∎

It follows the first main result.

**THEOREM 20.1**
Let $f \in C^N([a,b])$, $1 \leq n \leq N$, $x_0 \in (a,b)$ fixed, and $0 \leq h \leq b-a$. Then it holds

$$\omega_1(f, x_0, h) \leq \sum_{k=1}^{n} \frac{|f^{(k)}(x_0)|}{k!} \cdot h^k + \frac{h^n}{n!} \cdot \omega_1(f^{(n)}, x_0, h). \tag{20.4}$$

**PROOF**  From Lemma 20.1, we get the following: case of $x \geq x_0$, such that $x - x_0 \leq h$, then

$$|\mathcal{R}_n(x_0, x)| \leq \int_{x_0}^{x} |f^{(n)}(t) - f^{(n)}(x_0)| \cdot \frac{(x-t)^{n-1}}{(n-1)!} \cdot dt$$

$$\leq \omega_1(f^{(n)}, x_0, h) \cdot \int_{x_0}^{x} \frac{(x-t)^{n-1}}{(n-1)!} \cdot dt$$

$$= \omega_1(f^{(n)}, x_0, h) \cdot \frac{(x-x_0)^n}{n!} \leq \omega_1(f^{(n)}, x_0, h) \cdot \frac{h^n}{n!}.$$

I.e.,
$$|\mathcal{R}_n(x_0, x)| \leq \omega_1(f^{(n)}, x_0, h) \cdot \frac{h^n}{n!}, \quad \text{when } x \geq x_0. \tag{20.5}$$

Case of $x \leq x_0$, such that $x_0 - x \leq h$, then it holds

$$|\mathcal{R}_n(x_0, x)| \leq \int_{x}^{x_0} |f^{(n)}(t) - f^{(n)}(x_0)| \cdot \frac{|x-t|^{n-1}}{(n-1)!} \cdot dt$$

$$\leq \omega_1(f^{(n)}, x_0, h) \cdot \int_{x}^{x_0} \frac{(t-x)^{n-1}}{(n-1)!} \cdot dt$$

$$= \omega_1(f^{(n)}, x_0, h) \cdot \frac{(x_0-x)^n}{n!}$$

$$\leq \omega_1(f^{(n)}, x_0, h) \cdot \frac{h^n}{n!}.$$

I.e.,
$$|\mathcal{R}_n(x_0, x)| \leq \omega_1(f^{(n)}, x_0, h) \cdot \frac{h^n}{n!}, \quad \text{when } x_0 \geq x. \tag{20.6}$$

Thus, from (20.5) and (20.6), we get that
$$|\mathcal{R}_n(x_0, x)| \leq \omega_1(f^{(n)}, x_0, h) \cdot \frac{h^n}{n!}, \tag{20.7}$$

where $0 \leq |x - x_0| \leq h \leq b - a$. Therefore, from (20.2) and (20.7), we obtain
$$|f(x) - f(x_0)| \leq \sum_{k=1}^{n} \frac{|f^{(k)}(x_0)|}{k!} \cdot h^k + \frac{h^n}{n!} \cdot \omega_1(f^{(n)}, x_0, h), \tag{20.8}$$

where $0 \leq |x - x_0| \leq h \leq b - a$. Hence from Definition 20.1 and (20.8), we find inequality (20.4). ∎

To continue we use the next notion.

**DEFINITION 20.2** Let $f \in C([a,b])$ and $x_0 \in [a,b]$. Then the second local modulus of continuity of $f$ at $x_0$ is given by
$$\omega_2(f, x_0, h) := \sup_{\substack{\text{all } y: a \leq x_0 \pm y \leq b \\ \text{with } |y| \leq h}} |f(x_0 + y) + f(x_0 - y) - 2f(x_0)|, \tag{20.9}$$

where
$$0 \leq h \leq \frac{b-a}{2}.$$

The second main result follows.

**THEOREM 20.2**
Let $f \in C^N([a,b])$, $N \in \mathbf{N}$ and $n$ even such that $2 \leq n \leq N$, $x_0 \in (a,b)$ fixed. Then we get
$$\omega_2(f, x_0, h) \leq 2 \cdot \sum_{\rho=1}^{n/2} \frac{|f^{(2\rho)}(x_0)|}{(2\rho)!} \cdot h^{2\rho} + \frac{h^n}{n!} \cdot \omega_2(f^{(n)}, x_0, h), \tag{20.10}$$

where
$$0 \leq h \leq (b-a)/2.$$

**PROOF** Let $f$ as in the assumption and $y \in [a,b]$ such that $x_0 \pm y \in [a,b]$. Then by Lemma 20.1, we get

$$f(x_0 + y) - f(x_0) = \sum_{k=1}^{n} \frac{f^{(k)}(x_0)}{k!} \cdot y^k + \mathcal{R}_n(x_0, x_0 + y), \qquad (20.11)$$

where

$$\mathcal{R}_n(x_0, x_0 + y) := \int_{x_0}^{x_0+y} (f^{(n)}(t) - f^{(n)}(x_0))$$

$$\cdot \frac{(x_0 + y - t)^{n-1}}{(n-1)!} \cdot dt. \qquad (20.12)$$

Using the change of variable method, we obtain

$$\mathcal{R}_n(x_0, x_0 + y) = \int_0^y (f^{(n)}(x_0 + t) - f^{(n)}(x_0)) \cdot \frac{(y-t)^{n-1}}{(n-1)!} \cdot dt. \qquad (20.13)$$

Also, from Lemma 20.1, we have

$$f(x_0 - y) - f(x_0) = \sum_{k=1}^{n} \frac{f^{(k)}(x_0)}{k!} (-y)^k + \mathcal{R}_n(x_0, x_0 - y). \qquad (20.14)$$

Here it is

$$\mathcal{R}_n(x_0, x_0 - y) = \int_{x_0}^{x_0-y} (f^{(n)}(t) - f^{(n)}(x_0)) \cdot \frac{(x_0 - y - t)^{n-1}}{(n-1)!} \cdot dt$$

$$(n-1 \text{ is odd})$$

$$= -\int_0^{-y} (f^{(n)}(x_0 + t) - f^{(n)}(x_0)) \cdot \frac{(y+t)^{n-1}}{(n-1)!} \cdot dt$$

$$= \int_{-y}^{0} (f^{(n)}(x_0 + t) - f^{(n)}(x_0)) \cdot \frac{(y+t)^{n-1}}{(n-1)!} \cdot dt$$

## 20. Comparisons for Local Moduli of Continuity

$$= -\int_y^0 (f^{(n)}(x_0 - t) - f^{(n)}(x_0)) \cdot \frac{(y-t)^{n-1}}{(n-1)!} \cdot dt.$$

I.e.,

$$\mathcal{R}_n(x_0, x_0 - y) = \int_0^y (f^{(n)}(x_0 - t) - f^{(n)}(x_0)) \cdot \frac{(y-t)^{n-1}}{(n-1)!} \cdot dt. \quad (20.15)$$

Take now the central second-order differences

$$(\Delta_y^2 f)(x_0) := f(x_0 + y) + f(x_0 - y) - 2f(x_0) \quad (20.16)$$

and

$$(\Delta_t^2 f^{(n)})(x_0) = f^{(n)}(x_0 + t) + f^{(n)}(x_0 - t) - 2f^{(n)}(x_0). \quad (20.17)$$

Hence, from (20.13) and (20.15), we find

$$\mathcal{R}_n(x_0, x_0 + y) + \mathcal{R}_n(x_0, x_0 - y)$$

$$= \int_0^y (\Delta_t^2 f^{(n)})(x_0) \cdot \frac{(y-t)^{n-1}}{(n-1)!} \cdot dt. \quad (20.18)$$

Adding (20.11) and (20.14), we obtain

$$(\Delta_y^2 f)(x_0) = 2 \cdot \sum_{\rho=1}^{n/2} \frac{f^{(2\rho)}(x_0)}{(2\rho)!} \cdot y^{2\rho}$$

$$+ \int_0^y (\Delta_t^2 f^{(n)})(x_0) \cdot \frac{(y-t)^{n-1}}{(n-1)!} \cdot dt. \quad (20.19)$$

So that,

$$|(\Delta_y^2 f)(x_0)| \le 2 \cdot \sum_{\rho=1}^{n/2} \frac{|f^{(2\rho)}(x_0)|}{(2\rho)!} \cdot y^{2\rho} + \Gamma(y), \quad (20.20)$$

where

$$\Gamma(y) := \left| \int_0^y (\Delta_t^2 f^{(n)})(x_0) \cdot \frac{(y-t)^{n-1}}{(n-1)!} \cdot dt \right|. \quad (20.21)$$

Let $h \geq y \geq 0$, then it holds

$$\Gamma(y) \leq \int_0^y |(\Delta_t^2 f^{(n)})(x_0)| \cdot \frac{(y-t)^{n-1}}{(n-1)!} dt$$

$$\leq \int_0^y \omega_2(f^{(n)}, x_0, t) \cdot \frac{(y-t)^{n-1}}{(n-1)!} dt$$

$$\leq \omega_2(f^{(n)}, x_0, y) \cdot \int_0^y \frac{(y-t)^{n-1}}{(n-1)!} dt$$

$$\leq \omega_2(f^{(n)}, x_0, h) \cdot \frac{h^n}{n!}.$$

I.e.,

$$\Gamma(y) \leq \frac{h^n}{n!} \cdot \omega_2(f^{(n)}, x_0, h), \quad 0 \leq y \leq h. \tag{20.22}$$

Let $y \leq 0$ such that $-y \leq h$, then it holds

$$\Gamma(y) \leq \int_y^0 |(\Delta_t^2 f^{(n)})(x_0)| \cdot \frac{(t-y)^{n-1}}{(n-1)!} \cdot dt$$

$$\leq \int_y^0 \omega_2(f^{(n)}, x_0, |t|) \cdot \frac{(t-y)^{n-1}}{(n-1)!} \cdot dt$$

$$\leq \omega_2(f^{(n)}, x_0, |y|) \cdot \int_y^0 \frac{(t-y)^{n-1}}{(n-1)!} \cdot dt$$

$$= \omega_2(f^{(n)}, x_0, |y|) \cdot \frac{|y|^n}{n!} \leq \omega_2(f^{(n)}, x_0, h) \cdot \frac{h^n}{n!}.$$

I.e.,

$$\Gamma(y) \leq \frac{h^n}{n!} \cdot \omega_2(f^{(n)}, x_0, h), \quad y \leq 0 : -y \leq h. \tag{20.23}$$

That is

$$\Gamma(y) \leq \frac{h^n}{n!} \cdot \omega_2(f^{(n)}, x_0, h), \quad \text{all } |y| \leq h. \tag{20.24}$$

Consequently from (20.20) and (20.24) we obtain that

$$|(\Delta_y^2 f)(x_0)| \leq 2 \cdot \sum_{\rho=1}^{n/2} \frac{|f^{(2\rho)}(x_0)|}{(2\rho)!} \cdot h^{2\rho} + \frac{h^n}{n!} \cdot \omega_2(f^{(n)}, x_0, h),$$

$$0 \leq h \leq \frac{b-a}{2}. \qquad (20.25)$$

Taking the supremum in the left-hand side of (20.25) over all $y$: $|y| \leq h$, we find the desired inequality (20.10). ■

## 20.2  A Multivariate Result

We use the following.

**DEFINITION 20.3**  Let $(V_1, \|\cdot\|_1)$, $(V_2, \|\cdot\|_2)$ real normed vector spaces, $Q$ a compact subset of $V_1$ and $\vec{x}_0 \in Q$. For a continuous function $f : Q \to V_2$, we define the first local modulus of continuity at $\vec{x}_0$ as

$$\omega_1(f, \vec{x}_0, h) := \sup\{\|f(\vec{x}) - f(\vec{x}_0)\|_2 : \text{ all } \vec{x} \in Q, \|\vec{x} - \vec{x}_0\|_1 \leq h\}, \quad (20.26)$$

where $h \geq 0$ is fixed.

We give the final result of this chapter.

**THEOREM 20.3**

Let $Q$ be a compact and convex subset of $\mathbf{R}^k$, $k \geq 1$ and $\vec{x}_0 \in Q$ fixed. Let $f \in C^n(Q)$ and assume that each nth partial derivative $f_\alpha := \frac{\partial^\alpha f}{\partial x^\alpha}$, where $\alpha := (\alpha_1, \ldots, \alpha_k)$, $\alpha_i \in \mathbf{Z}^+$, $i = 1, \ldots, k$, $|\alpha| := \sum_{i=1}^k \alpha_i = n$, has, relative to $Q$ and the $\ell_1$-norm $\|\cdot\|$ and $\|\cdot\|_\infty$, a first local modulus of continuity at $\vec{x}_0$: $\omega_1(f_\alpha, \vec{x}_0, h)$. Here $h > 0$ is given, and we set

$$w(\vec{x}_0) := \max_{\alpha: |\alpha| = n} (\omega_1(f_\alpha, \vec{x}_0, h)),$$

(i.e., $\omega_1(f_\alpha, \vec{x}_0, h) \le w(\vec{x}_0)$, all $\alpha : |\alpha| = n$). Then it holds

$$\omega_1(f, \vec{x}_0, h) \le \sum_{j=1}^{n} \frac{[(\sum_{i=1}^{k} |\frac{\partial}{\partial x_i}|)^j f](\vec{x}_0)}{j!} \cdot h^j$$

$$+ \frac{w(\vec{x}_0) \cdot h^n \cdot k^n}{n!}. \tag{20.27}$$

Here the first local modulus of continuity of $f$ at $\vec{x}_0$ is taken with respect to $\|\cdot\|_1 := \|\cdot\|_{\ell_1}$-norm in $Q$ and $\|\cdot\|_2 := \|\cdot\|_\infty$-supremum norm in $C^n(Q)$. (See Definition 20.3).

**PROOF** Here $\vec{x}_0 = (x_{01}, \ldots, x_{0k}) \in Q$ is a fixed point and let

$$g_{\vec{z}}(t) := f(\vec{x}_0 + t \cdot (z - \vec{x}_0)), \quad t \ge 0. \tag{20.28}$$

Hence
$$g_{\vec{z}}(0) = f(\vec{x}_0).$$

The $j$th derivative of $g_{\vec{z}}(t)$ is given from

$$g_{\vec{z}}^{(j)}(t) = \left[\left(\sum_{i=1}^{k}(z_i - x_{0i}) \cdot \frac{\partial}{\partial x_i}\right)^j f\right](x_{01} + t \cdot (z_1 - x_{01}), \ldots, x_{0k}$$

$$+ t \cdot (z_k - x_{0k})), \tag{20.29}$$

and

$$g_{\vec{z}}^{(j)}(0) = \left[\left(\sum_{i=1}^{k}(z_i - x_{0i}) \cdot \frac{\partial}{\partial x_i}\right)^j f\right](\vec{x}_0), \quad j = 1, \ldots, n. \tag{20.30}$$

Here $\|z\|_{\ell_1} := \sum_{i=1}^{k} |z_i|$. Observe that

$$|g_{\vec{z}}^{(j)}(0)| \le h^j \cdot \left[\left(\sum_{i=1}^{k} \left|\frac{\partial}{\partial x_i}\right|\right)^j f\right](\vec{x}_0), \tag{20.31}$$

given that $\|z - x_0\|_{\ell_1} \le h$.

By Taylor's Theorem in $\mathbf{R}^k$, $k \geq 1$, we obtain

$$f(\vec{z}) - f(\vec{x}_0) = \sum_{j=1}^{n} \frac{g_{\vec{z}}^{(j)}(0)}{j!} + \mathcal{R}_n(\vec{z}, 0), \qquad (20.32)$$

where

$$\mathcal{R}_n(\vec{z}, 0) := \int_0^1 \left( \int_0^{t_1} \cdot \left( \int_0^{t_{n-1}} (g_{\vec{z}}^{(n)}(t_n) - g_{\vec{z}}^{(n)}(0)) dt_n \right) \cdot \right) \cdot dt_1. \qquad (20.33)$$

Here $0 \leq t \leq 1$, $\|\vec{z} - \vec{x}_0\|_{\ell_1} \leq h$ and

$$\omega_1(f_\alpha, \vec{x}_0, t \cdot \|\vec{z} - \vec{x}_0\|_{\ell_1}) \leq \omega_1(f_\alpha, \vec{x}_0, t \cdot h)$$

$$\leq \omega_1(f_\alpha, \vec{x}_0, h) \leq w(\vec{x}_0). \qquad (20.34)$$

Consequently,

$$|f_\alpha(\vec{x}_0 + t \cdot (\vec{z} - \vec{x}_0)) - f_\alpha(\vec{x}_0)| \leq \omega_1(f_\alpha, \vec{x}_0, t \cdot \|\vec{z} - \vec{x}_0\|_{\ell_1}) \leq w(\vec{x}_0). \qquad (20.35)$$

Thus,

$$|\mathcal{R}_n(\vec{z}, 0)| \leq \int_0^1 \left( \int_0^{t_1} \cdots \left( \int_0^{t_{n-1}} \left( \sum_{|\alpha|=n} \frac{n!}{\alpha_1! \cdots \alpha_k!} \right. \right. \right.$$

$$\left. \left. \left. \cdot |z_1 - x_{01}|^{\alpha_1} \cdots |z_k - x_{0k}|^{\alpha_k} \cdot w(\vec{x}_0) \right) \cdot dt_n \right) \cdots \right) dt_1$$

$$\leq w(\vec{x}_0) \cdot h^n \cdot \int_0^1 \left( \int_0^{t_1} \cdots \left( \int_0^{t_{n-1}} \left( \sum_{|\alpha|=n} \frac{n!}{\alpha_1! \cdots \alpha_k!} \right) \right. \right.$$

$$\left. \left. \cdot dt_n \right) \cdots \right) dt_1 = w(\vec{x}_0) \cdot \frac{h^n}{n!} \cdot k^n. \qquad (20.36)$$

I.e.,

$$|\mathcal{R}_n(\vec{z}, 0)| \leq \frac{w(\vec{x}_0) \cdot h^n \cdot k^n}{n!}. \qquad (20.37)$$

Therefore, from (20.26) in this setting, (20.31)–(20.33), and (20.37), we get inequality (20.27). ∎

# Chapter 21

## Convergence with Rates of Univariate Singular Integrals to the Unit

Here we study the $L^p$-approximation, $(1 \leq p < +\infty)$, by the Jackson-type generalizations of Picard and Gauss–Weierstrass singular integrals using the $L^p$-moduli of smoothness, and study both uniform and $L^p$-approximation (using the corresponding moduli of smoothness) by Jackson-type generalizations of the Poisson–Cauchy singular integrals. This chapter relies on [36].

## 21.1 Background

Let $f$ be a function from $\mathbf{R}$ to itself. For $r \in \mathbf{N}$, the $rth - L_p$-modulus of smoothness over $\mathbf{R}$ $(1 \leq p \leq +\infty)$ is defined by

$$\omega_r(f;\delta)_X = \sup_{|h| \leq \delta} \|\Delta_h^r f\|_X,$$

where

$$\Delta_h^r f(x) = \sum_{i=0}^{r} (-1)^i \binom{r}{i} f(x+ih), \quad r \in \mathbf{N},$$

$$X = L^p(\mathbf{R}) \quad \text{or} \quad X = L_{2\pi}^p(\mathbf{R}),$$

$$\|f\|_{L^p(\mathbf{R})} = \left( \int_{-\infty}^{+\infty} |f(x)|^p dx \right)^{1/p}, \quad \|f\|_{L_{2\pi}^p(\mathbf{R})} = \left( \int_{-\pi}^{\pi} |f(x)|^p dx \right)^{1/p}.$$

In the following, for $\xi > 0$ we study the Jackson-type generalizations of Picard, Poisson–Cauchy, and Gauss–Weierstrass singular integrals, introduced in [85], respectively,

$$P_{n,\xi}(f;x) = -\frac{1}{2\xi}\sum_{k=1}^{n+1}(-1)^k\binom{n+1}{k}\int_{-\infty}^{+\infty}f(x+kt)e^{-|t|/\xi}dt,$$

$$Q_{n,\xi}(f;x) = \frac{1}{-\left(\frac{2}{\xi}\right)\tan^{-1}\left(\frac{\pi}{\xi}\right)}\sum_{k=1}^{n+1}(-1)^k\binom{n+1}{k}\int_{-\pi}^{\pi}\frac{f(x+kt)}{t^2+\xi^2}dt,$$

and

$$W_{n,\xi}(f;x) = -\frac{1}{2C(\xi)}\sum_{k=1}^{n+1}(-1)^k\binom{n+1}{k}\int_{-\pi}^{\pi}f(x+kt)e^{-t^2/\xi^2}dt,$$

$$C(\xi) = \int_0^{\pi}e^{-t^2/\xi^2}dt,$$

(the above operators come by generalizing the usual Picard, Poisson–Cauchy and Gauss–Weierstrass singular integrals, by following the same idea which was used to define the Jackson's generalized operator in classical approximation theory).

Here we take only $f$ such that $P_{n,\xi}(f;x)$, $Q_{n,\xi}(f;x)$, $W_{n,\xi}(f;x) \in \mathbf{R}$, for all $x \in \mathbf{R}$.

Uniform convergences to the unit by $P_{n,\xi}$, $W_{n,\xi}$ operators (as $\xi \to 0$) have been given in [85] and can be stated in the next:

**THEOREM A** *Let* $f \in C_{2\pi}(\mathbf{R})$. *Then it holds*

(i) $\|f - P_{n,\xi}(f)\| \leq \left[\sum_{k=0}^{n+1}\binom{n+1}{k}k!\right]\omega_{n+1}(f;\xi)$, $\xi > 0$;

(ii) $\|f - W_{n,\xi}\| \leq \left[1/\int_0^{\pi}e^{-u^2}du\right]\left[\int_0^{+\infty}(u+1)^{n+1}e^{-u^2}du\right]\omega_{n+1}(f;\xi)$, $0 \leq \xi \leq 1$,

*where* $\|\cdot\|$ *is the uniform norm on* $C_{2\pi}(\mathbf{R})$ *and* $\omega_r(f;\xi)$ *is the rth uniform modulus of smoothness.*

In Section 21.2 we study the $L^p$-approximation, $1 \leq p < +\infty$, of $P_{n,\xi}$, $Q_{n,\xi}$, $W_{n,\xi}$ operators; while in Section 21.3 we study the uniform convergence of the $Q_{n,\xi}$ operator.

## 21.2 Quantitative $L^p$-Approximation, $1 \le p < +\infty$

The first main result has as follows:

**THEOREM 21.1**
Here consider $X = L^1(\mathbf{R})$ (for $P_{n,\xi}$), $X = L^1_{2\pi}(\mathbf{R})$ (for $W_{n,\xi}, Q_{n,\xi}$), $\xi > 0$, $n \in \mathbf{N}$, $f \in X$. Then it holds

$$\|f - P_{n,\xi}\|_X \le \left[\sum_{k=0}^{n+1} \binom{n+1}{k} k!\right] \omega_{n+1}(f;\xi)_X, \quad \xi > 0, \qquad (21.1)$$

$$\|f - W_{n,\xi}(f)\|_X \le \left[1/\int_0^\pi e^{-u^2} du\right]$$

$$\left[\int_0^{+\infty} (u+1)^{n+1} e^{-u^2} du\right] \omega_{n+1}(f;\xi)_{L^1_{2\pi}(\mathbf{R})};$$

$$0 < \xi \le 1, \qquad (21.2)$$

$$\|f - Q_{n,\xi}(f)\|_X \le K(n,\xi) \omega_{n+1}(f;\xi)_{L^1_{2\pi}(\mathbf{R})}, \quad \xi > 0, \qquad (21.3)$$

where $K(n,\xi) = \left[1/\tan^{-1}\frac{\pi}{\xi}\right] \int_0^{\pi/\xi} \frac{(u+1)^{n+1}}{u^2+1} du$.

**PROOF** We notice that (for $X = L^1(\mathbf{R})$)

$$f(x) - P_{n,\xi}(f;x) = (2\xi)^{-1} \int_{-\infty}^{+\infty} (-1)^{n+1} \Delta_t^{n+1} f(x) e^{-|t|/\xi} dt, \qquad (21.4)$$

which gives us

$$\|f - P_{n,\xi}(f)\|_{L^1(\mathbf{R})} = \int_{-\infty}^{+\infty} |f(x) - P_{n,\xi}(f;x)| dx$$

$$\le (2\xi)^{-1} \int_{-\infty}^{+\infty} \left[\int_{-\infty}^{+\infty} |\Delta_t^{n+1} f(x)| dx\right] e^{-|t|/\xi} dt$$

$$\leq (2\xi)^{-1} \int_{-\infty}^{+\infty} \omega_{n+1}(f; |t|)_{L^1(\mathbf{R})} e^{-|t|/\xi} dt$$

$$= (2\xi)^{-1} \int_{-\infty}^{+\infty} \omega_{n+1}(f; \xi(|t|/\xi))_{L^1(\mathbf{R})} e^{-|t|/\xi} dt$$

$$\leq (2\xi)^{-1} \omega_{n+1}(f; \xi)_{L^1(\mathbf{R})} \int_{-\infty}^{+\infty} (|t|/\xi + 1)^{n+1} e^{-|t|/\xi} dt$$

$$= \xi^{-1} \omega_{n+1}(f; \xi)_{L^1(\mathbf{R})} \int_{0}^{+\infty} [t/\xi + 1]^{n+1} e^{-t/\xi} dt$$

$$= \omega_{n+1}(f; \xi)_{L^1(\mathbf{R})} \int_{0}^{\infty} [u+1]^{n+1} e^{-u} du$$

$$= \left[ \sum_{k=0}^{n+1} \binom{n+1}{k} k! \right] \omega_{n+1}(f; \xi)_{L^1(\mathbf{R})},$$

and establishes (21.1). Also it holds

$$f(x) - W_{n,\xi}(f; x) = [1/2C(\xi)] \int_{-\pi}^{\pi} (-1)^{n+1} \Delta_t^{n+1} f(x) e^{-t^2/\xi^2} dt, \quad (21.5)$$

and working as above, we obtain (for $0 < \xi \leq 1$)

$$\|f - W_{n,\xi}(f)\|_{L^1_{2\pi}(\mathbf{R})} = \int_{-\pi}^{\pi} |f(x) - W_{n,\xi}(f; x)| dx$$

$$\leq [1/(2C(\xi))] \omega_{n+1}(f; \xi)_{L^1_{2\pi}(\mathbf{R})} \int_{0}^{\pi} [t/\xi + 1]^{n+1} e^{-t^2/\xi^2} dt$$

$$= \left[ \frac{\xi}{C(\xi)} \right] \omega_{n+1}(f; \xi)_{L^1_{2\pi}(\mathbf{R})} \int_{0}^{\pi/\xi} [u+1]^{n+1} e^{-u^2} du$$

(see also, e.g., [85], Lemma 3.2)

$$\leq \left[ 1/\int_{0}^{\pi} e^{-u^2} du \right] \left[ \int_{0}^{+\infty} (u+1)^{n+1} e^{-u^2} du \right] \omega_{n+1}(f; \xi)_{L^1_{2\pi}(\mathbf{R})},$$

which produces (21.2).

At the end,

$$f(x) - Q_{n,\xi}(f;x) = \frac{1}{\frac{2}{\xi}\tan^{-1}\frac{\pi}{\xi}} \int_{-\pi}^{\pi} \frac{(-1)^{n+1}}{t^2+\xi^2} \Delta_t^{n+1} f(x) dt, \qquad (21.6)$$

and working as above, we get

$$\|f - Q_{n,\xi}(f)\|_{L^1_{2\pi}(\mathbf{R})} = \int_{-\pi}^{\pi} |f(x) - Q_{n,\xi}(f;x)| dx$$

$$\leq \frac{\xi}{\tan^{-1}\frac{\pi}{\xi}} \omega_{n+1}(f;\xi)_{L^1_{2\pi}(\mathbf{R})} \int_0^{\pi} \frac{[t/\xi+1]^{n+1}}{t^2+\xi^2} dt$$

$$= \frac{1}{\tan^{-1}\frac{\pi}{\xi}} \omega_{n+1}(f;\xi)_{L^1_{2\pi}(\mathbf{R})} \int_0^{\pi/\xi} \frac{(u+1)^{n+1}}{u^2+1} du,$$

which establishes (21.3). ∎

**REMARK 21.1** For fixed $n \in \mathbf{N}$, by (21.1) and (21.2) we see that

$$\|f - P_{n,\xi}(f)\|_X \to 0, \quad \|f - W_{n,\xi}(f)\|_X \to 0, \quad \text{as } \xi \to 0.$$

Furthermore, since $K(n,\xi) \to +\infty$, as $\xi \to 0$, by (21.3) we do not get, in general, the convergence $\|f - Q_{n,\xi}(f)\|_X \to 0$, as $\zeta \to 0$. But, in some special cases this convergence is valid, as can be seen by the next. ∎

**COROLLARY 21.1**

If $f^{(n+1)} \in L^1_{2\pi}(\mathbf{R})$ and $f^{(n)}$ is absolutely continuous on $\mathbf{R}$, then it holds

$$\|f - Q_{n,\xi}(f)\|_{L^1_{2\pi}(\mathbf{R})} \leq C_n \xi, \quad 0 < \xi \leq 1$$

where $C_n > 0$ is a constant independent of $f$ and $\xi$.

**PROOF** We observe that

$$\omega_{n+1}(f;\xi)_{L^1_{2\pi}(\mathbf{R})} \leq C_1 \xi^{n+1} \|f^{(n+1)}\|_{L^1_{2\pi}(\mathbf{R})}$$

and for $0 < \xi \leq 1$,

$$\xi^{n+1} \int_0^{\pi/\xi} \frac{(u+1)^{n+1}}{u^2+1} du$$

$$= \xi^{n+1} \left[ \int_0^1 \frac{(u+1)^{n+1}}{u^2+1} du + \int_1^{\pi/\xi} \frac{(u+1)^{n+1}}{u^2+1} du \right]$$

$$= \xi^{n+1} \left[ C_2 + \int_1^{\pi/2} \frac{(u+1)^{n+1}}{u^2+1} du \right] \leq \xi^{n+1} \left[ C_2 + \int_1^{\pi/\xi} \frac{(u+1)^{n+1}}{u^2} du \right]$$

$$\leq \xi^{n+1} \left[ C_2 + \sum_{k=0}^{n+1} \binom{n+1}{k} \int_1^{\pi/\xi} u^{n-k-1} du \right]$$

$$= \xi^{n+1} \left\{ C_2 + \left[ \sum_{k=0}^{n-1} \binom{n+1}{k} \frac{u^{n-k}}{n-k} \right]_1^{\pi/\xi} \right.$$

$$\left. + (n+1) \ln u \Big|_1^{\pi/\xi} - \frac{1}{u} \Big|_1^{\pi/\xi} \right\} \leq C\xi,$$

which together with relation (21.3) establishes the corollary. ∎

The second main result follows

### THEOREM 21.2
Let us take $X = L^p(\mathbf{R})$ (for $P_{n,\xi}$), $X = L^p_{2\pi}(\mathbf{R})$ (for $W_{n,p}, Q_{n,\xi}$), $0 < \zeta \leq 1$, $n \in \mathbf{N}$, $1 < p < +\infty$, $\frac{1}{p} + \frac{1}{q} = 1$, $f \in X$. Then it holds

$$\|f - P_{n,\xi}(f)\|_X \leq (2/q)^{1/q} \|g\|_{L^p(\mathbf{R}_+)} \omega_{n+1}(f;\xi)_X,$$

where $g(u) = (u+1)^{n+1} e^{-u/2}$;

$$\|f - W_{n,\xi}(f)\|_X \leq \left( \sqrt{\frac{\pi}{2q}} \right)^{1/q} \frac{1}{\int_0^\pi e^{-u^2} du} \|h\|_{L^p(\mathbf{R}_+)} \omega_{n+1}(f;\xi)_X,$$

where $h(u) = (u+1)^{n+1} e^{-u^2/2}$;

$$\|f - Q_{n,\xi}(f)\|_X \le K_p(n,\xi) \omega_{n+1}(f;\xi)_{L^p_{2\pi}(\mathbf{R})},$$

where

$$K_p(n,\xi) = \left[ \frac{1}{\tan^{-1} \frac{\pi}{\xi}} \int_0^{\pi/\xi} (u+1)^{(n+1)p} \frac{1}{u^2+1} du \right]^{1/p}.$$

**PROOF**  Let $X = L^p(\mathbf{R})$, $\frac{1}{p} + \frac{1}{q} = 1$ and $C_1 = \frac{1}{(2\xi)^p} \left( \frac{4\xi}{q} \right)^{p/q}$. From (21.4) we get

$$\int_{-\infty}^{+\infty} |f(x) - P_{n,\xi}(f;x)|^p dx$$

$$= \frac{1}{(2\xi)^p} \int_{-\infty}^{+\infty} \left| \int_{-\infty}^{+\infty} (-1)^{n+1} \Delta_t^{n+1} f(x) e^{-|t|/(2\xi)} e^{-|t|/(2\xi)} dt \right|^p dx$$

(by Hölder's inequality, see [33], proof of Theorem 5)

$$\le C_1 \int_{-\infty}^{+\infty} \left( \int_{-\infty}^{+\infty} |\Delta_t^{n+1} f(x)|^p dx \right) e^{-|t|p/(2\xi)} dt$$

$$\le C_1 \int_{-\infty}^{+\infty} [\omega_{n+1}(f;|t|)_X e^{-|t|/(2\xi)}]^p dt$$

$$\le 2C_1 \omega_{n+1}^p(f;\xi)_X \int_0^{+\infty} [t/\xi + 1]^{(n+1)p} e^{-tp/(2\xi)} dt$$

$$= \frac{2^{p-1}}{q^{p/q}} \omega_{n+1}^p(f;\xi)_X \int_0^{+\infty} (u+1)^{(n+1)p} e^{-pu/2} du,$$

which implies

$$\|f - P_{n,\xi}(f)\|_X \le \left( \frac{2}{q} \right)^{1/q} \|g\|_{L^p(\mathbf{R}_+)} \omega_{n+1}(f;\xi)_X,$$

with $g(u) = (u+1)^{n+1} e^{-u/2}$, $u \in \mathbf{R}_+$.

Next, let $X = L^p_{2\pi}(\mathbf{R})$, $\frac{1}{p} + \frac{1}{q} = 1$, $Erf(x) = \frac{2}{\sqrt{\pi}} \int_0^x e^{-t^2} dt$. From (21.5) we obtain

$$\int_{-\pi}^{\pi} |f(x) - W_{n,\xi}(f;x)|^p dx$$

$$= \frac{1}{[2C(\xi)]^p} \int_{-\pi}^{\pi} \left| \int_{-\pi}^{\pi} (-1)^{n+1} \Delta_t^{n+1} f(x) e^{-t^2/(2\xi^2)} e^{-t^2/(2\xi^2)} dt \right|^p dx$$

(by Hölder's inequality, see [33], proof of Theorem 5)

$$\leq \frac{1}{[2C(\xi)]^p} \left( \sqrt{\frac{2\pi}{q}} \xi \right)^{p/q} \left( Erf\left(\pi \sqrt{\frac{q}{2}} \cdot \frac{1}{\xi}\right) \right)^{p/q}$$

$$\cdot \left[ \int_{-\pi}^{\pi} \left( \int_{-\pi}^{\pi} |\Delta_t^{n+1} f(x)|^p dx \right) e^{-t^2 p/(2\xi^2)} dt \right].$$

Setting $C_2 = \frac{1}{[2C(\xi)]^p} \left( \sqrt{\frac{2\pi}{q}} \xi \right)^{p/q} \left( Erf\left(\pi \sqrt{\frac{q}{2}} \cdot \frac{1}{\xi}\right) \right)^{p/q}$, we find

$$\int_{-\pi}^{\pi} |f(x) - W_{n,\xi}(f;x)|^p dx$$

$$\leq 2C_2 \omega_{n+1}^p(f;\xi)_X \int_0^{\pi} [t/\xi + 1]^{(n+1)p} e^{-t^2 p/(2\xi^2)} dt$$

$$\leq 2C_2 \omega_{n+1}^p(f;\xi)_X \xi \int_0^{\pi/\xi} [u+1]^{(n+1)p} e^{-u^2 p/2} du.$$

That is,

$$\|f - W_{n,\xi}\|_{L^p_{2\pi}(\mathbf{R})} \leq \left[ 2C_2 \xi \int_0^{\pi/\xi} [u+1]^{(n+1)p} e^{-u^2 p/2} du \right]^{1/p} \omega_{n+1}(f;\xi)_X.$$

However,

$$\left( Erf\left(\pi \sqrt{\frac{q}{2}} \cdot \frac{1}{\xi}\right) \right) = \frac{2}{\sqrt{\pi}} \int_0^{\pi\sqrt{\frac{q}{2}}/\xi} e^{-t^2} dt \leq \frac{2}{\sqrt{\pi}} \int_0^{+\infty} e^{-t^2} dt = 1,$$

## 21. Rates of Univariate Singular Integrals to the Unit

and from [85], Lemma 3.2, we see that

$$\frac{1}{C(\xi)} \leq \frac{1}{\xi \int_0^\pi e^{-u^2} du}, \quad 0 < \xi \leq 1,$$

which gives us

$$2C_2\xi \leq \frac{2}{2^p} \cdot \frac{1}{\xi^p \left(\int_0^\pi e^{-u^2} du\right)^p} \left(\sqrt{\frac{2\pi}{q}}\right)^{p/q} \xi^{p/q}\xi$$

$$= \frac{1}{2^{p-1} \left(\int_0^\pi e^{-u^2} du\right)^p} \left(\sqrt{\frac{2\pi}{q}}\right)^{p/q}.$$

And finally

$$\|f - W_{n,\xi}(f)\|_{L^p_{2\pi}(\mathbf{R})} \leq \left(\sqrt{\frac{\pi}{2q}}\right)^{1/q} \frac{1}{\int_0^\pi e^{-u^2} du} \|h\|_{L^p(\mathbf{R}_+)} \omega_{n+1}(f;\xi)_X,$$

where $h(u) = (u+1)^{n+1} e^{-u^2/2}$.

At the end, for $X = L^p_{2\pi}(\mathbf{R})$, $\frac{1}{p} + \frac{1}{q} = 1$, from (21.6) we observe that

$$\int_{-\pi}^\pi |f(x) - Q_{n,\xi}(f;x)|^p dx$$

$$= \frac{1}{\left[\frac{2}{\xi}\tan^{-1}\frac{\pi}{\xi}\right]^p} \int_{-\pi}^\pi \left| \int_{-\pi}^\pi (-1)^{n+1}\Delta_t^{n+1} f(x) \right.$$

$$\left. \cdot \frac{1}{(t^2+\xi^2)^{1/p}} \cdot \frac{1}{(t^2+\xi^2)^{1/q}} dt \right|^p dx$$

(by Hölder's inequality, see [33], proof of Theorem 5)

$$\leq \frac{1}{\left[\frac{2}{\xi}\tan^{-1}\frac{\pi}{\xi}\right]^p} \left(\frac{2}{\xi}\tan^{-1}\frac{\pi}{\xi}\right)^{p/q} \int_{-\pi}^\pi \left[\int_{-\pi}^\pi |\Delta_t^{n+1} f(x)|^p \frac{1}{t^2+\xi^2} dx\right] dt$$

$$\leq \frac{1}{\left[\frac{2}{\xi}\tan^{-1}\frac{\pi}{\xi}\right]} \int_{-\pi}^\pi \left[\omega_{n+1}(f;|t|)_X \frac{1}{(t^2+\xi^2)^{1/p}}\right]^p dt$$

$$= \frac{\xi}{\tan^{-1}\frac{\pi}{\xi}} \int_0^\pi \left[ \omega_{n+1}(f;t)_X \frac{1}{(t^2+\xi^2)^{1/p}} \right]^p dt$$

$$\leq \frac{\xi}{\tan^{-1}\frac{\pi}{\xi}} \omega_{n+1}^p(f;\xi)_X \int_0^\pi [t/\xi + 1]^{(n+1)p} \frac{1}{\xi^2} \cdot \frac{1}{(t/\xi)^2 + 1} dt$$

$$= \frac{1}{\tan^{-1}\frac{\pi}{\xi}} \omega_{n+1}^p(f;\xi)_X \int_0^{\pi/\xi} [u+1]^{(n+1)p} \frac{1}{u^2+1} du,$$

which establishes the theorem. ∎

**REMARK 21.2**  Theorem 21.2 proves that

$$\|f - P_{n,\xi}(f)\|_X \leq C_1 \omega_{n+1}(f;\xi)_X, \quad \|f - W_{n,\xi}(f)\|_X \leq C_2 \omega_{n+1}(f;\xi)_X$$

where $C_1, C_2 > 0$ are independent of $f, n$ and $\xi$; while $K_p(n,\xi)$ in the third estimate (in Theorem 21.2) tends to $+\infty$ as $\xi \to 0$. In that case, as in Corollary 21.1 we can improve the estimate of $\|f - Q_{n,\xi}(f)\|_X$. ∎

## 21.3  Quantitative Uniform Approximation by $Q_{n,\xi}$ Operator

From (21.6) we easily obtain (for $X = C_{2\pi}(\mathbf{R})$)

$$|f(x) - Q_{n,\xi}(f;x)| \leq \frac{1}{\frac{2}{\xi}\tan^{-1}\frac{\pi}{\xi}} \int_{-\pi}^\pi |\Delta_t^{n+1} f(x)| \frac{1}{t^2+\xi^2} dt$$

$$\leq \frac{1}{\frac{2}{\xi}\tan^{-1}\frac{\pi}{\xi}} \int_{-\pi}^\pi \omega_{n+1}(f;|t|)_X \frac{1}{t^2+\xi^2} dt$$

$$= \frac{\xi}{\tan^{-1}\frac{\pi}{\xi}} \omega_{n+1}(f;\xi)_X \int_0^\pi [t/\xi + 1]^{n+1} \frac{1}{t^2+\xi^2} dt,$$

which immediately produces

**THEOREM 21.3**
For $0 < \xi \leq 1$, $n \in \mathbf{N}$, $f \in X = C_{2\pi}(\mathbf{R})$, it holds

$$\|f - Q_{n,\xi}(f)\|_X \leq K(n,\xi)\omega_{n+1}(f;\xi)_X,$$

where $K(n,\xi)$ is given by Theorem 21.1.

Working exactly as in Corollary 21.1, we immediately get

**COROLLARY 21.2**
If $f^{(n+1)} \in C_{2\pi}(\mathbf{R}) = X$, then it holds

$$\|f - Q_{n,\xi}(f)\|_X \leq C_n \xi, \quad 0 < \xi \leq 1,$$

where $C_n > 0$ is independent of $f$ and $\xi$.

**REMARK 21.3** The results in this chapter show us that while the generalized operators $P_{n,\xi}$ and $W_{n,\xi}$ give better estimates than the classical operators of Picard and of Gauss–Weierstrass, the same idea of generalization applied for the Poisson–Cauchy singular integral, resulting in the $Q_{n,\xi}$-operator, does not give a better estimate. ∎

# Part VIII
# On Classical Analysis

# Chapter 22

# About Univariate Ostrowski-Type Inequalities

We show optimal upper bounds for the deviation of a function $f \in C^N([a,b])$, $N \in \mathbf{N}$, from its averages. These bounds are of the form $A \cdot \|f^{(N)}\|_\infty$, where $A$ is the smallest universal constant. That is, the given inequalities are sharp and sometimes are attained. The presented work has been greatly motivated by the articles of Ostrowski (1938), [136] and Fink (1992), [83]. This chapter also relies on [15].

More precisely, let $f \in C^{n+1}([a,b])$, $n \in \mathbf{Z}_+$, such that $f^{(k)}(x) = 0$, $k = 1, \ldots, n$, where $x$ is a fixed point in $[a,b]$. Then we prove that

$$\left| \frac{1}{b-a} \cdot \int_a^b f(y)\, dy - f(x) \right| \leq \varphi_n(x) \cdot \|f^{(n+1)}\|_\infty, \qquad (*)$$

where $\varphi_n(x)$ is a continuous function that depends only on $n$, $a$, $b$, it has a simple form and it is the smallest possible. That is, $(*)$ is sharp and in cases attained. The special case of $x = \frac{a+b}{2}$ is also met.

## 22.1 About Ostrowski's Inequality

Ostrowski's inequality (see Ostrowski [136]) has as follows:

$$\left| \frac{1}{b-a} \cdot \int_a^b f(y)\, dy - f(x) \right|$$

$$\leq \left( \frac{1}{4} + \frac{(x - \frac{a+b}{2})^2}{(b-a)^2} \right) \cdot (b-a) \cdot \|f'\|_\infty, \qquad (22.1)$$

where $f \in C^1([a,b])$, $x \in [a,b]$. Inequality (22.1) is sharp because the function in ( ) cannot be replaced by a smaller one. One can easily see that

$$\left(\frac{1}{4} + \frac{(x - \frac{a+b}{2})^2}{(b-a)^2}\right) \cdot (b-a) = \frac{(x-a)^2 + (b-x)^2}{2 \cdot (b-a)}. \tag{22.2}$$

In the following we present a different proof for (22.1) than the one of Ostrowski's initial proof of 1938 in [136].

**THEOREM 22.1**
Let $f \in C^1([a,b])$, $x \in [a,b]$. Then it holds

$$\left|\frac{1}{b-a} \cdot \int_a^b f(y)dy - f(x)\right| \leq \left(\frac{(x-a)^2 + (b-x)^2}{2 \cdot (b-a)}\right) \cdot \|f'\|_\infty. \tag{22.3}$$

Inequality (22.3) is sharp. In particular the optimal function is

$$f^*(y) := |y - x|^\alpha \cdot (b-a), \quad \alpha > 1. \tag{22.4}$$

**PROOF** See that

$$\left|\frac{1}{b-a} \cdot \int_a^b f(y)dy - f(x)\right| = \frac{1}{(b-a)} \cdot \left|\int_a^b (f(y) - f(x))dy\right|$$

$$\leq \frac{1}{(b-a)} \cdot \int_a^b |f(y) - f(x)| \cdot dy$$

$$\leq \frac{1}{(b-a)} \cdot \|f'\|_\infty \cdot \int_a^b |y - x| \cdot dy$$

$$= \frac{\|f'\|_\infty}{2 \cdot (b-a)} \cdot ((x-a)^2 + (b-x)^2).$$

So, we have proved inequality (22.3). Notice that

$$f^{*\prime}(y) = \alpha \cdot |y - x|^{\alpha - 1} \cdot \text{sign}(y - x) \cdot (b - a),$$

hence

$$|f^{*\prime}(y)| = \alpha \cdot |y - x|^{\alpha - 1} \cdot (b - a)$$

and
$$\|f^{*\prime}\|_\infty = \alpha \cdot (b-a) \cdot (\max(b-x, x-a))^{\alpha-1}.$$

Also we see that $f^*(x) = 0$.

Therefore we get for $f^*$ that

$$\text{L.H.S.}(22.3) = \int_a^b |y-x|^\alpha \cdot dy = \frac{(x-a)^{\alpha+1} + (b-x)^{\alpha+1}}{\alpha+1}$$

and

$$\lim_{\alpha \to 1} \text{L.H.S.}(22.3) = \frac{(x-a)^2 + (b-x)^2}{2}. \tag{22.5}$$

Also, we notice that

$$\text{R.H.S.}(22.3) = \left(\frac{(x-a)^2 + (b-x)^2}{2}\right) \cdot \alpha \cdot (\max(b-x, x-a))^{\alpha-1}$$

and

$$\lim_{\alpha \to 1} \text{R.H.S.}(22.3) = \frac{(x-a)^2 + (b-x)^2}{2}.$$

I.e.,

$$\lim_{\alpha \to 1} \text{L.H.S.}(22.3) = \lim_{\alpha \to 1} \text{R.H.S.}(22.3),$$

establishing (22.3) as sharp. ∎

Notice that when $x = a$ or $x = b$, inequality (22.3) can be attained by $f_a(y) := (y-a) \cdot (b-a)$, $f_b(y) := (y-b) \cdot (b-a)$, respectively (then both sides of (22.3) equal $(b-a)^2/2$).

## 22.2 About More General Univariate Ostrowski-Type Inequalities

The next material has been greatly motivated by the important work of Fink [83]. Let $f \in C^{n+1}([a,b])$, $n \in \mathbf{N}$, $x \in [a,b]$, be fixed. Then by Taylor's theorem we obtain

$$f(y) - f(x) = \sum_{k=1}^n \frac{f^{(k)}(x)}{k!} \cdot (y-x)^k + \mathcal{R}_n(x, y), \tag{22.6}$$

where
$$\mathcal{R}_n(x, y) := \int_x^y (f^{(n)}(t) - f^{(n)}(x)) \cdot \frac{(y-t)^{n-1}}{(n-1)!} \cdot dt; \qquad (22.7)$$

here $y$ can be $\geq x$ or $\leq x$.

Let $y \geq x$; then it holds

$$|\mathcal{R}_n(x, y)| \leq \int_x^y |f^{(n)}(t) - f^{(n)}(x)| \cdot \frac{(y-t)^{n-1}}{(n-1)!} \cdot dt$$

$$\leq \|f^{(n+1)}\|_\infty \cdot \int_x^y |t - x| \cdot \frac{|y-t|^{n-1}}{(n-1)!} \cdot dt$$

$$= \|f^{(n+1)}\|_\infty \cdot \frac{(y-x)^{n+1}}{(n+1)!}.$$

That is,
$$|\mathcal{R}_n(x, y)| \leq \frac{\|f^{(n+1)}\|_\infty}{(n+1)!} \cdot (y-x)^{n+1}, \quad y \geq x. \qquad (22.8)$$

Now let $x \geq y$; then it holds

$$|\mathcal{R}_n(x, y)| = \left| \int_y^x (f^{(n)}(t) - f^{(n)}(x)) \cdot \frac{(y-t)^{n-1}}{(n-1)!} \cdot dt \right|$$

$$\leq \int_y^x |f^{(n)}(t) - f^{(n)}(x)| \cdot \frac{|y-t|^{n-1}}{(n-1)!} \cdot dt$$

$$\leq \frac{\|f^{(n+1)}\|_\infty}{(n-1)!} \cdot \int_y^x (x-t) \cdot (t-y)^{n-1} \cdot dt$$

$$= \frac{\|f^{(n+1)}\|_\infty}{(n+1)!} \cdot (x-y)^{n+1}.$$

That is,
$$|\mathcal{R}_n(x, y)| \leq \frac{\|f^{(n+1)}\|_\infty}{(n+1)!} \cdot (x-y)^{n+1}, \quad x \geq y. \qquad (22.9)$$

By (22.8) and (22.9) we have that

$$|\mathcal{R}_n(x, y)| \leq \frac{\|f^{(n+1)}\|_\infty}{(n+1)!} \cdot |y-x|^{n+1}, \quad \text{for all } x, y \in [a, b]. \qquad (22.10)$$

## 22. About Univariate Ostrowski-Type Inequalities

In the following we treat

$$\left| \frac{1}{b-a} \cdot \int_a^b f(y)dy - f(x) \right|$$

$$= \frac{1}{b-a} \cdot \left| \int_a^b (f(y) - f(x)) \cdot dy \right|$$

$$= \frac{1}{b-a} \cdot \left| \int_a^b \left[ \sum_{k=1}^n \frac{f^{(k)}(x)}{k!} \cdot (y-x)^k + \mathcal{R}_n(x,y) \right] \cdot dy \right|$$

$$= \frac{1}{b-a} \cdot \left| \sum_{k=1}^n \frac{f^{(k)}(x)}{k!} \cdot \int_a^b (y-x)^k \cdot dy + \int_a^b \mathcal{R}_n(x,y) \cdot dy \right|$$

$$= \frac{1}{b-a} \cdot \left| \sum_{k=1}^n \frac{f^{(k)}(x)}{(k+1)!} \cdot [(b-x)^{k+1} - (a-x)^{k+1}] + \int_a^b \mathcal{R}_n(x,y) \cdot dy \right|$$

(from (22.10))

$$\leq \frac{1}{b-a} \cdot \left[ \sum_{k=1}^n \frac{|f^{(k)}(x)|}{(k+1)!} \cdot |(b-x)^{k+1} - (a-x)^{k+1}| \right.$$

$$\left. + \frac{\|f^{(n+1)}\|_\infty}{(n+1)!} \cdot \int_a^b |y-x|^{n+1} \cdot dy \right].$$

That is, we have established that

$$\left| \frac{1}{b-a} \cdot \int_a^b f(y)dy - f(x) \right|$$

$$\leq \frac{1}{b-a} \cdot \left[ \sum_{k=1}^n \frac{|f^{(k)}(x)|}{(k+1)!} \cdot |(b-x)^{k+1} - (a-x)^{k+1}| \right.$$

$$\left. + \frac{\|f^{(n+1)}\|_\infty}{(n+2)!} \cdot ((x-a)^{n+2} + (b-x)^{n+2}) \right], \qquad (22.11)$$

where $f \in C^{n+1}([a,b])$, $n \in \mathbb{N}$, $x \in [a,b]$, is fixed.
If we pick $x = \frac{a+b}{2}$, then

$$b - x = x - a = \frac{b-a}{2}.$$

Hence

$$\left| \frac{1}{b-a} \cdot \int_a^b f(y)dy - f\left(\frac{a+b}{2}\right) \right|$$

$$\leq \frac{1}{b-a} \cdot \left[ \sum_{1 \leq k \text{ even} \leq n} \frac{|f^{(k)}(\frac{a+b}{2})|}{(k+1)!} \cdot \frac{(b-a)^{k+1}}{2^k} \right.$$

$$\left. + \frac{\|f^{(n+1)}\|_\infty}{(n+2)!} \cdot \frac{(b-a)^{n+2}}{2^{n+1}} \right], \qquad (22.12)$$

where $f \in C^{n+1}([a,b])$, $n \in \mathbb{N}$.

The above considerations and the proved inequalities (22.11) and (22.12) lead to the next results.

**THEOREM 22.2**
Let $f \in C^{n+1}([a,b])$, $n \in \mathbb{N}$ and $x \in [a,b]$ be fixed, such that $f^{(k)}(x) = 0$, $k = 1, \ldots, n$. Then it holds

$$\left| \frac{1}{b-a} \cdot \int_a^b f(y)dy - f(x) \right| \leq \frac{\|f^{(n+1)}\|_\infty}{(n+2)!} \cdot \left( \frac{(x-a)^{n+2} + (b-x)^{n+2}}{b-a} \right). \qquad (22.13)$$

Inequality (22.13) is sharp. In particular, when $n$ is odd it is attained by $f^*(y) := (y-x)^{n+1} \cdot (b-a)$, while when $n$ is even the optimal function is

$$\tilde{f}(y) := |y-x|^{n+\alpha} \cdot (b-a), \quad \alpha > 1.$$

**PROOF** Inequality (22.13) comes immediately from (22.11). In the following we establish the sharpness of inequality (22.13).

When $n$ is odd: See that $f^{*(k)}(x) = 0$, $k = 0, 1, \ldots, n$, and $f^{*(n+1)}(y) = (n+1)! \cdot (b-a)$. Thus

$$\|f^{*(n+1)}\|_\infty = (n+1)! \cdot (b-a).$$

## 22. About Univariate Ostrowski-Type Inequalities

Plugging $f^*$ into (22.13) we obtain that

$$\text{L.H.S.}(22.13) = \frac{(b-x)^{n+2} + (x-a)^{n+2}}{n+2}. \tag{22.14}$$

Also, it holds

$$\text{R.H.S.}(22.13) = \frac{(x-a)^{n+2} + (b-x)^{n+2}}{n+2}. \tag{22.15}$$

From (22.14) and (22.15), when $n$ is odd, inequality (22.13) was established to be sharp, in particular is attained by $f^*$.

When $n$ is even: Observe that $\tilde{f}^{(k)}(x) = 0$, $k = 0, 1, \ldots, n$, and

$$\tilde{f}^{(n+1)}(y) = (n+\alpha)(n+\alpha-1)\cdots(\alpha+1) \cdot \alpha \cdot |y-x|^{\alpha-1} \cdot \text{sign}(y-x) \cdot (b-a).$$

Thus

$$|\tilde{f}^{(n+1)}(y)| = \left(\prod_{j=0}^{n}(n+\alpha-j)\right) \cdot |y-x|^{\alpha-1} \cdot (b-a)$$

and

$$\|\tilde{f}^{(n+1)}\|_\infty = \left(\prod_{j=0}^{n}(n+\alpha-j)\right) \cdot (\max(b-x, x-a))^{\alpha-1} \cdot (b-a).$$

Therefore we have

$$\text{R.H.S.}(22.13) = \frac{\left(\prod_{j=0}^{n}(n+\alpha-j)\right) \cdot (\max(b-x, x-a))^{\alpha-1}}{(n+2)!}$$

$$\cdot ((x-a)^{n+2} + (b-x)^{n+2}), \quad \alpha > 1.$$

Hence

$$\lim_{\alpha \to 1} \text{R.H.S.}(22.13) = \frac{(x-a)^{n+2} + (b-x)^{n+2}}{n+2} \tag{22.16}$$

and

$$\text{L.H.S.}(22.13) = \frac{(x-a)^{n+\alpha+1} + (b-x)^{n+\alpha+1}}{n+\alpha+1}.$$

Consequently,

$$\lim_{\alpha \to 1} \text{L.H.S.}(22.13) = \frac{(x-a)^{n+2} + (b-x)^{n+2}}{n+2}. \tag{22.17}$$

From (22.16) and (22.17) we obtain that (22.13) is sharp also when $n$ is even. ∎

Notice that when $x = a$ or $x = b$ and $n$ is even, inequality (22.13) can be attained by $\tilde{f}_a(y) := (y-a)^{n+1} \cdot (b-a)$, $\tilde{f}_b(y) := (y-b)^{n+1} \cdot (b-a)$, respectively (then both sides of (22.13) equal $(b-a)^{n+2}/n+2$). When $x = (a+b)/2$, we have a case of special interest that is presented next.

**THEOREM 22.3**
Let $f \in C^{n+1}([a,b])$, $n \in \mathbf{N}$ such that $f^{(k)}((a+b)/2) = 0$, all $k$ even $\in \{1,\ldots,n\}$. Then it holds

$$\left| \frac{1}{b-a} \cdot \int_a^b f(y)dy - f\left(\frac{a+b}{2}\right) \right| \leq \frac{\|f^{(n+1)}\|_\infty}{(n+2)!} \cdot \frac{(b-a)^{n+1}}{2^{n+1}}. \quad (22.18)$$

Inequality (22.18) is sharp. More precisely, when $n$ is odd it is attained by $f^*(y) := (y - \frac{a+b}{2})^{n+1} \cdot (b-a)$, while when $n$ is even the optimal function is

$$\tilde{f}(y) := \left| y - \frac{a+b}{2} \right|^{n+\alpha} \cdot (b-a), \quad \alpha > 1.$$

**COROLLARY 22.1**
Let $f \in C^2([a,b])$ such that $f''((a+b)/2) = 0$. Then it holds

$$\left| \frac{1}{b-a} \cdot \int_a^b f(y)dy - f\left(\frac{a+b}{2}\right) \right| \leq \|f''\|_\infty \cdot \frac{(b-a)^2}{24}, \quad (22.19)$$

which is sharp as in Theorem 22.3.

**PROOF** Apply Theorem 22.3 with $n = 1$. ∎

**PROOF** (of Theorem 22.3). Inequality (22.18) comes immediately by (22.12) and the assumption $f^{(k)}((a+b)/2) = 0$, all $k$ even in $\{1,\ldots,n\}$.
In the following we prove the sharpness of inequality (22.18).
When $n$ is odd: We observe that

$$f^{*(k)}\left(\frac{a+b}{2}\right) = 0, \quad \text{for } k = 0 \text{ and all } k \text{ even} \in \{1,\ldots,n\},$$

and moreover
$$f^{*(n+1)}(y) = (n+1)! \cdot (b-a), \quad \text{all } y \in [a,b].$$

Hence
$$\text{R.H.S.}(22.18) = \frac{(b-a)^{n+2}}{(n+2) \cdot 2^{n+1}}. \tag{22.20}$$

Also we get
$$\text{L.H.S.}(22.18) = \frac{(b-a)^{n+2}}{(n+2) \cdot 2^{n+1}}. \tag{22.21}$$

From (22.20) and (22.21) we obtain that (22.18) is attained by $f^*$, therefore (22.18) has been established sharp when $n$ is odd.

When $n$ is even: We see that $\tilde{f}^{(k)}((a+b)/2) = 0$, for $k = 0$ and for all $k$ even in $\{1, \ldots, n\}$, furthermore

$$\tilde{f}^{(n+1)}(y) = \prod_{j=0}^{n}(n+\alpha-j) \cdot \left|y - \frac{a+b}{2}\right|^{\alpha-1} \cdot \text{sign}\left(y - \frac{a+b}{2}\right) \cdot (b-a),$$

and
$$\|\tilde{f}^{(n+1)}\|_\infty = \left(\prod_{j=0}^{n}(n+\alpha-j)\right) \cdot \left(\frac{b-a}{2}\right)^{\alpha-1} \cdot (b-a).$$

Therefore,
$$\text{R.H.S.}(22.18) = \frac{\left(\prod_{j=0}^{n}(n+\alpha-j)\right) \cdot ((b-a)/2)^{\alpha-1} \cdot (b-a)}{(n+2)!}$$

$$\cdot \frac{(b-a)^{n+1}}{2^{n+1}}, \quad \alpha > 1.$$

Thus
$$\lim_{\alpha \to 1} \text{R.H.S.}(22.18) = \frac{(b-a)^{n+2}}{(n+2) \cdot 2^{n+1}}. \tag{22.22}$$

Also we obtain
$$\text{L.H.S.}(22.18) = \frac{2 \cdot ((b-a)/2)^{n+\alpha+1}}{n+\alpha+1}$$

and
$$\lim_{\alpha \to 1} \text{L.H.S.}(22.18) = \frac{(b-a)^{n+2}}{2^{n+1} \cdot (n+2)}. \tag{22.23}$$

From (22.22) and (22.23) we have proved that inequality (22.18) is sharp again when $n$ is even. ∎

# Chapter 23

## About Multidimensional Ostrowski-Type Inequalities

Here we exhibit optimal upper bounds on the deviation of a multidimensional function from its averages. These lead to sharp/attained inequalities. Namely, let $f \in C^{n+1}(Q)$, $n \in \mathbf{Z}_+$ be such that $f_\alpha(\vec{x}_0) = 0$, $|\alpha| = 1, \ldots, n$, where $f_\alpha$ is any partial derivative of $f$, and $\vec{x}_0$ is a fixed point of $Q$. Here $Q \subset \mathbf{R}^k$, $k \geq 1$ is a convex, compact subset. The special case of $Q := \prod_{i=1}^{k}[a_i, b_i]$ is taken care of extensively.

Then we prove that

$$\left| \frac{1}{\mathrm{Vol}(Q)} \int_Q f(\vec{z}) d\vec{z} - f(\vec{x}_0) \right| \leq \sum_{j=1}^{\ell} A_{jn}(\vec{x}_0, Q) \| f_j^{(n+1)} \|_\infty, \qquad (*)$$

where $A_{jn}(\vec{x}_0, Q)$ are continuous functions that depend only upon $n$, $Q$, have a simple/concrete form and give the optimal constants. Here $f_j^{(n+1)}$ stand for any partial derivative of order $n+1$. That is, $(*)$ is a sharp inequality and many times it is even attained. Several different cases of $(*)$ are treated here regarding different types of $n \in \mathbf{Z}_+$, $\mathbf{R}^k$, $Q$ and $f$. Here we follow the treatment [18].

## 23.1 General Results

The first result follows:

**THEOREM 23.1**
Let $f \in C^1(\prod_{i=1}^{k}[a_i, b_i])$, where $a_i < b_i$; $a_i, b_i \in \mathbf{R}$, $i = 1, \ldots, k$, and let

$\vec{x}_0 := (x_{01}, \ldots, x_{0k}) \in \prod_{i=1}^{k}[a_i, b_i]$ be fixed. Then it holds

$$\left| \frac{1}{\prod_{i=1}^{k}(b_i - a_i)} \int_{a_1}^{b_1} \cdots \int_{a_i}^{b_i} \cdots \int_{a_k}^{b_k} f(z_1, \ldots, z_k) dz_1 \cdots dz_k - f(\vec{x}_0) \right|$$

$$\leq \sum_{i=1}^{k} \left( \frac{(x_{0i} - a_i)^2 + (b_i - x_{0i})^2}{2(b_i - a_i)} \right) \left\| \frac{\partial f}{\partial z_i} \right\|_{\infty}. \qquad (23.1)$$

Inequality (23.1) is sharp, here the optimal function is

$$f^*(z_1, \ldots, z_k) := \sum_{i=1}^{k} |z_i - x_{0i}|^{\alpha_i}, \quad \alpha_i > 1. \qquad (23.2)$$

(23.1) generalizes the classical Ostrowski inequality to multidimension, see Ostrowski [136] and Anastassiou [15].

**PROOF** Set $\vec{z} := (z_1, \ldots, z_k)$. Consider $g_{\vec{z}}(t) := f(\vec{x}_0 + t(\vec{x} - \vec{x}_0))$, $t \geq 0$. Note that $g_{\vec{z}}(0) = f(\vec{x}_0)$, $g_{\vec{z}}(1) = f(\vec{z})$. Hence

$$f(\vec{z}) - f(\vec{x}_0) = g_{\vec{z}}(1) - g_{\vec{z}}(0) = g'_{\vec{z}}(\xi)(1 - 0) = g'_{\vec{z}}(\xi), \quad \text{where } \xi \in (0, 1).$$

I.e.,

$$f(\vec{z}) - f(\vec{x}_0) = \sum_{i=1}^{k}(z_i - x_{0i}) \frac{\partial f}{\partial z_i}(\vec{x}_0 + \xi(\vec{z} - \vec{x}_0)).$$

Thus

$$|f(\vec{z}) - f(\vec{x}_0)| \leq \sum_{i=1}^{k} |z_i - x_{0i}| \cdot \left| \frac{\partial f}{\partial z_i}(\vec{x}_0 + \xi(\vec{z} - \vec{x}_0)) \right|$$

$$\leq \sum_{i=1}^{k} |z_i - x_{0i}| \cdot \left\| \frac{\partial f}{\partial z_i} \right\|_{\infty}. \qquad (23.3)$$

Next we see that

$$\left| \frac{1}{\prod_{i=1}^{k}(b_i - a_i)} \int_{a_1}^{b_1} \cdots \int_{a_i}^{b_i} \cdots \int_{a_k}^{b_k} f(z_1, \ldots, z_k) dz_1 \cdots dz_k - f(\vec{x}_0) \right|$$

$$= \frac{1}{\prod_{i=1}^{k}(b_i - a_i)} \left| \int_{a_1}^{b_1} \cdots \int_{a_i}^{b_i} \cdots \int_{a_k}^{b_k} (f(\vec{z}) - f(\vec{x}_0)) d\vec{z} \right|$$

$$\leq \frac{1}{\prod_{i=1}^{k}(b_i - a_i)} \int_{a_1}^{b_1} \cdots \int_{a_i}^{b_i} \cdots \int_{a_k}^{b_k} |f(\vec{z}) - f(\vec{x}_0)| d\vec{z} \quad \text{(by (23.3))}$$

$$\leq \frac{1}{\prod_{i=1}^{k}(b_i - a_i)} \int_{a_1}^{b_1} \cdots \int_{a_i}^{b_i} \cdots \int_{a_k}^{b_k}$$

$$\cdot \left( \sum_{i=1}^{k} |z_i - x_{0i}| \left\| \frac{\partial f}{\partial z_i} \right\|_{\infty} \right) dz_1 \cdots dz_k = \frac{1}{\prod_{i=1}^{k}(b_i - a_i)}$$

$$\cdot \left[ \sum_{i=1}^{k} \left( \int_{a_1}^{b_1} \cdots \int_{a_i}^{b_i} \cdots \int_{a_k}^{b_k} |z_i - x_{0i}| dz_1 \cdots dz_k \right) \left\| \frac{\partial f}{\partial z_i} \right\|_{\infty} \right] =: *.$$

Here notice that

$$\int_{a_i}^{b_i} |z_i - x_{0i}| dz_i = \frac{(x_{0i} - a_i)^2 + (b_i - x_{0i})^2}{2}, \quad i = 1, \ldots, k. \quad (23.4)$$

Therefore, by (23.4) we obtain

$$* = \frac{1}{\prod_{j=1}^{k}(b_j - a_j)}$$

$$\cdot \left[ \sum_{i=1}^{k} \left\| \frac{\partial f}{\partial z_i} \right\|_{\infty} \left( \frac{(x_{0i} - a_i)^2 + (b_i - x_{0i})^2}{2} \right) \left( \prod_{\substack{j=1 \\ j \neq i}}^{k} (b_j - a_j) \right) \right]$$

$$= \sum_{i=1}^{k} \left( \frac{(x_{0i} - a_i)^2 + (b_i - x_{0i})^2}{2 \cdot (b_i - a_i)} \right) \left\| \frac{\partial f}{\partial z_i} \right\|_{\infty},$$

so that we establish inequality (23.1).

In the following we prove the sharpness of (23.1): Notice that $f^*(\vec{x}_0) = 0$ and

$$\frac{\partial f^*}{\partial z_i}(\vec{z}) = \alpha_i |z_i - x_{0i}|^{\alpha_i - 1} \operatorname{sgn}(z_i - x_{0i}), \quad \alpha_i > 1.$$

In particular we find

$$\left|\frac{\partial f^*(\vec{x})}{\partial z_i}\right| = \alpha_i |z_i - x_{0i}|^{\alpha_i - 1},$$

and $(a_i \leq z_i \leq b_i)$

$$\left\|\frac{\partial f^*}{\partial z_i}\right\|_\infty = \alpha_i (\max(b_i - x_{0i}, x_{0i} - a_i))^{\alpha_i - 1}.$$

Consequently, we observe

$$\text{R.H.S.(23.1)} = \sum_{i=1}^{k} \left(\frac{(x_{0i} - a_i)^2 + (b_i - x_{0i})^2}{2 \cdot (b_i - a_i)}\right) \left\|\frac{\partial f^*}{\partial z_i}\right\|_\infty$$

$$= \sum_{i=1}^{k} \left(\frac{(x_{0i} - a_i)^2 + (b_i - x_{0i})^2}{2(b_i - a_i)}\right) \alpha_i (\max(b_i - x_{0i}, x_{0i} - a_i))^{\alpha_i - 1},$$

and

$$\lim_{\substack{\alpha_i \to 1 \\ i=1,\ldots,k}} \text{R.H.S.(23.1)} = \sum_{i=1}^{k} \left(\frac{(x_{0i} - a_i)^2 + (b_i - x_{0i})^2}{2(b_i - a_i)}\right). \qquad (23.5)$$

Moreover, we get that

$$\text{L.H.S.(23.1)} = \frac{1}{\prod_{i=1}^{k}(b_i - a_i)} \int_{a_1}^{b_1} \cdots \int_{a_i}^{b_i} \cdots \int_{a_k}^{b_k} \left(\sum_{i=1}^{k} |z_i - x_{0i}|^{\alpha_i}\right) dz_1 \cdots dz_k$$

$$= \frac{1}{\prod_{i=1}^{k}(b_i - a_i)} \left[\sum_{i=1}^{k} \left(\int_{a_1}^{b_1} \cdots \int_{a_i}^{b_i} \cdots \int_{a_k}^{b_k} |z_i - x_{0i}|^{\alpha_i} dz_1 \cdots dz_k\right)\right]$$

$$= \frac{1}{\prod_{i=1}^{k}(b_i - a_i)} \left[\sum_{i=1}^{k} \left(\frac{(x_{0i} - a_i)^{\alpha_i + 1} + (b_i - x_{0i})^{\alpha_i + 1}}{\alpha_i + 1}\right) \left(\prod_{\substack{j=1 \\ j \neq i}}^{k}(b_j - a_j)\right)\right]$$

$$= \sum_{i=1}^{k} \left(\frac{(x_{0i} - a_i)^{\alpha_i + 1} + (b_i - x_{0i})^{\alpha_i + 1}}{(\alpha_i + 1)(b_i - a_i)}\right),$$

and

$$\lim_{\substack{\alpha_i \to 1 \\ i=1,\ldots,k}} \text{L.H.S.}(23.1) = \sum_{i=1}^{k} \left( \frac{(x_{0i} - a_i)^2 + (b_i - x_{0i})^2}{2(b_i - a_i)} \right). \quad (23.6)$$

At the end from (23.5) and (23.6) we obtain that

$$\lim_{\substack{\alpha_i \to 1 \\ i=1,\ldots,k}} \text{L.H.S.}(23.1) = \lim_{\substack{\alpha_i \to 1 \\ i=1,\ldots,k}} \text{R.H.S.}(23.1),$$

proving that inequality (23.1) is sharp. ∎

Regarding higher order derivatives we give the next results:

### THEOREM 23.2
Let $Q$ be a compact and convex subset of $\mathbf{R}^k$, $k \geq 1$. Let $f \in C^{n+1}(Q)$, $n \in \mathbf{N}$ and $\vec{x}_0 \in Q$ be fixed such that all partial derivatives $f_\alpha := \frac{\partial^\alpha f}{\partial z^\alpha}$, where $\alpha = (\alpha_1, \ldots, \alpha_k)$, $\alpha_i \in \mathbf{Z}^+$, $i = 1, \ldots, k$, $|\alpha| = \sum_{i=1}^{k} \alpha_i = j$, $j = 1, \ldots, n$ fulfill $f_\alpha(\vec{x}_0) = 0$. Then it holds

$$\left| \frac{1}{\text{Vol}(Q)} \int_Q f(\vec{z}) \, d\vec{z} - f(\vec{x}_0) \right|$$

$$\leq \frac{D_{n+1}(f)}{(n+1)! \text{Vol}(Q)} \int_Q (\|\vec{z} - \vec{x}_0\|_{\ell_1})^{n+1} \, d\vec{z}, \quad (23.7)$$

where

$$D_{n+1}(f) := \max_{\alpha : |\alpha| = n+1} \|f_\alpha\|_\infty \quad (23.8)$$

and

$$\|\vec{z} - \vec{x}_0\|_{\ell_1} := \sum_{i=1}^{k} |z_i - x_{0i}|. \quad (23.9)$$

**PROOF** Take $g_{\vec{z}}(t) := f(\vec{x}_0 + t(\vec{z} - \vec{x}_0))$, $0 \leq t \leq 1$. Notice that $g_{\vec{z}}(0) = f(\vec{x}_0)$ and $g_{\vec{z}}(1) = f(\vec{z})$. The $j$th derivative of $g_{\vec{z}}(t)$ is given from

$$g_{\vec{z}}^{(j)}(t) = \left[ \left( \sum_{i=1}^{k} (z_i - x_{0i}) \frac{\partial}{\partial z_i} \right)^j f \right]$$

$$\cdot(x_{01} + t(z_1 - x_{01}), \ldots, x_{0k} + t(z_k - x_{0k}))$$

and

$$g_{\vec{z}}^{(j)}(0) = \left[\left(\sum_{i=1}^{k}(z_i - x_{0i})\frac{\partial}{\partial z_i}\right)^j f\right](\vec{x}_0),$$

for $j = 1, \ldots, n+1$.

Let $f_\alpha$ be a partial derivative of $f \in C^{n+1}(Q)$. Because by assumption of the theorem we have $f_\alpha(\vec{x}_0) = 0$ for all $\alpha : |\alpha| = j$, $j = 1, \ldots, n$, we find that

$$g_{\vec{z}}^{(j)}(0) = 0, \quad j = 1, \ldots, n.$$

Hence by Taylor's theorem we see that

$$f(\vec{z}) - f(\vec{x}_0) = \sum_{j=1}^{n} \frac{g_{\vec{z}}^{(j)}(0)}{j!} + \mathcal{R}_n(\vec{z}, 0) = \mathcal{R}_n(\vec{z}, 0), \qquad (23.10)$$

where

$$\mathcal{R}_n(\vec{z}, 0) := \int_0^1 \left(\int_0^{t_1} \cdots \right.$$

$$\left.\left(\int_0^{t_{n-1}} (g_{\vec{z}}^{(n)}(t_n) - g_{\vec{z}}^{(n)}(0))dt_n\right) \cdots \right) dt_1. \qquad (23.11)$$

Therefore,

$$\mathcal{R}_n(\vec{z}, 0) = \int_0^1 \left(\int_0^{t_1} \cdots \left(\int_0^{t_{n-1}} g_{\vec{z}}^{(n+1)}(\xi(t_n))t_n \, dt_n\right) \cdots \right) dt_1,$$

by the mean value theorem applied on $g_{\vec{z}}^{(n)}$ over $(0, t_n)$. Moreover, we get

$$|\mathcal{R}_n(\vec{z}, 0)| \leq \|g_{\vec{z}}^{(n+1)}\|_\infty \int_0^1 \int_0^{t_1}$$

$$\cdots \left(\int_0^{t_{n-1}} t_n \, dt_n\right) \cdots dt_1 = \frac{\|g_{\vec{z}}^{(n+1)}\|_\infty}{(n+1)!}. \qquad (23.12)$$

However, there exists a $t_0 \in [0,1]$ such that $\|g_{\vec{z}}^{(n+1)}\|_\infty = |g_{\vec{z}}^{(n+1)}(t_0)|$. That is,

$$\|g_{\vec{z}}^{(n+1)}\|_\infty = \left|\left[\left(\sum_{i=1}^k (z_i - x_{0i})\frac{\partial}{\partial z_i}\right)^{n+1} f\right](\vec{x}_0 + t_0(\vec{z} - \vec{x}_0))\right|$$

$$\leq \left[\left(\sum_{i=1}^k |z_i - x_{0i}| \cdot \left|\frac{\partial}{\partial z_i}\right|\right)^{n+1} f\right](\vec{x}_0 + t_0(\vec{z} - \vec{x}_0)).$$

I.e.,

$$\|g_{\vec{z}}^{(n+1)}\|_\infty \leq \left[\left(\sum_{i=1}^k |z_i - x_{0i}| \cdot \left\|\frac{\partial}{\partial z_i}\right\|_\infty\right)^{n+1} f\right]. \qquad (23.13)$$

Hence from (23.10), (23.11), (23.12), and (23.13) we see that

$$|f(\vec{z}) - f(\vec{x}_0)| = |\mathcal{R}_n(\vec{z}, 0)|$$

$$\leq \frac{\left[\left(\sum_{i=1}^k |z_i - x_{0i}| \cdot \left\|\frac{\partial}{\partial z_i}\right\|_\infty\right)^{n+1} f\right]}{(n+1)!}. \qquad (23.14)$$

In the following we observe that

$$\left|\frac{1}{\text{Vol}(Q)} \int_Q f(\vec{z}) d\vec{z} - f(\vec{x}_0)\right|$$

$$= \frac{1}{\text{Vol}(Q)} \left|\int_Q (f(\vec{z}) - f(\vec{x}_0)) d\vec{z}\right|$$

$$\leq \frac{1}{\text{Vol}(Q)} \int_Q |f(\vec{z}) - f(\vec{x}_0)| d\vec{z} \quad \text{(by (23.14))}$$

$$\leq \frac{1}{(n+1)!\text{Vol}(Q)} \int_Q \left(\left(\sum_{i=1}^k |z_i - x_{0i}| \cdot \left\|\frac{\partial}{\partial z_i}\right\|_\infty\right)^{n+1} f\right) d\vec{z}$$

$$\leq \frac{D_{n+1}(f)}{(n+1)!\text{Vol}(Q)} \int_Q (\|\vec{z} - \vec{x}_0\|_{\ell_1})^{n+1} d\vec{z}. \qquad (23.15)$$

This establishes inequality (23.7). ■

**COROLLARY 23.1**
Under the assumptions of Theorem 23.2 we find that

$$\left| \frac{1}{\text{Vol}(Q)} \int_Q f(\vec{z}) d\vec{z} - f(\vec{x}_0) \right|$$

$$\leq \frac{1}{(n+1)!\text{Vol}(Q)} \int_Q \left[ \left( \sum_{i=1}^k |z_i - x_{0i}| \cdot \left\| \frac{\partial}{\partial z_i} \right\|_\infty \right)^{n+1} f \right] d\vec{z}. \quad (23.16)$$

Furthermore, (23.16) is sharp: when $n$ is odd it is attained by

$$f^*(z_1, \ldots, z_k) := \sum_{i=1}^k (z_i - x_{0i})^{n+1}, \quad (23.17)$$

while when $n$ is even the optimal function is

$$\tilde{f}(z_1, \ldots, z_k) := \sum_{i=1}^k |z_i - x_{0i}|^{n+\alpha_i}, \quad \alpha_i > 1. \quad (23.18)$$

**PROOF** Inequality (23.16) comes directly from (23.15).
Next we prove the sharpness of (23.16).
i) When $n$ is odd: Notice that $f^*(\vec{x}_0) = 0$ and

$$\left\| \frac{\partial^{n+1} f^*}{\partial z_i^{n+1}} \right\|_\infty = (n+1)!,$$

furthermore any mixed partial of $f^*$ equals zero. Thus by plugging $f^*$ into (23.16) we observe that

$$\text{R.H.S.}(23.16) = \frac{1}{(n+1)!\text{Vol}(Q)} \int_Q \left[ \sum_{i=1}^k |z_i - x_{0i}|^{n+1} (n+1)! \right] d\vec{z}$$

$$= \frac{1}{\text{Vol}(Q)} \int_Q \left( \sum_{i=1}^k (z_i - x_{0i})^{n+1} \right) d\vec{z}$$

$$= \frac{1}{\text{Vol}(Q)} \int_Q f^*(\vec{z}) d\vec{z} = \text{L.H.S.}(23.16).$$

proving the sharpness of (23.16) when $n$ is odd.

ii) When $n$ is even: Notice that $\tilde{f}(\vec{x}_0) = 0$ and any mixed partial of $\tilde{f}$ equals zero. Especially we observe that

$$\left|\frac{\partial^{n+1}\tilde{f}(\vec{z})}{\partial z_i^{n+1}}\right| = \left(\prod_{j=0}^{n}(n+\alpha_i - j)\right)|z_i - x_{0i}|^{\alpha_i - 1}, \quad \alpha_i > 1,$$

and

$$\left\|\frac{\partial^{n+1}\tilde{f}}{\partial z_i^{n+1}}\right\|_\infty = \left(\prod_{j=0}^{n}(n+\alpha_i - j)\right)\|z_i - x_{0i}\|_\infty^{\alpha_i - 1}$$

(here $\|z_i - x_{0i}\|_\infty < +\infty$), all $i = 1, \ldots, k$. Hence by plugging $\tilde{f}$ into (23.16) we obtain

$$\lim_{\text{all } \alpha_i \to 1} \text{R.H.S.}(23.16)$$

$$= \frac{1}{(n+1)!\text{Vol}(Q)} \lim_{\text{all } \alpha_i \to 1} \int_Q \left[\sum_{i=1}^{k} |z_i - x_{0i}|^{n+1}\right.$$

$$\left.\cdot \left(\prod_{j=0}^{n}(n+\alpha_i - j)\right)\|z_i - x_{0i}\|_\infty^{\alpha_i - 1}\right] d\vec{z}$$

$$= \frac{1}{\text{Vol}(Q)} \int_Q \left(\sum_{i=1}^{k} |z_i - x_{0i}|^{n+1}\right) d\vec{z}. \tag{23.19}$$

Furthermore,

$$\lim_{\text{all } \alpha_i \to 1} \text{L.H.S.}(23.16)$$

$$= \frac{1}{\text{Vol}(Q)} \lim_{\text{all } \alpha_i \to 1} \int_Q \left(\sum_{i=1}^{k} |z_i - x_{0i}|^{n+\alpha_i}\right) d\vec{z}$$

$$= \frac{1}{\text{Vol}(Q)} \int_Q \left( \sum_{i=1}^k |z_i - x_{0i}|^{n+1} \right) d\vec{z}. \tag{23.20}$$

By (23.19) and (23.20) we find that

$$\lim_{\text{all } \alpha_i \to 1} \text{R.H.S.}(23.16) = \lim_{\text{all } \alpha_i \to 1} \text{L.H.S.}(23.16),$$

proving the sharpness of (23.16) when $n$ is even. ∎

When $f$ is specialized we obtain an explicit upper bound.

### COROLLARY 23.2
*Under the assumptions of Theorem 23.2, consider only the particular functions $f(\vec{z}) = \sum_{i=1}^k \phi_i(z_i)$, where $\phi_i \in C^{n+1}(pr_i(Q))$. Then it holds*

$$\left| \frac{1}{\text{Vol}(Q)} \int_Q f(\vec{z}) d\vec{z} - f(\vec{x}_0) \right|$$

$$\leq \frac{1}{(n+1)! \text{Vol}(Q)} \sum_{i=1}^k \left( \int_Q |z_i - x_{0i}|^{n+1} d\vec{z} \right) \|\phi_i^{(n+1)}\|_\infty. \tag{23.21}$$

*Inequality (23.21) is sharp exactly in the same manner as inequality (23.16).*

**PROOF** From Corollary 23.1 and by noticing that all mixed partials of $f$ are zero and $\frac{\partial^{n+1} f}{\partial z_i^{n+1}} = \phi_i^{(n+1)}$. Here notice that we have $\phi_i^{(j)}(x_{0i}) = 0$, all $j = 1, \ldots, n$ and for all $i = 1, \ldots, k$; where $\vec{x}_0 := (x_{01}, \ldots, x_{0k})$. ∎

In the following we elaborate on inequality (23.16) when $Q = \prod_{i=1}^k [a_i, b_i]$ with $k = 2, 3$ and $a_i < b_i$, $a_i, b_i \in \mathbf{R}$. Especially we simplify its right-hand side. We obtain

### PROPOSITION 23.1
*Let $f \in C^{n+1}([a_1, b_1] \times [a_2, b_2])$, $n \in \mathbf{N}$, where $a_1 < b_1$, $a_2 < b_2$; $a_1, a_2, b_1, b_2 \in \mathbf{R}$ and let $\vec{x}_0 = (x_{01}, x_{02}) \in [a_1, b_1] \times [a_2, b_2]$ be fixed. We suppose here that all partial derivatives $f_\alpha := \frac{\partial^\alpha f}{\partial z^\alpha}$, where $\alpha = (\alpha_1, \alpha_2)$, $\alpha_1, \alpha_2 \in \mathbf{Z}^+$, $|\alpha| = \alpha_1 + \alpha_2 = j$, $j = 1, \ldots, n$ fulfill $f_\alpha(\vec{x}_0) = 0$. Then it*

## 23. Multidimensional Ostrowski-Type Inequalities

holds

$$\left| \frac{1}{(b_1 - a_1)(b_2 - a_2)} \int_{a_1}^{b_1} \int_{a_2}^{b_2} f(z_1, z_2) \, dz_1 dz_2 - f(\vec{x}_0) \right|$$

$$\leq \sum_{\ell=0}^{n+1} \left\{ \frac{[(x_{01} - a_1)^{n+2-\ell} + (b_1 - x_{01})^{n+2-\ell}][(x_{02} - a_2)^{\ell+1} + (b_2 - x_{02})^{\ell+1}]}{(n+2-\ell)!(\ell+1)!(b_1 - a_1)(b_2 - a_2)} \right\}$$

$$\cdot \left\| \frac{\partial^{n+1} f}{\partial z_1^{n+1-\ell} \partial z_2^{\ell}} \right\|_\infty. \tag{23.22}$$

*Inequality (23.22) is sharp, exactly the same manner as inequality (23.16).*

**PROOF** It is an application of Corollary 23.1, for $k = 2$ and $Q := [a_1, b_1] \times [a_2, b_2]$. Working on the right-hand side of (23.16) we observe the following: By the binomial theorem we obtain

$$\left( |z_1 - x_{01}| \cdot \left\| \frac{\partial}{\partial z_1} \right\|_\infty + |z_2 - x_{02}| \cdot \left\| \frac{\partial}{\partial x_2} \right\|_\infty \right)^{n+1} f \tag{23.23}$$

$$= \sum_{\ell=0}^{n+1} \binom{n+1}{\ell} |z_1 - x_{01}|^{n+1-\ell} |z_2 - x_{02}|^{\ell} \cdot \left\| \frac{\partial^{n+1} f}{\partial z_1^{n+1-\ell} \partial z_2^{\ell}} \right\|_\infty.$$

Moreover

$$\int_{a_1}^{b_1} |z_1 - x_{01}|^{n+1-\ell} dz_1 = \frac{(x_{01} - a_1)^{n+2-\ell} + (b_1 - x_{01})^{n+2-\ell}}{n+2-\ell}$$

and

$$\int_{a_1}^{b_2} |z_2 - x_{02}|^{\ell} dz_2 = \frac{(x_{02} - a_2)^{\ell+1} + (b_2 - x_{02})^{\ell+1}}{\ell + 1}. \tag{23.24}$$

Next integrate (23.23) over $Q$ and use (23.24). Therefore,

$$\text{R.H.S.}(23.16) = \frac{1}{(n+1)!(b_1 - a_1)(b_2 - a_2)} \sum_{\ell=0}^{n+1} \binom{n+1}{\ell}$$

$$\cdot \left[ \frac{((x_{01}-a_1)^{n+2-\ell} + (b_1-x_{01})^{n+2-\ell})((x_{02}-a_2)^{\ell+1} + (b_2-x_{02}))^{\ell+1})}{(n+2-\ell)(\ell+1)} \right]$$

$$\cdot \left\| \frac{\partial^{n+1} f}{\partial z_1^{n+1-\ell} \partial z_2^\ell} \right\|_\infty = \text{R.H.S.}(23.22).$$

∎

**COROLLARY 23.3**

Let $f \in C^2([a_1,b_1] \times [a_2,b_2])$, where $a_1 < b_1$, $a_2 < b_2$; $a_1, a_2, b_1, b_2 \in \mathbf{R}$ and let $\vec{x}_0 = (x_{01}, x_{02}) \in [a_1,b_1] \times [a_2,b_2]$ be fixed. We suppose that $\frac{\partial f}{\partial z_1}(\vec{x}_0) = \frac{\partial f}{\partial z_2}(\vec{x}_0) = 0$. Then it holds

$$\left| \frac{1}{(b_1-a_1)(b_2-a_2)} \int_{a_1}^{b_1} \int_{a_2}^{b_2} f(z_1,z_2) dz_1 dz_2 - f(\vec{x}_0) \right|$$

$$\leq \left( \frac{(x_{01}-a_1)^3 + (b_1-x_{01})^3}{6(b_1-a_1)} \right) \left\| \frac{\partial^2 f}{\partial z_1^2} \right\|_\infty$$

$$+ \left[ \frac{((x_{01}-a_1)^2 + (b_1-x_{01})^2)((x_{02}-a_2)^2 + (b_2-x_{02})^2)}{4(b_1-a_1)(b_2-a_2)} \right]$$

$$\cdot \left\| \frac{\partial^2 f}{\partial z_1 \partial z_2} \right\|_\infty + \left( \frac{(x_{02}-a_2)^3 + (b_2-x_{02})^3}{6(b_2-a_2)} \right) \left\| \frac{\partial^2 f}{\partial z_2^2} \right\|_\infty. \quad (23.25)$$

Inequality (23.25) is sharp; in fact it is attained by

$$f^*(z_1,z_2) = (z_1 - x_{01})^2 + (z_2 - x_{02})^2. \quad (23.26)$$

**PROOF**  By Proposition 23.1, directly, when $n=1$.  ∎

**PROPOSITION 23.2**

Let $f \in C^2(\prod_{i=1}^3 [a_i,b_i])$, where $a_i < b_i$, $i=1,2,3$; $a_i, b_i \in \mathbf{R}$ and let $\vec{x}_0 = (x_{01}, x_{02}, x_{03}) \in \prod_{i=1}^3 [a_i,b_i]$ is fixed. We suppose here that $\frac{\partial f}{\partial z_i}(\vec{x}_0) = 0$; $i=1,2,3$. Then it holds

$$\left| \frac{1}{\prod_{i=1}^3 (b_i-a_i)} \int_{a_1}^{b_1} \int_{a_2}^{b_2} \int_{a_3}^{b_3} f(z_1,z_2,z_3) dz_1 dz_2 dz_3 - f(\vec{x}_0) \right|$$

## 23. Multidimensional Ostrowski-Type Inequalities

$$\leq \left\{ \sum_{i=1}^{3} \left( \frac{(x_{0i} - a_i)^3 + (b_i - x_{0i})^3}{6(b_i - a_i)} \right) \left\| \frac{\partial^2 f}{\partial z_i^2} \right\|_\infty \right.$$

$$+ \sum_{i=1}^{2} \frac{((x_{0i} - a_i)^2 + (b_i - x_{0i})^2)((x_{0,i+1} - a_{i+1})^2 + (b_{i+1} - x_{0,i+1})^2)}{4(b_i - a_i)(b_{i+1} - a_{i+1})}$$

$$\cdot \left\| \frac{\partial^2 f}{\partial z_i \partial z_{i+1}} \right\|_\infty$$

$$+ \frac{((x_{03} - a_3)^2 + (b_3 - x_{03})^2)((x_{01} - a_1)^2 + (b_1 - x_{01})^2)}{4(b_3 - a_3)(b_1 - a_1)}$$

$$\left. \cdot \left\| \frac{\partial^2 f}{\partial z_3 \partial z_1} \right\|_\infty \right\}. \tag{23.27}$$

Inequality (23.27) is sharp and is attained by

$$f^*(z_1, z_2, z_3) = \sum_{i=1}^{3} (z_i - x_{0i})^2. \tag{23.28}$$

**PROOF** This is an application of Corollary 23.1, for $k = 3$, $n = 1$ and $Q := \prod_{i=1}^{3} [a_i, b_i]$. Working on the right-hand side of (23.16) we obtain the following:

$$\left[ \left( \sum_{i=1}^{3} |z_i - x_{0i}| \cdot \left\| \frac{\partial}{\partial z_i} \right\|_\infty \right)^2 f \right]$$

$$= \sum_{i=1}^{3} (z_i - x_{0i})^2 \left\| \frac{\partial^2 f}{\partial z_i^2} \right\|_\infty$$

$$+ 2 \sum_{i=1}^{2} |z_i - x_{0i}| \cdot |z_{i+1} - x_{0,i+1}| \cdot \left\| \frac{\partial^2 f}{\partial z_i \partial z_{i+1}} \right\|_\infty$$

$$+ 2|z_3 - x_{03}| \cdot |z_1 - x_{01}| \cdot \left\| \frac{\partial^2 f}{\partial z_3 \partial z_1} \right\|_\infty. \tag{23.29}$$

Then integrate (23.29) over $Q$ to get (23.27). ∎

# Chapter 24

## General Opial-Type Inequalities for Linear Differential Operators

A full collection of $L_r$ ($r \neq 0$) form very general Opial-type weighted inequalities is presented for a general linear differential operator $L$. These involve its related initial value problem solution $y$, $Ly$, the associated Green's function $H$ and initial conditions point $x_0 \in \mathbf{R}$. An application to establishing uniqueness in solutions of initial value problems is presented at the end. This treatment follows [37].

In 1960, Z. Opial [135] proved the next very important inequality: Let $f(x)$ belong to class $C^1$ on $0 \leq x \leq h$, and fulfill $f(0) = f(h) = 0$, $f(x) > 0$ on $(0, h)$. Then it holds

$$\int_0^h |f(x)f'(x)|dx \leq \frac{h}{4}\int_0^h f'^2(x)dx, \qquad (*)$$

where $h/4$ is the best constant.

Integral inequalities of type $(*)$ have a great interest by themselves and also have many important applications in the theory of ordinary differential equations and boundary value problems, see [2] and [175]. In fact, a tremendous number of articles have been written establishing and generalizing inequalities of the above form $(*)$, to all possible directions, e.g., see again [2] and [137]. Also a very large number of articles have been written in applying $(*)$-type inequalities in proving uniqueness of initial value problems, upper bounds of solutions, and uniqueness of boundary value problems, all for ordinary differential equations, see [2], Chapter 6, and [175].

In [2], pp. 200–204, Agarwal and Pang introduced/established some very general Opial-type inequalities involving generalized $\rho$-derivatives; that is, for the class of differential operators $D_\rho^{(n)}$, which properly contains the class of disconjugate linear operators

$$L := D^{(n)} + \sum_{i=1}^{n} a_i(t) D^{(n-i)}.$$

Then the author in 1998 (see [22]) proved for the first time Opial-type inequalities for general linear differential operators.

This chapter, among others, generalizes [22] to many different directions.

## 24.1 Setting

Here we use the notation from [111], pp. 145–154. Let $I$ be a closed interval of $\mathbf{R}$. Let $\alpha_i(x)$, $i = 0, 1, \ldots, n-1$ ($n \in \mathbf{N}$), $h(x)$ be continuous functions on $I$ and let $L = D^n + \alpha_{n-1}(x) D^{n-1} + \cdots + \alpha_0(x)$ be a fixed linear differential operator on $C^n(I)$. Let $y_1(x), \ldots, y_n(x)$ be a set of linear independent solutions to $Ly = 0$. Here the associated Green's function for $L$ is given by

$$H(x,t) := \begin{vmatrix} y_1(t) \cdots y_n(t) \\ y_1'(t) \cdots y_n'(t) \\ \vdots \\ y_1^{(n-2)}(t) \cdots y_n^{(n-2)}(t) \\ y_1(x) \cdots y_n(x) \end{vmatrix} \bigg/ \begin{vmatrix} y_1(t) \cdots y_n(t) \\ y_1'(t) \cdots y_n'(t) \\ \vdots \\ y_1^{(n-2)}(t) \cdots y_n^{(n-2)}(t) \\ y_1^{(n-1)}(t) \cdots y_n^{(n-1)}(t) \end{vmatrix},$$

which is a continuous function on $I^2$.

Take a fixed $x_0 \in I$; then

$$y(x) = \int_{x_0}^{x} H(x,t) h(t)\, dt, \quad \text{all } x \in I$$

is the unique solution of the initial value problem

$$Ly = h; \quad y^{(i)}(x_0) = 0, \quad i = 0, 1, \ldots, n-1.$$

## 24.2 General Results

The first result follows.

**THEOREM 24.1**
Let $x \geq x_0$; $x_0, x \in I$ and $r > 1$, $\alpha, \beta > 0$, $r > \alpha$ and let continuous functions $p(w) > 0$ and $q(w) \geq 0$ on $I$. Then it holds

$$\int_{x_0}^{x} q(w)|y(w)|^\beta |(Ly)(w)|^\alpha \, dw \leq K \left[ \int_{x_0}^{x} p(w)|(Ly)(w)|^r dw \right]^{(\frac{\alpha+\beta}{r})}. \quad (24.1)$$

Here

$$K := \left(\frac{\alpha}{\alpha+\beta}\right)^{\alpha/r} \cdot \left[ \int_{x_0}^{x} (q^r(w) p^{-\alpha}(w))^{\frac{1}{(r-\alpha)}} (P_1(w))^{\frac{\beta(r-1)}{r-\alpha}} dw \right]^{\frac{(r-\alpha)}{r}}, \quad (24.2)$$

and

$$P_1(w) := \int_{x_0}^{w} (p(t))^{-\frac{1}{(r-1)}} |H(w,t)|^{\frac{r}{r-1}} \, dt.$$

**PROOF** We have

$$|y(w)| \leq \int_{x_0}^{w} |H(w,t)| \, |h(t)| \, dt, \quad \text{all } x_0 \leq w \leq x.$$

In the following by Hölder's inequality with indices $r > 1$ and $\frac{r}{(r-1)}$ we obtain

$$|y(w)| \leq (P_1(w))^{(\frac{r-1}{r})} \left[ \int_{x_0}^{w} p(t)|h(t)|^r \, dt \right]^{1/r}. \quad (24.3)$$

We put

$$z(w) := \int_{x_0}^{w} p(t)|h(t)|^r \, dt, \quad x_0 \leq w \leq x \ (z(x_0) = 0), \quad (24.4)$$

so that

$$z'(w) = p(w)|h(w)|^r. \quad (24.5)$$

Thus for any $\alpha > 0$ it follows that

$$|h(w)|^\alpha = (p(w))^{-\alpha/r} (z'(w))^{\alpha/r}. \quad (24.6)$$

Hence, if $\beta > 0$ we obtain

$$q(w)|y(w)|^\beta |h(w)|^\alpha \le q(w)(P_1(w))^{\frac{\beta(r-1)}{r}} (p(w))^{-\alpha/r}(z(w))^{\beta/r}(z'(w))^{\alpha/r}. \tag{24.7}$$

Notice that $(z(x_0))^{(\alpha+\beta)/\alpha} = 0$. So we can integrate (24.7) over $[x_0, x]$, and apply Hölder's inequality with indices $r/\alpha$ and $r/(r-\alpha)$, $(r > \alpha)$, to find

$$\int_{x_0}^x q(w)|y(w)|^\beta |h(w)|^\alpha \, dw \le K_0 \left[ \int_{x_0}^x (z(w))^{\beta/\alpha}(z'(w)) \, dw \right]^{\alpha/r}, \tag{24.8}$$

where

$$K_0 := \left[ \int_{x_0}^x (q^r(w)p^{-\alpha}(w))^{\frac{1}{(r-\alpha)}} (P_1(w))^{\frac{\beta(r-1)}{r-\alpha}} \, dw \right]^{\frac{(r-\alpha)}{r}}. \tag{24.9}$$

Consequently we get the following result:

$$\int_{x_0}^x q(w)|y(w)|^\beta |h(w)|^\alpha \, dw \le K \left[ \int_{x_0}^x p(w)|h(w)|^r \, dw \right]^{\frac{(\alpha+\beta)}{r}}, \tag{24.10}$$

where

$$K := \left( \frac{\alpha}{\alpha + \beta} \right)^{\alpha/r} K_0.$$

We have proved (24.1). ∎

The counterpart of the previous comes next.

**THEOREM 24.2**
Let $x \le x_0$; $x, x_0 \in I$ and $r > 1$, $\alpha, \beta > 0$, $r > \alpha$. And let continuous functions $p(w) > 0$ and $q(w) \ge 0$ on $I$. Then it holds

$$\int_x^{x_0} q(w)|y(w)|^\beta |(Ly)(w)|^\alpha \, dw \le K \left[ \int_x^{x_0} p(w)|(Ly)(w)|^r \, dw \right]^{(\frac{\alpha+\beta}{r})}. \tag{24.11}$$

Here

$$K := \left( \frac{\alpha}{\alpha+\beta} \right)^{\alpha/r} \left[ \int_x^{x_0} (q^r(w)p^{-\alpha}(w))^{\frac{1}{(r-\alpha)}} (P_1(w))^{\frac{\beta(r-1)}{r-\alpha}} \, dw \right]^{\frac{(r-\alpha)}{r}}, \tag{24.12}$$

## 24. General Opial-Type Inequalities

and
$$P_1(w) := \int_w^{x_0} (p(t))^{-\frac{1}{(r-1)}} |H(w,t)|^{\frac{r}{r-1}} dt.$$

**PROOF** We see that
$$|y(w)| = \left| \int_w^{x_0} H(w,t) h(t) \, dt \right|$$
$$\leq \int_w^{x_0} |H(w,t)| \, |h(t)| \, dt, \quad \text{all } x \leq w \leq x_0.$$

Next by Hölder's inequality with indices $r > 1$ and $\frac{r}{(r-1)}$ we obtain
$$|y(w)| \leq (P_1(w))^{\frac{(r-1)}{r}} \left[ \int_w^{x_0} p(t)|h(t)|^r \, dt \right]^{1/r}. \tag{24.13}$$

We put
$$z(w) := \int_w^{x_0} p(t)|h(t)|^r \, dt, \quad x \leq w \leq x_0 \; (z(x_0) = 0). \tag{24.14}$$

That is
$$-z(w) = \int_{x_0}^w p(t)|h(t)|^r \, dt \leq 0,$$

so that
$$-z'(w) = p(w)|h(w)|^r. \tag{24.15}$$

Thus for any $\alpha > 0$ it follows that
$$|h(w)|^\alpha = (p(w))^{-\alpha/r} (-z'(w))^{\alpha/r}. \tag{24.16}$$

Hence, if $\beta > 0$ we obtain
$$q(w)|y(w)|^\beta |h(w)|^\alpha$$
$$\leq q(w)(P_1(w))^{\frac{\beta(r-1)}{r}} (p(w))^{-\alpha/r} (z(w))^{\beta/r} (-z'(w))^{\alpha/r}. \tag{24.17}$$

Notice that $(z(x_0))^{(\alpha+\beta)/\alpha} = 0$. Then we integrate (24.17) over $[x, x_0]$, and apply Hölder's inequality with indices $r/\alpha$ and $r/(r-\alpha)$, $(r > \alpha)$, to get

$$\int_x^{x_0} q(w)|y(w)|^\beta |h(w)|^\alpha \, dw \le K_0 \left[ \int_x^{x_0} (z(w))^{\beta/\alpha}(-z'(w)) \, dw \right]^{\alpha/r}, \quad (24.18)$$

where

$$K_0 := \left[ \int_x^{x_0} (q^r(w) p^{-\alpha}(w))^{\frac{1}{(r-\alpha)}} (P_1(w))^{\frac{\beta(r-1)}{r-\alpha}} \, dw \right]^{\frac{(r-\alpha)}{r}}. \quad (24.19)$$

Thus we find the next result

$$\int_x^{x_0} q(w)|y(w)|^\beta |h(w)|^\alpha \, dw \le K \left( \int_x^{x_0} p(w)|h(w)|^r \, dw \right)^{\left(\frac{\alpha+\beta}{r}\right)}, \quad (24.20)$$

where $K := \left(\frac{\alpha}{\alpha+\beta}\right)^{\alpha/r} K_0$. We have established (24.11). ∎

Extreme cases follow.

### PROPOSITION 24.1
Let $x \ge x_0$; $x_0, x \in I$, $\alpha, \beta > 0$. And let continuous functions $p(w) > 0$ and $q(w) \ge 0$ on $I$. Then it holds

$$\int_{x_0}^x q(w)|y(w)|^\beta |(Ly)(w)|^\alpha \, dw \quad (24.21)$$

$$\le \left( \int_{x_0}^x q(w) \left( \int_{x_0}^w (p(t))^{-1} |H(w,t)| \, dt \right)^\beta dw \right) \|p\|_\infty^\beta \|Ly\|_\infty^{\alpha+\beta}.$$

**PROOF** We see that $(x_0 \le w \le x)$

$$|y(w)|^\beta \le \left( \int_{x_0}^w |H(w,t)| \, |h(t)| \, dt \right)^\beta$$

$$= \left( \int_{x_0}^w p(t)(p(t))^{-1} |H(w,t)| \, |h(t)| \, dt \right)^\beta$$

## 24. General Opial-Type Inequalities

$$\le \left( \int_{x_0}^{w} (p(t))^{-1} |H(w,t)| \, dt \right)^{\beta} \|p\|_{\infty}^{\beta} \|h\|_{\infty}^{\beta}.$$

So that

$$q(w)|y(w)|^{\beta}|h(w)|^{\alpha} \le q(w) \left( \int_{x_0}^{w} (p(t))^{-1} |H(w,t)| \, dt \right)^{\beta} \|p\|_{\infty}^{\beta} \|h\|_{\infty}^{\alpha+\beta}.$$

Integrating that last inequality over $[x_0, x]$ we get

$$\int_{x_0}^{x} q(w)|y(w)|^{\beta}|h(w)|^{\alpha} \, dw$$

$$\le \left( \int_{x_0}^{x} q(w) \left( \int_{x_0}^{w} (p(t))^{-1} |H(w,t)| \, dt \right)^{\beta} dw \right) \|p\|_{\infty}^{\beta} \|h\|_{\infty}^{\alpha+\beta},$$

proving (24.21). ∎

### PROPOSITION 24.2

Let $x \le x_0$; $x, x_0 \in I$, $\alpha, \beta > 0$. And let continuous functions $p(w) > 0$ and $q(w) \ge 0$ on $I$. Then it holds

$$\int_{x}^{x_0} q(w)|y(w)|^{\beta}|Ly(w)|^{\alpha} \, dw \tag{24.22}$$

$$\le \left( \int_{x}^{x_0} q(w) \left( \int_{w}^{x_0} (p(t))^{-1} |H(w,t)| \, dt \right)^{\beta} dw \right) \|p\|_{\infty}^{\beta} \|Ly\|_{\infty}^{\alpha+\beta}.$$

**PROOF** We see that ($x \le w \le x_0$)

$$|y(w)|^{\beta} = \left| \int_{x_0}^{w} H(w,t)h(t) \, dt \right|^{\beta} = \left| \int_{w}^{x_0} H(w,t)h(t) \, dt \right|^{\beta}$$

$$\le \left( \int_{w}^{x_0} |H(w,t)| \, |h(t)| \, dt \right)^{\beta}$$

$$= \left( \int_{w}^{x_0} p(t)(p(t))^{-1} |H(w,t)| \, |h(t)| \, dt \right)^{\beta}$$

$$\leq \left( \int_w^{x_0} (p(t))^{-1} |H(w,t)| \, dt \right)^\beta \|p\|_\infty^\beta \|h\|_\infty^\beta.$$

So that

$$q(w)|y(w)|^\beta |h(w)|^\alpha \leq q(w) \left( \int_w^{x_0} (p(t))^{-1} |H(w,t)| \, dt \right)^\beta \|p\|_\infty^\beta \|h\|_\infty^{\alpha+\beta}.$$

Integrating the last inequality over $[x, x_0]$ we obtain

$$\int_x^{x_0} q(w)|y(w)|^\beta |h(w)|^\alpha \, dw$$

$$\leq \left( \int_x^{x_0} q(w) \left( \int_w^{x_0} (p(t))^{-1} |H(w,t)| \, dt \right)^\beta dw \right) \|p\|_\infty^\beta \|h\|_\infty^{\alpha+\beta},$$

proving (24.22). ∎

### COROLLARY 24.1

Let $x_0, x \in I$ and $r > 1$, $\alpha, \beta > 0$, $r > \alpha$. And let continuous functions $p(w) > 0$ and $q(w) \geq 0$ on $I$. Then it holds

$$\left| \int_{x_0}^x q(w)|y(w)|^\beta |(Ly)(w)|^\alpha \, dw \right| \leq \tilde{K} \left| \int_{x_0}^x p(w)|(Ly)(w)|^r \, dw \right|^{\left(\frac{\alpha+\beta}{r}\right)}.$$
(24.23)

Here

$$\tilde{K} := \left( \frac{\alpha}{\alpha+\beta} \right)^{\alpha/r} \left| \int_{x_0}^x (q^r(w) p^{-\alpha}(w))^{1/(r-\alpha)} (\tilde{P}_1(w))^{\frac{\beta(r-1)}{r-\alpha}} dw \right|^{\left(\frac{r-\alpha}{r}\right)},$$
(24.24)

and

$$\tilde{P}_1(w) := \left| \int_{x_0}^w (p(t))^{-\frac{1}{(r-1)}} |H(w,t)|^{\frac{r}{r-1}} \, dt \right|.$$

**PROOF** By Theorems 24.1 and 24.2. ∎

In particular, we obtain

### COROLLARY 24.2

Let $x_0, x \in I$, $\alpha, \beta > 0$, $\alpha < 2$. And let continuous functions $p(w) > 0$ and

$q(w) \geq 0$ on $I$. Then it holds

$$\left| \int_{x_0}^{x} q(w) |y(w)|^{\beta} |(Ly)(w)|^{\alpha} \, dw \right| \leq \tilde{K} \left| \int_{x_0}^{x} p(w) [(Ly)(w)]^2 \, dw \right|^{(\frac{\alpha+\beta}{2})}. \tag{24.25}$$

Here

$$\tilde{K} := \left( \frac{\alpha}{\alpha + \beta} \right)^{\alpha/2} \left| \int_{x_0}^{x} (q^2(w) p^{-\alpha}(w))^{\frac{1}{(2-\alpha)}} (\tilde{P}_1(w))^{\frac{\beta}{2-\alpha}} \, dw \right|^{(\frac{2-\alpha}{2})}, \tag{24.26}$$

and

$$\tilde{P}_1(w) := \left| \int_{x_0}^{w} (p(t))^{-1} [H(w,t)]^2 \, dt \right|.$$

**PROOF**  By Corollary 24.1, when $r = 2$.  ∎

Moreover, we have

### COROLLARY 24.3

Let $x_0, x \in I$, $\alpha, \beta > 0$ and continuous functions $p(w) > 0$ and $q(w) \geq 0$ on $I$. Then it holds

$$\left| \int_{x_0}^{x} q(w) |y(w)|^{\beta} |(Ly)(w)|^{\alpha} \, dw \right| \tag{24.27}$$

$$\leq \left( \int_{x_0}^{x} q(w) \left( \int_{x_0}^{w} (p(t))^{-1} |H(w,t)| \, dt \right)^{\beta} dw \right) \|p\|_{\infty}^{\beta} \|Ly\|_{\infty}^{\alpha+\beta}.$$

**PROOF**  By Propositions 24.1 and 24.2.  ∎

In the following we continue this study about several other interesting cases when $h(w)$ is of fixed sign and nowhere zero.

Furthermore, in the next we consider functions $p(w)$, $q(w)$ that are nonnegative and Lebesgue measurable over $I$, points $x_0, x \in I$ and real numbers $\alpha, \beta, r \neq 0$. Suppose that $\left( \frac{\alpha+\beta}{\alpha} \right) > 0$. Here either $x_0 \leq w \leq x$ or $x \leq w \leq x_0$.

We set

$$P_1(w) := \left| \int_{x_0}^{w} (p(t))^{-\frac{1}{r-1}} |H(w,t)|^{\frac{r}{r-1}} \, dt \right|, \tag{24.28}$$

$$z(w) := \left| \int_{x_0}^{w} p(t)|h(t)|^r \, dt \right|, \tag{24.29}$$

$$QY := \left| \int_{x_0}^{x} q(w)|y(w)|^\beta |h(w)|^\alpha \, dw \right|, \tag{24.30}$$

$$PQ := \left| \int_{x_0}^{x} (q^r(w) p^{-\alpha}(w))^{\frac{1}{r-\alpha}} (P_1(w))^{\frac{\beta(r-1)}{(r-\alpha)}} \, dw \right|, \tag{24.31}$$

and
$$P_1 := P_1(x), \quad z := z(x). \tag{24.32}$$

**THEOREM 24.3**
Let $x > x_0$. Suppose that $H(w,t) \geq 0$ for $x_0 \leq t \leq w$, $w \in I$. Also suppose that either $r > 1$, $\beta > 0$, $0 < \alpha < r$, or $r < \alpha < 0$, $\beta < 0$, or $-\alpha < \beta < 0$, $0 < \alpha < r < 1$, and $P_1(w)$ exists for all $w \in [x_0, x]$, $PQ < \infty$, $z < \infty$. Then it holds

$$\int_{x_0}^{x} q(w)|y(w)|^\beta |(Ly)(w)|^\alpha \, dw \leq K \left[ \int_{x_0}^{x} p(w)|(Ly)(w)|^r \, dw \right]^{\frac{(\alpha+\beta)}{r}}, \tag{24.33}$$

where
$$K := \left( \frac{\alpha}{\alpha+\beta} \right)^{\alpha/r} (PQ)^{\frac{(r-\alpha)}{r}}. \tag{24.34}$$

**PROOF** Because $Ly = h$ is of fixed sign it follows that

$$|y(w)| = \int_{x_0}^{w} H(w,t)|h(t)| \, dt =: g(w). \tag{24.35}$$

Now by Hölder's inequality with indices $r$ and $\frac{r}{r-1}$ we get

$$|y(w)| \stackrel{\leq}{\geq} (P_1(w))^{\frac{(r-1)}{r}} (z(w))^{1/r}. \tag{24.36}$$

In (24.36) it is clear that $\leq$ holds if $r > 1$, and $\geq$ holds for $r < 0$ or $0 < r < 1$. Here we notice that

$$z'(w) = p(w)|h(w)|^r, \quad z(x_0) = 0. \tag{24.37}$$

Thus it holds
$$|h(w)|^\alpha = (p(w))^{-\alpha/r} (z'(w))^{\alpha/r}. \tag{24.38}$$

Hence, if $\beta > 0$ we obtain

$$q(w)|y(w)|^\beta |h(w)|^\alpha$$

$$\overset{\leq}{\geq} q(w)(P_1(w))^{\beta(r-1)/r}(p(w))^{-\alpha/r}(z(w))^{\beta/r}(z'(w))^{\alpha/r}, \quad (24.39)$$

where $\leq$ holds if $r > 1$, and $\geq$ holds if $r < 0$ or $0 < r < 1$. On the other hand, if $\beta < 0$ we get that

$$q(w)|y(w)|^\beta |h(w)|^\alpha$$

$$\overset{\geq}{\leq} q(w)(P_1(w))^{\frac{\beta(r-1)}{r}}(p(w))^{-\alpha/r}(z(w))^{\beta/r}(z'(w))^{\alpha/r}, \quad (24.40)$$

where $\geq$ holds if $r > 1$, and $\leq$ holds if $r < 0$ or $0 < r < 1$. Here $(z(x_0))^{(\alpha+\beta)/\alpha} = 0$. Next, we integrate (24.39) and (24.40) over $[x_0, x]$ and apply Hölder's inequality with indices $r/\alpha$ and $r/(r-\alpha)$, to find

$$\int_{x_0}^x q(w)|y(w)|^\beta |h(w)|^\alpha \, dw \overset{\leq}{\geq} K_0 \left[ \int_{x_0}^x (z(w))^{\beta/\alpha}(z'(w)) \, dw \right]^{\alpha/r}, \quad (24.41)$$

where

$$K_0 := (PQ)^{(\frac{r-\alpha}{r})}. \quad (24.42)$$

Consequently, when $\leq$ holds in (24.39) or (24.40) and we require $r/\alpha > 1$ we get (24.33), which is clearly valid for all mentioned cases in the statement of the theorem. ■

The counterpart of the previous result comes next.

## THEOREM 24.4

Let $x > x_0$. Suppose again that $H(w,t) \geq 0$ for $x_0 \leq t \leq w$, $w \in I$. Also suppose that either

$$\begin{cases} \beta > 0, \ 0 < r < \min(\alpha, 1), \text{ or } \alpha < 0 < r < 1, \\ 0 < \beta < -\alpha \text{ and } P_1(w) \text{ exists for all} \\ w \in [x_0, x], \ PQ < \infty, \ QY < \infty; \end{cases} \quad (24.43)$$

or

$$\begin{cases} \beta < 0, \ \alpha < 0, \ r > 1, \text{ or } 1 < r < \alpha, \ -\alpha < \beta < 0, \\ \text{and } P_1(w), \ z(w) \text{ exist for all } w \in [x_0, x], \\ PQ < \infty, \ QY < \infty; \end{cases} \quad (24.44)$$

or

$$\left\{\begin{array}{l} \beta > 0,\ r < 0 < \alpha,\ or\ \alpha < r < 0,\ 0 < \beta < -\alpha, \\ and\ z(w)\ exists\ for\ all\ w \in [x_0, x], \\ PQ < \infty,\ QY < \infty. \end{array}\right\}. \quad (24.45)$$

Then it holds

$$\int_{x_0}^{x} q(w)|y(w)|^{\beta}|(Ly)(w)|^{\alpha}\,dw \geq K \left[\int_{x_0}^{x} p(w)|(Ly)(w)|^r\,dw\right]^{(\frac{\alpha+\beta}{r})}, \quad (24.46)$$

where

$$K := \left(\frac{\alpha}{\alpha+\beta}\right)^{\alpha/r} (PQ)^{(\frac{r-\alpha}{r})}. \quad (24.47)$$

**PROOF**  Clearly inequality (24.46) comes from the Proof of Theorem 24.3, when $\geq$ holds in (24.39) or (24.40) and we require $\frac{r}{\alpha} < 0$ or $0 < \frac{r}{\alpha} < 1$. Moreover, if $r < 0$, then from Hölder's inequality with indices $(1 - r)$ and $\left(\frac{r-1}{r}\right)$, we get

$$P_1(w) \leq \left(\int_{x_0}^{w} H(w,t)|h(t)|\,dt\right)^{\frac{r}{r-1}} (z(w))^{\frac{1}{1-r}}.$$

Thus the existence of $P_1(w)$ follows from that of $z(w)$ and $g(w)$ (see (24.35)). Similarly, for $0 < r < 1$, Hölder's inequality with indices $\frac{1}{r}$ and $\frac{1}{(1-r)}$ produces

$$z(w) \leq \left(\int_{x_0}^{w} H(w,t)|h(t)|\,dt\right)^{r} (P_1(w))^{1-r},$$

and hence the existence of $z(w)$ comes from that of $P_1(w)$ and $g(w)$. With the above remarks inequality (24.46) clearly makes sense in all cases.  ∎

As related results we give

**THEOREM 24.5**
Let $x < x_0$. Suppose that $H(w,t) \leq 0$ for $w \leq t \leq x_0$, $w \in I$. Also suppose that either $r > 1$, $\beta > 0$, $0 < \alpha < r$, or $r < \alpha < 0$, $\beta < 0$, or $-\alpha < \beta < 0$, $0 < \alpha < r < 1$, and $P_1(w)$ exists for all $w \in [x, x_0]$, $PQ < \infty$, $z < \infty$. Then it holds

$$\int_{x}^{x_0} q(w)|y(w)|^{\beta}|(Ly)(w)|^{\alpha}\,dw \leq K \left[\int_{x}^{x_0} p(w)|(Ly)(w)|^r\,dw\right]^{(\frac{\alpha+\beta}{r})}, \quad (24.48)$$

## 24. General Opial-Type Inequalities

where
$$K := \left(\frac{\alpha}{\alpha+\beta}\right)^{\alpha/r} (PQ)^{\left(\frac{r-\alpha}{r}\right)}. \tag{24.49}$$

**PROOF** Here for $x \le w \le x_0$ we observe that

$$|y(w)| = \left|\int_{x_0}^{w} H(w,t)h(t)\,dt\right| = \left|\int_{w}^{x_0} H(w,t)h(t)\,dt\right|$$

$$= \left|\int_{w}^{x_0} (-H(w,t))h(t)\,dt\right|$$

$$= \int_{w}^{x_0} (-H(w,t))|h(t)|\,dt =: g(w). \tag{24.50}$$

Next by Hölder's inequality with indices $r$ and $\frac{r}{r-1}$ we obtain

$$|y(w)| \underset{\ge}{\le} (P_1(w))^{\frac{r-1}{r}} (z(w))^{1/r}. \tag{24.51}$$

In (24.51) it is clear that $\le$ holds if $r > 1$, and $\ge$ holds for $r < 0$ or $0 < r < 1$. Here

$$-z'(w) = p(w)|h(w)|^r, \quad z(x_0) = 0. \tag{24.52}$$

Thus it holds
$$|h(w)|^\alpha = (p(w))^{-\alpha/r} (-z'(w))^{\alpha/r}. \tag{24.53}$$

Therefore, if $\beta > 0$ we get

$$g(w)|y(w)|^\beta |h(w)|^\alpha \tag{24.54}$$

$$\underset{\ge}{\le} q(w)(P_1(w))^{\frac{\beta(r-1)}{r}} (p(w))^{-\alpha/r} (z(w))^{\beta/r} (-z'(w))^{\alpha/r},$$

where $\le$ holds if $r > 1$, and $\ge$ holds if $r < 0$ or $0 < r < 1$. However, if $\beta < 0$ we obtain

$$q(w)|y(w)|^\beta |h(w)|^\alpha \tag{24.55}$$

$$\underset{\le}{\ge} q(w)(P_1(w))^{\frac{\beta(r-1)}{r}} (p(w))^{-\alpha/r} (z(w))^{\beta/r} (-z'(w))^{\alpha/r},$$

where $\geq$ holds if $r > 1$, and $\leq$ holds if $r < 0$ or $0 < r < 1$. Here $(z(x_0))^{(\frac{\alpha+\beta}{\alpha})} = 0$. Next, we integrate (24.54) and (24.55) over $[x, x_0]$ and apply Hölder's inequality with indices $r/\alpha$ and $\frac{r}{(r-\alpha)}$, to get

$$\int_x^{x_0} q(w)|y(w)|^\beta |h(w)|^\alpha \, dw \gtreqless K_0 \left[ \int_x^{x_0} (z(w))^{\beta/\alpha}(-z'(w)) \, dw \right]^{\alpha/r}$$

$$= \left( \frac{\alpha}{\alpha+\beta} \right)^{\alpha/r} K_0 (z(x))^{(\alpha+\beta)/r}, \quad (24.56)$$

where

$$K_0 := (PQ)^{(\frac{r-\alpha}{r})}. \quad (24.57)$$

Hence, when $\leq$ holds in (24.54) or (24.55) and we require $\frac{r}{\alpha} > 1$ we obtain (24.48), which is clearly valid for all mentioned cases in the statement of the theorem. ■

The counterpart of the previous theorem follows.

### THEOREM 24.6
Let $x < x_0$. Suppose that $H(w,t) \leq 0$ for $w \leq t \leq x_0$, $w \in I$. Also suppose that either

$$\begin{cases} \beta > 0, \ 0 < r < \min(\alpha, 1), \ \text{or } \alpha < 0 < r < 1, \\ 0 < \beta < -\alpha \text{ and } P_1(w) \text{ exists for all} \\ w \in [x, x_0], \ PQ < \infty, \ QY < \infty; \end{cases} \quad (24.58)$$

or

$$\begin{cases} \beta < 0, \ \alpha < 0, \ r > 1, \ \text{or } 1 < r < \alpha, \ -\alpha < \beta < 0, \\ \text{and } P_1(w), \ z(w) \text{ exist for all } w \in [x, x_0], \\ PQ < \infty, \ QY < \infty; \end{cases} \quad (24.59)$$

or

$$\begin{cases} \beta > 0, \ r < 0 < \alpha, \ \text{or } \alpha < r < 0, \ 0 < \beta < -\alpha, \\ \text{and } z(w) \text{ exists for all } w \in [x, x_0], \\ PQ < \infty, \ QY < \infty. \end{cases} \quad (24.60)$$

Then it holds

$$\int_x^{x_0} q(w)|y(w)|^\beta |(Ly)(w)|^\alpha \, dw \geq K \left[ \int_x^{x_0} p(w)|(Ly)(w)|^r \, dw \right]^{(\frac{\alpha+\beta}{r})}, \quad (24.61)$$

where
$$K := \left(\frac{\alpha}{\alpha+\beta}\right)^{\alpha/r} (PQ)^{(\frac{r-\alpha}{r})}. \qquad (24.62)$$

**PROOF** Evidently inequality (24.61) comes from the Proof of Theorem 24.5, when $\geq$ holds in (24.54) or (24.55) and we require $\frac{r}{\alpha} < 0$ or $0 < \frac{r}{\alpha} < 1$. Moreover if $r < 0$, then from Hölder's inequality with indices $(1-r)$ and $\left(\frac{r-1}{r}\right)$, we have

$$P_1(w) \leq \left(\int_w^{x_0} |H(w,t)||h(t)|\,dt\right)^{\frac{r}{r-1}} (z(w))^{\frac{1}{1-r}}.$$

Thus the existence of $P_1(w)$ comes from the one of $z(w)$ and $g(w)$ (see (24.50)). Similarly, for $0 < r < 1$, by Hölder's inequality with indices $\frac{1}{r}$ and $\frac{1}{1-r}$ we get

$$z(w) \leq \left(\int_w^{x_0} |H(w,t)||h(t)|\,dt\right)^{r} (P_1(w))^{1-r},$$

and therefore the existence of $z(w)$ comes from the one of $P_1(w)$ and $g(w)$. Having made all the above remarks, inequality (24.61) is true in all cases. ∎

In the next $Ly = h$ still is of fixed sign and nowhere zero and $\left(\frac{\alpha+\beta}{\alpha}\right) > 0$. So putting things together we get

**COROLLARY 24.4**
Let $x_0, x \in I$ such that $x \neq x_0$. Suppose that

$$(w - x_0)H(w,t) \geq 0,$$

for all $t$ between $x_0$, $w \in I$. Also suppose the rest of common assumptions of Theorems 24.3 and 24.5. Then it holds

$$\left|\int_{x_0}^{x} q(w)|y(w)|^{\beta}|(Ly)(w)|^{\alpha}\,dw\right| \leq K \left[\left|\int_{x_0}^{x} p(w)|(Ly)(w)|^{r}\,dw\right|\right]^{(\frac{\alpha+\beta}{r})}, \qquad (24.63)$$

where
$$K := \left(\frac{\alpha}{\alpha+\beta}\right)^{\alpha/r} (PQ)^{(\frac{r-\alpha}{r})}. \qquad (24.64)$$

**PROOF** From Theorems 24.3 and 24.5. ∎

**COROLLARY 24.5**
Let $x_0, x \in I$ such that $x \neq x_0$. Suppose that

$$(w - x_0)H(w, t) \geq 0,$$

for all $t$ between $x_0, w \in I$. Also suppose the rest of common assumptions of Theorems 24.4 and 24.6. Then it holds

$$\left| \int_{x_0}^{x} q(w)|y(w)|^\beta |(Ly)(w)|^\alpha \, dw \right| \geq K \left[ \left| \int_{x_0}^{x} p(w) |(Ly)(w)|^r \, dw \right| \right]^{(\frac{\alpha+\beta}{r})}, \tag{24.65}$$

where

$$K := \left( \frac{\alpha}{\alpha + \beta} \right)^{\alpha/r} (PQ)^{(\frac{r-\alpha}{r})}. \tag{24.66}$$

**PROOF** By Theorems 24.4 and 24.6. ∎

## 24.3 An Application to Differential Equations

We use the next corollary of Theorem 24.1.

**COROLLARY 24.6**
Let $x_0, x \in I$ such that $x \geq x_0$. Then it holds

$$\int_{x_0}^{x} |y(w)| \, |(Ly)(w)| dw \leq 2^{-1/2} \left( \int_{x_0}^{x} \left( \int_{x_0}^{w} (H(w,t))^2 dt \right) dw \right)^{1/2}$$

$$\cdot \left( \int_{x_0}^{x} ((Ly)(w))^2 dw \right). \tag{24.67}$$

So we will use (24.67) in establishing uniqueness of the solution of the following general initial value problem (IVP):

$$(Ly)'(w) = a(w) \cdot y(w), \quad w \in I := [x_0, b], \ 0 \neq a \in C(I), \tag{24.68}$$

## 24. General Opial-Type Inequalities

such that $y^{(i)}(x_0) = \gamma_i$, $i = 0, 1, \ldots, n$. Here $\alpha_i \in C^1(I)$, $i = 0, 1, \ldots, n-1$. Suppose that there are two solutions $y_1, y_2$ to IVP (24.68):

$$(Ly_1)' = ay_1, \quad (Ly_2)' = ay_2.$$

Hence for $z := y_1 - y_2$ we find that

$$(Lz)' = az \quad \text{with } z^{(i)}(x_0) = 0, \quad i = 0, 1, \ldots, n.$$

That is $Lz(x_0) = 0$. Moreover

$$(Lz)(Lz)' = az(Lz),$$

and

$$\int_{x_0}^{x} (Lz)(Lz)' dw = \int_{x_0}^{x} az(Lz) dw.$$

Thus

$$\left. \frac{(Lz(w))^2}{2} \right|_{x_0}^{x} = \int_{x_0}^{x} az(Lz) dw.$$

And

$$(Lz(x))^2 = 2 \left| \int_{x_0}^{x} a(w) z(w)(Lz)(w) dw \right| \leq 2 \|a\|_\infty \int_{x_0}^{x} |z(w)| |Lz(w)| dw.$$

Therefore, by (24.67) we obtain

$$((Lz)(x))^2 \leq 2^{1/2} \cdot \|a\|_\infty \cdot \left( \int_{x_0}^{x} \left( \int_{x_0}^{w} (H(w,t))^2 dt \right) dw \right)^{1/2}$$

$$\cdot \left( \int_{x_0}^{x} ((Lz)(w))^2 dw \right), \quad \text{true for all } I \ni x \geq x_0.$$

I.e., there exists $M > 0$ such that

$$((Lz)(x))^2 \leq M \cdot \left( \int_{x_0}^{x} ((Lz)(w))^2 dw \right), \quad \text{valid for all } I \ni x \geq x_0.$$

Consequently, from Gronwall's inequality results that $(Lz)(x) = 0$, all $x \geq x_0$, and $((Lz)(x))' = 0$, all $x \geq x_0$. Moreover,

$$a(w) z(w) = 0, \quad \text{all } w \geq x_0,$$

*and*
$$z(w) = 0, \quad \text{all } w \geq x_0.$$

*At the end we get*
$$y_1(w) = y_2(w), \quad \text{all } I \ni w \geq x_0,$$

*establishing uniqueness.*

# Chapter 25

## $L_p$-Opial-Type Inequalities Engaging Fractional Derivatives of Functions

A large variety of $L_p$ form Opial-type inequalities are presented involving generalized fractional derivatives of functions of different orders. These rely on a generalization of Taylor's formula for generalized fractional derivatives. Many of the exposed results are applied in treating very general fractional differential equations. More precisely, they are used in proving uniqueness and giving upper bounds to solutions of important fractional initial value problems. Other related upper bounds to fractional derivatives are also shown. This treatment follows [28].

## 25.1 Background

Here we follow [60]. Let $g \in C([0,1])$, $n := [\nu]$, $\nu > 0$, and $\alpha := \nu - n$ ($0 < \alpha < 1$). Define

$$(J_\nu g)(x) := \frac{1}{\Gamma(\nu)} \int_0^x (x-t)^{\nu-1} g(t) dt, \quad 0 \le x \le 1, \quad (25.1)$$

the *Riemann–Liouville integral*, where $\Gamma$ is the gamma function. We consider the subspace $C^\nu([0,1])$ of $C^n([0,1])$:

$$C^\nu([0,1]) := \{g \in C^n([0,1]) : J_{1-\alpha} D^n g \in C^1([0,1])\},$$

where $D := \frac{d}{dx}$. So let $g \in C^\nu([0,1])$, we define the $\nu$-*fractional derivative* of $g$ as

$$D^\nu g := D J_{1-\alpha} D^n g. \quad (25.2)$$

539

When $\nu \geq 1$ we obtain the Taylor's formula

$$g(t) = g(0) + g'(0)t + g''(0)\frac{t^2}{2!} + \cdots + g^{(n-1)}(0)\frac{t^{n-1}}{(n-1)!}$$

$$+ (J_\nu D^\nu g)(t), \quad \forall t \in [0,1]. \tag{25.3}$$

When $0 < \nu < 1$ we get

$$g(t) = (J_\nu D^\nu g)(t), \quad \forall t \in [0,1]. \tag{25.4}$$

In the following we carry above notions over to arbitrary $[a,b] \subseteq \mathbf{R}$. Let $x, x_0 \in [a,b]$ such that $x \geq x_0$, $x_0$ is fixed. Let $f \in C([a,b])$ and define

$$(J_\nu^{x_0} f)(x) := \frac{1}{\Gamma(\nu)} \int_{x_0}^{x} (x-t)^{\nu-1} f(t) dt, \quad x_0 \leq x \leq b, \tag{25.5}$$

the *generalized Riemann–Liouville integral*. We consider the subspace $C_{x_0}^\nu([a,b])$ of $C^n([a,b])$:

$$C_{x_0}^\nu([a,b]) := \{f \in C^n([a,b]) : J_{1-\alpha}^{x_0} D^n f \in C^1([x_0,b])\}.$$

Hence, let $f \in C_{x_0}^\nu([a,b])$, we define the *generalized $\nu$-fractional derivative of $f$ over $[x_0, b]$* as

$$D_{x_0}^\nu f := D J_{1-\alpha}^{x_0} f^{(n)} \quad (f^{(n)} := D^n f). \tag{25.6}$$

Notice that

$$(J_{1-\alpha}^{x_0} f^{(n)})(x) = \frac{1}{\Gamma(1-\alpha)} \int_{x_0}^{x} (x-t)^{-\alpha} f^{(n)}(t) dt$$

exists for $f \in C_{x_0}^\nu([a,b])$.

We present the following generalization of Taylor's formula (also see [60]):

**THEOREM 25.1**
Let $f \in C_{x_0}^\nu([a,b])$, $x_0 \in [a,b]$ fixed.
(i) If $\nu \geq 1$ then it holds

$$f(x) = f(x_0) + f'(x_0)(x - x_0) + f''(x_0)\frac{(x-x_0)^2}{2}$$

## 25. $L_p$-Opial-Type Inequalities

$$\cdots + f^{(n-1)}(x_0)\frac{(x-x_0)^{n-1}}{(n-1)!}$$

$$+ (J_\nu^{x_0} D_{x_0}^\nu f)(x), \quad \text{all } x \in [a,b] : x \geq x_0. \tag{25.7}$$

*(ii)* If $0 < \nu < 1$ we have

$$f(x) = (J_\nu^{x_0} D_{x_0}^\nu f)(x), \quad \text{all } x \in [a,b] : x \geq x_0. \tag{25.8}$$

**PROOF** For $x = x_0$, (25.7) holds trivially. Otherwise, suppose that $x \neq x_0$ and define

$$w := \varphi(t) := x_0 + t(x - x_0), \quad 0 \leq t \leq 1.$$

Hence $\varphi(0) = x_0$ and $\varphi(1) = x$. Let $g(t) := f(x_0 + t(x - x_0))$, i.e. $g(0) = f(x_0)$ and $g(1) = f(x)$. Here we observe that $(\nu \geq 1)$

$$g^{(i)}(t) = f^{(i)}(x_0 + t(x - x_0))(x - x_0)^i, \quad i = 0, 1, \ldots, n-1.$$

In particular, $g^{(i)}(0) = f^{(i)}(x_0)(x - x_0)^i$. Moreover, $g \in C^\nu([0,1])$. Consequently from (25.3) we obtain

$$f(x) = g(1) = g(0) + g'(0) + \frac{g''(0)}{2} + \cdots + \frac{g^{(n-1)}(0)}{(n-1)!} + (J_\nu D^\nu g)(1)$$

$$= f(x_0) + f'(x_0)(x - x_0) + \frac{f''(x_0)}{2}(x - x_0)^2$$

$$+ \cdots + \frac{f^{(n-1)}(x_0)}{(n-1)!}(x - x_0)^{n-1} + (J_\nu D^\nu g)(1), \quad \nu \geq 1.$$

Next we translate $(J_\nu D^\nu g)(1)$ for any $\nu > 0$.
Notice that

$$t = \frac{w - x_0}{x - x_0}; \quad t = 0 \text{ iff } w = x_0,$$

and

$$dt = \frac{dw}{x - x_0}.$$

Denote

$$z := x_0 + y(x - x_0) = \varphi(y), \quad y \in [0, 1],$$

therefore
$$y = \frac{z - x_0}{x - x_0}, \quad \text{and} \quad y - t = \frac{z - w}{x - x_0},$$

in particular $t = y$ iff $w = z$.

We see that
$$(J_{1-\alpha} g^{(n)})(y) = \frac{1}{\Gamma(1-\alpha)} \int_0^y (y-t)^{-\alpha} \cdot f^{(n)}(x_0 + t(x - x_0))(x - x_0)^n dt$$

$$= (x - x_0)^n \frac{(J_{1-\alpha}^{x_0} f^{(n)})(z)}{(x - x_0)^{(1-\alpha)}}.$$

That is,
$$(J_{1-\alpha} g^{(n)})(y) = (x - x_0)^{n-(1-\alpha)} \cdot (J_{1-\alpha}^{x_0} f^{(n)})(z)$$

or
$$(J_{1-\alpha} g^{(n)})(y) = (x - x_0)^{\nu-1} (J_{1-\alpha}^{x_0} f^{(n)})(x_0 + y(x - x_0)).$$

Then it holds
$$(D(J_{1-\alpha} g^{(n)}))(y) = (x - x_0)^{\nu} D(J_{1-\alpha}^{x_0} f^{(n)})(x_0 + y(x - x_0))$$

$$= (x - x_0)^{\nu} (D_{x_0}^{\nu} f)(z).$$

I.e.,
$$(D^{\nu} g)(y) = (x - x_0)^{\nu} (D_{x_0}^{\nu} f)(z). \qquad (25.9)$$

Especially
$$(D^{\nu} g)(1) = (x - x_0)^{\nu} (D_{x_0}^{\nu} f)(x).$$

Next by using (25.9) we treat
$$(J_{\nu} D^{\nu} g)(1) = \frac{1}{\Gamma(\nu)} \int_0^1 (1-t)^{\nu-1} (D^{\nu} g)(t) dt$$

$$= \frac{(x - x_0)^{\nu}}{\Gamma(\nu)} \int_0^1 (1-t)^{\nu-1} (D_{x_0}^{\nu} f)(w) dt$$

$$= \frac{(x - x_0)^{\nu}}{\Gamma(\nu)} \int_0^1 \frac{(x - w)^{\nu-1}}{(x - x_0)^{\nu-1}} (D_{x_0}^{\nu} f)(w) \frac{dw}{(x - x_0)}$$

$$= \frac{1}{\Gamma(\nu)} \int_{x_0}^x (x - w)^{\nu-1} (D_{x_0}^{\nu} f)(w) dw$$

## 25. $L_p$-Opial-Type Inequalities

$$= (J_\nu^{x_0} D_{x_0}^\nu f)(x).$$

That is,
$$(J_\nu D^\nu g)(1) = (J_\nu^{x_0} D_{x_0}^\nu f)(x). \tag{25.10}$$

So that (25.7) becomes clear. Also (25.8) is obvious by (25.4) and (25.10). ∎

**REMARK 25.1** 1) $(D_{x_0}^n f) = f^{(n)}$, $n \in \mathbf{N}$.
2) Let $f \in C_{x_0}^\nu([a,b])$, $\nu \geq 1$ and $f^{(i)}(x_0) = 0$, $i = 0, 1, \ldots, n-1$, $n := [\nu]$. Then by (25.7)
$$f(x) = (J_\nu^{x_0} D_{x_0}^\nu f)(x).$$

That is,
$$f(x) = \frac{1}{\Gamma(\nu)} \int_{x_0}^x (x-t)^{\nu-1} (D_{x_0}^\nu f)(t)dt, \tag{25.11}$$

all $x \in [a,b] : x \geq x_0$. Notice that (25.11) is true, also when $0 < \nu < 1$. ∎

We use

**LEMMA 25.1**
Let $f \in C([a,b])$, $\mu, \nu > 0$. Then it holds
$$J_\mu^{x_0}(J_\nu^{x_0} f) = J_{\mu+\nu}^{x_0}(f). \tag{25.12}$$

**PROOF** Observe that

$$((J_\mu^{x_0} \circ J_\nu^{x_0})(f))(x) = (J_\mu^{x_0}(J_\nu^{x_0}(f)))(x)$$

$$= \frac{1}{\Gamma(\mu)} \int_{x_0}^x (x-s)^{\mu-1} \left( \frac{1}{\Gamma(\nu)} \int_{x_0}^s (s-t)^{\nu-1} f(t) dt \right) ds$$

$$= \frac{1}{\Gamma(\mu)\Gamma(\nu)} \cdot \int_{x_0}^b (x-s)_+^{\mu-1} \left( \int_{x_0}^b (s-t)_+^{\nu-1} f(t) dt \right) ds$$

$$= \frac{1}{\Gamma(\mu)\Gamma(\nu)} \int_{x_0}^b \int_{x_0}^b (x-s)_+^{\mu-1} (s-t)_+^{\nu-1} f(t) dt ds$$

$$= \frac{1}{\Gamma(\mu)\Gamma(\nu)} \int_{x_0}^{b} \left( \int_{x_0}^{b} (x-s)_+^{\mu-1}(s-t)_+^{\nu-1} ds \right) f(t) dt$$

$$= \frac{1}{\Gamma(\mu)\Gamma(\nu)} \int_{x_0}^{x} \left( \int_{t}^{x} (x-s)^{\mu-1}(s-t)^{\nu-1} ds \right) f(t) dt =: \otimes.$$

By [174], p. 256, for $\mu, \nu > 0$ we find that

$$\int_{t}^{x} (x-s)^{\mu-1}(s-t)^{\nu-1} ds = \frac{\Gamma(\mu)\Gamma(\nu)}{\Gamma(\mu+\nu)} (x-t)^{\mu+\nu-1}.$$

Hence

$$\otimes = \frac{1}{\Gamma(\mu+\nu)} \int_{x_0}^{x} (x-t)^{\mu+\nu-1} f(t) dt$$

$$= (J_{\mu+\nu}^{x_0} f)(x).$$

∎

**REMARK 25.2** Let $\nu, \gamma \geq 1$ such that $\nu - \gamma \geq 1$, that is $\gamma < \nu$. Call $n := [\nu]$, $\alpha := \nu - n$; $m := [\gamma]$, $\rho := \gamma - m$. See that $\nu - m \geq 1$ and $n - m \geq 1$. Let $f \in C_{x_0}^{\nu}([a,b])$ such that $f^{(i)}(x_0) = 0$, $i = 0, 1, \ldots, n-1$. Thus by (25.7),

$$f(x) = (J_{\nu}^{x_0} D_{x_0}^{\nu} f)(x), \quad \text{all } x \in [a,b] : x \geq x_0.$$

Then by Leibnitz's formula and $\Gamma(p+1) = p\Gamma(p)$, $p > 0$, we obtain that

$$f^{(m)}(x) = (J_{\nu-m}^{x_0} D_{x_0}^{\nu} f)(x), \quad \text{all } x \geq x_0. \tag{25.13}$$

∎

It follows next that $f \in C_{x_0}^{\gamma}([a,b])$ and hence

$$(D_{x_0}^{\gamma} f)(x) := (D J_{1-\rho}^{x_0} f^{(m)})(x) \text{ exists for all } x \geq x_0. \tag{25.14}$$

So by application of (25.13) we get (over $x \geq x_0$)

$$J_{1-\rho}^{x_0}(f^{(m)}) = J_{1-\rho}^{x_0}(J_{\nu-m}^{x_0} D_{x_0}^{\nu} f)$$

$$= (J_{1-\rho}^{x_0} \circ J_{\nu-m}^{x_0})(D_{x_0}^{\nu}f) \overset{(25.12)}{=} (J_{\nu-m+1-\rho}^{x_0})(D_{x_0}^{\nu}f)$$

$$= J_{\nu-\gamma+1}^{x_0}(D_{x_0}^{\nu}f).$$

I.e.,

$$(J_{1-\rho}^{x_0}f^{(m)})(x) = \frac{1}{\Gamma(\nu-\gamma+1)}\int_{x_0}^{x}(x-t)^{\nu-\gamma}(D_{x_0}^{\nu}f)(t)dt.$$

Consequently,

$$(D_{x_0}^{\gamma}f)(x) = D((J_{1-\rho}^{x_0}f^{(m)})(x))$$

$$= \frac{1}{\Gamma(\nu-\gamma)} \cdot \int_{x_0}^{x}(x-t)^{(\nu-\gamma)-1}(D_{x_0}^{\nu}f)(t)dt. \quad (25.15)$$

That is,

$$(D_{x_0}^{\gamma}f)(x) = (J_{\nu-\gamma}^{x_0}(D_{x_0}^{\nu}f))(x) \text{ exists for all } x \in [a,b] : x \geq x_0. \quad (25.16)$$

## 25.2  General Results

We give the following Opial-type inequality:

### THEOREM 25.2

Let $\nu, \gamma \geq 1$ such that $\nu - \gamma \geq 1$ and $f \in C_{x_0}^{\nu}([a,b])$ with $f^{(i)}(x_0) = 0$, $i = 0, 1, \ldots, n-1$, $n := [\nu]$. Here $x, x_0 \in [a,b] : x \geq x_0$. Let $p, q > 1$ such that $\frac{1}{p} + \frac{1}{q} = 1$. Then it holds

$$\int_{x_0}^{x} |(D_{x_0}^{\gamma}f)(w)| |(D_{x_0}^{\nu}f)(w)| dw$$

$$\leq \frac{(x-x_0)^{\frac{p\nu-p\gamma-p+2}{p}}}{(\sqrt[q]{2})\Gamma(\nu-\gamma)((p\nu-p\gamma-p+1)(p\nu-p\gamma-p+2))^{1/p}}$$

$$\cdot \left(\int_{x_0}^{x}|(D_{x_0}^{\nu}f)(w)|^q dw\right)^{2/q}. \quad (25.17)$$

**PROOF**  From (25.15) and Hölder's inequality we obtain ($x \geq x_0$)

$$|(D_{x_0}^\gamma f)(x)| \leq \frac{1}{\Gamma(\nu-\gamma)} \int_{x_0}^x (x-t)^{(\nu-\gamma)-1} |(D_{x_0}^\nu f)(t)| dt$$

$$\leq \frac{1}{\Gamma(\nu-\gamma)} \left( \int_{x_0}^x (x-t)^{p\nu-p\gamma-p} dt \right)^{1/p} \cdot \left( \int_{x_0}^x |(D_{x_0}^\nu f)(t)|^q dt \right)^{1/q}$$

$$= \frac{1}{\Gamma(\nu-\gamma)} \cdot \frac{(x-x_0)^{\frac{p\nu-p\gamma-p+1}{p}}}{(p\nu-p\gamma-p+1)^{1/p}} \cdot (z(x))^{1/q},$$

where

$$z(x) := \int_{x_0}^x |(D_{x_0}^\nu f)(t)|^q dt \geq 0,$$

($z(x_0) = 0$). That is,
$$z'(x) = |(D_{x_0}^\nu f)(x)|^q,$$

and
$$|(D_{x_0}^\nu f)(x)| = (z'(x))^{1/q}, \quad \text{all } x \geq x_0.$$

So here for $x_0 \leq w \leq x$ we see that

$$|(D_{x_0}^\gamma f)(w)| \, |(D_{x_0}^\nu f)(w)| \leq \frac{(w-x_0)^{\frac{p\nu-p\gamma-p+1}{p}} (z(w) \cdot z'(w))^{1/q}}{\Gamma(\nu-\gamma)(p\nu-p\gamma-p+1)^{1/p}}.$$

Integrating the last inequality we get

$$\int_{x_0}^x |(D_{x_0}^\gamma f)(w)| \, |(D_{x_0}^\nu f)(w)| dw$$

$$\leq \frac{\int_{x_0}^x (w-x_0)^{\frac{p\nu-p\gamma-p+1}{p}} (z(w) z'(w))^{1/q} dw}{\Gamma(\nu-\gamma)(p\nu-p\gamma-p+1)^{1/p}}$$

(by Hölder's inequality),

$$\leq \frac{\left( \int_{x_0}^x (w-x_0)^{p\nu-p\gamma-p+1} \right)^{1/p} \left( \int_{x_0}^x z(w) z'(w) dw \right)^{1/q}}{\Gamma(\nu-\gamma)(p\nu-p\gamma-p+1)^{1/p}}$$

$$= \frac{(x-x_0)^{\frac{p\nu-p\gamma-p+2}{p}} (z(x))^{2/q}}{\sqrt[q]{2} \, \Gamma(\nu-\gamma)(p\nu-p\gamma-p+1)^{1/p} (p\nu-p\gamma-p+2)^{1/p}},$$

establishing (25.17). ∎

A related extreme case comes next.

**PROPOSITION 25.1**
Under the assumptions of Theorem 25.2 when $p = 1$ and $q = \infty$ we find

$$\int_{x_0}^{x} |D_{x_0}^{\gamma} f|(w) \cdot |D_{x_0}^{\nu} f|(w) dw \leq \frac{(x - x_0)^{(\nu - \gamma + 1)}}{\Gamma(\nu - \gamma + 2)} \cdot (\|D_{x_0}^{\nu} f\|_{\infty})^2. \quad (25.18)$$

**PROOF** When $x_0 \leq w \leq x$ from (25.15) we get

$$|(D_{x_0}^{\gamma} f)(w)| \leq \frac{1}{\Gamma(\nu - \gamma)} \left( \int_{x_0}^{w} (w - t)^{(\nu - \gamma) - 1} dt \right) \cdot \|D_{x_0}^{\nu} f\|_{\infty}.$$

Hence,

$$|(D_{x_0}^{\gamma} f)(w)| \leq \frac{(w - x_0)^{(\nu - \gamma)}}{\Gamma(\nu - \gamma + 1)} \cdot \|D_{x_0}^{\nu} f\|_{\infty},$$

and

$$|(D_{x_0}^{\gamma} f)(w)| |(D_{x_0}^{\nu} f)(w)| \leq \frac{(w - x_0)^{(\nu - \gamma)}}{\Gamma(\nu - \gamma + 1)} \cdot (\|D_{x_0}^{\nu} f\|_{\infty})^2. \quad (25.19)$$

Integrating (25.19) over $[x_0, x]$ we find (25.18). ∎

The counterpart of Theorem 25.2 follows.

**THEOREM 25.3**
Let $\nu, \gamma \geq 1$ such that $\nu - \gamma \geq 1$ and $f \in C_{x_0}^{\nu}([a, b])$ with $f^{(i)}(x_0) = 0$, $i = 0, 1, \ldots, n - 1$, $n := \lceil \nu \rceil$. Here $x, x_0 \in [a, b] : x > x_0$. Suppose that $(D_{x_0}^{\nu} f)(t) \neq 0$ and of fixed sign over $[x_0, b]$. Let $p, q$ such that $0 < p < 1$ and $\frac{1}{p} + \frac{1}{q} = 1$. Then it holds

$$\int_{x_0}^{x} |D_{x_0}^{\gamma} f|(w) \cdot |D_{x_0}^{\nu} f|(w) dw$$

$$\geq \frac{(x - x_0)^{\frac{p\nu - p\gamma - p + 2}{p}}}{\sqrt[q]{2} \Gamma(\nu - \gamma)((p\nu - p\gamma - p + 1)(p\nu - p\gamma - p + 2))^{1/p}}$$

$$\cdot \left( \int_{x_0}^{x} (|D_{x_0}^{\nu} f|(w))^q dw \right)^{2/q}. \tag{25.20}$$

**PROOF**  From (25.15) and assumption on $(D_{x_0}^{\nu} f)$ we observe that

$$|D_{x_0}^{\gamma} f|(x) = \frac{1}{\Gamma(\nu - \gamma)} \int_{x_0}^{x} (x - t)^{(\nu - \gamma) - 1} |D_{x_0}^{\nu} f|(t) dt.$$

Here $q = \frac{p}{p-1} < 0$. By Hölder's inequality $(0 < p < 1)$ we obtain for $x > x_0$ that

$$|D_{x_0}^{\gamma} f|(x) \geq \frac{1}{\Gamma(\nu - \gamma)} \left( \int_{x_0}^{x} (x - t)^{p\nu - p\gamma - p} dt \right)^{1/p}$$

$$\cdot \left( \int_{x_0}^{x} (|D_{x_0}^{\nu} f|(t))^q dt \right)^{1/q}$$

$$= \frac{(x - x_0)^{\frac{p\nu - p\gamma - p + 1}{p}}}{\Gamma(\nu - \gamma)(p\nu - p\gamma - p + 1)^{1/p}} \cdot (z(x))^{1/q}.$$

Here set

$$z(x) := \int_{x_0}^{x} (|D_{x_0}^{\nu} f|(t))^q dt \geq 0, \quad z(x_0) = 0.$$

So that

$$z'(x) = (|D_{x_0}^{\nu} f|(x))^q,$$

and

$$|D_{x_0}^{\nu} f|(x) = (z'(x))^{1/q}, \quad \text{for all } x \geq x_0.$$

Then for $x_0 < w \leq x$ we have

$$|D_{x_0}^{\gamma} f|(w) \cdot |D_{x_0}^{\nu} f|(w) \geq \frac{(w - x_0)^{\frac{p\nu - p\gamma - p + 1}{p}}}{\Gamma(\nu - \gamma)(p\nu - p\gamma - p + 1)^{1/p}}$$

$$\cdot (z(w) z'(w))^{1/q}. \tag{25.21}$$

At the end integrating (25.21) we observe that

$$\int_{x_0}^{x} |D_{x_0}^{\gamma} f|(w) \cdot |D_{x_0}^{\nu} f|(w) dw = \lim_{\theta \downarrow x_0} \int_{\theta}^{x} |D_{x_0}^{\gamma} f|(w) \cdot |D_{x_0}^{\nu} f|(w) dw$$

$$\geq \lim_{\theta \downarrow x_0} \frac{\int_\theta^x (w-x_0)^{\frac{p\nu-p\gamma-p+1}{p}} \cdot (z(w)z'(w))^{1/q} dw}{\Gamma(\nu-\gamma)(p\nu-p\gamma-p+1)^{1/p}} \quad \text{(Hölder)}$$

$$\geq \lim_{\theta \downarrow x_0} \frac{\left(\int_\theta^x (w-x_0)^{p\nu-p\gamma-p+1} dw\right)^{1/p} \cdot \left(\int_\theta^x z(w)z'(w) dw\right)^{1/q}}{\Gamma(\nu-\gamma)(p\nu-p\gamma-p+1)^{1/p}}$$

$$= \frac{(x-x_0)^{\frac{p\nu-p\gamma-p+2}{p}} (z(x))^{2/q}}{\sqrt[q]{2}\,\Gamma(\nu-\gamma)(p\nu-p\gamma-p+1)^{1/p}(p\nu-p\gamma-p+2)^{1/p}},$$

establishing (25.20). ∎

The following inequality involves fractional derivatives of three different orders.

### THEOREM 25.4
Let $\gamma, k, \nu \geq 1$ such that $\nu - k > 1$, $\nu - \gamma \geq 1$ and $\gamma = k+1$, also $f \in C_{x_0}^\nu([a,b])$ with $f^{(i)}(x_0) = 0$, $i = 0, 1, \ldots, n-1$, $n := [\nu]$. Here $x, x_0 \in [a,b] : x \geq x_0$. Let $p, q > 1$ such that $\frac{1}{p} + \frac{1}{q} = 1$. Then it holds

$$\int_{x_0}^x |(D_{x_0}^k f)(w)|\,|(D_{x_0}^\gamma f)(w)|\,dw$$

$$\leq \frac{(x-x_0)^{\frac{2(p\nu-pk-p+1)}{p}}}{2(\Gamma(\nu-k))^2 (p\nu-pk-p+1)^{2/p}}$$

$$\cdot \left(\int_{x_0}^x |(D_{x_0}^\nu f)(w)|^q dw\right)^{2/q}. \qquad (25.22)$$

**PROOF** From (25.15) we see that

$$(D_{x_0}^\gamma f)(x) = \frac{1}{\Gamma(\nu-\gamma)} \int_{x_0}^x (x-t)^{\nu-\gamma-1} (D_{x_0}^\nu f)(t)\,dt$$

and

$$(D_{x_0}^k f)(x) = \frac{1}{\Gamma(\nu-k)} \int_{x_0}^x (x-t)^{\nu-k-1} (D_{x_0}^\nu f)(t)\,dt.$$

Because $\gamma = k+1$, we get from the above that

$$((D_{x_0}^k f)(x))' = (D_{x_0}^\gamma f)(x). \qquad (25.23)$$

Next notice that

$$|(D_{x_0}^\gamma f)(x)| \le A(x) := \frac{1}{\Gamma(\nu-\gamma)} \cdot \int_{x_0}^x (x-t)^{\nu-\gamma-1} \cdot |D_{x_0}^\nu f|(t) dt,$$

and

$$|(D_{x_0}^k f)(x)| \le B(x) := \frac{1}{\Gamma(\nu-k)} \int_{x_0}^x (x-t)^{\nu-k-1} \cdot |D_{x_0}^\nu f|(t) dt$$

$(B(x_0) = 0)$. Moreover, observe that

$$B'(x) = A(x). \qquad (25.24)$$

Hence

$$\int_{x_0}^x |(D_{x_0}^k f)(w)| \, |(D_{x_0}^\gamma f)(w)| dw$$

$$\le \int_{x_0}^x B(w) A(w) dw \stackrel{(25.24)}{=} \int_{x_0}^x B(w) B'(w) dw$$

$$= \int_{x_0}^x B(w) dB(w) = \frac{B^2(x)}{2}$$

$$= \frac{1}{2(\Gamma(\nu-k))^2} \left( \int_{x_0}^x (x-w)^{\nu-k-1} |D_{x_0}^\nu f|(w) dw \right)^2 \quad \text{(Hölder)}$$

$$\le \frac{1}{2(\Gamma(\nu-k))^2} \left( \int_{x_0}^x (x-w)^{p\nu-pk-p} dw \right)^{2/p} \left( \int_{x_0}^x |D_{x_0}^\nu f(w)|^q dw \right)^{2/q}$$

$$= \frac{1}{2(\Gamma(\nu-k))^2} \cdot \frac{(x-x_0)^{\frac{2(p\nu-pk-p+1)}{p}}}{(p\nu-pk-p+1)^{2/p}} \cdot \left( \int_{x_0}^x |D_{x_0}^\nu f(w)|^q dw \right)^{2/q}.$$

■

A related extreme case comes next.

## PROPOSITION 25.2
Under the assumptions of Theorem 25.4 when $p = 1$, $q = \infty$ we find

$$\int_{x_0}^{x} |D_{x_0}^k f(w)| |D_{x_0}^{k+1} f(w)| dw \leq \frac{(x-x_0)^{2(\nu-k)}(\|D_{x_0}^\nu f\|_\infty)^2}{2(\Gamma(\nu-k))^2(\nu-k)^2}. \quad (25.25)$$

**PROOF** From (25.15) we get that

$$|(D_{x_0}^\gamma f)(w)| \leq \frac{(w-x_0)^{(\nu-\gamma)}}{\Gamma(\nu-\gamma+1)} \|D_{x_0}^\nu f\|_\infty,$$

$$|(D_{x_0}^k f)(w)| \leq \frac{(w-x_0)^{(\nu-k)}}{\Gamma(\nu-k+1)} \|D_{x_0}^\nu f\|_\infty, \quad \text{all } x_0 \leq w \leq x.$$

Therefore,

$$|(D_{x_0}^\gamma f)(w)| |(D_{x_0}^k f)(w)| \leq \frac{(w-x_0)^{2\nu-k-\gamma}}{\Gamma(\nu-k+1)\Gamma(\nu-\gamma+1)} (\|D_{x_0}^\nu f\|_\infty)^2.$$

Integrating the last over $[x_0, x]$ we observe that

$$\int_{x_0}^{x} |(D_{x_0}^\gamma f)(w)| |(D_{x_0}^k f)(w)| dw$$

$$\leq \frac{(x-x_0)^{2\nu-k-\gamma+1}(\|D_{x_0}^\nu f\|_\infty)^2}{(2\nu-k-\gamma+1)\Gamma(\nu-k+1)\Gamma(\nu-\gamma+1)}. \quad (25.26)$$

Using $\gamma = k+1$ and $\Gamma(p+1) = p\Gamma(p)$, $p > 0$ in (25.26) we find (25.25).
∎

## PROPOSITION 25.3
Inequality (25.25) is sharp! In fact it is attained.

**PROOF** Let $\nu \geq 1$, $n := [\nu]$, $\alpha := \nu - n$ and $f$ such that $f^{(n)}(t) = (t-x_0)^\alpha$. Then it holds

$$(J_{1-\alpha}^{x_0} f^{(n)})(x) = \frac{1}{\Gamma(1-\alpha)} \int_{x_0}^{x} (x-t)^{(1-\alpha)-1}(t-x_0)^{(\alpha+1)-1} dt$$

$$= \frac{1}{\Gamma(1-\alpha)} \frac{\Gamma(1-\alpha)\Gamma(\alpha+1)}{\Gamma(2)}(x-x_0), \text{ by [174], p. 256.}$$

I.e.,
$$(J_{1-\alpha}^{x_0} f^{(n)})(x) = \Gamma(\alpha+1)(x-x_0).$$

And
$$(D_{x_0}^{\nu} f)(x) = \Gamma(\alpha+1). \tag{25.27}$$

A convenient choice here for $f$ is
$$f(t) = \frac{(t-x_0)^{\alpha+n}}{\prod_{j=1}^{n}(\alpha+j)}.$$

One sees that (from (25.15) and (25.27))
$$(D_{x_0}^{\gamma} f)(x) = \frac{1}{\Gamma(\nu-\gamma)} \int_{x_0}^{x} (x-t)^{\nu-\gamma-1} \Gamma(\alpha+1) dt.$$

That is,
$$(D_{x_0}^{\gamma} f)(x) = \frac{\Gamma(\alpha+1)}{\Gamma(\nu-\gamma+1)}(x-x_0)^{\nu-\gamma} = \frac{\Gamma(\alpha+1)}{\Gamma(\nu-k)}(x-x_0)^{\nu-k-1},$$

by $\gamma = k+1$, and
$$(D_{x_0}^{k} f)(x) = \frac{\Gamma(\alpha+1)}{\Gamma(\nu-k+1)}(x-x_0)^{\nu-k}, \quad \text{all } x \geq x_0.$$

Consequently, working on (25.25) we observe
$$L := \int_{x_0}^{x} |(D_{x_0}^{k} f)(w)| |(D_{x_0}^{k+1} f)(w)| dw$$

$$= \int_{x_0}^{x} \frac{\Gamma(\alpha+1)}{\Gamma(\nu-k+1)}(w-x_0)^{\nu-k} \cdot \frac{\Gamma(\alpha+1)}{\Gamma(\nu-k)}(w-x_0)^{\nu-k-1} \cdot dw$$

$$= \frac{(\Gamma(\alpha+1))^2}{(\Gamma(\nu-k))^2(\nu-k)} \int_{x_0}^{x} (w-x_0)^{2\nu-2k-1} dw$$

$$= \frac{(\Gamma(\alpha+1))^2}{(\Gamma(\nu-k))^2(\nu-k)} \cdot \frac{(x-x_0)^{2(\nu-k)}}{2(\nu-k)}.$$

25. $L_p$-Opial-Type Inequalities

That is,
$$L = \frac{(\Gamma(\alpha+1))^2 (x-x_0)^{2(\nu-k)}}{2(\Gamma(\nu-k))^2 (\nu-k)^2}. \tag{25.28}$$

Next we work on $R$, the right-hand side of (25.25), we see that
$$R \overset{(25.27)}{=} \frac{(x-x_0)^{2(\nu-k)} (\Gamma(\alpha+1))^2}{2(\Gamma(\nu-k))^2 (\nu-k)^2}. \tag{25.29}$$

From (25.28) and (25.29) we see that
$$L = R,$$

that is inequality (25.25) is attained, i.e., it is sharp! ∎

The counterpart of Theorem 25.4 comes next.

### THEOREM 25.5
Let $\gamma, k, \nu \geq 1$ such that $\nu - k > 1$, $\nu - \gamma \geq 1$ and $\gamma = k+1$, also $f \in C_{x_0}^{\nu}([a,b])$ with $f^{(i)}(x_0) = 0$, $i = 0, 1, \ldots, n-1$, $n := [\nu]$. Here $x, x_0 \in [a,b] : x > x_0$. Suppose that $(D_{x_0}^{\nu} f)(t) \neq 0$ and of fixed sign over $[x_0, b]$. Let $p, q$ such that $0 < p < 1$ and $\frac{1}{p} + \frac{1}{q} = 1$. Then it holds

$$\int_{x_0}^{x} |D_{x_0}^k f|(w)|D_{x_0}^{\gamma} f|(w) dw \leq \frac{(x-x_0)^{\frac{2(p\nu-pk-p+1)}{p}}}{2(\Gamma(\nu-k))^2 (p\nu - pk - p + 1)^{2/p}}$$

$$\cdot \left( \int_{x_0}^{x} |D_{x_0}^{\nu} f|^q(w) \, dw \right)^{2/q}. \tag{25.30}$$

**PROOF** From (25.15) we get $(D_{x_0}^{\gamma} f)(x) \neq 0$, $(D_{x_0}^k f)(x) \neq 0$ and of the same fixed sign as $(D_{x_0}^{\nu} f)(x)$. Also from (25.23),

$$(D_{x_0}^k f)'(x) = (D_{x_0}^{\gamma} f)(x), \quad x > x_0.$$

Hence $((D_{x_0}^k f)(x_0) = 0)$

$$\int_{x_0}^{x} |(D_{x_0}^k f)(w)| \, |(D_{x_0}^{\gamma} f)(w)| dw = \int_{x_0}^{x} (D_{x_0}^k f)(w)(D_{x_0}^{\gamma} f)(w) dw$$

$$= \int_{x_0}^{x} (D_{x_0}^k f)(w)(D_{x_0}^k f)'(w) dw = \int_{x_0}^{x} (D_{x_0}^k f)(w) d(D_{x_0}^k f)(w)$$

$$= \frac{((D_{x_0}^k f)(x))^2}{2} = \frac{1}{2(\Gamma(\nu-k))^2} \left( \int_{x_0}^{x} (x-w)^{\nu-k-1} (D_{x_0}^\nu f)(w) dw \right)^2$$

$$= \frac{1}{2(\Gamma(\nu-k))^2} \left( \int_{x_0}^{x} (x-w)^{\nu-k-1} |D_{x_0}^\nu f|(w) dw \right)^2$$

$$\geq \frac{1}{2(\Gamma(\nu-k))^2} \left( \int_{x_0}^{x} (x-w)^{p\nu-pk-p} dw \right)^{2/p}$$

$$\cdot \left( \int_{x_0}^{x} (|(D_{x_0}^\nu f)(w)|)^q dw \right)^{2/q}$$

$$= \frac{1}{2(\Gamma(\nu-k))^2} \cdot \frac{(x-x_0)^{\frac{2(p\nu-pk-p+1)}{p}}}{(p\nu-pk-p+1)^{2/p}}$$

$$\cdot \left( \int_{x_0}^{x} (|(D_{x_0}^\nu f)(w)|)^q dw \right)^{2/q}.$$

∎

Especially we obtain

### THEOREM 25.6

Let $\nu, k \geq 1$ such that $\nu - k \geq 1$, also $f \in C_{x_0}^\nu([a,b])$ with $f^{(i)}(x_0) = 0$, $i = 0, 1, \ldots, n-1$, $n := \lceil \nu \rceil$. Here $x, x_0 \in [a,b] : x \geq x_0$. Let $p, q > 1$ such that $\frac{1}{p} + \frac{1}{q} = 1$. Then it holds ($\ell > 0$)

$$\int_{x_0}^{x} |(D_{x_0}^k f)(w)|^\ell dw$$

$$\leq \frac{(x-x_0)^{(\ell\nu-\ell k-\ell+1+\frac{\ell}{p})}}{(\Gamma(\nu-k))^\ell \cdot (\ell\nu-\ell k-\ell+1+\frac{\ell}{p}) \cdot (p\nu-pk-p+1)^{\ell/p}}$$

$$\cdot \left( \int_{x_0}^{x} |(D_{x_0}^\nu f)(w)|^q dw \right)^{\ell/q}. \qquad (25.31)$$

## 25. $L_p$-Opial-Type Inequalities

**PROOF** From (25.15) and Hölder's inequality we obtain ($x_0 \le w \le x$)

$$|(D_{x_0}^k f)(w)| \le \frac{1}{\Gamma(\nu - k)} \int_{x_0}^w (w - t)^{\nu - k - 1} |(D_{x_0}^\nu f)(t)| dt$$

$$\le \frac{1}{\Gamma(\nu - k)} \left( \int_{x_0}^w (w - t)^{p\nu - pk - p} dt \right)^{1/p} \left( \int_{x_0}^w |D_{x_0}^\nu f(t)|^q dt \right)^{1/q}$$

$$= \frac{1}{\Gamma(\nu - k)} \frac{(w - x_0)^{\frac{p\nu - pk - p + 1}{p}}}{(p\nu - pk - p + 1)^{1/p}} \left( \int_{x_0}^w |(D_{x_0}^\nu f)(t)|^q dt \right)^{1/q}.$$

That is,

$$|(D_{x_0}^k f)(w)|^\ell \le \frac{1}{(\Gamma(\nu - k))^\ell} \frac{(w - x_0)^{\frac{\ell p\nu - \ell pk - \ell p + \ell}{p}}}{(p\nu - pk - p + 1)^{\ell/p}}$$

$$\cdot \left( \int_{x_0}^x (|(D_{x_0}^\nu f)(t)|)^q dt \right)^{\ell/q}.$$

Integrating the last we observe

$$\int_{x_0}^x |(D_{x_0}^k f)(w)|^\ell dw \le \frac{\left( \int_{x_0}^x (|(D_{x_0}^\nu f)(t)|)^q dt \right)^{\ell/q}}{(\Gamma(\nu - k))^\ell (p\nu - pk - p + 1)^{\ell/p}}$$

$$\cdot \frac{(x - x_0)^{(\ell\nu - \ell k - \ell + 1 + \frac{\ell}{p})}}{(\ell\nu - \ell k - \ell + 1 + \frac{\ell}{p})},$$

establishing (25.31). ∎

Another extreme case comes next.

### PROPOSITION 25.4
*Same assumptions as in Theorem 25.6, however $p = 1$ and $q = \infty$. Then it holds ($\ell > 0$)*

$$\int_{x_0}^x |(D_{x_0}^k f)(w)|^\ell dw \le \frac{(x - x_0)^{\ell\nu - \ell k + 1} (\|D_{x_0}^\nu f\|_\infty)^\ell}{(\ell\nu - \ell k + 1)(\Gamma(\nu - k + 1))^\ell}. \tag{25.32}$$

**PROOF**  From (25.15) we obtain

$$|(D_{x_0}^k f)(w)| \le \frac{1}{\Gamma(\nu - k)} \left( \int_{x_0}^w (w - t)^{\nu - k - 1} dt \right) \|D_{x_0}^\nu f\|_\infty$$

$$= \frac{1}{\Gamma(\nu - k)} \frac{(w - x_0)^{\nu - k}}{(\nu - k)} \|D_{x_0}^\nu f\|_\infty.$$

That is,

$$|(D_{x_0}^k f)(w)| \le \frac{(w - x_0)^{\nu - k}}{\Gamma(\nu - k + 1)} \|D_{x_0}^\nu f\|_\infty.$$

And

$$|(D_{x_0}^k f)(w)|^\ell \le \frac{(w - x_0)^{\ell\nu - \ell k}}{(\Gamma(\nu - k + 1))^\ell} (\|D_{x_0}^\nu f\|_\infty)^\ell,$$

all $x_0 \le w \le x$. Integrating the last inequality we get

$$\int_{x_0}^x |(D_{x_0}^k f)(w)|^\ell dw \le \frac{(\|D_{x_0}^\nu f\|_\infty)^\ell}{(\Gamma(\nu - k + 1))^\ell} \int_{x_0}^x (w - x_0)^{\ell\nu - \ell k} dw$$

$$= \frac{(\|D_{x_0}^\nu f\|_\infty)^\ell}{(\Gamma(\nu - k + 1))^\ell} \cdot \frac{(x - x_0)^{\ell\nu - \ell k + 1}}{(\ell\nu - \ell k + 1)},$$

establishing (25.32).  ∎

---

## 25.3  Applications to Differential Equations

i) Uniqueness of the solution to fractional initial value problem (see also [1]).

$$\begin{cases} \text{Let } \nu, \gamma_i \ge 1 \text{ such that } \nu - \gamma_i \ge 1,\, i = 1, \ldots, r \in \mathbf{N}, \\ n := [\nu],\, f \in C_a^\nu([a, b]);\, f^{(i)}(a) = a_i \in \mathbf{R},\, i = 0, 1, \ldots, n - 1. \\ \text{Furthermore} \\ (D_a^\nu f)(t) = F(t, \{(D_a^{\gamma_i} f)(t)\}_{i=1}^r), \\ \text{all } t \in [a, b]. \end{cases} \quad (25.33)$$

## 25. $L_p$-Opial-Type Inequalities

Here $F$ is a continuous function on $[a,b] \times \mathbf{R}^r$, $q_i(t) \geq 0$, $1 \leq i \leq r$ all continuous over $[a,b]$. Moreover, $F$ fulfills the Lipschitz condition:

$$|F(t, z_1, z_2, \ldots, z_r) - F(t, z'_1, z'_2, \ldots, z'_r)| \leq \sum_{i=1}^{r} q_i(t)|z_i - z'_i|. \qquad (25.34)$$

Suppose that

$$\phi^*(b) := \sum_{i=1}^{r} \|q_i\|_\infty \frac{(b-a)^{(\nu-\gamma_i)}}{2\Gamma(\nu-\gamma_i)\sqrt{(\nu-\gamma_i)(2\nu-2\gamma_i-1)}} < 1. \qquad (25.35)$$

Notice that an $f_* \in C_a^\nu([a,b])$ such that $f_*^{(i)}(a) = 0$, $i = 0, 1, \ldots, n-1$, implies that $f_* \in C_a^{\gamma_i}([a,b])$, $i = 1, \ldots, r$. From (25.15) we find that

$$(D_a^{\gamma_i} f_*)(x) = \frac{1}{\Gamma(\nu-\gamma_i)} \int_a^x (x-t)^{(\nu-\gamma_i)-1} (D_a^\nu f_*)(t) dt, \qquad (25.36)$$

all $a \leq x \leq b$, $i = 1, \ldots, r$. When $p = q = 2$ from (25.17) we get (all $i = 1, \ldots, r$)

$$\int_a^b |(D_a^{\gamma_i} f_*)(w)| \, |(D_a^\nu f_*)(w)| dw \leq \frac{(b-a)^{\nu-\gamma_i}}{2\Gamma(\nu-\gamma_i)\sqrt{(\nu-\gamma_i)(2\nu-2\gamma_i-1)}}$$

$$\cdot \int_a^b ((D_a^\nu f_*)(w))^2 dw. \qquad (25.37)$$

Let $f_1, f_2$ solve (25.33), i.e.,

$$(D_a^\nu f_j)(t) = F\left(t, \{(D_a^{\gamma_i} f_j)(t)\}_{i=1}^r\right),$$

$t \in [a,b]$, $j = 1, 2$: $f_j^{(i)}(a) = a_i$; $i = 0, 1, \ldots, n-1$. Put $g := f_1 - f_2$, then

$$(D_a^\nu g)(t) = F(t, \{(D_a^{\gamma_i} f_1)(t)\}_{i=1}^r) - F(t, \{(D_a^{\gamma_i} f_2)(t)\}_{i=1}^r), \qquad (25.38)$$

such that
$$g^{(i)}(a) = 0, \quad 0 \leq i \leq n-1.$$

Here we see that

$$|F(t, (D_a^{\gamma_1} f_1(t)), \ldots, (D_a^{\gamma_r} f_1(t)) - F(t, (D_a^{\gamma_1} f_2)(t), \ldots, (D_a^{\gamma_r} f_2(t))|$$

$$\leq \sum_{i=1}^{r} q_i(t)|(D_a^{\gamma_i} f_1)(t) - (D_a^{\gamma_i} f_2)(t)| = \sum_{i=1}^{r} q_i(t)|(D_a^{\gamma_i}(f_1 - f_2))(t)|$$

$$= \sum_{i=1}^{r} q_i(t)|(D_a^{\gamma_i} g)(t)| \leq \sum_{i=1}^{r} \|q_i\|_\infty |(D_a^{\gamma_i} g)(t)|. \tag{25.39}$$

So that we obtain

$$\begin{aligned}
(D_a^\nu g(t))^2 &= (D_a^\nu g)(t) \cdot \left\{ F(t, \{(D_a^{\gamma_i} f_1)(t)\}_{i=1}^r) - F(t, \{(D_a^{\gamma_i} f_2)(t)\}_{i=1}^r) \right\} \\
&\leq |(D_a^\nu g)(t)| \, |F(t, \{(D_a^{\gamma_i} f_1)(t)\}_{i=1}^r) - F(t, \{(D_a^{\gamma_i} f_2)(t)\}_{i=1}^r)| \\
&\overset{(25.39)}{\leq} |(D_a^\nu g)(t)| \left( \sum_{i=1}^{r} \|q_i\|_\infty |(D_a^{\gamma_i} g)(t)| \right) \\
&= \sum_{i=1}^{r} \|q_i\|_\infty |(D_a^{\gamma_i} g)(t)| \, |(D_a^\nu g)(t)|.
\end{aligned}$$

Consequently,

$$\int_a^b (D_a^\nu g(t))^2 dt \leq \sum_{i=1}^{r} \|q_i\|_\infty \cdot \int_a^b |(D_a^{\gamma_i} g)(t)| \, |(D_a^\nu g)(t)| dt$$

$$\overset{(25.37)}{\leq} \phi^*(b) \int_a^b ((D_a^\nu g)(t))^2 dt.$$

If $\int_a^b (D_a^\nu g(t))^2 dt \neq 0$, then by last inequality $\phi^*(b) \geq 1$, a contradiction by the assumption that $\phi^*(b) < 1$. Hence $\int_a^b (D_a^\nu g(t))^2 dt = 0$. Thus $(D_a^\nu g(t))^2 = 0$, a.e. in $[a,b]$. I.e., $D_a^\nu g(t) = 0$, a.e. in $[a,b]$. But $g^{(i)}(a) = 0$, $0 \leq i \leq n-1$ by assumption.

Consequently from (25.7) we find $g(t) \equiv 0$ on $[a,b]$. That is $f_1 = f_2$ on $[a,b]$, establishing the uniqueness of solution to initial value problem (25.33).

ii) **Upper bounds on** $(D_a^\nu f)$, **solution** $f$, **etc. (see also [1]).**

$$\begin{cases} \text{Consider the initial value problem } (a \le t \le b) \\ (D_a^\nu f)'(t) = F(t, \{(D_a^{\gamma_i} f)(t)\}_{i=1}^r, D_a^\nu f(t)), \\ \text{here } \nu, \gamma_i \ge 1 \text{ such that } \nu - \gamma_i \ge 1, \, i = 1, \ldots, r \in \mathbf{N}, \\ n := [\nu], f \in C_a^\nu([a, b]); f^{(i)}(a) = 0, i = 0, 1, \ldots, n - 1, \\ \text{and } (D_a^\nu f)(a) = A \in \mathbf{R}. \end{cases} \quad (25.40)$$

Here $F$ is a continuous function on $[a, b] \times \mathbf{R}^{r+1}$; $q_i(t) \ge 0$, $1 \le i \le r$ continuous functions over $[a, b]$ such that

$$|F(t, x_1, x_2, \ldots, x_r, x_{r+1})| \le \sum_{i=1}^{r} q_i(t)|x_i|. \quad (25.41)$$

We see that

$$(D_a^\nu f)(t)(D_a^\nu f)'(t) = (D_a^\nu f)(t) \cdot F(t, \{(D_a^{\gamma_i} f)(t)\}_{i=1}^r, (D_a^\nu f)(t)),$$

and for $a \le x \le b$ we notice that

$$\int_a^x (D_a^\nu f)(t) \cdot (D_a^\nu f)'(t) dt = \int_a^x (D_a^\nu f)(t)$$

$$\cdot F(t, \{(D_a^{\gamma_i} f)(t)\}_{i=1}^r, (D_a^\nu f)(t)) dt.$$

I.e.,

$$\frac{((D_a^\nu f)(t))^2}{2} \bigg|_a^x \le \int_a^x |(D_a^\nu f)(t)| \cdot |F(t, \{(D_a^{\gamma_i} f)(t)\}_{i=1}^r, (D_a^\nu f)(t))| dt$$

$$\stackrel{(25.41)}{\le} \int_a^x |(D_a^\nu f)(t)| \left(\sum_{i=1}^r q_i(t)|(D_a^{\gamma_i} f)(t)|\right) dt$$

$$\le \sum_{i=1}^r \|q_i\|_\infty \left(\int_a^x |(D_a^{\gamma_i} f)(t)| \, |(D_a^\nu f)(t)| dt\right).$$

Thus

$$((D_a^\nu f)(x))^2 \le A^2 + 2 \cdot \sum_{i=1}^r \|q_i\|_\infty \left(\int_a^x |(D_a^{\gamma_i} f)(t)| \, |(D_a^\nu f)(t)| dt\right)$$

$$\stackrel{(25.17)}{\leq} A^2 + \left( \sum_{i=1}^{r} \|q_i\|_\infty \left( \frac{(x-a)^{\nu-\gamma_i}}{\Gamma(\nu-\gamma_i)\sqrt{(\nu-\gamma_i)(2\nu-2\gamma_i-1)}} \right) \right)$$

$$\cdot \left( \int_a^x ((D_a^\nu f)(t))^2 dt \right).$$

That is, for $a \leq x \leq b$

$$((D_a^\nu f)(x))^2 \leq A^2$$

$$+ \left\{ \sum_{i=1}^{r} \|q_i\|_\infty \cdot \left( \frac{(x-a)^{\nu-\gamma_i}}{\Gamma(\nu-\gamma_i)\sqrt{(\nu-\gamma_i)(2\nu-2\gamma_i-1)}} \right) \right\}$$

$$\cdot \left( \int_a^x ((D_a^\nu f)(t))^2 dt \right). \tag{25.42}$$

Set $\theta(x) := ((D_a^\nu f)(x))^2$, $\rho := A^2$,

$$Q(x) := \sum_{i=1}^{r} \|q_i\|_\infty \cdot \left( \frac{(x-a)^{\nu-\gamma_i}}{\Gamma(\nu-\gamma_i)\sqrt{(\nu-\gamma_i)(2\nu-2\gamma_i-1)}} \right). \tag{25.43}$$

That is, we obtain

$$\theta(x) \leq \rho + Q(x) \int_a^x \theta(t) dt. \tag{25.44}$$

Here $\rho \geq 0$, $Q(x) \geq 0$, $Q(a) = 0$, $\theta(x) \geq 0$, all $a \leq x \leq b$.
Put

$$w(x) := \int_a^x \theta(t) dt \quad (w(a) = 0). \tag{25.45}$$

Therefore,

$$w'(x) = \theta(x) \leq \rho + Q(x)w(x).$$

And

$$w'(x) - Q(x)w(x) \leq \rho, \quad \text{any } a \leq x \leq b. \tag{25.46}$$

That is,

$$w'(t) - Q(t)w(t) \leq \rho, \quad \text{all } a \leq t \leq x.$$

And observe that

$$w'(t)e^{-\int_a^t Q(s)ds} - e^{-\int_a^t Q(s)ds} Q(t)w(t) \leq \rho e^{-\int_a^t Q(s)ds}.$$

Moreover, we see that

$$\left(w(t)e^{-\int_a^t Q(s)ds}\right)' \leq \rho e^{-\int_a^t Q(s)ds}, \quad a \leq t \leq x$$

and

$$\left(w(t)e^{-\int_a^t Q(s)ds}\right)\bigg|_a^x \leq \rho \left(\int_a^x \left(e^{-\int_a^t Q(s)ds}\right)dt\right).$$

Consequently,

$$w(x)e^{-\int_a^x Q(s)ds} \leq \rho \left(\int_a^x \left(e^{-\int_a^t Q(s)ds}\right)dt\right),$$

and

$$w(x) \leq \rho e^{\int_a^x Q(s)ds} \left(\int_a^x \left(e^{-\int_a^t Q(s)ds}\right)dt\right).$$

Hence ($a \leq x \leq b$)

$$\theta(x) \leq \rho \left(1 + Q(x) \cdot e^{\int_a^x Q(s)ds} \cdot \left(\int_a^x \left(e^{-\int_a^t Q(s)ds}\right)dt\right)\right). \quad (25.47)$$

We have shown that

$$|(D_a^\nu f)(x)| \leq |A| \cdot \left\{1 + Q(x) \cdot e^{\int_a^x Q(s)ds}\right.$$

$$\left. \cdot \left(\int_a^x \left(e^{-\int_a^t Q(s)ds}\right)dt\right)\right\}^{1/2} =: \chi(x), \quad (25.48)$$

all $a \leq x \leq b$.

From (25.11) we see that

$$|f(x)| \leq \frac{1}{\Gamma(\nu)} \int_a^x (x-t)^{\nu-1} |D_a^\nu f(t)| dt$$

and

$$|f(x)| \stackrel{(25.48)}{\leq} \frac{1}{\Gamma(\nu)} \int_a^x (x-t)^{\nu-1} \chi(t) dt, \quad \text{all } a \leq x \leq b. \quad (25.49)$$

Also from (25.15) we get

$$|(D_a^{\gamma_i} f)(x)| \le \frac{1}{\Gamma(\nu - \gamma_i)} \int_a^x (x-t)^{\nu - \gamma_i - 1} |(D_a^\nu f)(t)| dt$$

and

$$|(D_a^{\gamma_i} f)(x)| \stackrel{(25.48)}{\le} \frac{1}{\Gamma(\nu - \gamma_i)} \int_a^x (x-t)^{\nu - \gamma_i - 1} \chi(t) dt, \qquad (25.50)$$

all $i = 1, \ldots, r$, any $a \le x \le b$.

iii) **Consider again the fractional initial value problem (25.40) with $a \ne b$.** Here for fixed $i_* \in \{1, \ldots, r\}$ we suppose that $\gamma_{i_*+1} = \gamma_{i_*} + 1$, where $\gamma_{i_*}, \gamma_{i_*+1} \in \{\gamma_1, \ldots, \gamma_r\}$. Also suppose that $F$ is continuous on $[a,b] \times \mathbf{R}^{r+1}$ such that

$$|F(t, x_1, x_2, \ldots, x_{r+1})| \le q(t) |x_{i_*}| |x_{i_*+1}|, \qquad (25.51)$$

with $q \ne 0$ and $q(t) \ge 0$ continuous over $[a,b]$. Set $k := \gamma_{i_*}$, $\gamma := \gamma_{i_*+1}$, i.e., $\gamma = k+1$. Let $f \in C_a^\nu([a,b])$, $n := \lceil \nu \rceil$, such that $f^{(i)}(a) = 0$, $i = 0, 1, \ldots, n-1$. For $a \le x \le b$ we find (use of (25.22), $p = q = 2$)

$$\int_a^x |(D_a^k f)(t)||(D_a^\gamma f)(t)| dt \le \frac{(x-a)^{(2\nu - 2k - 1)}}{2(\Gamma(\nu - k))^2 (2\nu - 2k - 1)}$$

$$\cdot \left( \int_a^x ((D_a^\nu f)(t))^2 dt \right). \qquad (25.52)$$

Set $\rho := |A| \ge 0$,

$$Q(x) := \|q\|_\infty \cdot \frac{(x-a)^{2\nu - 2k - 1}}{2(\Gamma(\nu - k))^2 (2\nu - 2k - 1)} \ge 0, \qquad (25.53)$$

($Q(a) = 0$).

Put $\sigma := \|Q(x)\|_\infty > 0$ and suppose that

$$(b-a)\rho\sigma < 1. \qquad (25.54)$$

Here we get again upper bounds to $f$, $(D_a^\nu f)$, $(D_a^{\gamma_i} f)$. For that we do the following:

$$\int_a^x (D_a^\nu f)'(t) dt \stackrel{(25.40)}{=} \int_a^x F(t, \{(D_a^{\gamma_i} f)(t)\}_{i=1}^r, (D_a^\nu f)(t)) dt,$$

that is,
$$(D_a^\nu f)(x) = A + \int_a^x F(t, \{(D_a^{\gamma_i} f)(t)\}_{i=1}^r, (D_a^\nu f)(t))dt.$$

Thus we have
$$|(D_a^\nu f)(x)| \le |A| + \int_a^x |F(t, \{(D_a^{\gamma_i} f)(t)\}_{i=1}^r, (D_a^\nu f)(t))|dt$$

$$\le |A| + \int_a^x q(t)|(D_a^k f)(t)| \, |(D_a^\gamma f)(t)|dt$$

$$\le |A| + \|q\|_\infty \cdot \int_a^x |(D_a^k f)(t)| \, |(D_a^\gamma f)(t)|dt.$$

I.e., by (25.52), (25.53) and the above we have
$$\theta(x) \le \rho + Q(x) \cdot \int_a^x (\theta(t))^2 dt, \qquad (25.55)$$

where
$$\theta(x) := |(D_a^\nu f)(x)| \ge 0, \quad a \le x \le b.$$

Set
$$w(x) := \int_a^x (\theta(t))^2 dt \ge 0, \quad w(a) = 0. \qquad (25.56)$$

I.e.,
$$w'(x) = (\theta(x))^2$$

and
$$\theta(x) = (w'(x))^{1/2}, \quad \text{all } a \le x \le b.$$

We see that (use of (25.55))
$$(w'(x))^{1/2} \le \rho + Q(x)w(x), \quad a \le x \le b. \qquad (25.57)$$

Consequently,
$$(w'(x))^{1/2} \le \rho + \sigma w(x) < \rho + \varepsilon + \sigma w(x),$$

where $\varepsilon > 0$ arbitrarily small. I.e.,
$$w'(x) < (\rho + \varepsilon + \sigma w(x))^2, \quad a \le x \le b.$$

Here $(\rho + \varepsilon + \sigma w(x))^2 > 0$. Especially
$$w'(t) < (\rho + \varepsilon + \sigma w(t))^2, \quad a \le t \le x,$$
and
$$\frac{w'(t)}{(\rho + \varepsilon + \sigma w(t))^2} < 1. \qquad (25.58)$$

Inequality (25.58) is the same as
$$\left(-\frac{1}{\sigma(\rho + \varepsilon + \sigma w(t))}\right)' < 1. \qquad (25.59)$$

I.e.,
$$-\int_a^x \left(\frac{1}{\rho + \varepsilon + \sigma w(t)}\right)' dt < \sigma(x - a),$$
and
$$-\left(\frac{1}{\rho + \varepsilon + \sigma w(t)}\right)\bigg|_a^x < \sigma(x - a).$$

Hence it holds
$$\frac{1}{(\rho + \varepsilon)} - \frac{1}{(\rho + \varepsilon + \sigma w(x))} < \sigma(x - a),$$
and
$$\frac{1}{(\rho + \varepsilon)} - \sigma(x - a) < \frac{1}{\rho + \varepsilon + \sigma w(x)}.$$

So that
$$\frac{1 - \sigma(\rho + \varepsilon)(x - a)}{\rho + \varepsilon} < \frac{1}{(\rho + \varepsilon) + \sigma w(x)}. \qquad (25.60)$$

By assumption (25.54) we see that $(x - a)\rho\sigma < 1$, all $a \le x \le b$. There exists $\varepsilon > 0$ small enough such that
$$(x - a)(\rho + \varepsilon)\sigma < 1.$$

That is,
$$1 - \sigma(\rho + \varepsilon)(x - a) > 0.$$

From (25.60) we obtain that
$$(\rho + \varepsilon) + \sigma w(x) < \frac{\rho + \varepsilon}{1 - \sigma(\rho + \varepsilon)(x - a)}, \qquad (25.61)$$

all $a \leq x \leq b$.

Following (25.61) we find that

$$w(x) < \frac{(\rho+\varepsilon)^2(x-a)}{1-\sigma(\rho+\varepsilon)(x-a)},$$

where $\varepsilon > 0$ very small, all $a \leq x \leq b$. Taking in the last inequality $\varepsilon \to 0$ we get

$$w(x) \leq \frac{\rho^2(x-a)}{1-\sigma\rho(x-a)}, \quad \text{all } a \leq x \leq b. \tag{25.62}$$

From (25.55), (25.56) and (25.62), we see that

$$\theta(x) \leq \rho + Q(x)\left(\frac{\rho^2(x-a)}{1-\sigma\rho(x-a)}\right), \quad \text{all } a \leq x \leq b.$$

That is

$$|(D_a^\nu f)(x)| \leq |A| + Q(x)\left(\frac{A^2(x-a)}{1-|A|\,\|Q\|_\infty(x-a)}\right)$$

$$= |A|\left(1 + \left(\frac{|A|Q(x)(x-a)}{1-|A|\,\|Q\|_\infty(x-a)}\right)\right)$$

$$= |A|\left(\frac{1+|A|(x-a)(Q(x)-\|Q\|_\infty)}{1-|A|(x-a)\|Q\|_\infty}\right).$$

We have proved for any $a \leq x \leq b$ that

$$|(D_a^\nu f)(x)| \leq |A|\left(\frac{1+|A|(x-a)(Q(x)-\|Q\|_\infty)}{1-|A|(x-a)\|Q\|_\infty}\right) =: \tilde{\varphi}(x). \tag{25.63}$$

That is,

$$|(D_a^\nu f)(x)| \leq \tilde{\varphi}(x), \quad \text{all } a \leq x \leq b. \tag{25.64}$$

Again from (25.11) we observe that

$$|f(x)| \stackrel{(25.63)}{\leq} \frac{1}{\Gamma(\nu)}\int_a^x (x-t)^{\nu-1}\tilde{\varphi}(t)dt, \tag{25.65}$$

all $a \leq x \leq b$, and from (25.15) we obtain

$$|(D_a^{\gamma_i} f)(x)| \stackrel{(25.63)}{\leq} \frac{1}{\Gamma(\nu-\gamma_i)}\int_a^x (x-t)^{\nu-\gamma_i-1}\tilde{\varphi}(t)dt, \tag{25.66}$$

all $i = 1, \ldots, r$, any $a \leq x \leq b$.

# Chapter 26

# $L_p$-General Fractional Opial Inequalities

This is a continuation of Chapter 25. Several $L_p$ $(p > 0)$ form very general Opial-type inequalities are given involving different order generalized fractional derivatives of a function. These are based again on a generalization of Taylor's formula for generalized fractional derivatives. In the last results of this chapter we use a monotonicity property of the involved function/highest order generalized fractional derivative. This treatment follows [24].

In this chapter we make use of Section 25.1, i.e., we are on the same setting as in Chapter 25, and we use the same notions and notations.

## 26.1 General Results

The following theorem and its proof resemble Theorem 1.3 and proof from [140] which treats ordinary derivatives. So this result is a very general Opial-type inequality involving fractional derivatives.

**THEOREM 26.1**

Let $\gamma_i \geq 1$, $\nu \geq 2$ such that $\nu - \gamma_i \geq 1$; $i = 1, \ldots, \ell$ and $f \in C_{x_0}^\nu([a,b])$ with $f^{(j)}(x_0) = 0$, $j = 0, 1, \ldots, n-1$, $n := [\nu]$. Here $x, x_0 \in [a,b] : x \geq x_0$. Let $q_1, q_2 > 0$ continuous functions on $[a,b]$ and $r_i > 0 : \sum_{i=1}^{\ell} r_i = r$. Let $s_1, s_1' > 1$: $\frac{1}{s_1} + \frac{1}{s_1'} = 1$ and $s_2, s_2' > 1$: $\frac{1}{s_2} + \frac{1}{s_2'} = 1$, and $p > s_2$. Furthermore suppose that

$$Q_1 := \left( \int_{x_0}^{x} (q_1(w))^{s_1'} dw \right)^{1/s_1'} < +\infty \qquad (26.1)$$

and
$$Q_2 := \left(\int_{x_0}^{x} (q_2(w))^{-s_2'/p} dw\right)^{r/s_2'} < +\infty. \quad (26.2)$$

Call $\sigma := \frac{p-s_2}{ps_2}$. Then it holds

$$\int_{x_0}^{x} q_1(w) \prod_{i=1}^{\ell} |(D_{x_0}^{\gamma_i}(f))(w)|^{r_i} dw \le Q_1 Q_2$$

$$\cdot \prod_{i=1}^{\ell} \left\{ \frac{\sigma^{r_i \sigma}}{(\Gamma(\nu-\gamma_i))^{r_i}(\nu-\gamma_i-1+\sigma)^{r_i \sigma}} \right\}$$

$$\cdot \frac{(x-x_0)^{\left(\sum_{i=1}^{\ell}(\nu-\gamma_i-1)r_i + \sigma r + \frac{1}{s_1}\right)}}{\left(\left(\sum_{i=1}^{\ell}(\nu-\gamma_i-1)r_i s_1\right) + r s_1 \sigma + 1\right)^{1/s_1}}$$

$$\cdot \left(\int_{x_0}^{x} q_2(w) |(D_{x_0}^{\nu} f)(w)|^p dw\right)^{r/p}. \quad (26.3)$$

**PROOF** From (25.15) we have ($x_0 \le w \le x$)

$$(D_{x_0}^{\gamma_i} f)(w) = \frac{1}{\Gamma(\nu-\gamma_i)} \int_{x_0}^{w} (w-t)^{\nu-\gamma_i-1} (D_{x_0}^{\nu} f)(t) dt$$

$$= \frac{1}{\Gamma(\nu-\gamma_i)} \int_{x_0}^{x} (w-t)_+^{\nu-\gamma_i-1} (D_{x_0}^{\nu} f)(t) dt, \quad (26.4)$$

all $i = 1, \ldots, \ell$. Hence it holds

$$\int_{x_0}^{x} q_1(w) \prod_{i=1}^{\ell} |(D_{x_0}^{\gamma_i}(f))(w)|^{r_i} dw \le Q_1 \left(\int_{x_0}^{x} \prod_{i=1}^{\ell} |(D_{x_0}^{\gamma_i}(f))(w)|^{r_i s_1} dw\right)^{1/s_1}$$

$$\stackrel{(26.4)}{\le} Q_1 \left\{\int_{x_0}^{x} \prod_{i=1}^{\ell} \left(\frac{1}{\Gamma(\nu-\gamma_i)} \int_{x_0}^{x} (q_2(t))^{-1/p} (q_2(t))^{1/p} (w-t)_+^{\nu-\gamma_i-1}\right.\right.$$

$$\left.\left.\cdot |(D_{x_0}^{\nu} f)(t)| dt\right)^{r_i s_1} dw\right\}^{1/s_1}$$

$$\leq Q_1 \left\{ \int_{x_0}^{x} \prod_{i=1}^{\ell} \frac{1}{(\Gamma(\nu-\gamma_i))^{r_i s_1}} \cdot \left( \int_{x_0}^{x} (q_2(t))^{-s_2'/p} dt \right)^{\frac{r_i s_1}{s_2'}} \right.$$

$$\left. \cdot \int_{x_0}^{x} (q_2(t))^{s_2/p} |(D_{x_0}^{\nu} f)(t)|^{s_2} (w-t)_+^{(\nu-\gamma_i-1)s_2} dt \right)^{\frac{r_i s_1}{s_2}} dw \right\}^{1/s_1}$$

$$= Q_1 Q_2 \prod_{i=1}^{\ell} \frac{1}{(\Gamma(\nu-\gamma_i))^{r_i}} \left\{ \int_{x_0}^{x} \prod_{i=1}^{\ell} \left( \int_{x_0}^{x} (q_2(t))^{s_2/p} \right. \right.$$

$$\left. \left. \cdot |(D_{x_0}^{\nu} f)(t)|^{s_2} (w-t)_+^{(\nu-\gamma_i-1)s_2} dt \right)^{\frac{r_i s_1}{s_2}} dw \right\}^{1/s_1}$$

$$\leq Q_1 Q_2 \cdot \prod_{i=1}^{\ell} \frac{1}{(\Gamma(\nu-\gamma_i))^{r_i}} \cdot \left\{ \int_{x_0}^{x} \prod_{i=1}^{\ell} \left( \int_{x_0}^{x} q_2(t) |(D_{x_0}^{\nu} f)(t)|^p dt \right)^{\frac{r_i s_1}{p}} \right.$$

$$\left. \cdot \left( \int_{x_0}^{x} (w-t)_+^{(\nu-\gamma_i-1)/\sigma} dt \right)^{r_i s_1 \sigma} dw \right\}^{1/s_1}$$

$$= Q_1 Q_2 \prod_{i=1}^{\ell} \frac{1}{(\Gamma(\nu-\gamma_i))^{r_i}} \cdot \left( \int_{x_0}^{x} q_2(w) |D_{x_0}^{\nu} f(w)|^p dw \right)^{r/p}$$

$$\cdot \left( \int_{x_0}^{x} \prod_{i=1}^{\ell} \left( \int_{x_0}^{x} (w-t)_+^{(\nu-\gamma_i-1)/\sigma} dt \right)^{r_i s_1 \sigma} dw \right)^{1/s_1}$$

$$= Q_1 Q_2 \cdot \prod_{i=1}^{\ell} \frac{1}{(\Gamma(\nu-\gamma_i))^{r_i}} \cdot \left( \int_{x_0}^{x} q_2(w) |D_{x_0}^{\nu} f(w)|^p dw \right)^{r/p}$$

$$\cdot \left( \int_{x_0}^{x} \prod_{i=1}^{\ell} \frac{(w-x_0)^{(\nu-\gamma_i-1+\sigma)r_i s_1}}{\left( \frac{\nu-\gamma_i-1+\sigma}{\sigma} \right)^{r_i s_1 \sigma}} dw \right)^{1/s_1}$$

$$= Q_1 Q_2 \cdot \prod_{i=1}^{\ell} \left\{ \frac{\sigma^{r_i \sigma}}{(\Gamma(\nu-\gamma_i))^{r_i} (\nu-\gamma_i-1+\sigma)^{r_i \sigma}} \right\}$$

$$\cdot \left( \int_{x_0}^{x} q_2(w) |(D_{x_0}^{\nu} f)(w)|^p dw \right)^{r/p}$$

$$\cdot \left( \frac{(x-x_0)^{\sum_{i=1}^{\ell}(\nu-\gamma_i-1)r_i + \frac{1}{s_1}+\sigma r}}{\left( \left( \sum_{i=1}^{\ell}(\nu-\gamma_i-1)r_i s_1 \right) + 1 + r\sigma s_1 \right)^{1/s_1}} \right),$$

establishing (26.3). ∎

The counterpart of the last theorem comes next.

### THEOREM 26.2

Let $\gamma_i \geq 1$, $\nu \geq 2$ such that $\nu - \gamma_i \geq 1$; $i = 1, \ldots, \ell$ and $f \in C_{x_0}^{\nu}([a,b])$ with $f^{(j)}(x_0) = 0$, $j = 0, 1, \ldots, n-1$, $n := [\nu]$. Here $x, x_0 \in [a,b] : x > x_0$. Let $q_1, q_2 > 0$ continuous functions on $[a,b]$ and $r_i > 0$: $\sum_{i=1}^{\ell} r_i = r$. Let $0 < s_1, s_2 < 1$ and $s_1', s_2' < 0$ such that $\frac{1}{s_1} + \frac{1}{s_1'} = 1$, $\frac{1}{s_2} + \frac{1}{s_2'} = 1$. Assume that $(D_{x_0}^{\nu} f)(t)$ is never zero and of fixed sign over $[x_0, b]$. Furthermore suppose that

$$Q_1 := \left( \int_{x_0}^{x} (q_1(w))^{s_1'} dw \right)^{1/s_1'} < +\infty \qquad (26.5)$$

and

$$Q_2 := \left( \int_{x_0}^{x} (q_2(w))^{-s_2'} dw \right)^{r/s_2'} < +\infty. \qquad (26.6)$$

Set $\lambda := \frac{s_1 s_2}{s_1 s_2 - 1}$. Then it holds

$$\int_{x_0}^{x} q_1(w) \left( \prod_{i=1}^{\ell} (|D_{x_0}^{\gamma_i}(f)|(w))^{r_i} \right) dw$$

$$\geq \frac{Q_1 Q_2}{\prod_{i=1}^{\ell} \left\{ (\Gamma(\nu-\gamma_i))^{r_i} \cdot ((\nu-\gamma_i-1)s_2^2 s_1 + 1)^{\left(\frac{r_i}{s_2^2 s_1}\right)} \right\}}$$

$$\cdot \frac{(x-x_0)^{\{(\sum_{i=1}^{\ell} r_i((\nu-\gamma_i-1)s_1 + s_2^{-2})) + 1\}/s_1}}{\{(\sum_{i=1}^{\ell} r_i((\nu-\gamma_i-1)s_1 + s_2^{-2})) + 1\}^{1/s_1}}$$

$$\cdot \left( \int_{x_0}^{x} q_2^{\lambda s_2}(w) |(D_{x_0}^{\nu} f)(w)|^{\lambda s_2} dw \right)^{r/\lambda s_2}. \qquad (26.7)$$

## 26. $L_p$-General Fractional Opial Inequalities

**PROOF** From (25.15) we get ($x_0 \le w \le x$)

$$(D_{x_0}^{\gamma_i} f)(w) = \frac{1}{\Gamma(\nu - \gamma_i)} \int_{x_0}^{w} (w-t)^{\nu-\gamma_i-1} (D_{x_0}^{\nu} f)(t) dt$$

$$= \frac{1}{\Gamma(\nu - \gamma_i)} \int_{x_0}^{x} (w-t)_+^{\nu-\gamma_i-1} (D_{x_0}^{\nu} f)(t) dt.$$

By assumption we have

$$|(D_{x_0}^{\gamma_i} f)(w)| = \frac{1}{\Gamma(\nu - \gamma_i)} \int_{x_0}^{x} (w-t)_+^{\nu-\gamma_i-1} |(D_{x_0}^{\nu} f)(t)| dt. \qquad (26.8)$$

Therefore, we have

$$\int_{x_0}^{x} q_1(w) \left( \prod_{i=1}^{\ell} (|D_{x_0}^{\gamma_i}(f)|(w))^{r_i} \right) dw$$

$$\ge Q_1 \cdot \left( \int_{x_0}^{x} \prod_{i=1}^{\ell} (|D_{x_0}^{\gamma_i}(f)|(w))^{r_i s_1} dw \right)^{1/s_1}$$

$$\stackrel{(26.8)}{=} Q_1 \left\{ \int_{x_0}^{x} \prod_{i=1}^{\ell} \left( \frac{1}{\Gamma(\nu - \gamma_i)} \int_{x_0}^{x} q_2^{-1}(t) q_2(t) (w-t)_+^{\nu-\gamma_i-1} \right. \right.$$

$$\left. \left. \cdot |(D_{x_0}^{\nu} f)(t)| dt \right)^{r_i s_1} dw \right\}^{1/s_1}$$

$$\ge Q_1 \cdot \left\{ \int_{x_0}^{x} \prod_{i=1}^{\ell} \frac{1}{(\Gamma(\nu - \gamma_i))^{r_i s_1}} \cdot \left( \int_{x_0}^{x} (q_2(t))^{-s_2'} dt \right)^{\frac{r_i s_1}{s_2'}} \right.$$

$$\left. \cdot \left( \int_{x_0}^{x} (q_2(t))^{s_2} |(D_{x_0}^{\nu} f)(t)|^{s_2} (w-t)_+^{(\nu-\gamma_i-1)s_2} dt \right)^{\frac{r_i s_1}{s_2}} dw \right\}^{1/s_1}$$

$$= Q_1 Q_2 \cdot \prod_{i=1}^{\ell} \frac{1}{(\Gamma(\nu - \gamma_i))^{r_i}} \cdot \left\{ \int_{x_0}^{x} \prod_{i=1}^{\ell} \left( \int_{x_0}^{x} (q_2(t))^{s_2} \right. \right.$$

$$\cdot |(D_{x_0}^\nu f)(t)|^{s_2} (w-t)_+^{(\nu-\gamma_i-1)s_2} dt \Big)^{\frac{r_i s_1}{s_2}} dw \Big\}^{1/s_1}$$

$$\geq Q_1 Q_2 \cdot \prod_{i=1}^{\ell} \frac{1}{(\Gamma(\nu-\gamma_i))^{r_i}}$$

$$\cdot \Big\{ \int_{x_0}^{x} \prod_{i=1}^{\ell} \Big( \int_{x_0}^{x} (q_2(t))^{\lambda s_2} |(D_{x_0}^\nu f)(t)|^{\lambda s_2} dt \Big)^{\frac{r_i s_1}{\lambda s_2}}$$

$$\cdot \Big( \int_{x_0}^{x} (w-t)_+^{(\nu-\gamma_i-1)s_2^2 s_1} dt \Big)^{\frac{r_i}{s_2^2}} dw \Big\}^{1/s_1}$$

$$= Q_1 Q_2 \cdot \prod_{i=1}^{\ell} \frac{1}{(\Gamma(\nu-\gamma_i))^{r_i}}$$

$$\cdot \Big( \int_{x_0}^{x} (q_2(t))^{\lambda s_2} |(D_{x_0}^\nu f)(t)|^{\lambda s_2} dt \Big)^{\frac{r}{\lambda s_2}} \cdot \chi. \tag{26.9}$$

Here we put

$$\chi := \Big\{ \int_{x_0}^{x} \prod_{i=1}^{\ell} \Big( \int_{x_0}^{x} (w-t)_+^{(\nu-\gamma_i-1)s_2^2 s_1} dt \Big)^{\frac{r_i}{s_2^2}} dw \Big\}^{1/s_1}.$$

We obtain

$$\Delta(w) := \prod_{i=1}^{\ell} \Big( \int_{x_0}^{x} (w-t)_+^{(\nu-\gamma_i-1)s_2^2 s_1} dt \Big)^{\frac{r_i}{s_2^2}}$$

$$= \prod_{i=1}^{\ell} \Big( \int_{x_0}^{w} (w-t)^{(\nu-\gamma_i-1)s_2^2 s_1} dt \Big)^{\frac{r_i}{s_2^2}}$$

$$= \frac{(w-x_0)^{\sum_{i=1}^{\ell} r_i((\nu-\gamma_i-1)s_1+s_2^{-2})}}{\prod_{i=1}^{\ell}((\nu-\gamma_i-1)s_2^2 s_1+1)^{r_i/s_2^2}}. \tag{26.10}$$

So that

$$\chi = \left(\int_{x_0}^{x} \Delta(w) dw\right)^{1/s_1} \stackrel{(26.10)}{=} \frac{1}{\prod_{i=1}^{\ell}((\nu - \gamma_i - 1)s_2^2 s_1 + 1)^{r_i/s_2^2 s_1}}$$

$$\cdot \left(\int_{x_0}^{x} (w - x_0)^{\sum_{i=1}^{\ell} r_i((\nu - \gamma_i - 1)s_1 + s_2^{-2})} dw\right)^{1/s_1},$$

producing

$$\chi = \frac{1}{\prod_{i=1}^{\ell}((\nu - \gamma_i - 1)s_2^2 s_1 + 1)^{r_i/s_2^2 s_1}}$$

$$\cdot \frac{(x - x_0)^{\{(\sum_{i=1}^{\ell} r_i((\nu - \gamma_i - 1)s_1 + s_2^{-2}))+1\}/s_1}}{\{(\sum_{i=1}^{\ell} r_i((\nu - \gamma_i - 1)s_1 + s_2^{-2})) + 1\}^{1/s_1}}. \qquad (26.11)$$

Using (26.11) into (26.9) we establish (26.7). ■

A related extreme case comes next for $p = 1$ and $q = \infty$.

**PROPOSITION 26.1**
Let $\gamma_i \geq 1$, $\nu \geq 2$ such that $\nu - \gamma_i \geq 1$; $i = 1, \ldots, \ell$ and $f \in C_{x_0}^{\nu}([a,b])$ with $f^{(j)}(x_0) = 0$, $j = 0, 1, \ldots, n-1$, $n := \lceil \nu \rceil$. Here $x, x_0 \in [a,b] : x \geq x_0$. Let $\tilde{q}(w) \geq 0$ continuous on $[a,b]$ and $r_i > 0$: $\sum_{i=1}^{\ell} r_i = r$. Then it holds

$$\int_{x_0}^{x} \tilde{q}(w) \cdot \prod_{i=1}^{\ell} (|D_{x_0}^{\gamma_i} f|(w))^{r_i} dw \qquad (26.12)$$

$$\leq \left\{ \frac{\|\tilde{q}\|_{\infty} (\|D_{x_0}^{\nu} f\|_{\infty})^r}{\prod_{i=1}^{\ell} (\Gamma(\nu - \gamma_i + 1))^{r_i}} \right\} \cdot \left\{ \frac{(x - x_0)^{r\nu - \sum_{i=1}^{\ell} r_i \gamma_i + 1}}{(r\nu - \sum_{i=1}^{\ell} r_i \gamma_i + 1)} \right\}.$$

**PROOF** From (25.15) we have ($x_0 \leq w \leq x$)

$$(D_{x_0}^{\gamma_i} f)(w) = \frac{1}{\Gamma(\nu - \gamma_i)} \int_{x_0}^{w} (w - t)^{\nu - \gamma_i - 1} (D_{x_0}^{\nu} f)(t) dt.$$

Hence

$$|D_{x_0}^{\gamma_i} f|(w) \le \frac{\|D_{x_0}^{\nu} f\|_\infty}{\Gamma(\nu - \gamma_i)} \cdot \frac{(w - x_0)^{\nu - \gamma_i}}{(\nu - \gamma_i)}$$

$$= \frac{\|D_{x_0}^{\nu} f\|_\infty (w - x_0)^{\nu - \gamma_i}}{\Gamma(\nu - \gamma_i + 1)}.$$

And

$$(|D_{x_0}^{\gamma_i} f|(w))^{r_i} \le \frac{(w - x_0)^{r_i \nu - r_i \gamma_i}}{(\Gamma(\nu - \gamma_i + 1))^{r_i}} \cdot (\|D_{x_0}^{\nu} f\|_\infty)^{r_i}.$$

Moreover, for all $x_0 \le w \le x$ we have

$$\tilde{q}(w) \cdot \prod_{i=1}^{\ell} (|D_{x_0}^{\gamma_i} f|(w))^{r_i} \le \frac{(w - x_0)^{(r\nu - \sum_{i=1}^{\ell} r_i \gamma_i)} \cdot \|\tilde{q}\|_\infty \cdot (\|D_{x_0}^{\nu} f\|_\infty)^r}{\prod_{i=1}^{\ell} (\Gamma(\nu - \gamma_i + 1))^{r_i}}. \tag{26.13}$$

Integrating (26.13) over $[x_0, x]$ we find (26.12). ∎

A quite sharp fractional Opial-type estimate comes next. The statement and proof of this theorem resembles the corresponding theorem and proof in the ordinary derivatives case, see Theorem 1.1 and its proof from [140] and [84].

### THEOREM 26.3

Let $k \ge 1$, $\gamma \ge 2$, $\nu \ge 3$ such that $\nu - \gamma \ge 1$, $\nu - k \ge 1$, $\gamma - k \ge 1$ and $f \in C_{x_0}^{\nu}([a,b])$ with $f^{(i)}(x_0) = 0$, $i = 1, \ldots, n-1$, $n := [\nu]$. Here $x, x_0 \in [a,b] : x \ge x_0$. Let $p, q > 1$: $\frac{1}{p} + \frac{1}{q} = 1$. Then it holds

$$\int_{x_0}^{x} |D_{x_0}^{\gamma} f|(w) \cdot |D_{x_0}^{k} f|(w) dw \tag{26.14}$$

$$\le \frac{2^{-1/p} (x - x_0)^{(2\nu - k - \gamma - 1 + \frac{2}{q})} \cdot \left( \int_{x_0}^{x} |D_{x_0}^{\nu} f(w)|^p dw \right)^{2/p}}{\Gamma(\nu - k) \Gamma(\nu - \gamma + 1)((\nu - \gamma)q + 1)^{1/q} (2\nu q - kq - \gamma q - q + 2)^{1/q}}.$$

**PROOF** From (25.15) we have ($x_0 \le w \le x$)

$$(D_{x_0}^{\gamma} f)(w) = \frac{1}{\Gamma(\nu - \gamma)} \int_{x_0}^{x} (w - t)_{+}^{\nu - \gamma - 1} (D_{x_0}^{\nu} f)(t) dt \tag{26.15}$$

26. $L_p$-General Fractional Opial Inequalities

and

$$(D_{x_0}^k f)(w) = \frac{1}{\Gamma(\nu - k)} \int_{x_0}^{x} (w - t)_+^{\nu-k-1} (D_{x_0}^\nu f)(t) dt.$$

Using (26.15), Hölder's inequality and calculus techniques we get

$$\int_{x_0}^{x} |D_{x_0}^\gamma f|(w) \cdot |D_{x_0}^k f|(w) dw \leq \frac{1}{\Gamma(\nu - \gamma)\Gamma(\nu - k)} \cdot \int_{x_0}^{x} |D_{x_0}^\nu f|(t)$$

$$\cdot \left( \int_t^x |D_{x_0}^\nu f|(s) \cdot \left( \int_{x_0}^x \{(w-t)_+^{\nu-k-1}(w-s)_+^{\nu-\gamma-1} \right. \right.$$

$$\left. \left. + (w-s)_+^{\nu-k-1}(w-t)_+^{\nu-\gamma-1}\} dw \right) ds \right) dt$$

$$\leq \frac{1}{\Gamma(\nu - \gamma)\Gamma(\nu - k)} \cdot \int_{x_0}^{x} |D_{x_0}^\nu f|(t) \cdot \left( \int_t^x |D_{x_0}^\nu f|(s) \cdot (x-t)^{\gamma-k-1} \right.$$

$$\left. \cdot \left( \int_{x_0}^x \{(w-t)_+^{\nu-\gamma}(w-s)_+^{\nu-\gamma-1} + (w-s)_+^{\nu-\gamma}(w-t)_+^{\nu-\gamma-1}\} dw \right) ds \right) dt$$

$$= \frac{1}{\Gamma(\nu - \gamma)\Gamma(\nu - k)} \int_{x_0}^{x} |D_{x_0}^\nu f(t)|$$

$$\cdot \left( \int_t^x |D_{x_0}^\nu f(s)| \cdot (x-t)^{\gamma-k-1} \cdot \frac{(x-t)^{\nu-\gamma}(x-s)^{\nu-\gamma}}{\nu - \gamma} ds \right) dt$$

$$= \frac{1}{\Gamma(\nu - \gamma + 1)\Gamma(\nu - k)} \int_{x_0}^{x} |D_{x_0}^\nu f(t)|(x-t)^{\nu-k-1}$$

$$\cdot \left( \int_t^x |D_{x_0}^\nu f(s)|(x-s)^{\nu-\gamma} ds \right) dt$$

$$\leq \frac{1}{\Gamma(\nu - \gamma + 1)\Gamma(\nu - k)} \cdot \int_{x_0}^{x} |D_{x_0}^\nu f(t)|(x-t)^{\nu-k-1}$$

$$\cdot \left( \int_t^x |D_{x_0}^\nu f(s)|^p ds \right)^{1/p} \cdot \left( \int_t^x (x-s)^{q\nu - q\gamma} ds \right)^{1/q} dt$$

$$= \frac{1}{\Gamma(\nu-\gamma+1)\Gamma(\nu-k)((\nu-\gamma)q+1)^{1/q}} \cdot \int_{x_0}^{x} |D_{x_0}^{\nu}f(t)|$$

$$\cdot (x-t)^{2\nu-k-\gamma-1+\frac{1}{q}} \cdot \left( \int_{t}^{x} |D_{x_0}^{\nu}f(s)|^p ds \right)^{1/p} \cdot dt$$

$$\leq \frac{1}{\Gamma(\nu-\gamma+1)\Gamma(\nu-k)((\nu-\gamma)q+1)^{1/q}} \cdot \left( \int_{x_0}^{x} |D_{x_0}^{\nu}f(t)|^p \right.$$

$$\cdot \left. \left( \int_{t}^{x} |D_{x_0}^{\nu}f(s)|^p ds \right) dt \right)^{1/p} \cdot \left( \int_{x_0}^{x} (x-t)^{2\nu q - kq - \gamma q - q + 1} \right)^{1/q}$$

$$= \frac{2^{-1/p}(x-x_0)^{(2\nu-k-\gamma-1+\frac{2}{q})} \cdot \left( \int_{x_0}^{x} |D_{x_0}^{\nu}f(t)|^p dt \right)^{2/p}}{\Gamma(\nu-\gamma+1)\Gamma(\nu-k)((\nu-\gamma)q+1)^{1/q}(2\nu q - kq - \gamma q - q + 2)^{1/q}},$$

that is establishing (26.14). ∎

Another extreme related case is next when $p=1$ and $q=\infty$.

### PROPOSITION 26.2

Let $\nu \geq 2$, $k, \gamma \geq 1$ such that $\nu - \gamma \geq 1$, $\nu - k \geq 1$ and $f \in C_{x_0}^{\nu}([a,b])$ with $f^{(j)}(x_0) = 0$, $j = 0, 1, \ldots, n-1$, $n := \lceil \nu \rceil$. Here $x, x_0 \in [a,b] : x \geq x_0$. Then it holds

$$\int_{x_0}^{x} |D_{x_0}^{k}f|(w) \cdot |D_{x_0}^{\gamma}f|(w) \cdot dw$$

$$\leq \frac{(x-x_0)^{2\nu-k-\gamma+1}(\|D_{x_0}^{\nu}f\|_{\infty})^2}{\Gamma(\nu-\gamma+1) \cdot \Gamma(\nu-k+1) \cdot (2\nu-k-\gamma+1)}. \quad (26.16)$$

**PROOF** From (25.15) we have ($x_0 \leq w \leq x$)

$$(D_{x_0}^{k}f)(w) = \frac{1}{\Gamma(\nu-k)} \int_{x_0}^{w} (w-t)^{(\nu-k-1)} \cdot (D_{x_0}^{\nu}f)(t) dt,$$

$$(D_{x_0}^{\gamma}f)(w) = \frac{1}{\Gamma(\nu-\gamma)} \int_{x_0}^{w} (w-t)^{(\nu-\gamma-1)} \cdot (D_{x_0}^{\nu}f)(t) dt.$$

Moreover, we notice

$$|(D_{x_0}^k f)(w)| \le \frac{(w-x_0)^{\nu-k}}{\Gamma(\nu-k+1)} \|D_{x_0}^\nu f\|_\infty,$$

$$|(D_{x_0}^\gamma f)(w)| \le \frac{(w-x_0)^{\nu-\gamma}}{\Gamma(\nu-\gamma+1)} \|D_{x_0}^\nu f\|_\infty.$$

Hence

$$|(D_{x_0}^k f)(w)| \cdot |(D_{x_0}^\gamma f)(w)| \le \frac{(w-x_0)^{2\nu-k-\gamma} \cdot (\|D_{x_0}^\nu f\|_\infty)^2}{\Gamma(\nu-k+1) \cdot \Gamma(\nu-\gamma+1)}. \quad (26.17)$$

Integrating (26.17) over $[x_0, x]$ we obtain (26.16). ∎

Next we use

**THEOREM 26.4**
(Chebyshev's inequality for integrals, see [95], p. 1099). *Let $f_1, f_2, \ldots, f_n$ be nonnegative integrable functions on $[a,b]$, $n \in \mathbf{N}$. These are either all increasing or all decreasing. Then it holds*

$$\prod_{i=1}^n \int_a^b f_i(x)dx \le (b-a)^{n-1} \int_a^b \prod_{i=1}^n f_i(x)dx. \quad (26.18)$$

The following is a fractional integral Opial-type result involving derivatives of generalized Riemann–Liouville integral. Let $x_0 \in [a,b] \subset \mathbf{R}$, $f \in C([a,b])$ and the generalized Riemann–Liouville integral

$$(J_\nu^{x_0} f)(w) := \frac{1}{\Gamma(\nu)} \int_{x_0}^w (w-t)^{\nu-1} f(t)dt, \quad (26.19)$$

all $x_0 \le w \le b$, and $\nu > 1$. Let $k \in \mathbf{N}$ then

$$(J_\nu^{x_0} f)^{(k)}(w) = (J_{\nu-k}^{x_0} f)(w). \quad (26.20)$$

**THEOREM 26.5**
*Let $f \in C([a,b])$ such that $|f|$ is decreasing. Let $\nu \ge 2$, $n := [\nu]$ and $\ell_i \in \mathbf{N}$ such that $\nu - \ell_i \ge 1$, $i = 1, \ldots, k$, $\ell_i \in \{1, \ldots, n-1\}$, $1 \le k \le n-1$,*

$\ell := \sum_{i=1}^{k} \ell_i$. Here $p, q > 1$: $\frac{1}{p} + \frac{1}{q} = 1$ and $x, x_0 \in [a, b]$ such that $x \geq x_0$. Then it holds

$$\int_{x_0}^{x} \prod_{i=1}^{k} |(J_{\nu-\ell_i}^{x_0} f)(w)| dw \leq \frac{\left(\int_{x_0}^{x} |f(t)|^{qk} dt\right)^{1/q}}{\prod_{i=1}^{k} \Gamma(\nu - \ell_i)}$$

$$\cdot \frac{p(x - x_0)^{\left(\frac{1+pk\nu - p\ell}{p}\right)}}{(pk\nu - p\ell - pk + 1)^{1/p}(pk\nu - p\ell - pk + p + 1)}. \quad (26.21)$$

**PROOF** Set

$$G(w) := \frac{1}{\Gamma(\nu)} \int_{x_0}^{w} (w - t)^{\nu - 1} |f(t)| dt.$$

We see that $|(J_{\nu}^{x_0} f)(w)| \leq G(w)$, all $x_0 \leq w \leq x$. And furthermore,

$$G^{(\ell_i)}(w) = \frac{1}{\Gamma(\nu - \ell_i)} \int_{x_0}^{w} (w - t)^{\nu - \ell_i - 1} |f(t)| dt$$

$$= \frac{1}{\Gamma(\nu - \ell_i)} \int_{x_0}^{x} (w - t)_+^{\nu - \ell_i - 1} |f(t)| dt, \quad (26.22)$$

all $i = 1, \ldots, k$.

Also observe the following: Let $t_1 \leq t_2$ when $w - t_1 \geq w - t_2$. If $w - t_2 \geq 0$, then $(w - t_1)_+ \geq (w - t_2)_+$. If $w - t_1 \leq 0$, then $(w - t_1)_+ \geq (w - t_2)_+$. If $w - t_1 \geq 0$ and $w - t_2 \leq 0$, then $(w - t_1)_+ \geq (w - t_2)_+$. That is, $(w - t)_+$ is nonincreasing in $t$. Also $(w - t)_+^{\nu - \ell_i - 1}$ is nonincreasing in $t$. Hence $|f|(w - t)_+^{\nu - \ell_i - 1}$ is decreasing in $t$.

From (26.20) we get that

$$(J_{\nu}^{x_0} f)^{(\ell_i)}(w) = (J_{\nu - \ell_i}^{x_0} f)(w)$$

and

$$|(J_{\nu - \ell_i}^{x_0} f)(w)| \leq G^{(\ell_i)}(w),$$

any $i = 1, \ldots, k$ and all $x_0 \leq w \leq x$. Therefore,

$$\prod_{i=1}^{k} |(J_{\nu - \ell_i}^{x_0} f)(w)| \leq \prod_{i=1}^{k} G^{(\ell_i)}(w)$$

$$= \frac{1}{\prod_{i=1}^{k}\Gamma(\nu-\ell_i)} \cdot \prod_{i=1}^{k}\int_{x_0}^{x}(w-t)_+^{(\nu-\ell_i-1)}|f(t)|dt$$

(by Theorem 26.4)
$$\leq \frac{(x-x_0)^{k-1}}{\prod_{i=1}^{k}\Gamma(\nu-\ell_i)}\int_{x_0}^{x}(w-t)_+^{k\nu-\ell-k}|f(t)|^k dt$$

$$= \frac{(x-x_0)^{k-1}}{\prod_{i=1}^{k}\Gamma(\nu-\ell_i)} \cdot \int_{x_0}^{w}(w-t)^{k\nu-\ell-k}|f(t)|^k dt$$

(by Hölder's inequality)
$$\leq \frac{(x-x_0)^{k-1}}{\prod_{i=1}^{k}\Gamma(\nu-\ell_i)} \cdot \left(\int_{x_0}^{w}(w-t)^{pk\nu-p\ell-pk}dt\right)^{1/p}$$

$$\cdot \left(\int_{x_0}^{x}|f(t)|^{qk}dt\right)^{1/q}.$$

Thus

$$\prod_{i=1}^{k}|(J_{\nu-\ell_i}^{x_0}f)(w)| \leq \frac{(w-x_0)^{\frac{pk\nu-p\ell-pk+1}{p}}(x-x_0)^{k-1}}{\prod_{i=1}^{k}\Gamma(\nu-\ell_i)\cdot(pk\nu-p\ell-pk+1)^{1/p}}$$

$$\cdot \left(\int_{x_0}^{x}|f(t)|^{qk}dt\right)^{1/q}. \tag{26.23}$$

Integrating (26.23) over $[x_0, x]$ we get (26.21). ∎

**COROLLARY 26.1**

Same assumptions as in Theorem 26.5. Here $\nu = n \in \mathbf{N} - \{1\}$. Set

$$(\mathcal{R}_{n-\ell_i}^{x_0}(f))(w) := (J_{n-\ell_i}^{x_0}f)(w) = \frac{1}{(n-\ell_i-1)!} \cdot \int_{x_0}^{w}(w-t)^{n-\ell_i-1}f(t)dt, \tag{26.24}$$

all $x_0 \leq w \leq x$. Then it holds

$$\int_{x_0}^{x}\prod_{i=1}^{k}|(\mathcal{R}_{n-\ell_i}^{x_0}(f))(w)|dw \leq \frac{\left(\int_{x_0}^{x}|f(t)|^{qk}dt\right)^{1/q}}{\prod_{i=1}^{k}(n-\ell_i-1)!}$$

$$\cdot \frac{p(x-x_0)^{(\frac{1+pkn-p\ell}{p})}}{(pkn-p\ell-pk+1)^{1/p} \cdot (pkn-p\ell-pk+p+1)}. \quad (26.25)$$

An extreme case follows.

**PROPOSITION 26.3**
Same assumptions as in Theorem 26.5. However, here $p=1$, $q=\infty$. Then it holds

$$\int_{x_0}^{x} \left( \prod_{i=1}^{k} |(J_{\nu-\ell_i}^{x_0} f)(w)| \right) dw \quad (26.26)$$

$$\leq \frac{(x-x_0)^{(k\nu-\ell+1)} \cdot \|f\|_\infty^k}{\left(\prod_{i=1}^{k} \Gamma(\nu-\ell_i)\right) \cdot (k\nu-\ell-k+1)(k\nu-\ell+1)}, \quad \forall x \geq x_0.$$

**PROOF** As in the proof of Theorem 26.5 we find that

$$\prod_{i=1}^{k} |(J_{\nu-\ell_i}^{x_0} f)(w)| \leq \frac{1}{\prod_{i=1}^{k} \Gamma(\nu-\ell_i)} \cdot \left( \prod_{i=1}^{k} \int_{x_0}^{w} (w-t)^{\nu-\ell_i-1} |f(t)| dt \right)$$

(by Theorem 26.4)
$$\leq \frac{(w-x_0)^{k-1}}{\prod_{i=1}^{k} \Gamma(\nu-\ell_i)} \cdot \int_{x_0}^{w} (w-t)^{k\nu-\ell-k} \cdot |f(t)|^k dt$$

$$\leq \frac{(w-x_0)^{k-1}}{\prod_{i=1}^{k} \Gamma(\nu-\ell_i)} \cdot \frac{(w-x_0)^{k\nu-\ell-k+1}}{(k\nu-\ell-k+1)} \cdot \|f\|_\infty^k.$$

That is,

$$\prod_{i=1}^{k} |(J_{\nu-\ell_i}^{x_0} f)(w)| \leq \frac{(w-x_0)^{(k\nu-\ell)} \cdot \|f\|_\infty^k}{\left(\prod_{i=1}^{k} \Gamma(\nu-\ell_i)\right) \cdot (k\nu-\ell-k+1)}. \quad (26.27)$$

Integrating (26.27) over $[x_0, x]$ we find (26.26). ∎

The counterpart of Theorem 26.5 for fractional derivatives comes next.

**THEOREM 26.6**
Let $[x_0, x] \subset [a,b]$, $f \in C_{x_0}^{\nu}([a,b])$, $\nu \geq 2$, $n := [\nu]$, $\gamma_i \geq 1$ such that $\nu - \gamma_i \geq 1$, all $i = 1, \ldots, k$, $k \in \mathbf{N} - \{1\}$. Here suppose that $f^{(j)}(x_0) = 0$,

## 26. $L_p$-General Fractional Opial Inequalities

$j = 0, 1, \ldots, n-1$. Set $\gamma := \sum_{i=1}^{k} \gamma_i$. Let $p, q > 1$ such that $\frac{1}{p} + \frac{1}{q} = 1$. Furthermore, suppose that $|(D_{x_0}^{\nu} f)(t)|$ is decreasing on $[x_0, x]$. Then it holds

$$\int_{x_0}^{x} \prod_{i=1}^{k} |(D_{x_0}^{\gamma_i} f)(w)| dw \tag{26.28}$$

$$\leq \frac{p(x-x_0)^{(\frac{1+k\nu p - \gamma p}{p})} \left( \int_{x_0}^{x} |D_{x_0}^{\nu} f(t)|^{kq} dt \right)^{1/q}}{\prod_{i=1}^{k} \Gamma(\nu - \gamma_i) \cdot (k\nu p - \gamma p - kp + 1)^{1/p} \cdot (k\nu p - \gamma p - kp + p + 1)}.$$

**PROOF** From (25.15) ($x_0 \leq w \leq x$) we have

$$(D_{x_0}^{\gamma_i} f)(w) = \frac{1}{\Gamma(\nu - \gamma_i)} \int_{x_0}^{x} (w-t)_+^{(\nu - \gamma_i - 1)} \cdot (D_{x_0}^{\nu} f)(t) dt,$$

and

$$|(D_{x_0}^{\gamma_i} f)(w)| \leq \frac{1}{\Gamma(\nu - \gamma_i)} \cdot \int_{x_0}^{x} (w-t)_+^{(\nu - \gamma_i - 1)} \cdot |(D_{x_0}^{\nu} f)(t)| dt.$$

Here clearly by assumption $(w-t)_+^{(\nu - \gamma_i - 1)} \cdot |D_{x_0}^{\nu} f(t)| \geq 0$ is decreasing and integrable over $[x_0, x]$, all $i = 1, \ldots, k$. Hence by Theorem 26.4 we get

$$\prod_{i=1}^{k} |(D_{x_0}^{\gamma_i} f)(w)| \leq \frac{1}{\prod_{i=1}^{k} \Gamma(\nu - \gamma_i)} \cdot \prod_{i=1}^{k} \int_{x_0}^{x} (w-t)_+^{(\nu - \gamma_i - 1)} |D_{x_0}^{\nu} f(t)| dt$$

$$\leq \frac{(x-x_0)^{k-1}}{\prod_{i=1}^{k} \Gamma(\nu - \gamma_i)} \cdot \int_{x_0}^{x} (w-t)_+^{k\nu - \gamma - k} \cdot |D_{x_0}^{\nu} f(t)|^k dt$$

$$\leq \frac{(x-x_0)^{k-1}}{\prod_{i=1}^{k} \Gamma(\nu - \gamma_i)} \cdot \left( \int_{x_0}^{x} (w-t)_+^{k\nu p - \gamma p - kp} dt \right)^{1/p} \cdot \left( \int_{x_0}^{x} |D_{x_0}^{\nu} f(t)|^{kq} dt \right)^{1/q}$$

$$= \frac{(x-x_0)^{k-1}}{\prod_{i=1}^{k} \Gamma(\nu - \gamma_i)} \cdot \left( \int_{x_0}^{w} (w-t)^{k\nu p - \gamma p - kp} dt \right)^{1/p} \left( \int_{x_0}^{x} |D_{x_0}^{\nu} f(t)|^{kq} dt \right)^{1/q}.$$

So that

$$\prod_{i=1}^{k} |(D_{x_0}^{\gamma_i} f)(w)| \leq \frac{(x-x_0)^{k-1}}{\prod_{i=1}^{k} \Gamma(\nu - \gamma_i)} \cdot \frac{(w-x_0)^{(\frac{k\nu p - \gamma p - kp + 1}{p})}}{(k\nu p - \gamma p - kp + 1)^{1/p}}$$

$$\cdot \left( \int_{x_0}^{x} |D_{x_0}^{\nu} f(t)|^{kq} dt \right)^{1/q}. \tag{26.29}$$

Integrating (26.29) over $[x_0, x]$ we establish (26.28). ∎

### COROLLARY 26.2

Same assumptions as in Theorem 26.6. Here $\nu = n \in \mathbf{N} - \{1\}$, also $\gamma_i \in \mathbf{N}$, $i = 1, \ldots, k$. Then it holds

$$\int_{x_0}^{x} \prod_{i=1}^{k} |f^{(\gamma_i)}(w)| dw \tag{26.30}$$

$$\leq \frac{p(x - x_0)^{\frac{(1+knp-\gamma p)}{p}} \cdot \left( \int_{x_0}^{x} |f^{(n)}(t)|^{kq} dt \right)^{1/q}}{\prod_{i=1}^{k}((n - \gamma_i - 1)!) \cdot (knp - \gamma p - kp + 1)^{1/p} \cdot (knp - \gamma p - kp + p + 1)}.$$

A final extreme case comes next.

### PROPOSITION 26.4

Same assumptions as in Theorem 26.6; however, here we have that $p = 1$ and $q = \infty$. Then it holds

$$\int_{x_0}^{x} \prod_{i=1}^{k} |(D_{x_0}^{\gamma_i} f)(w)| dw \leq \frac{(x - x_0)^{k\nu - \gamma + 1} \cdot (\|D_{x_0}^{\nu} f\|_{\infty})^k}{\left( \prod_{i=1}^{k} \Gamma(\nu - \gamma_i) \right) \cdot (k\nu - \gamma - k + 1) \cdot (k\nu - \gamma + 1)}. \tag{26.31}$$

**PROOF**   As in the proof of Theorem 26.6 we observe that

$$\prod_{i=1}^{k} |(D_{x_0}^{\gamma_i} f)(w)| \leq \frac{1}{\prod_{i=1}^{k} \Gamma(\nu - \gamma_i)} \cdot \prod_{i=1}^{k} \int_{x_0}^{w} (w - t)^{(\nu - \gamma_i - 1)} |D_{x_0}^{\nu} f(t)| dt$$

$$\overset{\text{(by Theorem 26.4)}}{\leq} \frac{(w - x_0)^{k-1}}{\prod_{i=1}^{k} \Gamma(\nu - \gamma_i)}$$

$$\cdot \left( \int_{x_0}^{w} (w - t)^{k\nu - \gamma - k} dt \right) \cdot (\|D_{x_0}^{\nu} f\|_{\infty}^{k})$$

## 26. $L_p$-General Fractional Opial Inequalities

$$= \frac{(w - x_0)^{k-1} \cdot (w - x_0)^{k\nu-\gamma-k+1} \cdot (\|D_{x_0}^{\nu} f\|_{\infty}^{k})}{\left(\prod_{i=1}^{k} \Gamma(\nu - \gamma_i)\right) \cdot (k\nu - \gamma - k + 1)}$$

$$= \frac{(w - x_0)^{k\nu-\gamma} \cdot (\|D_{x_0}^{\nu} f\|_{\infty}^{k})}{\left(\prod_{i=1}^{k} \Gamma(\nu - \gamma_i)\right) \cdot (k\nu - \gamma - k + 1)}.$$

That is, for all $x_0 \leq w \leq x$ we obtain

$$\prod_{i=1}^{k} |(D_{x_0}^{\gamma_i} f)(w)| \leq \frac{(w - x_0)^{k\nu-\gamma} \cdot (\|D_{x_0}^{\nu} f\|_{\infty}^{k})}{\left(\prod_{i=1}^{k} \Gamma(\nu - \gamma_i)\right) \cdot (k\nu - \gamma - k + 1)}. \tag{26.32}$$

Integrating (26.32) over $[x_0, x]$ we prove (26.31). ■

# References

[1] R. P. Agarwal, Sharp Opial-type inequalities involving $r$-derivatives and their applications, *Tôhoku Math. J.* **47** (1995), 567–593.

[2] R. P. Agarwal and P. Y. H. Pang, *Opial Inequalities with Applications in Differential and Difference Equations*, Kluwer Academic Publishers, Dordrecht, Netherlands, 1995.

[3] Ch. D. Aliprantis and O. Burkinshaw, *Positive Operators*, Academic Press, New York, 1985.

[4] C. D. Aliprantis and O. Burkinshaw, *Principles of Real Analysis*, Second Edition, Academic Press, Boston, 1990.

[5] G. A. Anastassiou, A study of positive linear operators by the method of moments, one-dimensional case, *J. Approx. Theory* **45** (1985), 247–270.

[6] G. A. Anastassiou, Korovkin type inequalities in real normed vector spaces, *Approx. Theory Appl.* **2**(2) (1986), 39–53.

[7] G. A. Anastassiou, Multi-dimensional quantitative results for probability measures approximating the unit measure, *Approx. Theory Appl.* **2**(4) (1986), 93–103.

[8] G. A. Anastassiou, Spline monotone approximation with linear differential operators, *J. Approx. Theory Appl.* **5** (1989), 61–67.

[9] G. A. Anastassiou, Korovkin inequalities for stochastic processes, *J. Math. Anal. Appl.*, **157**(2) (1991), 366–384.

[10] G. A. Anastassiou, Monotone approximation by pseudopolynomials, *Approximation Theory*, Academic Press, New York, 1991, 5–11.

[11] G. A. Anastassiou, Bivariate monotone approximation, *Proc. A.M.S.* **112**(4) (1991), 959–964.

[12] G. A. Anastassiou, *Moments in Probability and Approximation The-*

*ory*, Pitman Research Notes in Math., Vol. 287, Longman Sci. & Tech., Harlow, U.K., 1993.

[13] G. A. Anastassiou, Higher order monotone approximation with linear differential operators, *Indian J. Pure Appl. Math.* **24**(4) (1993), 263–266.

[14] G. A. Anastassiou, $L_p$-Korovkin type inequalities for positive linear operators, *Approximation Probability and Related Fields*, edited by G. Anastassiou and S. T. Rachev, Plenum, Proc. U.C.S.B. Conference, May 20–22, 1993, 19–40, 1994.

[15] G. A. Anastassiou, Ostrowski type inequalities, *Proc. A.M.S.* **123** (1995), 3775–3781.

[16] G. A. Anastassiou, Central limit theorem, weak law of large numbers for martingales in Banach spaces and weak invariance principle a quantitative study, *J. Multivariate Anal.* **52** (1995), 158–180.

[17] G. A. Anastassiou, Rate of convergence of some neural network operators to the unit-univariate case, *J. Math. Anal. Appl.* **212** (1997), 237–262.

[18] G. A. Anastassiou, Multivariate Ostrowski type inequalities, *Acta Math. Hungarica* **76**(4) (1997), 267–278.

[19] G. A. Anastassiou, Lattice homomoprhism–Korovkin type inequalities for vector valued functions, *Hokkaido Math. J.* **26** (1997), 337–364.

[20] G. A. Anastassiou, Shape and probability preserving univariate wavelet type operators, *Commun. Appl. Anal.* **1**(3) (1997), 303–314.

[21] G. A. Anastassiou, Weak convergence of squashing neural network operators studied asymptotically, *Neural Parallel Sci. Computation* **5** (1997), 439–448.

[22] G. A. Anastassiou, Opial type inequalities for linear differential operators, *Math. Inequalities Appl.* **1**(2) (1998), 193–200.

[23] G. A. Anastassiou, Higher order univariate wavelet type approximation, *Approximation Theory, In Memory of A.K. Varma*, Marcel Dekker, New York, 1998, 43–60.

[24] G. A. Anastassiou, General fractional Opial type inequalities, *Acta Applicandae Mathematicae* **54** (1998), 303–317.

[25] G. A. Anastassiou, Weak convergence of Cardaliaguet Euvrard neural network operators studied asymptotically, *Results Math.* **34** (1998), 214–223.

[26] G. A. Anastassiou, Differentiated shift-invariant integral operators, univariate case, *Applicable Anal.* **68**(3–4) (1998), 281–311.

[27] G. A. Anastassiou, Inequalities for local moduli of continuity, *Appl. Math. Lett.* **12** (1999), 7–12.

[28] G. A. Anastassiou, Opial type inequalities involving fractional derivatives of functions, *Nonlinear Studies* **6**(2) (1999), 207–230.

[29] G. A. Anastassiou, Higher order multivariate wavelet type approximation, to appear, Proc. #941 AMS Meeting, *Wavelet Analysis and Multiresolution Methods*, edited by Tian He, Marcel Dekker, New York, 2000.

[30] G. A. Anastassiou, Rate of convergence of some multivariate neural network operators to the unit, to appear, *J. Comp. Math. Appl.* 2000.

[31] G. A. Anastassiou and A. Bendikov, A discrete analog of Kac's formula and optimal approximation of the solution of the heat equation, *Indian J. Pure Appl. Math.* **28**(10) (1997), 1367–1389.

[32] G. A. Anastassiou and S. Cambanis, Non-orthogonal wavelet approximation with rates of deterministic signals, to appear, *J. Comp. Math. Appl.* 2000.

[33] G. A. Anastassiou and S. G. Gal, General theory of global smoothness preservation by singular integrals, univariate case, *J. Comp. Anal. Appl.* **1**(3) (1999), 289–317.

[34] G. A. Anastassiou and S. Gal, On some differentiated shift-invariant integral operators, univariate case revisited, *Adv. Nonlinear Var. Inequal.* **2**(2) (1999), 71–83.

[35] G. A. Anastassiou and S. G. Gal, Some shift invariant integral operators, univariate case, revisited, *J. Comp. Anal. Appl.* **1**(1) (1999), 3–23.

[36] G. A. Anastassiou and S. Gal, Convergence of generalized singular iintegrals to the unit, univariate case, submitted.

[37] G. A. Anastassiou and J. Pecaric, General weighted Opial inequalities for linear differential operators, *J. Math. Anal. Appl.* **239** (1999), 402–418.

[38] G. A. Anastassiou and S. T. Rachev, *Approximation, Probability, and Related Fields*, Proceeding Intern. Conf. U.C.S.B., 1993, Plenum, New York, 1994.

[39] G. A. Anastassiou and T. Rychlik, Refined rates of bias convergence

for generalized $L$-statistics in the I.I.D. case, to appear, *Applicationes Mathematicae* Poland, 2000.

[40] G. A. Anastassiou and O. Shisha, Monotone approximation with linear differential operators, *J. Approx. Theory* **44** (1985), 391–393.

[41] G. A. Anastassiou and X. M. Yu, Bivariate probabilistic wavelet approximation, *Proceedings of the 6th S.E. Approximation Theory International Conference*, held in Memphis, 1991, Marcel Dekker, New York, 1992, 79–92.

[42] G. A. Anastassiou and X. M. Yu, Monotone and probabilistic wavelet approximation, *Stochastic Anal. Appl.* **10**(3) (1992), 251–264.

[43] G. A. Anastassiou and X. M. Yu, Convex and coconvex—probabilistic wavelet approximation, *Stochastic Anal. Appl.* **10**(5) (1992), 507–521.

[44] G. A. Anastassiou and X. M. Yu, Probabilistic discrete wavelet approximation, *Num. Func. Anal. Opt.* **13** (1992), 117–121.

[45] G. A. Anastassiou and X. M. Yu, Multivariable probabilistic scale approximation, *J. Fundam. Sci. Appl.* Plovdiv, Bulgaria, **5** (1997), 41–57.

[46] I. Badea and C. Badea, On the order of simultaneous approximation of bivariate functions by Bernstein operators, *Anal. Numer. Theor. Approx.* **16** (1987), 11–17.

[47] D. Bainov and P. Simeonov, *Integral Inequalities and Applications*, Kluwer Academic Publishers, Dordrecht, Netherlands, 1992.

[48] N. Balakrishnan and A. C. Cohen, *Order Statistics and Inference*, Academic Press, Boston, 1991.

[49] R. M. Blumenthal and R. K. Getoor, *Markov Processes and Potential Theory*, Academic Press, New York, 1968.

[50] M. Bogdan, Asymptotic distributions of linear combinations of order statistics, *Zastos. Math.* **24** (1994), 201–225.

[51] D. Boos, A differential for $L$-statistics, *Ann. Statist.* **7** (1979), 955–959.

[52] B. Büttgenbach, H. Esser, and R. S. Nessel, On the sharpness of error bounds in connection with finite difference schemes on uniform grids for boundary value problems of ordinary differential equtions, *Numer. Funct. Anal. Optim.* **12** (1991), 285–298.

[53] B. Büttgenbach, H. Esser, G. Lüttgens, and R. J. Nessel, A sharp error estimate for the numerical solution of a Dirichlet problem for

the Poisson equation, *J. Comp. Appl. Math.* **44** (1992), 331–337.

[54] P. L. Butzer and H. Berens, *Semigroups of Operators and Approximation*, Springer-Verlag, New York, 1967.

[55] P. L. Butzer, L. Hahn, and M. Th. Roeckerath, Central limit theorem and weak law of large numbers with rates for martingales in Banach spaces, *J. Multivariate Anal.* **13** (1983), 287–301.

[56] P. L. Butzer and H. J. W. Kirschfink, Donsker's weak invariance principle with rates for $C[0,1]$-valued, dependent random functions, *Approx. Theory Appl.* **2**(4) (1986), 55–77.

[57] P. L. Butzer and D. Schulz, The weak invariance principle with rates for $C[0,1]$-valued random functions. *Anniversary Volume on Approximation Theory and Functional Analysis*, edited by P. L. Butzer, R. L. Sterns, and B. Sz.-Nagy, ISNM 65, 567–584, Birkhaüser, Basel, Switzerland, 1984.

[58] S. Cambanis and E. Masry, Wavelet approximation of deterministic and random signals: Convergence properties and rates, Center for Stochastic Processes, University of North Carolina, Technical Report No. 352, November 1991.

[59] S. Cambanis and E. Masry, Wavelet approximation of deterministic and random signals: Convergence properties and rates, *IEEE Trans. Inf. Theory* **40** (July 1994), 1013–1029.

[60] J. A. Canavati, The Riemann-Liouville integral, *Nieuw Archief Voor Wiskunde* **5**(1) (1987), 53–75.

[61] J. D. Cao and H. H. Gonska, Pointwise estimates for modified positive linear operators, *Portugal. Math.* **46** (1989), 402–430.

[62] P. Cardaliaguet and G. Euvrard, Approximation of a function and its derivative with a neural network, *Neural Networks* **5** (1992), 207–220.

[63] W. Z. Chen and M. Zhou, Freud-Butzer-Hahn type quantitative theorem for probabilistic representations of $(C_0)$ operator semigroups, *Approx. Theory Appl.* **9** (1993), 1–8.

[64] W. Z. Chen and M. Zhou, Asymptotic formulae for probabilistic representation of $(C_0)$ operator semigroups, *Northeastern Math. J.* **10** (1994), 159–166.

[65] H. Chernoff, J. L. Gastwirth, and M. V. Johns, Asymptotic distribution of linear combinations of order statistics, with applications to estimation, *Ann. Math. Statist.* **38** (1967), 52–72.

[66] G. Choquet, *Lectures on Analysis, Vol. I, Advanced Book Program*, W.A. Benjamin, 1969.

[67] C. K. Chui, *An Introduction to Wavelets*, Academic Press, New York, 1992.

[68] C. K. Chui and J. Z. Wang, A cardinal spline approach to wavelets, *CAT Report* **211** (1990).

[69] K. L. Chung, On the exponential formulas of semi-group theory, *Math. Scand.* **10** (1962), 153–162.

[70] I. Daubechies, Orthonormal bases of compactly supported wavelets, *Comm. Pure Appl. Math.* **41** (1988), 909–996.

[71] I. Daubechies, *Ten Lectures on Wavelets*, SIAM, Philadelphia, 1992.

[72] H. A. David, *Order Statistics*, 2nd ed., Wiley, New York, 1981.

[73] R. A. DeVore, Monotone approximation by polynomials, *SIAM J. Math. Anal.* **8** (1977), 906–921.

[74] R. A. DeVore and G. G. Lorentz, *Constructive Approximation*, Grundlehren der mathematischen Wissenschaften, Vol. 303, Springer-Verlag, Berlin, 1993.

[75] R. A. DeVore and X. M. Yu, Pointwise estimates for monotone polynomial approximation, *Const. Approx.* **1** (1985), 323–331.

[76] W. Dickmeis, R. J. Nessel, and E. Van Wickeren, Quantitative extensions of the uniform boundedness principle, *Jahr. Deutsch. Math. Verein.* **89** (1987), 105–134.

[77] Z. Ditzian and K. Ivanov, Bernstein-type operators and their derivatives, *J. Approx. Theory* **56** (1989), 72–90.

[78] M. D. Donsker, An invariance principle for certain probability limit theorems, *Mem. Amer. Math. Soc.* No. 6 (1951).

[79] J. L. Doob, *Classical Potential Theory and its Probabilistic Counterpart*, Springer-Verlag, New York, 1984.

[80] E. B. Dynkin, *Markov Processes*, Springer-Verlag, Berlin, 1965.

[81] E. B. Dynkin and A. A. Yushkevich, *Markov Processes: Theorems and Problems*, Plenum Press, New York, 1969.

[82] H. Esser, St. J. Goebbels, G. Lüttgens, and R. J. Nessel, Sharp error bounds for the Crank-Nicolson and Saulyev difference scheme in connection with an initial boundary value problem for the inhomogeneous heat equation, Special Issue on *Concrete Analysis*, *J. Comp. Math. Appl.* **30**(3–6) (1995), 59–68.

[83] A. M. Fink, Bounds on the deviation of a function from its averages, *Czechoslovak Math. J.* **42**(117) (1992), 289–310.

[84] A. M. Fink, On Opial's inequality for $f^{(n)}$, *Proc. AMS* **115**(1) (1992), 177–181.

[85] S. G. Gal, Degree of approximation of continuous functions by some singular integrals, *Rev. Anal. Numér. Théor. Approx. (Cluj)*, Tome XXVII, No. 2 (1998), 251–261.

[86] I. Gavrea and D. H. Mache, Generalization of Bernstein-type approximation methods. *Approximation Theory, Proc. IDoMAT95*, edited by M. W. Müller, M. Felten, and D. H. Mache, Math. Research, Vol. 86, Akademie Verlag, Berlin, 1995, 115–126.

[87] E. Giné, Some remarks on the central limit theorem in $C(S)$. *International Conference on Probability in Banach Spaces, First, Oberwolfach, July 20–26, 1975. Probability in Banach Spaces*, edited by A. Beck, Lecture Notes in Math., Vol. 526, Springer-Verlag, Berlin, 1976, 101–106.

[88] H. Gonska, On approximation of continuously differentiable functions by positive linear operators, *Bull. Austral. Math. Soc.* **27** (1983), 73–81.

[89] H. H. Gonska, Simultaneous approximation by algebraic blending functions. *Alfred Haar Memorial Conference, Budapest 1985, Colloquia Mathematica Soc. János Bolyai*, **49**, North-Holand, Amsterdam, 1985, 363–382.

[90] H. Gonska and E. Hinnemann, Pointwise estimates for approximation by algebraic polynomials, *Acta Math. Hungar.* **46** (1985), 243–254.

[91] H. H. Gonska and R. K. Kovacheva, The second order modulus revisited: Remarks, Applications, Problems, *Conf. Sem. Mat. Univ. Bari* **257** (1994), 1–32.

[92] H. H. Gonska and J. Meier, Quantitative theorems on approximation by Bernstein-Stancu operators, *Estratto da Calcolo* **21**, fasc. IV (1984), 317–335.

[93] H. H. Gonska and D.-X. Zhou, Local smoothness of functions and Bernstein-Durrmeyer operators, special issue *Concrete Analysis*, G. A. Anastassiou (Ed.), *Comput. Math. Appl.* **30**(3–6) (1995), 83–101.

[94] H. H. Gonska and X.-L. Zhou, The strong converse inequality for the Bernstein-Kantorovich operators, special issue *Concrete Analysis*, G. A. Anastassiou (Ed.), *Comput. Math. Appl.* **30**(3–6) (1995), 103–

128.

[95] I. S. Gradshteyn and I. M. Ryzhik, *Table of Integrals, Series, and Products*, Academic Press, New York, 1980.

[96] L. M. Graves, Riemann integration and Taylor's theorem in general analysis, *Trans. A.M.S.* **29** (1927), 163–177.

[97] M. Heilmann, $L_p$-saturation of some modified Bernstein operators, *J. Approx. Theory* **54** (1988), 260–273.

[98] R. Helmers, P. Janssen, and R. Serfling, Berry-Essen and bootstrap results for generalized $L$-statistics, *Scand. J. Statist.* **17** (1990), 65–77.

[99] R. Helmers and H. Ruymgaart, Asymptotic normality of generalized $L$-statistics with unbounded scores, *J. Statist. Planning Infer.* **19** (1988), 43–53.

[100] E. Hewitt and K. Stromberg, *Real and Abstract Analysis*, Vol. 25, Springer-Verlag, New York, 1965.

[101] E. Hille and R. S. Phillips, *Functional Analysis and Semi-Groups*, Amer. Math. Soc. Colloq. Publ. 31, Providence, RI, 1957.

[102] K. Ito and M. P. McKean, *Diffusion Processes and Their Sample Path*, Springer-Verlag, Berlin, 1974.

[103] M. Kac, On some connection between probability theory and differential and integral equations, *Proc. Second Berkeley Symposium on Math. Stat. Probability*, University of California Press, 1951, 189–215.

[104] U. Kamps, *A Concept of Generalized Order Statistics*, Teubner Texts on Mathematical Stochastics, B.G. Teubner, Stuttgart, Germany, 1995.

[105] L. V. Kantorovich and G. P. Akilov, *Functional Analysis*, 2nd ed., Pergamon Press, London, 1982.

[106] R. A. Khan, Some probabilistic methods in the theory of approximation operators, *Acta Math. Acad. Sci. Hungar.* **35** (1980), 193–203.

[107] M. K. Khan and M. A. Peters, Lipschitz constants for some approximation operators of a Lipschitz continuous function, *J. Approx. Theory* **59** (1989), 307–315.

[108] H.-B. Knoop and X.-L. Zhou, The lower estimate for linear positive operators, Schritfenreihe des Fachbereichs Mathematik, Duisburg, SM-DU-201, (First part in *Const. Approx.* **11** (1995), 53–66; second part in *Results in Math.* **25** (1994), 300–315), 1992.

[109] W. Köhnen, Einige Saturationssätze für $n$-Parametrige Halbgruppen von Operatoren, *L'Anal. Numer. Th. L'Appro.* **9**(1) (1980), 65–73.

[110] P. P. Korovkin, *Linear Operators and Approximation Theory*, Hindustan Publ. Corp., Delhi, India, 1960.

[111] D. Kreider, R. Kuller, D. Ostberg, and F. Perkins, *An Introduction to Linear Analysis*, Addison-Wesley, Reading, MA, 1966.

[112] M. K. Kwong and A. Zettl, *Norm Inequalities for Derivatives and Differences*, Lecture Notes in Mathematics, No. 1536, Springer-Verlag, Berlin, 1992.

[113] R. G. Laha and V. K. Rohatgi, *Probability Theory*, Wiley, New York, 1979.

[114] M. J. Lai, Asymptotic formulae of multivariate Bernstein approximation, *J. Approx. Theory* **70** (1992), 229–242.

[115] C.-D. Lea and M. L. Puri, Asymptotic properties of linear functions of order statistics, *J. Statist. Planning Infer.* **18** (1988), 203–223.

[116] D. Leviatan, Pointwise estimates for convex polynomial approximation, *Proc. Amer. Math. Soc.* **98** (1986), 471–474.

[117] D. Leviatan, Monotone and comonotone polynomial approximation revisited, *J. Approx. Theory* **53** (1988), 1–16.

[118] G. G. Lorentz, Monotone approximation, *Inequalities III*, edited by O. Shisha, Academic Press, New York, 1972, 201–215.

[119] G. G. Lorentz and K. Zeller, Degree of approximation by monotone polynomials I, *J. Approx. Theory* **1** (1968), 501–504.

[120] D. H. Mache, A link between Bernstein polynomials and Durrmeyer polynomials with Jacobi weights. *Approximation Theory VIII*, Vol. 1: *Approximation and Interpolation*, edited by C. K. Chui and L. L. Schumaker, World Scientific Publishing Co., Singapore, 1995, 403–410.

[121] V. Maier, $L_p$ approximation by Kantorovich operators, *Anal. Math.* **4** (1978), 289–295.

[122] V. Maier, The $L_1$ saturation class of the Kantorovich operator, *J. Approx. Theory* **22** (1978), 223–232.

[123] S. G. Mallat, Multiresolution approximations and wavelet orthonormal bases of $L^2(\mathbf{R})$, *Trans. Amer. Math. Soc.* **315** (1989), 69–87.

[124] R. G. Mamedov, On the order of the approximation of functions

by linear positive operators, *Dokl. Akad. Nauk USSR* **128** (1959), 674–676.

[125] A. Marchaud, Différences et deérivées d'une fonction de deux variables, *C.R. Acad. Sci.* **178** (1924), 1467–1470.

[126] A. Marchaud, Sur les dérivées et sur les différences des fonctions de variables réelles, *J. Math. Pures Appl.* **6** (1927), 337–425.

[127] D. M. Mason, Asymptotic normality of linear combinations of order statistics with a smooth score function, *Ann. Statist.* **9** (1981), 899–908.

[128] D. M. Mason, Some characteristics of stong laws of linear functions of order statistics, *Ann. Prob.* **10** (1982), 1051–1057.

[129] D. M. Mason and G. R. Shorack, Necessary and sufficient conditions for asymptotic normality of $L$-statistics, *Ann. Prob.* **20** (1992), 1779–1804.

[130] Y. Meyer, *Ondelettes et Opérateurs I, Ondelettes*, Hermann, Paris, 1990.

[131] D. S. Mitrinovic, *Analytic Inequalities*, Springer-Verlag, Berlin, 1970.

[132] D. J. Newman, Efficient co-monotone approximation, *J. Approx. Theory* **25** (1979), 189–192.

[133] R. Norvsaiša, Laws of large numbers for $L$-statistics, *J. Appl. Math. Stochastic Anal.* **7** (1994), 125–143.

[134] R. Norvaiša and R. Zitikis, Asymptotic behavior of linear combinations of functions of order statistics, *J. Statist. Planning Infer.* **28** (1991), 305–317.

[135] Z. Opial, Sur une inégalité, *Ann. Polon. Math.* **8** (1960), 29–32.

[136] A. Ostrowski, Über die Absolutabweichung einer differentiebaren Funktion von ihrem Integralmittelwert, *Comment. Math. Helv.* **10** (1938), 226–227.

[137] B. G. Pachpatte, On Opial-type integral inequalities, *J. Math. Anal. Appl.* **120** (1986), 547–556.

[138] R. Paltanea, Best constants in estimates with second order moduli of continuity. *Approximation Theory, Proc. IDoMAT95*, edited by M. W. Müller, M. Felten, and D. H. Mache, Math. Research, Vol. 86, Akademie Verlag, Berlin, 1995, 251–275.

[139] R. Paltanea, On an optimal constant in approximation by Bernstein operators, *Rend. Circ. Mat. Palermo*, to appear, 2000.

[140] P. Y. H. Pang and R. P. Agarwal, On an Opial type inequality due to Fink, *J. Math. Anal. Appl.* **196** (1995), 748–753.

[141] E. Passow and L. Raymon, Monotone and comonotone approximation, *Proc. Amer. Math. Soc.* **42** (1974), 340–349.

[142] E. Passow, L. Raymon, and J. A. Roulier, Comonotone polynomial approximation, *J. Approx. Theory* **11** (1974), 221–224.

[143] D. Pfeifer, Probabilistic representations of operator semigroups—a unifying approach, *Semigroup Forum* **30** (1984), 17–34.

[144] D. Pfeifer, Approximation-theoretic aspects of probabilistic representations for operator semigroups, *J. Approx. Theory* **43** (1985), 271–296.

[145] D. Pfeifer, Probabilistic concepts of approximation theory in connexion with operator semigroups, *Approx. Theory Appl.* **1** (1985), 93–118.

[146] V. A. Popov, On the quantitative Korovkin theorems in $L_p$, *C.R. Acad. Bulgare Sci., Mathématiques Théorie des Approx.* **35**(7) (1982), 897–900.

[147] E. Quak, $L_p$-error estimates for positive linear operators using the second-order $\tau$-modulus, *Anal. Math.* **14** (1988), 259–272.

[148] E. Quak, Multivariate $L_p$-error estimates for positive linear operators via the first-order $\tau$ modulus, *J. Approx. Theory* **56** (1989), 277–286.

[149] L. B. Rall, *Computational Solution of Nonlinear Operator Equations*, John Wiley & Sons, New York, 1969.

[150] S. D. Riemenschneider, The $L_p$ saturation of the Bernstein Kantorovich polynomials, *J. Approx. Theory* **23** (1978), 158–162.

[151] V. K. Rohatgi and A. K. M. D. E. Saleh, A class of distributions connected to order statistics with nonintegral sample size, *Commun. Statist. Theory Meth.* **17** (1988), 2005–2012.

[152] J. A. Roulier, Monotone approximation of certain classes of functions, *J. Approx. Theory* **1** (1968), 319–324.

[153] H. Y. Schaefer, *Banach Lattices and Positive Operators*, Springer-Verlag, New York, #**215**, 1974.

[154] L. L. Schumaker, *Spline Functions: Basic Theory*, Wiley, New York, 1981.

[155] P. K. Sen, An invariance principle for linear combinations of order

statistics, *Z. Wahrsch. verw. Gebiete* **42** (1978), 327–340.

[156] S. Y. Shaw, Some exponential formulas for $m$-parameter semigroups, *Bull. Inst. Math. Acad. Sinica* **9** (1981), 221–228.

[157] A. Shiryaev, *Probability*, Springer-Verlag, New York, 1984.

[158] O. Shisha, Monotone approximation, *Pacific J. Math.* **15** (1965), 667–671.

[159] O. Shisha and B. Mond, The degree of convergence of sequences of linear positive operators, *Nat. Acad. of Sci.* **60** (1968), 1196–1200.

[160] G. R. Shorack, Asymptotic normality of linear combinations of functions of order statistics, *Ann. Math. Statist.* **40** (1969), 2041–2050.

[161] G. R. Shorack, Functions of order statistics, *Ann. Math. Statist.* **43** (1972), 412–427.

[162] P. C. Sikkema, Der Wert einiger Konstanten in der Theorie der Approximation mit Bernstein-Polynomen, *Numer. Math.* **3** (1961), 107–116.

[163] F. Spitzer, *Principles of Random Walk*, Graduate Texts in Math., Vol. **34**, Springer-Verlag, New York, 1976.

[164] D. D. Stancu, Sur quelques polynomes de type Bernstein, *Studii Si Cercetari Stiintifice* **XI** (2) (1960), 221–233.

[165] D. D. Stancu, Probabilistic methods in the theory of approximation of functions of several variables by linear positive operators. *Approximation Theory*, edited by A. Talbot, Academic Press, New York, 1970, 329–342.

[166] S. M. Stigler, Linear functions of order statistics with smooth weight functions, *Ann. Statist.* **2** (1974), 676–693.

[167] S. M. Stigler, Fractional order statistics, with applications, *J. Amer. Statist. Ass.* **72** (1977), 544–550.

[168] A. F. Timan, *Theory of Approximation of Functions of a Real Variable*, Macmillan, New York, 1963.

[169] V. Totik, $L_p$ ($p > 1$)-approximation by Kantorovich polynomials, *Analysis* **3** (1983), 79–100.

[170] V. Totik, An interpolation theorem and its application to positive operators, *Pacific J. Math.* **111** (1984), 447–481.

[171] X. C. Wang, Note on Bernstein polynomials and Kantorovich polynomials, *Approx. Theory Appl.* **7** (1991), 99–105.

[172] J. A. Wellner, A Glivenko-Cantelli theorm and strong laws of large numbers for functions of order statistics, *Ann. Statist.* **5** (1977), 473–480.

[173] J. A. Wellner, A law of the iterated logarithm for functions of order statistics, *Ann. Statist.* **5** (1977), 481–494.

[174] E. T. Whittaker and G. N. Watson, *A Course in Modern Analysis*, Cambridge University Press, 1927.

[175] D. Willett, The existence-uniqueness theorem for an $n$th order linear ordinary differential equation, *Amer. Math. Monthly* **75** (1968), 174–178.

[176] Z. C. Wu, Linear combinations of Bernstein operators on a simplex, *Approx. Theory Appl.* **7** (1991), 81–90.

[177] X. Xiang, A note on the bias of $L$-estimators and a bias reducing procedure, *Statist. Prob. Lett.* **23** (1995), 123–127.

[178] M. Zhou, On the study of probabilistic approximation processes, Ph.D. Thesis, The University of Memphis, Memphis, TN, 1997.

[179] M. Zhou and G. A. Anastassiou, Asymptotic expansions of the $(C_0)$ $m$-parameter operator semigroups, *Num. Funct. Anal. Optim.* **16**(9 & 10) (1995), 1273–1291.

[180] M. Zhou and G. A. Anastassiou, Representation formulae for $(C_0)$ $m$-parameter operator semigroups, *Annales Polonici Mathematici* **LXIII.3** (1996), 247–272.

[181] W. R. van Zwet, A stong law of linear functions of order statistics, *Ann. Prob.* **8** (1980), 986–990.

# List of Symbols

$R$ real numbers, 60
$b$, 60
$F_n$, 60
$\omega_1$, 60
$\lceil \cdot \rceil$, 60
$\lfloor \cdot \rfloor$, 60
$S$, 76
$G_n$, 77
$\mathbf{R}^d$ $d$-dimensional Euclidean space, 90
$\mathbf{Z}$ integers, 96
$f^{(N)}$, 121
$e_n$, 123
$K_n$, 136
$\gamma_n$, 136
$A_k$, 146
$B_k$, 156
$\varphi$, 146
$\omega$, 149
$F$, 155
$C_k$, 174
$D_k$, 177
$C(\mathbf{R})$, 180
$\|\cdot\|_\infty$ supremum norm, 203
$\Phi$, 229
$\delta_{x_i}$, 229
$L_k$, 230
$\omega(\cdot,\cdot)_p$, 231
$\|\cdot\|_p$, 231
$L_2(\mathbf{R})$, 235
$N_m$, 236
$e_k$, 237

$C^{(i)}(\mathbf{R})$, 259
$\omega_p$, 259
$\Delta_h^p$, 259
$\ell_k$, 260
$L_k^{(i)}$, 261
$I_{k,q}$, 265
$A_{k,j}$, 270
$B_{k,j}$, 270
$C_{k,j}$, 270
$\Gamma_{k,j}$, 270
$\dot{\Omega}$, 275
$\dot{\mathbf{R}}^\ell$, 275
$E_{\dot{x}}$, 275
$\tau_{\dot{\Omega}}$, 275
$\dot{\Delta}$, 275
$\Delta$, 275
$\dot{\mathbf{Z}}^\ell$, 279
$\dot{\Delta}_h$, 294
$\dot{\Omega}_h$, 294
$\omega_{2,i}$, 297
$\overline{\dot{\Omega}}_{h,T}$, 297
$\mathbf{R}_+^m$, 305
$\mathbf{Z}_+^m$, 305
$T$, 305
$D^r$, 307
$E$, 307
$S_n$, 308
$\omega(T, r, g; \delta)$, 312
$BUC(\mathbf{R}^m)$, 327
$X$, 330
$MDS$, 338
$B$, 338
$\|g\|_{\mathcal{L}_j}$, 338
$[x]^j$, 339
$C_B^r$, 339
$x^v$, 339
$S_n(\omega, t)$, 348
$W(t)$, 348
$C_c$, 350
$C_c^r$, 350
$\overline{C}_c^r$, 350
$E[X_i \mid \mathcal{A}(X_1, \ldots, X_n)]$, 358
$CLT$, 360

# List of Symbols

$X_{i:n}$, 361
$L$-statistics, 361
$L_n(f, g)$, 362
$\mu(f, g, F)$, 362
$\omega_2(\cdot, \cdot)$, 363
$K(f, t)$, 364
$\omega_2^\varphi(f, t)$, 364
$K^*(f, t)_p$, 367
$E_0(f)_\infty$, 367
**N** natural numbers, 367
$(D_n f)(x)$, 369
$M_n(f, g)$, 370
$\omega_2^*(f, t)$, 372
$(M_n^\alpha f)(x)$, 374
$T_{\alpha, k, n} f$, 374
$M_n^\alpha(f, g)$, 376
$(L_{m0}^{\langle 0\beta\gamma\rangle} f)(x)$, 377
$p_{m,k}(x)$, 377
$L_n^{\beta\gamma}(f, g)$, 378
$\mu_n$, 384
$C(M, Y)$, 384
$T$, 384
$T(f \vee g)$, 387
$T(f \wedge g)$, 387
$B(M, Y)$, 384
$P^{(k)}$, 391
$B_n(f; x)$, 416
$\phi_n$, 392
$\mu$, 417
$L^1(\underline{0}, \mathcal{A}, P)$, 444
$C_\Omega(K)$, 445
$B_\Omega(K)$, 445
$EX$, 445
$\Delta_n(x)$, 452
$I$, 451
$L$, 452
$\mathbf{Z}_+$ nonnegative integers, 455
$P_\theta$, 455
$P_n \otimes C(I) + C(I) \otimes P_m$, 455
$C^{r,p}(I^2)$, 456
$\omega_{s,q}(f; \delta_1, \delta_2)$, 456
$Q_{n,m}$, 456
$\omega(f, \delta_1, \delta_2)$, 461

$P_{m,n}$, 463
$\ell_{ij}$, 463
$S_m(\Delta_n)$, 465
$Q_n(f)$, 465
$B[a,b]$, 465
$\dot{S}_m(\Delta_n)$, 469
$\dot{Q}_n(f)$, 469
$S_m(\Delta;h)$, 471
$D_h^j f(x)$, 471
$\|f\|_{\ell_\infty[a,b]_h}$, 471
$\omega_1(f;t)_{\ell_\infty[a,b]_h}$, 471
$B[a,b]_h$, 471
$[a,b]_h$, 471
$\omega_1(f,x_0,h)$, 473
$C^n[a,b]$, 473
$\omega_2(f,x_0,h)$, 475
$\omega_1(f,\vec{x}_0,h)$, 479
$\omega_r(f;\delta)_X$, 483
$\Delta_h^r f(x)$, 483
$P_{n,\xi}(f;x)$, 484
$Q_{n,\xi}(f;x)$, 484
$W_{n,\xi}(f;x)$, 484
$L^1_{2\pi}(\mathbf{R})$, 485
$L^1(\mathbf{R})$, 485
$L^p(\mathbf{R})$, 488
$L^p_{2\pi}(\mathbf{R})$, 488
$C^1[a,b]$, 498
$C^{n+1}[a,b]$, 497
$\mathrm{Vol}(Q)$, 507
$C^1(\prod_{i=1}^k [a_i,b_i])$, 507
$\|\cdot\|_{\ell_1}$, 511
$L_r$, 521
$H(x,t)$, 522
$Ly$, 522
$P_1(w), z(w), QY, PQ$, 529–530
$(J_\nu g)(x)$, 539
$C^\nu([0,1])$, 539
$(J_\nu D^\nu g)(t)$, 540
$(J_\nu^{x_0} f)(x)$, 540
$\Gamma(\nu)$, 540
$C^\nu_{x_0}([a,b])$, 540
$D^\nu_{x_0} f$, 540
$(J_{1-\alpha}^{x_0} f^{(n)})(x)$, 540

# List of Symbols

$D_{x_0}^\gamma f$, 545
$\phi^*(b)$, 557
$Q_1$, 567
$Q_2$, 568

# Index

absolutely continuous, 487
approximation on the grid, 294
asymptotic expansion, 125
asymptotic probabilistic expansion, 323
asymptotic weak convergence, 121
average, 497

Banach lattice, 384
Banach space, 335
    Banach space-valued martingales, 335
bell-shaped function, 60
Bernstein–Durrmeyer operator, 369
Bernstein operator, 365
bias convergence, 361
bivariate monotone approximation, 461
Borel measurable function, 415
bounded function, 384
Brownian motion, 276
B-spline, 236

Cardaliaguet–Euvrard operator, 60
Cauchy–Schwartz's inequality 317
central limit theorem, 360
Chebyshev's inequality for integrals, 577
compact and convex subset, 433
compact support, 146
continuous and bounded, 60

convergence to the unit with rates, 59
convex function, 181
covariance functional, 348

$d$-dimensional bell-shaped function, 90
dependent from above/below, 358
deterministic signals, 235
Dirichlet problem, 275
discrete Dirichlet problem, 294
discrete measure, 229
discrete parabolic Laplacian, 294
discrete polynomial spline, 471
Ditzian–Totik modulus of smoothness, 364

$E$-commutative operator, 445
endomorphism, 305
even function, 269
expectation, 307
extension operator, 411
extended Pettis integral, 307

Feller operator, 327–328
$\varphi$-decomposability, 340
first local modulus of continuity, 473
fractional derivative, 539
fractional initial value problem, 556
Fréchet derivative, 338

generalized $L$-statistics, 361

Green's function, 522

heat equation, 275
higher order approximation, 165
Hölder's inequality, 326

infinite divisibility, 340
infinitesimal generator, 306

Jackson-type generalizations of Picard, Poisson–Cauchy and Gauss–Weierstrass singular integrals, 484
Jackson-type inequality, 145
Jacobi weights, 375

$K$-functional, 364
Kantorovich operator, 367
Korovkin's theorem, 383

Laplacian operator, 275
lattice homomorphism, 387
lattice norm, 386
linear differential operator, 451
$L_p$-approximation, 428
$L_p$-general fractional Opial inequality, 567
$L_p$-modulus of continuity, 230
$L_p$-norm, 230
$L_p$-Opial inequality, 539
$L_p$-result for positive linear operator, 413
$L_p$-wavelet type approximation, 231

Mache operator, 374
martingale difference sequence, 338
mixed difference, 456
mixed modulus of smoothness, 456
modulus of continuity, 149
monotone approximation, 149, 451
multidimensional high order approximation, 201
multidimensional modulus of continuity, 201

multidimensional Ostrowski-type inequality, 507
multilinear continuous function, 338
multiresolution decomposition, 235
multivariate Cardaliaguet–Euvrard neural network operators, 90
multivariate function, 201
multivariate squashing function, 111
multivariate squashing operators, 112

neural network operator, 59
nonorthogonal wavelet type operators, 237
normed bounded function, 405

operator semigroup, 305
Opial-type inequality, 521
optimal upper bound, 419
order statistic, 361
orthomorphism, 412
Ostrowski-type inequality, 497

periodic polynomial spline, 469
pointwise convergence, 62
Poisson distribution, 330
polygonal line, 348
polynomial spline, 465
positive finite measure, 417
positive operator, 383
potential, 281
probability distribution function, 155
probability space, 348
pseudopolynomial, 456

quadrature wavelet type operator, 177

random vector, 312
random walk, 275
Riemann–Liouville integral, 539
right-continuous function, 146

scale function, 149
score functions, 361
second local modulus of continuity, 475
second modulus of continuity, 363
shape and probability preserving, 185
sharp inequality, 181
shift invariant integral operator, 263
Shisha–Mond type, 384
Sikkema best constant, 366
singular integral, 483
spline monotone approximation, 465
squashing operator, 77
Stancu modulus of continuity, 461
Stancu operator, 377
star-shaped, 336
statistical inference, 361
stochastic process, 445
Szász–Mirakjan operator, 330

Taylor's formula, 204
two-dimensional Bernstein polynomial, 461

uniform convergence, 166
uniformly continuous, 60
uniqueness of solution of initial value problem, 536

vector lattice, 385

wavelet type approximation, 145
weak convergence, 121
weak invariance principle, 348
weak law of large numbers, 346
weight function, 362
Wiener-measure, 348